D0151614

Finite Element Modeling and Simulation with *ANSYS Workbench*

Finite Element Modeling and Simulation with *ANSYS Workbench*

Xiaolin Chen • Yijun Liu

CRC Press is an imprint of the
Taylor & Francis Group, an **informa** business

CRC Press
Taylor & Francis Group
6000 Broken Sound Parkway NW, Suite 300
Boca Raton, FL 33487-2742

Printed on acid-free paper
Version Date: 20140709

International Standard Book Number-13: 978-1-4398-7384-7 (Hardback)

Visit the Taylor & Francis Web site at
http://www.taylorandfrancis.com

and the CRC Press Web site at
http://www.crcpress.com

Contents

Preface

This book presents an integrated introduction to the basic theories of finite element analysis (FEA) and the use of *ANSYS® Workbench* in the modeling and simulation of engineering problems. It is intended for undergraduate students in engineering and for engineers who have no previous experience with the FEA and wish to learn the basic theory and software usage from a single volume. The book covers the basic concepts in the FEA using simple mechanics problems as examples, and leads to discussions and applications of the one-dimensional bar and beam elements, two-dimensional plane stress and plane strain elements, plate and shell elements, and three-dimensional solid elements in the analyses of structural stresses, vibrations and dynamics, thermal responses, fluid flows, optimizations, and failures. In each of these application areas, *ANSYS Workbench* is introduced through hands-on case studies. Clear understanding of the FEA principles, element behaviors, solution procedures, and correct usage of the FEA software is emphasized throughout the book.

The materials in this book are good for teaching undergraduate FEA courses in engineering to students who need a general background in the FEA so that they can apply FEA in their design and analysis of structures using *ANSYS Workbench* or other FEA software. Presentations of FEA theories and formulations in the book are kept to a minimum, aimed solely for readers to understand the behaviors of the various elements and FEA solutions. Emphases are placed on the techniques in FEA modeling and simulation, such as in building a quality FEA mesh, verifying the results, and employing the symmetry of a structure. The book includes 12 chapters and can be used for an undergraduate FEA course in one semester (15 weeks with three 50-minute sessions each week) or in two quarters (10 weeks per quarter with three 50-minute sessions each week). It is recommended that computer lab sessions on using *ANSYS Workbench* be arranged for each chapter. Case studies included in the book provide step-by-step instructions on solving a variety of problems of moderate size and level of sophistication in *ANSYS Workbench*, and can be covered in the lab sessions. Modeling tips are available in the case studies wherever applicable to give students an immediate opportunity to use the skills they learn in a problem-solving context. Homework problems and projects using *ANSYS Workbench* software are provided at the end of each chapter.

The Solutions Manual contains solutions to all end-of-chapter problems. The manual features a compact presentation of solutions for over 100 problems. Each end-of-chapter problem that requires the use of an FEA tool is solved in *ANSYS Workbench*. For such a problem, the *Workbench* solution, which contains crucial properties such as geometry, material, mesh, boundary conditions, and simulation results, is saved using a *Workbench* project file (.wbpj) as well as a project folder that contains the supporting files for the *Workbench* project. The Solutions Manual, along with the *Workbench* solution files for over 80 end-of-chapter simulation problems, is available to instructors from CRC Press. Microsoft PowerPoint slides for each chapter are provided to instructors using this book. In addition, *Workbench* files for the case studies are available to instructors for use in lectures or computer lab sessions. The additional material is available from the CRC website: http://www.crcpress.com/product/isbn/9781439873847.

The materials in the book have been developed by the authors for undergraduate courses on FEA at the Washington State University Vancouver and at the University of Cincinnati for the last 10 years or so. The authors would like to thank many of their current and former students for their help in developing these lecture materials and their contributions to many of the examples used in the book, and in particular we thank students Qikai Xie, Fan Yang, and Bin He for their help in developing the supplementary materials for this book. We are grateful for the students we met in our classes at Vancouver and Cincinnati. This book could never have been written without their collective interest, feedback, and enthusiasm. Our special thanks go to the editorial staff at CRC Press, Jonathan Plant, the Senior Editor, and Amber Donley, the Project Coordinator, for their patience, cooperation, and professionalism in the production of this book. Much gratitude also goes to the reviewers for their valuable comments and suggestions. Of course, we take full responsibility for any errors or mistakes in the book. We also welcome any suggestions from readers that could help us improve the book in the future.

Xiaolin Chen
Vancouver, Washington
Yijun Liu
Cincinnati, Ohio

Authors

Dr. Xiaolin Chen is an associate professor of mechanical engineering at the Washington State University Vancouver. She earned her BS in engineering mechanics from Shanghai Jiao Tong University, MS in mechanical design and theory from the State Key Laboratory of Mechanical System and Vibration affiliated with Shanghai Jiao Tong University, and her PhD in mechanical engineering from the University of Cincinnati. Her research interests include computational mechanics, finite element analysis, boundary element analysis, multiphysics phenomena and coupled-field problems, inverse problems, and model order reduction techniques.

Dr. Yijun Liu is a professor of mechanical engineering at the University of Cincinnati. He earned his BS and MS in aerospace engineering from Northwestern Polytechnical University (China), and his PhD in theoretical and applied mechanics from the University of Illinois at Urbana-Champaign. Prior to joining the faculty, he conducted postdoctoral research at the Center of Nondestructive Evaluation of Iowa State University and worked at Ford Motor Company as a CAE analyst. Dr. Liu's interests are in computational mechanics, finite element method, boundary element method, and fast multipole method in modeling composite materials, fracture, fatigue, structural dynamics, and acoustics problems. He is the author of the book *Fast Multipole Boundary Element Method* (Cambridge University Press, 2009) and has had more than 100 papers published in technical journals and conference proceedings in areas of computational mechanics.

1

Introduction

1.1 Some Basic Concepts

The finite element method (FEM), or finite element analysis (FEA), is based on the idea of building a complicated object with simple blocks, or, dividing a complicated object into smaller and manageable pieces. Application of this simple idea can be found everywhere in everyday life (Figure 1.1), as well as in engineering. For example, children play with LEGO® toys by using many small pieces, each of very simple geometry, to build various objects such as trains, ships, or buildings. With more and more smaller pieces, these objects will look more realistic.

In mathematical terms, this is simply the use of the limit concept, that is, to approach or represent a smooth object with a finite number of simple pieces and increasing the number of such pieces in order to improve the accuracy of this representation.

1.1.1 Why FEA?

Computers have revolutionized the practice of engineering. Design of a product that used to be done by tedious hand drawings has been replaced by computer-aided design (CAD) using computer graphics. Analysis of a design used to be done by hand calculations and many of the testing have been replaced by computer simulations using computer-aided engineering (CAE) software. Together, CAD, CAE, and computer-aided manufacturing (CAM) have dramatically changed the landscape of engineering (Figure 1.2). For example, a car, that used to take five to six years from design to product, can now be produced starting from the concept design to the manufacturing within a year using the CAD/CAE/CAM technologies.

Among all the computational tools for CAE, the FEM is the most widely applied method or one of the most powerful modern "calculators" available for engineering students and professionals. FEA provides a way of virtually testing a product design. It helps users understand their designs and implement appropriate design changes early in the product development process. The adoption of FEA in the design cycle is driven by market pressure since it brings many benefits that will help companies make better products with reduced development costs and time-to-market.

1.1.2 Finite Element Applications in Engineering

The FEM can be applied in solving the mathematical models of many engineering problems, from stress analysis of truss and frame structures or complicated machines, to dynamic responses of automobiles, trains, or airplanes under different mechanical,

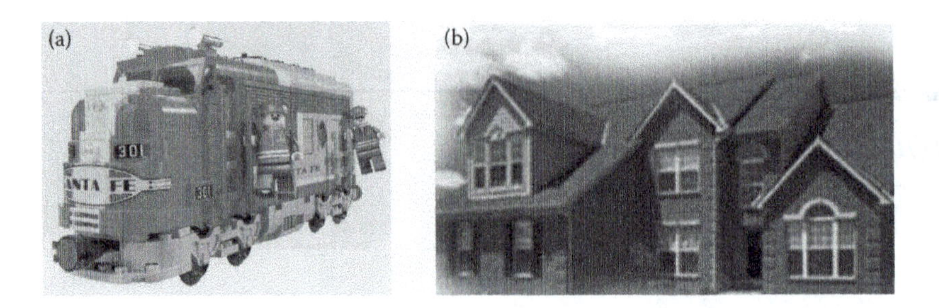

FIGURE 1.1
Objects built with simple and small pieces: (a) a fire engine built with LEGO® (http://lego.wikia.com/wiki/10020_Santa_Fe_Super_Chief); and (b) a house built with many elements—bricks, beams, columns, panels, and so on.

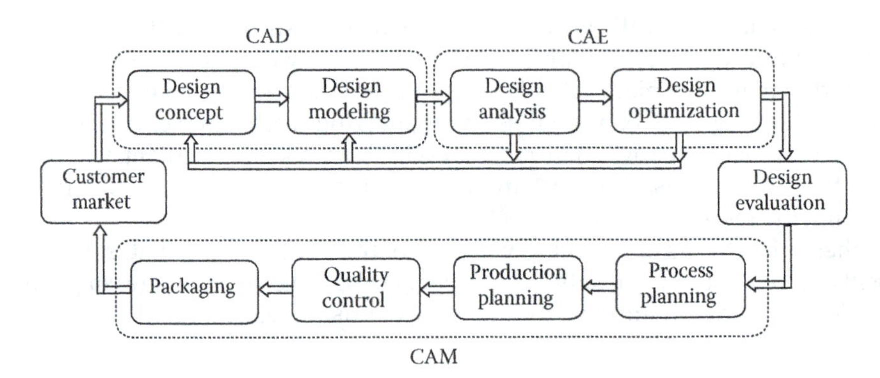

FIGURE 1.2
A sketch of the computer-aided product development process.

thermal, or electromagnetic loading. There are numerous finite element applications in industries, ranging from automotive, aerospace, defense, consumer products, and industrial equipment to energy, transportation and construction, as shown by some examples in Table 1.1. The applications of the FEA have also been extended to materials science, biomedical engineering, geophysics, and many other emerging fields in recent years.

TABLE 1.1
Examples of Engineering Applications Using FEA

Field of Study	Examples of Engineering Applications
Structural and solid mechanics	Offshore structure reliability analysis, vehicle crash simulation, nuclear reactor component integrity analysis, wind turbine blade design optimization
Heat transfer	Electronics cooling modeling, casting modeling, combustion engine heat-transfer analysis
Fluid flow	Aerodynamic analysis of race car designs, modeling of airflow patterns in buildings, seepage analysis through porous media
Electrostatics/electromagnetics	Field calculations in sensors and actuators, performance prediction of antenna designs, electromagnetic interference suppression analysis

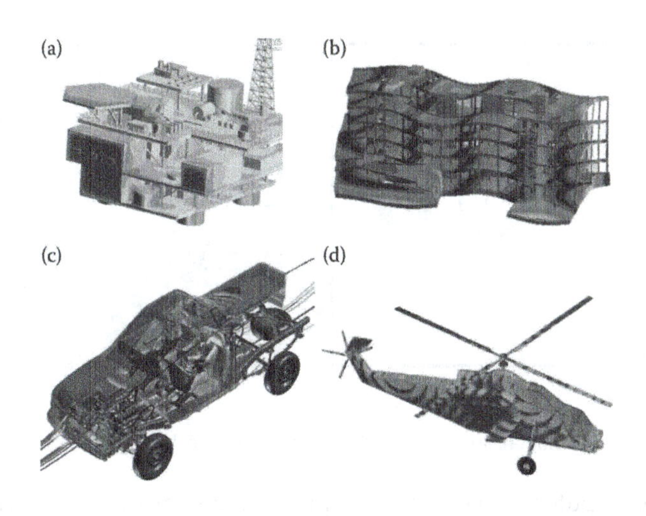

FIGURE 1.3
Examples of FEA using *ANSYS Workbench*: (a) wind load simulation of an offshore platform (Courtesy of ANSYS, Inc., http://www.ansys.com/Industries/Energy/Oil+&+Gas); (b) modal response of a steel frame building with concrete slab floors (http://www.isvr.co.uk/modelling/); (c) underhood flow and thermal management (Courtesy of ANSYS, Inc., http://www.ansys.com/Industries/Automotive/Application+Highlights/Underhood); and (d) electric field pattern of antenna mounted on helicopter (Courtesy of ANSYS, Inc., http://www.ansys.com/Industries/Electronics+&+Semiconductor/Defense+&+Aerospace + Electronics).

1.1.3 FEA with *ANSYS Workbench*

Over the last few decades, many commercial programs have become available for conducting the FEA. Among a comprehensive range of finite element simulation solutions provided by leading CAE companies, *ANSYS® Workbench* is a user-friendly platform designed to seamlessly integrate *ANSYS, Inc.*'s suite of advanced engineering simulation technology. It offers bidirectional connection to major CAD systems. The *Workbench* environment is geared toward improving productivity and ease of use among engineering teams. It has evolved as an indispensible tool for product development at a growing number of companies, finding applications in many diverse engineering fields (Figure 1.3).

1.1.4 A Brief History of FEA

An account of the historical development of FEM and the computational mechanics in general was given by O. C. Zienkiewicz recently, which can be found in Reference [1]. The foundation of the FEM was first developed by Courant in the early 1940s. The stiffness method, a prelude of the FEM, was developed by Turner, Clough et al., in 1956. The name "finite element" was coined by Clough in 1960. Computer implementation of FEM programs emerged during the early 1970s. To date, FEM has become one of the most widely used and versatile analysis techniques. A few major milestones are as follows:

1943—Courant (Variational methods which laid the foundation for FEM)
1956—Turner, Clough, Martin, and Topp (Stiffness method)
1960—Clough (Coined "Finite Element," solved plane problems)
1970s—Applications on "mainframe" computers

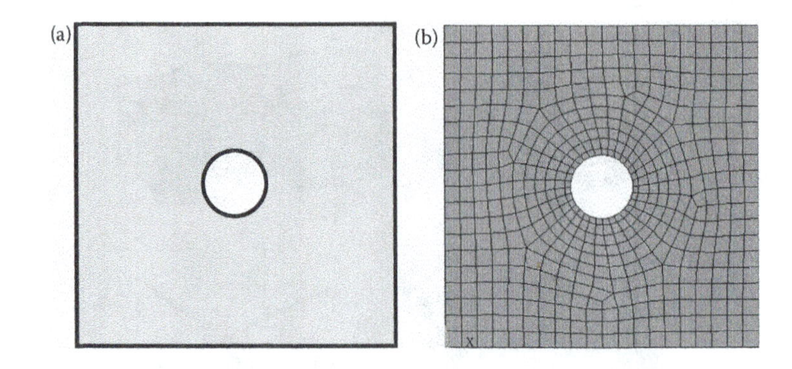

FIGURE 1.4
(a) A plate with a hole (CAD model); and (b) A FEM discretization (mesh).

1980s—Microcomputers, development of pre- and postprocessors (GUI)
1990s—Analysis of large structural systems, nonlinear, and dynamic problems
2000s—Analysis of multiphysics and multiscale problems

1.1.5 A General Procedure for FEA

To conduct an FEA, the following procedure is required in general:

- Divide the CAD/geometric model into pieces to create a "mesh" (a collection of elements with nodes, Figure 1.4).
- Describe the behavior of the physical quantities on each element.
- Connect (assemble) the elements at the nodes to form an approximate system of equations for the entire model.
- Apply loads and boundary conditions (e.g., to prevent the model from moving).
- Solve the system of equations involving unknown quantities at the nodes (e.g., the displacements).
- Calculate the desired quantities (e.g., strains and stresses) at elements or nodes.

In commercial FEA software, this procedure is typically rearranged into the following phases:

- Preprocessing (build FEM models, define element properties, and apply loads and constraints)
- FEA solver (assemble and solve the FEM system of equations, calculate element results)
- Postprocessing (sort and display the results)

1.2 An Example in FEA: Spring System

A glimpse into the steps involved in an FEA is provided through a simple example in this section. We will look at a spring element and a spring system to gain insight into the basic concepts of the FEM.

1.2.1 One Spring Element

For the single element shown in Figure 1.5, we have:

Two nodes	i, j
Nodal displacements	u_i, u_j (m, mm)
Nodal forces	f_i, f_j (Newton)
Spring constant (stiffness)	k (N/m, N/mm)

Relationship between spring force F and elongation Δ is shown in Figure 1.6.

In the linear portion of the curve shown in Figure 1.6, we have

$$F = k\Delta, \quad \text{with } \Delta = u_j - u_i \tag{1.1}$$

where $k = F/\Delta(>0)$ is the stiffness of the spring (the force needed to produce a unit stretch).

Consider the equilibrium of forces for the spring. At node i, we have

$$f_i = -F = -k(u_j - u_i) = ku_i - ku_j$$

and at node j

$$f_j = F = k(u_j - u_i) = -ku_i + ku_j$$

In matrix form,

$$\begin{bmatrix} k & -k \\ -k & k \end{bmatrix} \begin{Bmatrix} u_i \\ u_j \end{Bmatrix} = \begin{Bmatrix} f_i \\ f_j \end{Bmatrix} \tag{1.2}$$

FIGURE 1.5
One spring element.

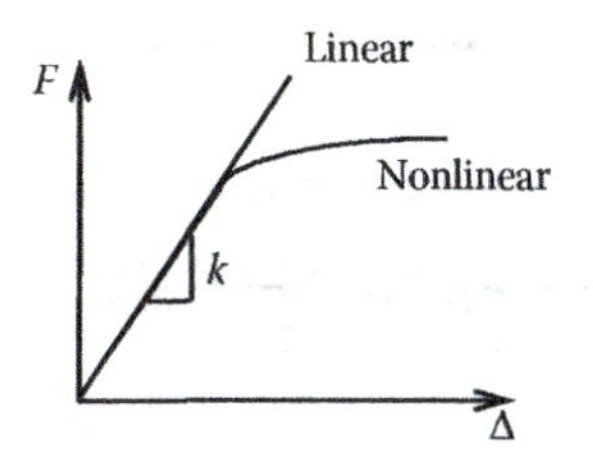

FIGURE 1.6
Force–displacement relation in a spring.

or,

$$\mathbf{ku} = \mathbf{f} \tag{1.3}$$

where

$\mathbf{k}$ = element stiffness matrix
$\mathbf{u}$ = element nodal displacement vector
$\mathbf{f}$ = element nodal force vector

From the derivation, we see that the first equation in Equation 1.2 represents the equilibrium of forces at node *i*, while the second equation in Equation 1.2 represents the equilibrium of forces at node *j*. Note also that **k** is symmetric. Is **k** singular or nonsingular? That is, can we solve the equation in Equation 1.2? If not, why?

1.2.2 A Spring System

For a system of multiple spring elements, we first write down the stiffness equation for each spring and then "assemble" them together to form the stiffness equation for the whole system. For example, for the two-spring system shown in Figure 1.7, we proceed as follows:

For element 1, we have

$$\begin{bmatrix} k_1 & -k_1 \\ -k_1 & k_1 \end{bmatrix} \begin{Bmatrix} u_1 \\ u_2 \end{Bmatrix} = \begin{Bmatrix} f_1^1 \\ f_2^1 \end{Bmatrix} \tag{1.4}$$

and for element 2,

$$\begin{bmatrix} k_2 & -k_2 \\ -k_2 & k_2 \end{bmatrix} \begin{Bmatrix} u_2 \\ u_3 \end{Bmatrix} = \begin{Bmatrix} f_1^2 \\ f_2^2 \end{Bmatrix} \tag{1.5}$$

where f_i^m is the (internal) force acting on *local* node *i* of element *m* ($i = 1, 2$).

1.2.2.1 Assembly of Element Equations: Direct Approach

Consider the equilibrium of forces at node 1,

$$F_1 = f_1^1$$

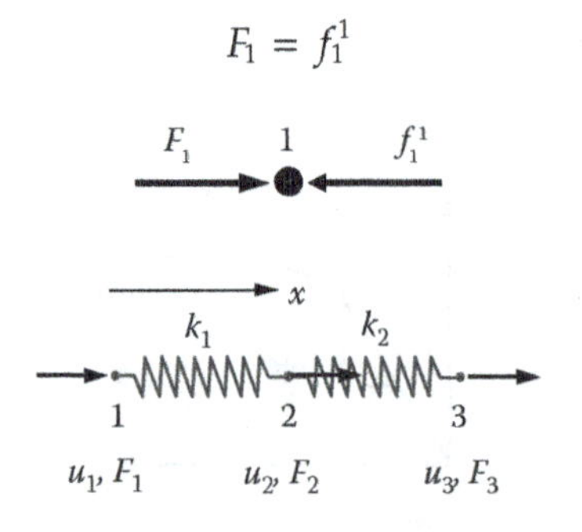

FIGURE 1.7
A system of two spring elements.

at node 2,

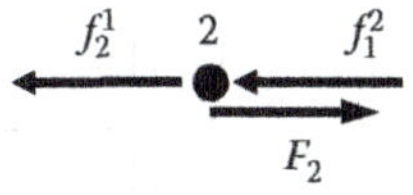

$$F_2 = f_2^1 + f_1^2$$

and, at node 3,

$$F_3 = f_2^2$$

Using Equations 1.4 and 1.5, we obtain

$$\begin{aligned} F_1 &= k_1u_1 - k_1u_2 \\ F_2 &= -k_1u_1 + (k_1 + k_2)u_2 - k_2u_3 \\ F_3 &= -k_2u_2 + k_2u_3 \end{aligned}$$

In matrix form, we have

$$\begin{bmatrix} k_1 & -k_1 & 0 \\ -k_1 & k_1 + k_2 & -k_2 \\ 0 & -k_2 & k_2 \end{bmatrix} \begin{Bmatrix} u_1 \\ u_2 \\ u_3 \end{Bmatrix} = \begin{Bmatrix} F_1 \\ F_2 \\ F_3 \end{Bmatrix} \tag{1.6}$$

or

$$\mathbf{Ku} = \mathbf{F} \tag{1.7}$$

in which, **K** is the stiffness matrix (structure matrix) for the entire spring system.

1.2.2.1.1 An Alternative Way of Assembling the Whole Stiffness Matrix

"Enlarging" the stiffness matrices for elements 1 and 2, we have

$$\begin{bmatrix} k_1 & -k_1 & 0 \\ -k_1 & k_1 & 0 \\ 0 & 0 & 0 \end{bmatrix} \begin{Bmatrix} u_1 \\ u_2 \\ u_3 \end{Bmatrix} = \begin{Bmatrix} f_1^1 \\ f_2^1 \\ 0 \end{Bmatrix}$$

and

$$\begin{bmatrix} 0 & 0 & 0 \\ 0 & k_2 & -k_2 \\ 0 & -k_2 & k_2 \end{bmatrix} \begin{Bmatrix} u_1 \\ u_2 \\ u_3 \end{Bmatrix} = \begin{Bmatrix} 0 \\ f_1^2 \\ f_2^2 \end{Bmatrix}$$

Adding the two matrix equations (i.e., using *superposition*), we have

$$\begin{bmatrix} k_1 & -k_1 & 0 \\ -k_1 & k_1+k_2 & -k_2 \\ 0 & -k_2 & k_2 \end{bmatrix} \begin{Bmatrix} u_1 \\ u_2 \\ u_3 \end{Bmatrix} = \begin{Bmatrix} f_1^1 \\ f_2^1 + f_1^2 \\ f_2^2 \end{Bmatrix}$$

This is the same equation we derived by using the concept of equilibrium of forces.

1.2.2.2 Assembly of Element Equations: Energy Approach

We can also obtain the result using an energy method, for example, the principle of minimum potential energy. In fact, the energy approach is more general and considered the foundation of the FEM. To proceed, we consider the strain energy U stored in the spring system shown in Figure 1.5.

$$U = \frac{1}{2}k_1\Delta_1^2 + \frac{1}{2}k_2\Delta_2^2 = \frac{1}{2}\Delta_1^T k_1 \Delta_1 + \frac{1}{2}\Delta_2^T k_2 \Delta_2$$

However,

$$\Delta_1 = u_2 - u_1 = [-1 \quad 1]\begin{Bmatrix} u_1 \\ u_2 \end{Bmatrix}, \qquad \Delta_2 = u_3 - u_2 = [-1 \quad 1]\begin{Bmatrix} u_2 \\ u_3 \end{Bmatrix}$$

We have

$$U = \frac{1}{2}[u_1 \quad u_2]\begin{bmatrix} k_1 & -k_1 \\ -k_1 & k_1 \end{bmatrix}\begin{Bmatrix} u_1 \\ u_2 \end{Bmatrix} + \frac{1}{2}[u_2 \quad u_3]\begin{bmatrix} k_2 & -k_2 \\ -k_2 & k_2 \end{bmatrix}\begin{Bmatrix} u_2 \\ u_3 \end{Bmatrix} = \text{(enlarging...)}$$

$$= \frac{1}{2}[u_1 \quad u_2 \quad u_3]\begin{bmatrix} k_1 & -k_1 & 0 \\ -k_1 & k_1+k_2 & -k_2 \\ 0 & -k_2 & k_2 \end{bmatrix}\begin{Bmatrix} u_1 \\ u_2 \\ u_3 \end{Bmatrix} \tag{1.8}$$

The potential of the external forces is

$$\Omega = -F_1u_1 - F_2u_2 - F_3u_3 = -[u_1 \quad u_2 \quad u_3]\begin{Bmatrix} F_1 \\ F_2 \\ F_3 \end{Bmatrix} \tag{1.9}$$

Thus, the total potential energy of the system is

$$\Pi = U + \Omega = \frac{1}{2}[u_1 \quad u_2 \quad u_3]\begin{bmatrix} k_1 & -k_1 & 0 \\ -k_1 & k_1+k_2 & -k_2 \\ 0 & -k_2 & k_2 \end{bmatrix}\begin{Bmatrix} u_1 \\ u_2 \\ u_3 \end{Bmatrix} - [u_1 \quad u_2 \quad u_3]\begin{Bmatrix} F_1 \\ F_2 \\ F_3 \end{Bmatrix} \tag{1.10}$$

which is a function of the three nodal displacements (u_1, u_2, u_3). According to the principle of minimum potential energy, for a system to be in equilibrium, the total potential energy must be minimum, that is, $d\Pi = 0$, or equivalently,

$$\frac{\partial \Pi}{\partial u_1} = 0, \quad \frac{\partial \Pi}{\partial u_2} = 0, \quad \frac{\partial \Pi}{\partial u_3} = 0, \tag{1.11}$$

which yield the same three equations as in Equation 1.6.

1.2.3 Boundary and Load Conditions

Assuming that node 1 is fixed, and same force P is applied at node 2 and node 3, that is

$$u_1 = 0 \quad \text{and} \quad F_2 = F_3 = P$$

we have from Equation 1.6

$$\begin{bmatrix} k_1 & -k_1 & 0 \\ -k_1 & k_1 + k_2 & -k_2 \\ 0 & -k_2 & k_2 \end{bmatrix} \begin{Bmatrix} 0 \\ u_2 \\ u_3 \end{Bmatrix} = \begin{Bmatrix} F_1 \\ P \\ P \end{Bmatrix}$$

which reduces to

$$\begin{bmatrix} k_1 + k_2 & -k_2 \\ -k_2 & k_2 \end{bmatrix} \begin{Bmatrix} u_2 \\ u_3 \end{Bmatrix} = \begin{Bmatrix} P \\ P \end{Bmatrix}$$

and

$$F_1 = -k_1 u_2$$

Unknowns are

$$\mathbf{u} = \begin{Bmatrix} u_2 \\ u_3 \end{Bmatrix}$$

and the reaction force F_1 (if desired).

Solving the equations, we obtain the displacements

$$\begin{Bmatrix} u_2 \\ u_3 \end{Bmatrix} = \begin{Bmatrix} 2P/k_1 \\ 2P/k_1 + P/k_2 \end{Bmatrix}$$

and the reaction force

$$F_1 = -2P$$

1.2.4 Solution Verification

It is very important in FEA to verify the results you obtained through either hand calculations or analytical solutions in the literature. The following is a list of items to check based on common sense or intuition, or analytical solutions if they are available.

- Deformed shape of the structure
- Equilibrium of the external forces (Reaction forces should balance with the applied loads.)
- Order of magnitudes of the obtained values

Notes about the Spring Elements:

- Spring elements are only suitable for stiffness analysis.
- They are not suitable for stress analysis of the spring itself.
- There are spring elements with stiffness in the lateral direction, spring elements for torsion, and so on.

1.2.5 Example Problems

EXAMPLE 1.1

k_1 k_2 P k_3

1 2 3 4 x

Given: For the spring system shown above,

$$k_1 = 100\ \text{N/mm},\ k_2 = 200\ \text{N/mm},\ k_3 = 100\ \text{N/mm}$$
$$P = 500\ \text{N},\ u_1 = u_4 = 0$$

Find:

a. The global stiffness matrix
b. Displacements of nodes 2 and 3
c. The reaction forces at nodes 1 and 4
d. The force in the spring 2

Solution

a. The element stiffness matrices are (make sure to put proper unit after each number)

$$\mathbf{k}_1 = \begin{bmatrix} 100 & -100 \\ -100 & 100 \end{bmatrix} (\text{N/mm})$$

$$\mathbf{k}_2 = \begin{bmatrix} 200 & -200 \\ -200 & 200 \end{bmatrix} (\text{N/mm})$$

$$\mathbf{k}_3 = \begin{bmatrix} 100 & -100 \\ -100 & 100 \end{bmatrix} \text{(N/mm)}$$

Applying the superposition concept, we obtain the global stiffness matrix for the spring system

$$\mathbf{K} = \begin{bmatrix} 100 & -100 & 0 & 0 \\ -100 & 100+200 & -200 & 0 \\ 0 & -200 & 200+100 & -100 \\ 0 & 0 & -100 & 100 \end{bmatrix} \quad (\text{columns } u_1, u_2, u_3, u_4)$$

or

$$\mathbf{K} = \begin{bmatrix} 100 & -100 & 0 & 0 \\ -100 & 300 & -200 & 0 \\ 0 & -200 & 300 & -100 \\ 0 & 0 & -100 & 100 \end{bmatrix}$$

which is *symmetric* and *banded*.

Equilibrium (FE) equation for the whole system is

$$\begin{bmatrix} 100 & -100 & 0 & 0 \\ -100 & 300 & -200 & 0 \\ 0 & -200 & 300 & -100 \\ 0 & 0 & -100 & 100 \end{bmatrix} \begin{Bmatrix} u_1 \\ u_2 \\ u_3 \\ u_4 \end{Bmatrix} = \begin{Bmatrix} F_1 \\ F_2 \\ F_3 \\ F_4 \end{Bmatrix}$$

b. Applying the BCs $u_1 = u_4 = 0$, $F_2 = 0$, and $F_3 = P$, and "deleting" the first and fourth rows and columns, we have

$$\begin{bmatrix} 300 & -200 \\ -200 & 300 \end{bmatrix} \begin{Bmatrix} u_2 \\ u_3 \end{Bmatrix} = \begin{Bmatrix} 0 \\ P \end{Bmatrix}$$

Solving this equation, we obtain

$$\begin{Bmatrix} u_2 \\ u_3 \end{Bmatrix} = \begin{Bmatrix} P/250 \\ 3P/500 \end{Bmatrix} = \begin{Bmatrix} 2 \\ 3 \end{Bmatrix} \text{(mm)}$$

c. From the first and fourth equations in the system of FE equations, we obtain the reaction forces

$$F_1 = -100u_2 = -200 \text{ (N)}$$

$$F_4 = -100u_3 = -300 \text{ (N)}$$

d. The FE equation for spring (element) 2 is

$$\begin{bmatrix} 200 & -200 \\ -200 & 200 \end{bmatrix} \begin{Bmatrix} u_i \\ u_j \end{Bmatrix} = \begin{Bmatrix} f_i \\ f_j \end{Bmatrix}$$

Here $i = 2, j = 3$ for element 2. Thus we can calculate the spring force as

$$F = f_j = -f_i = [-200 \quad 200] \begin{Bmatrix} u_2 \\ u_3 \end{Bmatrix} = [-200 \quad 200] \begin{Bmatrix} 2 \\ 3 \end{Bmatrix} = 200 \text{ (N)}$$

Check the results:

Draw the free-body diagram (FBD) of the system and consider the equilibrium of the forces.

$F_1 = -200$ N

$F_4 = -300$ N

$P = 500$ N

Equilibrium of the forces is satisfied!

EXAMPLE 1.2

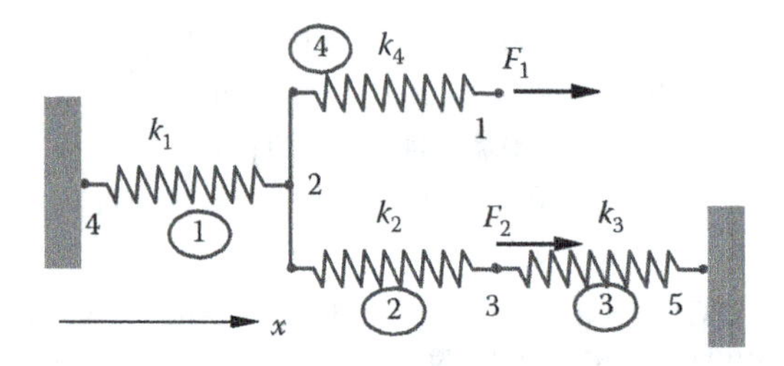

Problem

For the spring system with arbitrarily numbered nodes and elements, as shown above, find the global stiffness matrix.

Solution

First, we construct the following *element connectivity table*:

Element	Node *i* (1)	Node *j* (2)
1	4	2
2	2	3
3	3	5
4	2	1

This table specifies the *global* node numbers corresponding to the *local* node numbers for each element.

Then we write the element stiffness matrix for each element

$$\mathbf{k}_1 = \begin{bmatrix} k_1 & -k_1 \\ -k_1 & k_1 \end{bmatrix}, \quad (u_4, u_2)$$

$$\mathbf{k}_2 = \begin{bmatrix} k_2 & -k_2 \\ -k_2 & k_2 \end{bmatrix}, \quad (u_2, u_3)$$

$$\mathbf{k}_3 = \begin{bmatrix} k_3 & -k_3 \\ -k_3 & k_3 \end{bmatrix}, \quad (u_3, u_5)$$

$$\mathbf{k}_4 = \begin{bmatrix} k_4 & -k_4 \\ -k_4 & k_4 \end{bmatrix} \quad (u_2, u_1)$$

Finally, applying the superposition method, we obtain the global stiffness matrix as follows:

$$\mathbf{K} = \begin{bmatrix} k_4 & -k_4 & 0 & 0 & 0 \\ -k_4 & k_1 + k_2 + k_4 & -k_2 & -k_1 & 0 \\ 0 & -k_2 & k_2 + k_3 & 0 & -k_3 \\ 0 & -k_1 & 0 & k_1 & 0 \\ 0 & 0 & -k_3 & 0 & k_3 \end{bmatrix} \quad (u_1, u_2, u_3, u_4, u_5)$$

The matrix is *symmetric, banded,* but *singular,* as it should be.

After introducing the basic concepts, the section below introduces you to one of the general-purpose finite element software tools—*ANSYS Workbench.*

1.3 Overview of *ANSYS Workbench*

ANSYS Workbench is a simulation platform that enables users to model and solve a wide range of engineering problems using the FEA. It provides access to the *ANSYS* family of design and analysis modules in an integrated simulation environment. This section gives a brief overview of the different elements in the *ANSYS Workbench* simulation environment or the graphical-user interface (GUI). Readers are referred to *ANSYS Workbench* user's guide [2] for more detailed information.

1.3.1 The User Interface

The *Workbench* interface is composed primarily of a *Toolbox* region and a *Project Schematic* region (Figure 1.8). The main use of the two regions is described next.

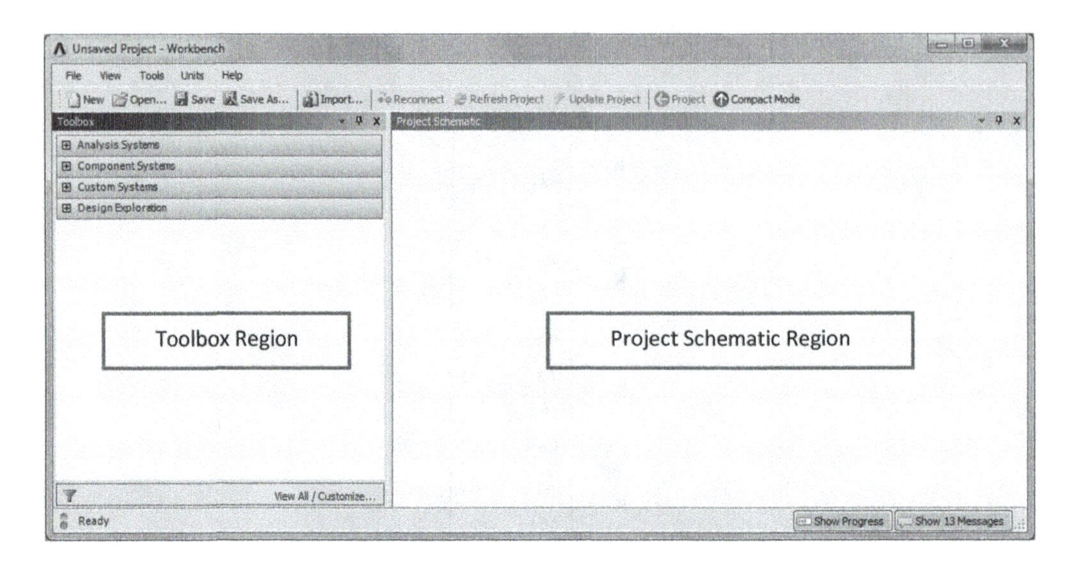

FIGURE 1.8
ANSYS Workbench user interface.

1.3.2 The Toolbox

The *Toolbox* contains the following four groups of systems:

Analysis Systems: Predefined analysis templates to be used to build your project, including static structural, steady-state thermal, transient thermal, fluid flow, modal, shape optimization, linear buckling, and many others.

Component Systems: Component applications that can be used to build or expand an analysis system, including geometry import, engineering data, mesh, postprocessing, and others.

Custom Systems: Coupled-field analysis systems such as fluid solid interaction, prestress modal, thermal-stress, and others.

Design Exploration: Parametric optimization studies such as response surface optimization, parameters correlation, six sigma analysis, and others.

1.3.3 The Project Schematic

A project schematic, that is, a graphical representation of the workflow, can be built by dragging predefined analysis templates or other components from the *Toolbox* and dropping them into the *Project Schematic* window. "Drag" here means to move the mouse while holding down the left mouse button, and "drop" means to release the mouse button.

To build a project for static structural analysis, for instance, drag the *Static Structural* template from the *Toolbox* and drop it into the rectangular box that appears in the *Project Schematic* window. A standalone analysis system that contains the components needed for static structural analysis is added to the project schematic as shown in Figure 1.9a. The system consists of seven individual components called cells.

Alternatively, a standalone analysis can be created by double-clicking. For example, double-click the *Steady-State Thermal* template from the *Toolbox*, and an independent

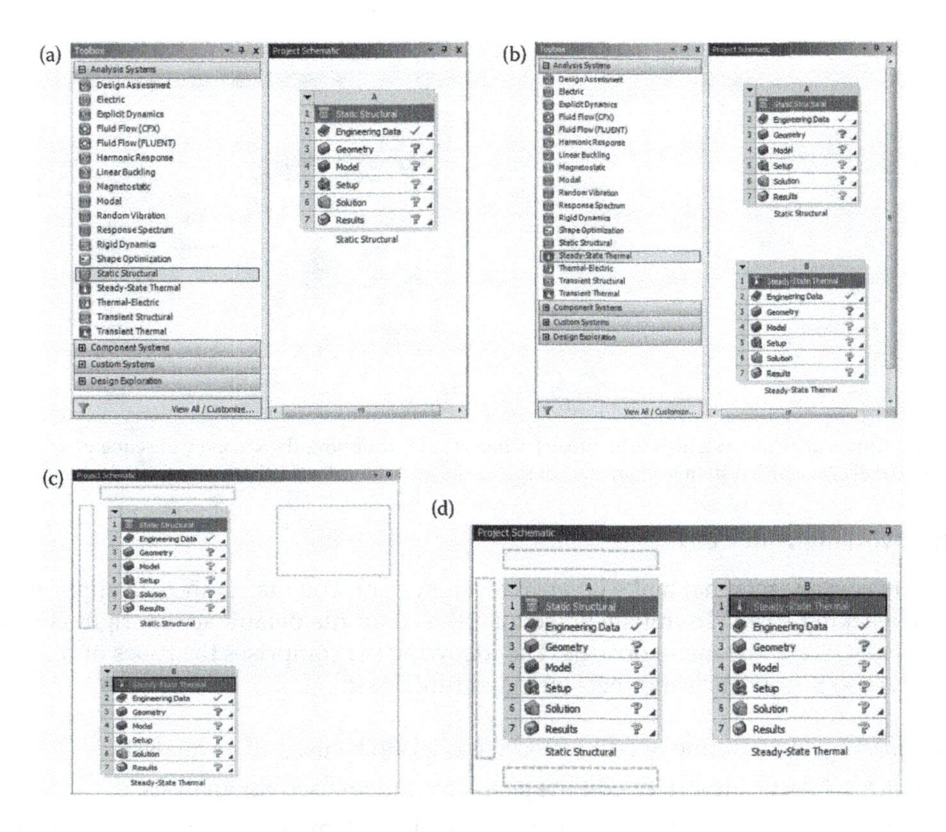

FIGURE 1.9
Defining standalone analysis systems in the project schematic: (a) a standalone system; (b) two independent standalone systems; (c) moving a system in a top-bottom configuration; and (d) moving a system in a side-by-side configuration.

Steady-State Thermal system will be placed in the default location below the existing *Static Structural* system, as shown in Figure 1.9b.

A system can be moved around another system in the project schematic. To move a system, click on the header cell (i.e., the cell titled *Steady-State Thermal* for the thermal system) and drag it to a new place. Once you drag the header cell, dashed rectangles appear for the possible new locations to drop the system. This is illustrated in Figures 1.9c and d for two systems with initial top–bottom and side-by-side configurations, respectively.

To delete a system, click on the down arrow button at the upper left corner of the system from the *Project Schematic* window, and then choose *Delete* from the drop-down context menu.

In some cases, a project may contain two or more analysis systems that share data. For example, a downstream modal analysis may use the same material, geometry, and model data from the preceding structural analysis. To build such a project, create a standalone system for *Static Structural* analysis. Then, drag the *Modal* analysis template from the *Toolbox* and drop it onto the *Model* cell of the *Static Structural* system. Immediately before the subsequent system is dropped, bounding boxes will appear on the *Engineering Data, Geometry,* and *Model* cells of the first system, as shown in Figure 1.10a. After the system is released, a project including two linked systems is created, as shown in Figure 1.10b, where the linked cells indicate data sharing at the *Model* and above levels.

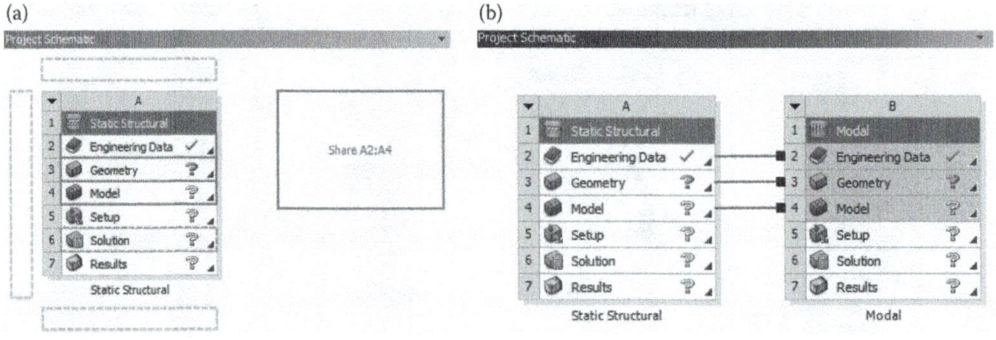

(a) (b)

FIGURE 1.10
Defining linked analysis systems in the project schematic: (a) dropping the second (subsequent) system onto the *Model* cell of the first system to share data at the model and above levels; (b) two systems that are linked.

1.3.4 Working with Cells

Cells are components that make up an analysis system. You may launch an application by double-clicking a cell. To initiate an action other than the default action, right-click on a cell to view its context menu options. The following list comprises the types of cells available in *ANSYS Workbench* and their intended functions:

Engineering Data: Define or edit material models to be used in an analysis.

Geometry: Create, import, or edit the geometry model used for analysis.

Model/Mesh: Assign material, define coordinate system, and generate mesh for the model.

Setup: Apply loads, boundary conditions, and configure the analysis settings.

Solution: Access the model solution or share solution data with other downstream systems.

Results: Indicate the results availability and status (also referred to as postprocessing).

As the data flows through a system, a cell's state can quickly change. *ANSYS Workbench* provides a state indicator icon placed on the right side of the cell. Table 1.2 describes the indicator icons and the various cell states available in *ANSYS Workbench*. For more information, please refer to *ANSYS Workbench* user's guide [2].

1.3.5 The Menu Bar

The menu bar is the horizontal bar anchored at the top of the *Workbench* user interface. It provides access to the following functions:

File Menu: Create a new project, open an existing project, save the current project, and so on.

View Menu: Control the window/workspace layout, customize the toolbox, and so on.

Tools Menu: Update the project and set the license preferences and other user options.

Units Menu: Select the unit system and specify unit display options.

Help Menu: Get help for *ANSYS Workbench*.

TABLE 1.2

Indicator Icons and Descriptions of the Various Cell States

Cell State	Indicator	Description
Unfulfilled		Need upstream data to proceed
Refresh required		A refresh action is needed as a result of changes made on upstream data
Attention required		User interaction with the cell is needed to proceed
Update required		An update action is needed as a result of changes made on upstream data
Up to date		Data are up to date and no attention is required
Input changes pending		An update or refresh action is needed to recalculate based on changes made to upstream cells
Interrupted		Solution has been interrupted. A resume or update action will make the solver continue from the interrupted point
Pending		Solution is in progress

Source: Courtesy of *ANSYS Workbench User's Guide*, Release 14.5, ANSYS, Inc., 2012.

In the chapters that follow, the use of *ANSYS Workbench* will be presented in a step-by-step fashion in the context of real-world problem solving that may involve different modeling concepts or different physics. It is worth noting that, although a commercial finite element program enables you to deal with a wide range of engineering problems with complex geometry, constraints, and material behaviors, it is your responsibility to understand the underlying physics to be able to setup a problem correctly. Your FEA results, no matter how pretty they may look, are only as good as the assumptions and decisions you have made in building your model.

1.4 Summary

In this chapter, the basic concepts in the FEM are introduced. The spring system is used as an example to show how to establish the element stiffness matrices, to assemble the finite element equations for a system from element stiffness matrices, and to solve the FE equations. Verifying the FE results is emphasized. *ANSYS Workbench* environment is briefly introduced. The concepts and procedures introduced in this chapter are very simple and yet very important for studying the finite element analyses of other problems.

PROBLEMS

1.1 Answer the following questions briefly:

a. What is the physical meaning of the FE equations (for either an element or the whole structure)?

b. What is the procedure in using the FEM?

1.2 Answer the following questions briefly:

a. What is a finite element discretization or a finite element mesh?

b. Define the terms: node and element.

1.3 Using *ANSYS Workbench,* solve the following problems:

a. Create a project schematic that includes structural analysis of two design parts of the same vehicle.

b. Create a project schematic that allows sequential thermal-stress coupling analysis.

1.4 Consider the spring system shown below. Find the global stiffness matrix for the entire system.

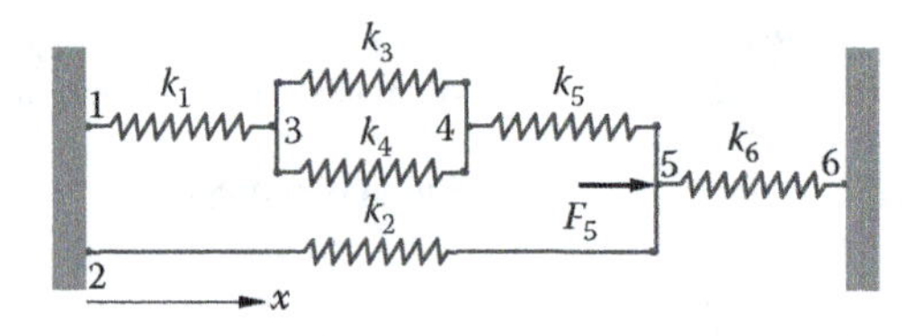

1.5 Consider the spring system shown below. Find the global stiffness matrix for the entire system.

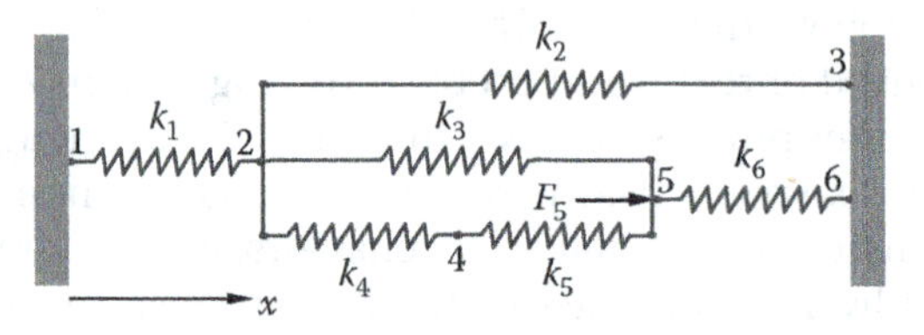

1.6 A spring system is shown below

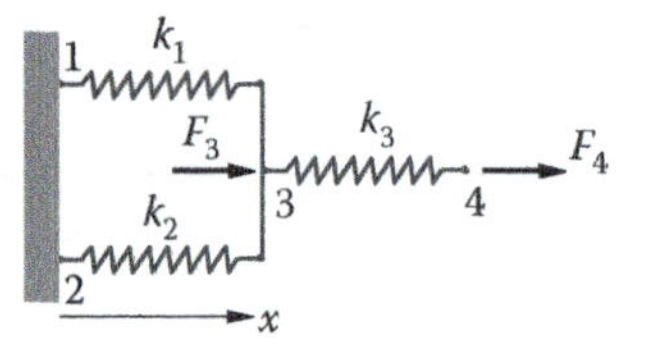

Given: $k_1 = 80$ N/mm, $k_2 = 100$ N/mm, $k_3 = 160$ N/mm, $F_3 = 200$ N, $F_4 = 100$ N, and nodes 1 and 2 are fixed;

Find:

a. Global stiffness matrix

b. Displacements of nodes 3 and 4

c. Reaction forces at nodes 1 and 2

d. Forces in springs 1 and 2

1.7 A spring system is shown below

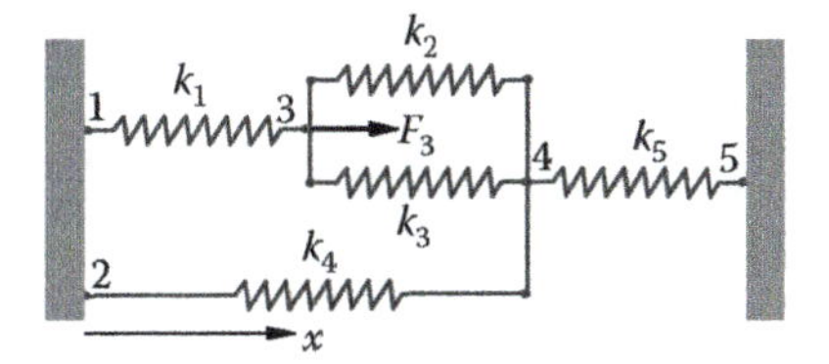

Given: $k_1 = k_2 = k_3 = k_4 = k_5 = k$, $F_3 = P$, and nodes 1, 2, and 5 are fixed;

Find:

a. Global stiffness matrix

b. Displacements of nodes 3 and 4

c. Reaction forces at nodes 1, 2, and 5

d. Forces in springs 2, 4, and 5

2

Bars and Trusses

2.1 Introduction

This chapter introduces you to the simplest one-dimensional (1-D) structural element, namely the bar element, and the FEA of truss structures using such element. Trusses are commonly used in the design of buildings, bridges, and towers (Figure 2.1). They are triangulated frameworks composed of slender bars whose ends are connected through bolts, pins, rivets, and so on. Truss structures create large, open, and uninterrupted space, and offer lightweight and economical solutions to many engineering situations. If a truss, along with the applied load, lies in a single plane, it is called a planar truss. If it has members and joints extending into the three-dimensional (3-D) space, it is then a space truss.

Most structural analysis problems such as stress and strain analysis can be treated as *linear static* problems, based on the following assumptions:

1. *Small deformations* (loading pattern is not changed due to the deformed shape)
2. *Elastic materials* (no plasticity or failures)
3. *Static loads* (the load is applied to the structure in a slow or steady fashion)

Linear analysis can provide most of the information about the behavior of a structure, and can be a good approximation for many analyses. It is also the basis of nonlinear FEA in most of the cases. In Chapters 2 through 7, only linear static responses of structures are considered.

2.2 Review of the 1-D Elasticity Theory

We begin by examining the problem of an axially loaded bar based on 1-D linear elasticity. Consider a uniform prismatic bar shown in Figure 2.2. The parameters L, A, and E are the length, cross-sectional area, and elastic modulus of the bar, respectively.

Let u, ε, and σ be the displacement, strain, and stress, respectively (all in the axial direction and functions of x only), we have the following basic relations:

Strain–displacement relation:

$$\varepsilon(x) = \frac{du(x)}{dx} \tag{2.1}$$

FIGURE 2.1
Truss examples: (a) Montreal Biosphere Museum (http://en.wikipedia.org/wiki/Montreal_Biosph%C3%A8re). (b) Betsy Ross Bridge (http://en.wikipedia.org/wiki/Betsy_Ross_Bridge).

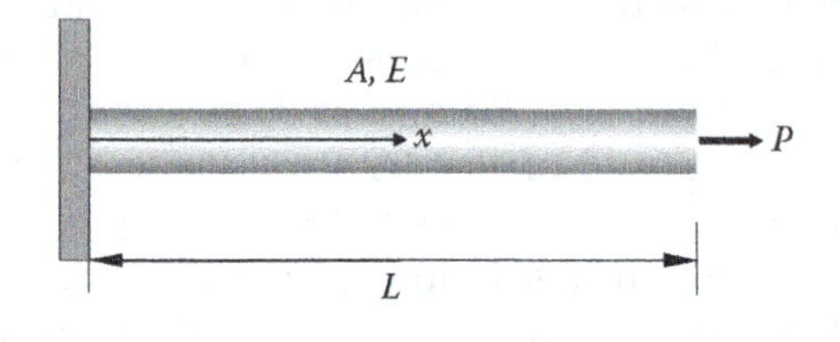

FIGURE 2.2
An axially loaded elastic bar.

Stress–strain relation:

$$\sigma(x) = E\varepsilon(x) \tag{2.2}$$

Equilibrium equation:

$$\frac{d\sigma(x)}{dx} + f(x) = 0 \tag{2.3}$$

where $f(x)$ is the body force (force per unit volume, such as gravitational and magnetic forces) inside the bar. To obtain the displacement, strain, and stress field in a bar, Equations 2.1 through 2.3 need to be solved under given boundary conditions, which can be done readily for a single bar, but can be tedious for a network of bars or a truss structure made of many bars.

2.3 Modeling of Trusses

For the truss analysis, it is often assumed that: (1) the bar members are of uniform cross sections and are joined together by frictionless pins, and (2) loads are applied to joints only and not in between joints along the truss members. It is based on these assumptions that the truss members are considered to carry only axial loads and have negligible bending resistance.

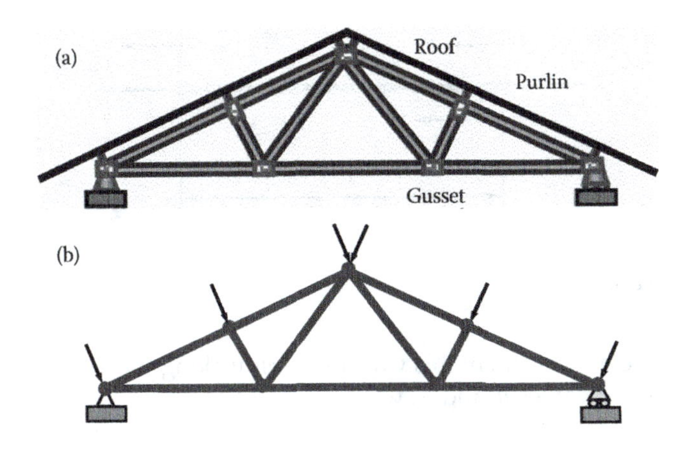

FIGURE 2.3
Modeling of a planar roof truss: (a) Physical structure. (b) Discrete model.

Given the fact that a truss is an assembly of axial bars, it is generally satisfactory to treat each truss member as a discrete element—bar element. A bar element is a 1-D finite element to describe the deformation, strain, and stress in a slender structure member that often has a uniform cross section and is loaded in tension or compression only along its axis.

Let us take a planar roof structure as an example to illustrate the truss model idealization process. In an actual structure pictured in Figure 2.3a, purlins are used to support the roof and they further transmit the load to gusset plates that connect the members. In the idealized model shown in Figure 2.3b, each truss member is simply replaced by a two-node bar element. A joint condition is modeled by node sharing between elements connected end-to-end, and loads are imposed only on nodes. Components such as purlins and gussets are neglected to avoid construction details that do not contribute significantly to the overall truss load–deflection behavior. Other factors such as the weight of the truss members can be considered negligible, compared to the loads they carry. If it were to be included, a member's weight can always be applied to its two ends, half the weight at each end node. Experience has shown that the deformation and stress calculations for such idealized models are often sufficient for the overall design and analysis purposes of truss structures [3].

2.4 Formulation of the Bar Element

In this section, we will formulate the equations for the bar element based on the 1-D elasticity theory. The structural behavior of the element, or the element stiffness matrix, can be established using two approaches: the direct approach and the energy approach. The two methods are discussed in the following.

2.4.1 Stiffness Matrix: Direct Method

A bar element with two end nodes is presented in Figure 2.4.

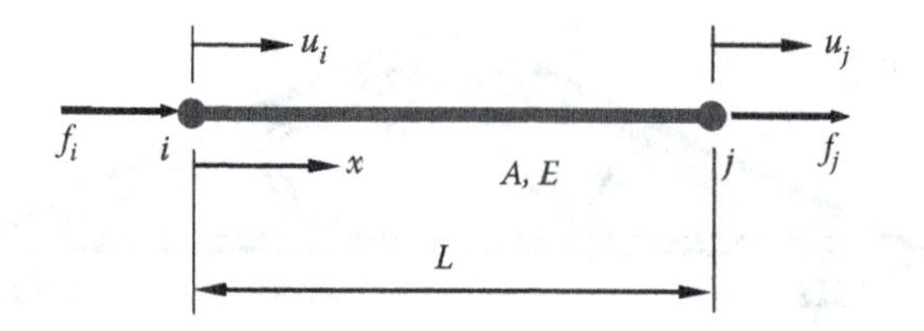

FIGURE 2.4
Notation for a bar element.

Assuming that the displacement u is varying *linearly* along the axis of the bar, that is, in terms of the two nodal values u_i and u_j, we can write

$$u(x) = \left(1 - \frac{x}{L}\right)u_i + \frac{x}{L}u_j \tag{2.4}$$

We have

$$\varepsilon = \frac{u_j - u_i}{L} = \frac{\Delta}{L} \quad (\Delta = \text{elongation}) \tag{2.5}$$

$$\sigma = E\varepsilon = \frac{E\Delta}{L} \tag{2.6}$$

We also have

$$\sigma = \frac{F}{A} \quad (F = \text{force in bar}) \tag{2.7}$$

Thus, Equations 2.6 and 2.7 lead to

$$F = \frac{EA}{L}\Delta = k\Delta$$

where $k = EA/L$ is the stiffness of the bar. That is, the bar behaves like a spring in this case and we conclude that the element stiffness matrix is

$$\mathbf{k} = \begin{bmatrix} k & -k \\ -k & k \end{bmatrix} = \begin{bmatrix} \frac{EA}{L} & -\frac{EA}{L} \\ -\frac{EA}{L} & \frac{EA}{L} \end{bmatrix}$$

or

$$\mathbf{k} = \frac{EA}{L}\begin{bmatrix} 1 & -1 \\ -1 & 1 \end{bmatrix} \tag{2.8}$$

This can be verified by considering the equilibrium of the forces at the two nodes.

The *Element equilibrium equation* is

$$\frac{EA}{L}\begin{bmatrix} 1 & -1 \\ -1 & 1 \end{bmatrix}\begin{Bmatrix} u_i \\ u_j \end{Bmatrix} = \begin{Bmatrix} f_i \\ f_j \end{Bmatrix} \tag{2.9}$$

Degree of Freedom (DOF): Number of components of the displacement vector at a node. For 1-D bar element along the x-axis, we have one DOF at each node.

Physical Meaning of the Coefficients in **k**: The jth column of **k** (here $j = 1$ or 2) represents the forces applied to the bar to maintain a deformed shape with unit displacement at node j and zero displacement at the other node.

2.4.2 Stiffness Matrix: Energy Approach

We derive the same stiffness matrix for the bar using a formal approach which can be applied to many other more complicated situations.

First, we define two *linear shape functions* as follows (Figure 2.5):

$$N_i(\xi) = 1 - \xi, \quad N_j(\xi) = \xi \tag{2.10}$$

where

$$\xi = \frac{x}{L}, \quad 0 \le \xi \le 1 \tag{2.11}$$

From Equation 2.4, we can write the displacement as

$$u(x) = u(\xi) = N_i(\xi)u_i + N_j(\xi)u_j$$

or

$$u = \begin{bmatrix} N_i & N_j \end{bmatrix}\begin{Bmatrix} u_i \\ u_j \end{Bmatrix} = \mathbf{Nu} \tag{2.12}$$

Strain is given by Equations 2.1 and 2.12 as

$$\varepsilon = \frac{du}{dx} = \left[\frac{d}{dx}\mathbf{N}\right]\mathbf{u} = \mathbf{Bu} \tag{2.13}$$

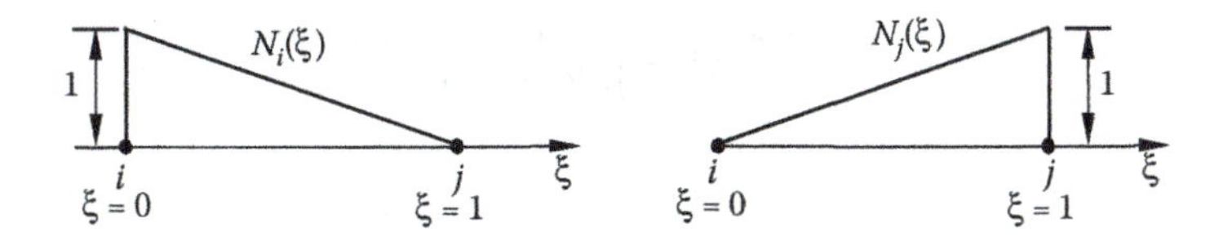

FIGURE 2.5
The shape functions for a bar element.

where **B** is the element *strain–displacement matrix,* which is

$$\mathbf{B} = \frac{d}{dx}\left[N_i(\xi) \quad N_j(\xi)\right] = \frac{d}{d\xi}\left[N_i(\xi) \quad N_j(\xi)\right] \cdot \frac{d\xi}{dx}$$

that is,

$$\mathbf{B} = \left[-1/L \quad 1/L\right] \tag{2.14}$$

Stress can be written as

$$\sigma = E\varepsilon = E\mathbf{Bu} \tag{2.15}$$

Consider the *strain energy* stored in the bar

$$\begin{aligned} \mathrm{U} &= \frac{1}{2}\int_V \sigma^{\mathrm{T}}\varepsilon dV = \frac{1}{2}\int_V (\mathbf{u}^{\mathrm{T}}\mathbf{B}^{\mathrm{T}}E\mathbf{Bu})dV \\ &= \frac{1}{2}\mathbf{u}^{\mathrm{T}}\left[\int_V (\mathbf{B}^{\mathrm{T}}E\mathbf{B})dV\right]\mathbf{u} \end{aligned} \tag{2.16}$$

where Equations 2.13 and 2.15 have been used.

The *potential* of the external forces is written as (this is by definition, and remember the negative sign)

$$\Omega = -f_i u_i - f_j u_j = -\mathbf{u}^{\mathrm{T}}\mathbf{f} \tag{2.17}$$

The total potential of the system is

$$\Pi = \mathrm{U} + \Omega$$

which yields by using Equations 2.16 and 2.17

$$\Pi = \frac{1}{2}\mathbf{u}^{\mathrm{T}}\left[\int_V (\mathbf{B}^{\mathrm{T}}E\mathbf{B})dV\right]\mathbf{u} - \mathbf{u}^{\mathrm{T}}\mathbf{f} \tag{2.18}$$

Setting $d\Pi = 0$ by the principle of minimum potential energy, we obtain (verify this)

$$\left[\int_V \left(\mathbf{B}^{\mathrm{T}}E\mathbf{B}\right)dV\right]\mathbf{u} = \mathbf{f}$$

or

$$\mathbf{ku} = \mathbf{f} \tag{2.19}$$

where

$$\mathbf{k} = \int_V (\mathbf{B}^T E \mathbf{B}) dV \tag{2.20}$$

is the *element stiffness matrix*.

Equation 2.20 is a general result which can be used for the construction of other types of elements.

Now, we evaluate Equation 2.20 for the bar element by using Equation 2.14

$$\mathbf{k} = \int_0^L \begin{Bmatrix} -1/L \\ 1/L \end{Bmatrix} E \begin{bmatrix} -1/L & 1/L \end{bmatrix} A dx = \frac{EA}{L}\begin{bmatrix} 1 & -1 \\ -1 & 1 \end{bmatrix}$$

which is the same as we derived earlier using the direct method.

Note that from Equations 2.16 and 2.20, the strain energy in the element can be written as

$$U = \frac{1}{2}\mathbf{u}^T\mathbf{k}\mathbf{u} \tag{2.21}$$

In the future, once we obtain an expression like Equation 2.16, we can immediately recognize that the matrix in between the displacement vectors is the stiffness matrix. Recall that for a spring, the strain energy can be written as

$$U = \frac{1}{2}k\Delta^2 = \frac{1}{2}\Delta^T k \Delta$$

Thus, result (2.21) goes back to the simple spring case again.

2.4.3 Treatment of Distributed Load

Distributed axial load q (N/mm, N/m, lb/in) (Figure 2.6) can be converted into two equivalent nodal forces using the shape functions. Consider the work done by the distributed load q,

$$\begin{aligned} W_q &= \frac{1}{2}\int_0^L u(x)q(x)dx = \frac{1}{2}\int_0^L (\mathbf{N}\mathbf{u})^T q(x)dx = \frac{1}{2}\begin{bmatrix} u_i & u_j \end{bmatrix}\int_0^L \begin{bmatrix} N_i(x) \\ N_j(x) \end{bmatrix} q(x)dx \\ &= \frac{1}{2}\mathbf{u}^T\int_0^L \mathbf{N}^T q(x)dx \end{aligned} \tag{2.22}$$

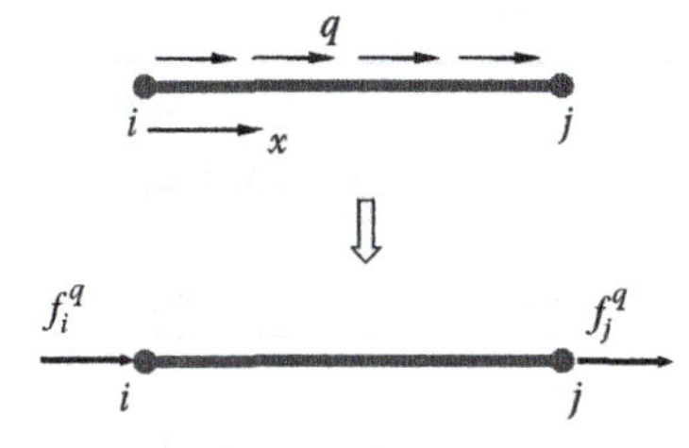

FIGURE 2.6
Conversion of a distributed load on one element.

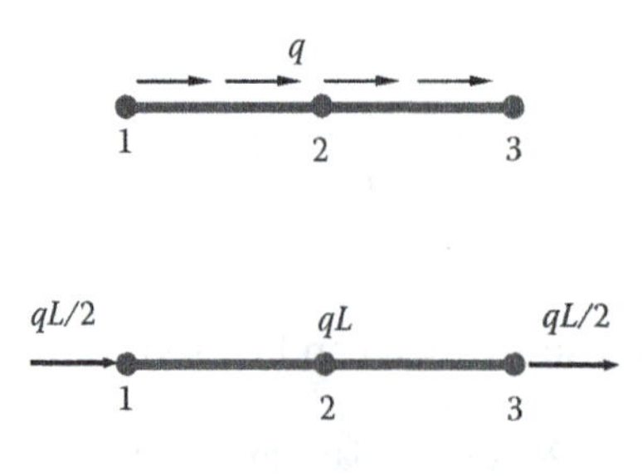

FIGURE 2.7
Conversion of a distributed load with constant intensity q on two elements.

The work done by the equivalent nodal forces are

$$W_{f_q} = \frac{1}{2} f_i^q u_i + \frac{1}{2} f_j^q u_j = \frac{1}{2} \mathbf{u}^T \mathbf{f}_q \tag{2.23}$$

Setting $W_q = W_{f_q}$ and using Equations 2.22 and 2.23, we obtain the equivalent nodal force vector

$$\mathbf{f}_q = \begin{Bmatrix} f_i^q \\ f_j^q \end{Bmatrix} = \int_0^L \mathbf{N}^T q(x)dx = \int_0^L \begin{bmatrix} N_i(x) \\ N_j(x) \end{bmatrix} q(x)dx \tag{2.24}$$

which is valid for any distributions of q. For example, if q is a *constant*, we have

$$\mathbf{f}_q = q \int_0^L \begin{bmatrix} 1 - x/L \\ x/L \end{bmatrix} dx = \begin{Bmatrix} qL/2 \\ qL/2 \end{Bmatrix} \tag{2.25}$$

that is, equivalent nodal forces can be added to replace the distributed load as shown in Figure 2.7.

2.4.4 Bar Element in 2-D and 3-D

To analyze the truss structures in 2-D or 3-D, we need to extend the 1-D bar element formulation to 2-D or 3-D. In the following, we take a look at the formulation for the 2-D case.

2.4.4.1 2-D Case

Local	Global
x, y	X, Y
u_i', v_i'	u_i, v_i
1 DOF at each node	2 DOFs at each node

Note that lateral displacement v_i' does not contribute to the stretch of the bar within the linear theory (Figure 2.8). Displacement vectors in the local and global coordinates are related as follows:

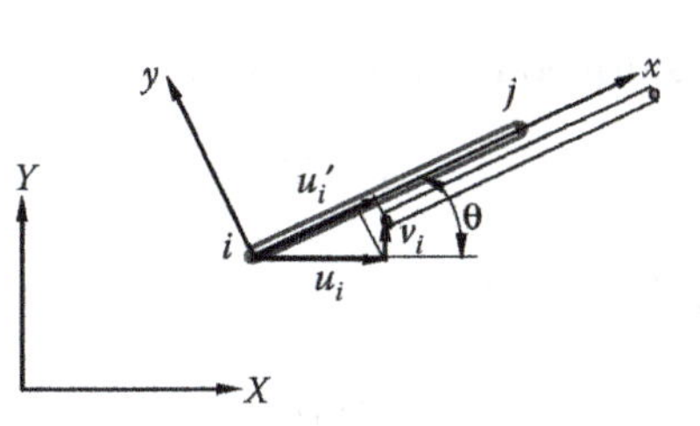

FIGURE 2.8
Local and global coordinates for a bar in 2-D space.

$$u_i' = u_i \cos\theta + v_i \sin\theta = \begin{bmatrix} l & m \end{bmatrix} \begin{Bmatrix} u_i \\ v_i \end{Bmatrix}$$

$$v_i^i = -u_i \sin\theta + v_i \cos\theta = \begin{bmatrix} -m & l \end{bmatrix} \begin{Bmatrix} u_i \\ v_i \end{Bmatrix}$$

where $l = \cos\theta$, $m = \sin\theta$.
In matrix form,

$$\begin{Bmatrix} u_i' \\ v_i' \end{Bmatrix} = \begin{bmatrix} l & m \\ -m & l \end{bmatrix} \begin{Bmatrix} u_i \\ v_i \end{Bmatrix} \tag{2.26}$$

or,

$$\mathbf{u}_i' = \tilde{\mathbf{T}} \mathbf{u}_i$$

where the *transformation matrix*

$$\tilde{\mathbf{T}} = \begin{bmatrix} l & m \\ -m & l \end{bmatrix} \tag{2.27}$$

is orthogonal, that is, $\tilde{\mathbf{T}}^{-1} = \tilde{\mathbf{T}}^T$.
For the two nodes of the bar element, we have

$$\begin{Bmatrix} u_i' \\ v_i' \\ u_j' \\ v_j' \end{Bmatrix} = \begin{bmatrix} l & m & 0 & 0 \\ -m & l & 0 & 0 \\ 0 & 0 & l & m \\ 0 & 0 & -m & l \end{bmatrix} \begin{Bmatrix} u_i \\ v_i \\ u_j \\ v_j \end{Bmatrix} \tag{2.28}$$

or,

$$\mathbf{u}' = \mathbf{T}\mathbf{u} \quad \text{with } \mathbf{T} = \begin{bmatrix} \tilde{\mathbf{T}} & \mathbf{0} \\ \mathbf{0} & \tilde{\mathbf{T}} \end{bmatrix} \tag{2.29}$$

The nodal forces are transformed in the same way,

$$\mathbf{f}' = \mathbf{T}\mathbf{f} \tag{2.30}$$

In the local coordinate system, we have

$$\frac{EA}{L}\begin{bmatrix} 1 & -1 \\ -1 & 1 \end{bmatrix}\begin{Bmatrix} u_i' \\ u_j' \end{Bmatrix} = \begin{Bmatrix} f_i' \\ f_j' \end{Bmatrix}$$

Augmenting this equation, we write

$$\frac{EA}{L}\begin{bmatrix} 1 & 0 & -1 & 0 \\ 0 & 0 & 0 & 0 \\ -1 & 0 & 1 & 0 \\ 0 & 0 & 0 & 0 \end{bmatrix}\begin{Bmatrix} u_i' \\ v_i' \\ u_j' \\ v_j' \end{Bmatrix} = \begin{Bmatrix} f_i' \\ 0 \\ f_j' \\ 0 \end{Bmatrix}$$

or,

$$\mathbf{k}'\,\mathbf{u}' = \mathbf{f}'$$

Using transformations given in Equations 2.29 and 2.30, we obtain

$$\mathbf{k}'\,\mathbf{T}\mathbf{u} = \mathbf{T}\mathbf{f}$$

Multiplying both sides by $\mathbf{T}^T$ and noticing that $\mathbf{T}^T\mathbf{T} = \mathbf{I}$, we obtain

$$\mathbf{T}^T\,\mathbf{k}'\,\mathbf{T}\mathbf{u} = \mathbf{f} \tag{2.31}$$

Thus, the element stiffness matrix $\mathbf{k}$ in the global coordinate system is

$$\mathbf{k} = \mathbf{T}^T\,\mathbf{k}'\,\mathbf{T} \tag{2.32}$$

which is a 4×4 symmetric matrix.

Explicit form is

$$\mathbf{k} = \frac{EA}{L}\begin{array}{c} \begin{array}{cccc} \;\;u_i\;\; & \;\;v_i\;\; & \;\;u_j\;\; & \;\;v_j\;\; \end{array} \\ \begin{bmatrix} l^2 & lm & -l^2 & -lm \\ lm & m^2 & -lm & -m^2 \\ -l^2 & -lm & l^2 & lm \\ -lm & -m^2 & lm & m^2 \end{bmatrix} \end{array} \tag{2.33}$$

Calculation of the *directional cosines* l and m:

$$l = \cos\theta = \frac{X_j - X_i}{L}, \quad m = \sin\theta = \frac{Y_j - Y_i}{L}$$

The structure stiffness matrix is assembled by using the element stiffness matrices in the usual way as in the 1-D case.

2.4.4.2 3-D Case

Local	Global
x, y, z	X, Y, Z
u_i', v_i', w_i'	u_i, v_i, w_i
1 DOF at each node	3 DOFs at each node

Similar to the 2-D case, element stiffness matrices in the 3-D case are calculated in the local coordinate systems first and then transformed into the global coordinate system (X, Y, and Z) where they are assembled (Figure 2.9). The transformation relation is

$$\begin{Bmatrix} u_i' \\ v_i' \\ w_i' \end{Bmatrix} = \begin{bmatrix} l_X & l_Y & l_Z \\ m_X & m_Y & m_Z \\ n_X & n_Y & n_Z \end{bmatrix} \begin{Bmatrix} u_i \\ v_i \\ w_i \end{Bmatrix} \tag{2.34}$$

where (l_X,l_Y,l_Z), (m_X,m_Y,m_Z), and (n_X,n_Y,n_Z) are the direction cosines of the local x, y, and z coordinate axis in the global coordinate system, respectively. FEM software packages will do this transformation automatically.

Therefore, the input data for bar elements are simply:

- Coordinates (X, Y, Z) for each node
- E and A for each element (length L can be computed from the coordinates of the two nodes)

2.4.5 Element Stress

Once the nodal displacement is obtained for an element, the stress within the element can be calculated using the basic relations. For example, for 2-D cases, we proceed as follows:

$$\sigma = E\varepsilon = E\mathbf{B}\begin{Bmatrix} u_i' \\ u_j' \end{Bmatrix} = E\begin{bmatrix} -\frac{1}{L} & \frac{1}{L} \end{bmatrix}\begin{bmatrix} l & m & 0 & 0 \\ 0 & 0 & l & m \end{bmatrix}\begin{Bmatrix} u_i \\ v_i \\ u_j \\ v_j \end{Bmatrix}$$

FIGURE 2.9
Local and global coordinates for a bar in 3-D space.

That is,

$$\sigma = \frac{E}{L}\begin{bmatrix} -l & -m & l & m \end{bmatrix}\begin{Bmatrix} u_i \\ v_i \\ u_j \\ v_j \end{Bmatrix} \tag{2.35}$$

which is a general formula for 2-D bar elements.

2.5 Examples with Bar Elements

We now look at examples of using the bar element in the analysis of 1-D stress problems and plane truss problems.

EXAMPLE 2.1

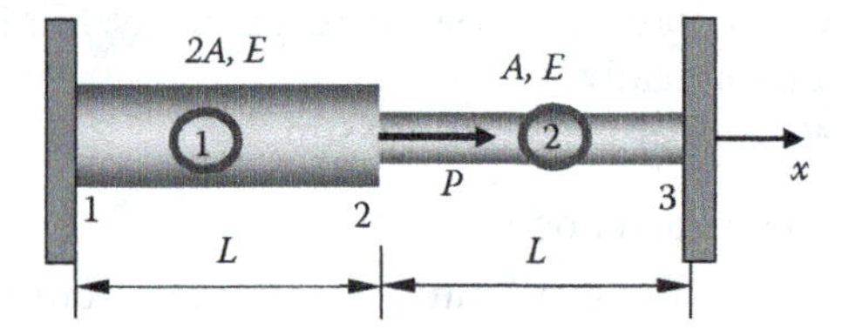

Problem

Find the stresses in the two-bar assembly which is loaded with force *P*, and constrained at the two ends, as shown in the above figure.

Solution

Use two 1-D bar elements.

For element 1,

$$\mathbf{k}_1 = \frac{2EA}{L}\begin{bmatrix} 1 & -1 \\ -1 & 1 \end{bmatrix} \quad (u_1, u_2)$$

For element 2,

$$\mathbf{k}_2 = \frac{EA}{L}\begin{bmatrix} 1 & -1 \\ -1 & 1 \end{bmatrix} \quad (u_2, u_3)$$

Imagine a frictionless pin at node 2, which connects the two elements. We can assemble the global FE equation as follows:

$$\frac{EA}{L}\begin{bmatrix} 2 & -2 & 0 \\ -2 & 3 & -1 \\ 0 & -1 & 1 \end{bmatrix}\begin{Bmatrix} u_1 \\ u_2 \\ u_3 \end{Bmatrix} = \begin{Bmatrix} F_1 \\ F_2 \\ F_3 \end{Bmatrix}$$

Load and boundary conditions (BCs) are

$$u_1 = u_3 = 0, \quad F_2 = P$$

FE equation becomes

$$\frac{EA}{L}\begin{bmatrix} 2 & -2 & 0 \\ -2 & 3 & -1 \\ 0 & -1 & 1 \end{bmatrix}\begin{Bmatrix} 0 \\ u_2 \\ 0 \end{Bmatrix} = \begin{Bmatrix} F_1 \\ P \\ F_3 \end{Bmatrix}$$

"Deleting" the first row and column, and the third row and column, we obtain

$$\frac{EA}{L}[3]\{u_2\} = \{P\}$$

Thus,

$$u_2 = \frac{PL}{3EA}$$

and

$$\begin{Bmatrix} u_1 \\ u_2 \\ u_3 \end{Bmatrix} = \frac{PL}{3EA}\begin{Bmatrix} 0 \\ 1 \\ 0 \end{Bmatrix}$$

Stress in element 1 is

$$\begin{aligned} \sigma_1 &= E\varepsilon_1 = E\mathbf{B}_1\mathbf{u}_1 = E\begin{bmatrix} -1/L & 1/L \end{bmatrix}\begin{Bmatrix} u_1 \\ u_2 \end{Bmatrix} \\ &= E\frac{u_2 - u_1}{L} = \frac{E}{L}\left(\frac{PL}{3EA} - 0\right) = \frac{P}{3A} \end{aligned}$$

Similarly, stress in element 2 is

$$\begin{aligned} \sigma_2 &= E\varepsilon_2 = E\mathbf{B}_2\mathbf{u}_2 = E\begin{bmatrix} -1/L & 1/L \end{bmatrix}\begin{Bmatrix} u_2 \\ u_3 \end{Bmatrix} \\ &= E\frac{u_3 - u_2}{L} = \frac{E}{L}\left(0 - \frac{PL}{3EA}\right) = -\frac{P}{3A} \end{aligned}$$

which indicates that bar 2 is in compression.

Check the results: Draw the FBD as follows and check the equilibrium of the structures.

Notes:

- In this case, the calculated stresses in elements 1 and 2 are exact. It will not help if we further divide element 1 or 2 into smaller elements.
- For tapered bars, averaged values of the cross-sectional areas should be used for the elements.
- We need to find the displacements first in order to find the stresses, and this approach is called the *displacement-based FEM*.

EXAMPLE 2.2

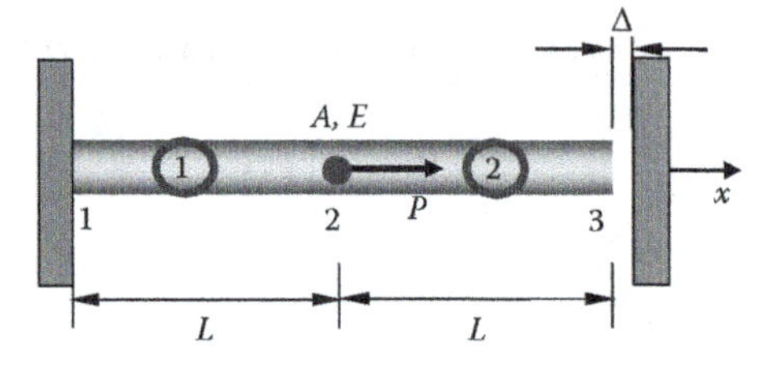

Problem

Determine the support reaction forces at the two ends of the bar shown above, given the following:

$$P = 6.0 \times 10^4 \text{ N}, \quad E = 2.0 \times 10^4 \text{ N/mm}^2,$$
$$A = 250 \text{ mm}^2, \quad L = 150 \text{ mm}, \quad \Delta = 1.2 \text{ mm}$$

Solution

We first check to see if contact of the bar with the wall on the right will occur or not. To do this, we imagine the wall on the right is removed and calculate the displacement at the right end

$$\Delta_0 = \frac{PL}{EA} = \frac{(6.0 \times 10^4)(150)}{(2.0 \times 10^4)(250)} = 1.8 \text{ mm} > \Delta = 1.2 \text{ mm}$$

Thus, contact occurs and the wall on the right should be accounted for in the analysis. The global FE equation is found to be

$$\frac{EA}{L}\begin{bmatrix} 1 & -1 & 0 \\ -1 & 2 & -1 \\ 0 & -1 & 1 \end{bmatrix}\begin{Bmatrix} u_1 \\ u_2 \\ u_3 \end{Bmatrix} = \begin{Bmatrix} F_1 \\ F_2 \\ F_3 \end{Bmatrix}$$

The load and boundary conditions are

$$F_2 = P = 6.0 \times 10^4 \text{ N}$$
$$u_1 = 0, \quad u_3 = \Delta = 1.2 \text{ mm}$$

The FE equation becomes

$$\frac{EA}{L}\begin{bmatrix} 1 & -1 & 0 \\ -1 & 2 & -1 \\ 0 & -1 & 1 \end{bmatrix}\begin{Bmatrix} 0 \\ u_2 \\ \Delta \end{Bmatrix} = \begin{Bmatrix} F_1 \\ P \\ F_3 \end{Bmatrix}$$

The second equation gives

$$\frac{EA}{L}\begin{bmatrix} 2 & -1 \end{bmatrix}\begin{Bmatrix} u_2 \\ \Delta \end{Bmatrix} = \{P\}$$

that is,

$$\frac{EA}{L}[2]\{u_2\} = \left\{P + \frac{EA}{L}\Delta\right\}$$

Solving this, we obtain

$$u_2 = \frac{1}{2}\left(\frac{PL}{EA} + \Delta\right) = 1.5 \text{ mm}$$

and

$$\begin{Bmatrix} u_1 \\ u_2 \\ u_3 \end{Bmatrix} = \begin{Bmatrix} 0 \\ 1.5 \\ 1.2 \end{Bmatrix} (\text{mm})$$

To calculate the support reaction forces, we apply the first and third equations in the global FE equation.

The first equation gives

$$F_1 = \frac{EA}{L}\begin{bmatrix} 1 & -1 & 0 \end{bmatrix}\begin{Bmatrix} u_1 \\ u_2 \\ u_3 \end{Bmatrix} = \frac{EA}{L}(-u_2) = -5.0 \times 10^4 \text{ N}$$

and the third equation gives,

$$F_3 = \frac{EA}{L}\begin{bmatrix} 0 & -1 & 1 \end{bmatrix}\begin{Bmatrix} u_1 \\ u_2 \\ u_3 \end{Bmatrix} = \frac{EA}{L}(-u_2 + u_3) = -1.0 \times 10^4 \text{ N}$$

Check the results: Again, we can draw the free-body diagram to verify that the equilibrium of the forces is satisfied.

EXAMPLE 2.3

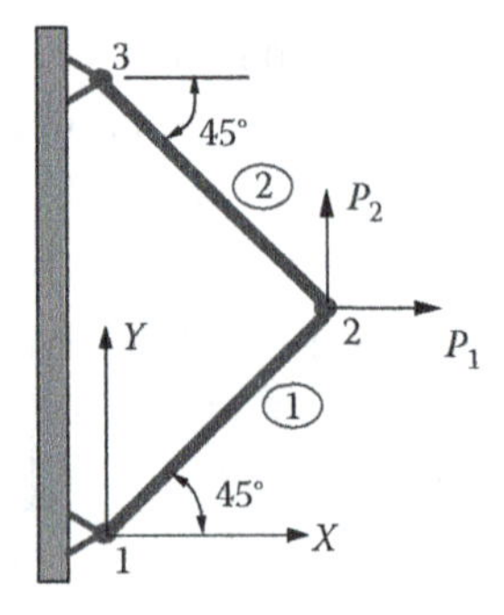

A simple plane truss is made of two identical bars (with *E*, *A*, and *L*), and loaded as shown in the above figure.

Find

a. displacement of node 2;
b. stress in each bar.

Solution

This simple structure is used here to demonstrate the FEA procedure using the bar element in 2-D space. In local coordinate systems, we have

$$\mathbf{k}_1' = \frac{EA}{L}\begin{bmatrix} 1 & -1 \\ -1 & 1 \end{bmatrix} = \mathbf{k}_2'$$

These two matrices cannot be assembled together, because they are in different coordinate systems. We need to convert them into global coordinate system *OXY*.

Element 1:

$$\theta = 45^\circ, \ l = m = \frac{\sqrt{2}}{2}$$

Using Equation 2.32 or 2.33, we obtain the stiffness matrix in the global system

$$\mathbf{k}_1 = \mathbf{T}_1^T \mathbf{k}_1' \mathbf{T}_1 = \frac{EA}{2L}\begin{bmatrix} 1 & 1 & -1 & -1 \\ 1 & 1 & -1 & -1 \\ -1 & -1 & 1 & 1 \\ -1 & -1 & 1 & 1 \end{bmatrix} \quad (\text{columns: } u_1 \ v_1 \ u_2 \ v_2)$$

Element 2:

$$\theta = 135^\circ, \ l = -\frac{\sqrt{2}}{2}, \ m = \frac{\sqrt{2}}{2}$$

$$\mathbf{k}_2 = \mathbf{T}_2^T \mathbf{k}_2' \mathbf{T}_2 = \frac{EA}{2L}\begin{bmatrix} 1 & -1 & -1 & 1 \\ -1 & 1 & 1 & -1 \\ -1 & 1 & 1 & -1 \\ 1 & -1 & -1 & 1 \end{bmatrix} \quad (\text{columns: } u_2 \ v_2 \ u_3 \ v_3)$$

Assemble the structure FE equation,

$$
\begin{array}{c}
\begin{array}{cccccc} \quad u_1 & \quad v_1 & \quad u_2 & \quad v_2 & \quad u_3 & \quad v_3 \end{array} \\
\dfrac{EA}{2L}\begin{bmatrix} 1 & 1 & -1 & -1 & 0 & 0 \\ 1 & 1 & -1 & -1 & 0 & 0 \\ -1 & -1 & 2 & 0 & -1 & 1 \\ -1 & -1 & 0 & 2 & 1 & -1 \\ 0 & 0 & -1 & 1 & 1 & -1 \\ 0 & 0 & 1 & -1 & -1 & 1 \end{bmatrix} \begin{Bmatrix} u_1 \\ v_1 \\ u_2 \\ v_2 \\ u_3 \\ v_3 \end{Bmatrix} = \begin{Bmatrix} F_{1X} \\ F_{1Y} \\ F_{2X} \\ F_{2Y} \\ F_{3X} \\ F_{3Y} \end{Bmatrix}
\end{array}
$$

Load and boundary conditions (BCs):

$$u_1 = v_1 = u_3 = v_3 = 0, \quad F_{2X} = P_1, \quad F_{2Y} = P_2$$

Condensed FE equation,

$$\frac{EA}{2L}\begin{bmatrix} 2 & 0 \\ 0 & 2 \end{bmatrix} \begin{Bmatrix} u_2 \\ v_2 \end{Bmatrix} = \begin{Bmatrix} P_1 \\ P_2 \end{Bmatrix}$$

Solving this, we obtain the displacement of node 2,

$$\begin{Bmatrix} u_2 \\ v_2 \end{Bmatrix} = \frac{L}{EA}\begin{Bmatrix} P_1 \\ P_2 \end{Bmatrix}$$

Using Equation 2.34, we calculate the stresses in the two bars,

$$\sigma_1 = \frac{E}{L}\frac{\sqrt{2}}{2}\begin{bmatrix} -1 & -1 & 1 & 1 \end{bmatrix} \frac{L}{EA}\begin{Bmatrix} 0 \\ 0 \\ P_1 \\ P_2 \end{Bmatrix} = \frac{\sqrt{2}}{2A}(P_1 + P_2)$$

$$\sigma_2 = \frac{E}{L}\frac{\sqrt{2}}{2}\begin{bmatrix} 1 & -1 & -1 & 1 \end{bmatrix} \frac{L}{EA}\begin{Bmatrix} P_1 \\ P_2 \\ 0 \\ 0 \end{Bmatrix} = \frac{\sqrt{2}}{2A}(P_1 - P_2)$$

Check the results: Again, we need to check the equilibrium conditions, symmetry, antisymmetry, and so on, of the FEA results.

EXAMPLE 2.4: (MULTIPOINT CONSTRAINT)

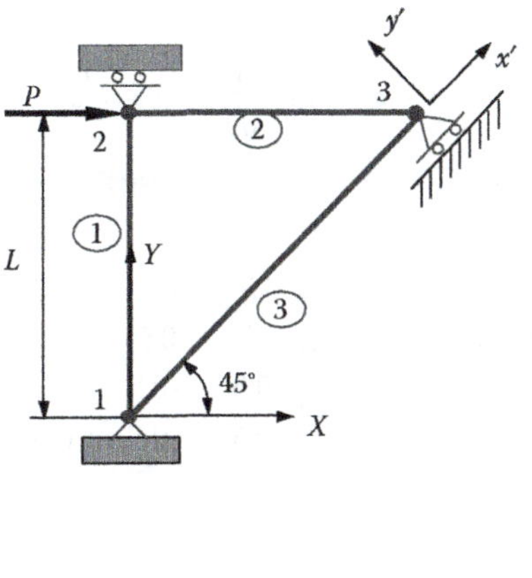

For the plane truss shown above,

$$P = 1000 \text{ kN}, \quad L = 1m, \quad E = 210\, GPa,$$

$$A = 6.0 \times 10^{-4} m^2 \text{ for elements 1 and 2,}$$

$$A = 6\sqrt{2} \times 10^{-4} m^2 \text{ for element 3.}$$

Determine the displacements and reaction forces.

Solution

We have an inclined roller at node 3, which needs special attention in the FE solution. We first assemble the global FE equation for the truss.

Element 1:

$$\theta = 90^\circ, \; l = 0, \; m = 1$$

$$\mathbf{k}_1 = \frac{(210 \times 10^9)(6.0 \times 10^{-4})}{1} \begin{array}{c} \begin{matrix} u_1 & v_1 & u_2 & v_2 \end{matrix} \\ \begin{bmatrix} 0 & 0 & 0 & 0 \\ 0 & 1 & 0 & -1 \\ 0 & 0 & 0 & 0 \\ 0 & -1 & 0 & 1 \end{bmatrix} \end{array} \text{(N/m)}$$

Element 2:

$$\theta = 0^\circ, \; l = 1, \; m = 0$$

$$\mathbf{k}_2 = \frac{(210 \times 10^9)(6.0 \times 10^{-4})}{1} \begin{array}{c} \begin{matrix} u_2 & v_2 & u_3 & v_3 \end{matrix} \\ \begin{bmatrix} 1 & 0 & -1 & 0 \\ 0 & 0 & 0 & 0 \\ -1 & 0 & 1 & 0 \\ 0 & 0 & 0 & 0 \end{bmatrix} \end{array} \text{(N/m)}$$

Element 3:

$$\theta = 45^\circ, \; l = \frac{1}{\sqrt{2}}, \; m = \frac{1}{\sqrt{2}}$$

$$\mathbf{k}_3 = \frac{(210 \times 10^9)(6\sqrt{2} \times 10^{-4})}{\sqrt{2}} \begin{array}{c} \begin{matrix} u_1 & v_1 & u_3 & v_3 \end{matrix} \\ \begin{bmatrix} 0.5 & 0.5 & -0.5 & -0.5 \\ 0.5 & 0.5 & -0.5 & -0.5 \\ -0.5 & -0.5 & 0.5 & 0.5 \\ -0.5 & -0.5 & 0.5 & 0.5 \end{bmatrix} \end{array} \text{(N/m)}$$

The global FE equation is,

$$1260 \times 10^5 \begin{bmatrix} 0.5 & 0.5 & 0 & 0 & -0.5 & -0.5 \\ & 1.5 & 0 & -1 & -0.5 & -0.5 \\ & & 1 & 0 & -1 & 0 \\ & & & 1 & 0 & 0 \\ & & & & 1.5 & 0.5 \\ Sym. & & & & & 0.5 \end{bmatrix} \begin{Bmatrix} u_1 \\ v_1 \\ u_2 \\ v_2 \\ u_3 \\ v_3 \end{Bmatrix} = \begin{Bmatrix} F_{1X} \\ F_{1Y} \\ F_{2X} \\ F_{2Y} \\ F_{3X} \\ F_{3Y} \end{Bmatrix}$$

Load and boundary conditions (BCs):

$$u_1 = v_1 = v_2 = 0, \quad \text{and} \quad v_3' = 0,$$
$$F_{2X} = P, \; F_{3x'} = 0.$$

From the transformation relation and the BCs, we have

$$v_3' = \begin{bmatrix} -\frac{\sqrt{2}}{2} & \frac{\sqrt{2}}{2} \end{bmatrix} \begin{Bmatrix} u_3 \\ v_3 \end{Bmatrix} = \frac{\sqrt{2}}{2}(-u_3 + v_3) = 0,$$

that is,

$$u_3 - v_3 = 0$$

This is a *multipoint constraint* (MPC).
Similarly, we have a relation for the force at node 3,

$$F_{3x'} = \begin{bmatrix} \frac{\sqrt{2}}{2} & \frac{\sqrt{2}}{2} \end{bmatrix} \begin{Bmatrix} F_{3X} \\ F_{3Y} \end{Bmatrix} = \frac{\sqrt{2}}{2}(F_{3X} + F_{3Y}) = 0,$$

that is,

$$F_{3X} + F_{3Y} = 0$$

Applying the load and BCs in the structure FE equation by "deleting" the first, second, and fourth rows and columns, we have

$$1260 \times 10^5 \begin{bmatrix} 1 & -1 & 0 \\ -1 & 1.5 & 0.5 \\ 0 & 0.5 & 0.5 \end{bmatrix} \begin{Bmatrix} u_2 \\ u_3 \\ v_3 \end{Bmatrix} = \begin{Bmatrix} P \\ F_{3X} \\ F_{3Y} \end{Bmatrix}$$

Further, from the MPC and the force relation at node 3, the equation becomes,

$$1260 \times 10^5 \begin{bmatrix} 1 & -1 & 0 \\ -1 & 1.5 & 0.5 \\ 0 & 0.5 & 0.5 \end{bmatrix} \begin{Bmatrix} u_2 \\ u_3 \\ u_3 \end{Bmatrix} = \begin{Bmatrix} P \\ F_{3X} \\ -F_{3X} \end{Bmatrix}$$

which is

$$1260 \times 10^5 \begin{bmatrix} 1 & -1 \\ -1 & 2 \\ 0 & 1 \end{bmatrix} \begin{Bmatrix} u_2 \\ u_3 \end{Bmatrix} = \begin{Bmatrix} P \\ F_{3X} \\ -F_{3X} \end{Bmatrix}$$

The third equation yields,

$$F_{3X} = -1260 \times 10^5 u_3$$

Substituting this into the second equation and rearranging, we have

$$1260 \times 10^5 \begin{bmatrix} 1 & -1 \\ -1 & 3 \end{bmatrix} \begin{Bmatrix} u_2 \\ u_3 \end{Bmatrix} = \begin{Bmatrix} P \\ 0 \end{Bmatrix}$$

Solving this, we obtain the displacements,

$$\begin{Bmatrix} u_2 \\ u_3 \end{Bmatrix} = \frac{1}{2520 \times 10^5} \begin{Bmatrix} 3P \\ P \end{Bmatrix} = \begin{Bmatrix} 0.01191 \\ 0.003968 \end{Bmatrix} \text{ (m)}$$

From the global FE equation, we can calculate the reaction forces,

$$\begin{Bmatrix} F_{1X} \\ F_{1Y} \\ F_{2Y} \\ F_{3X} \\ F_{3Y} \end{Bmatrix} = 1260 \times 10^5 \begin{bmatrix} 0 & -0.5 & -0.5 \\ 0 & -0.5 & -0.5 \\ 0 & 0 & 0 \\ -1 & 1.5 & 0.5 \\ 0 & 0.5 & 0.5 \end{bmatrix} \begin{Bmatrix} u_2 \\ u_3 \\ v_3 \end{Bmatrix} = \begin{Bmatrix} -500 \\ -500 \\ 0.0 \\ -500 \\ 500 \end{Bmatrix} \text{ (kN)}$$

Check the results!

A general MPC can be described as,

$$\sum_j A_j u_j = 0$$

where A_j's are constants and u_j's are nodal displacement components. In FE software, users only need to specify this relation to the software. The software will take care of the solution process.

2.6 Case Study with *ANSYS Workbench*

Problem Description: Truss bridges can span long distances and support heavy weights without intermediate supports. They are economical to construct and are available in a wide variety of styles. Consider the following planar truss, constructed of wooden timbers, which can be used in parallel to form bridges. Determine the deflections at each joint of the truss under the given loading conditions.

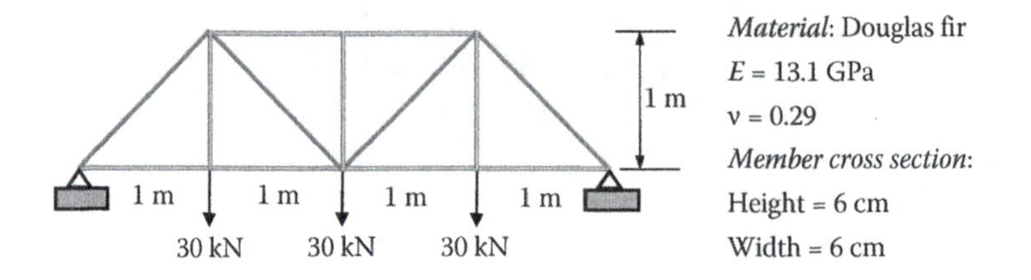

Solution

To solve the problem with *ANSYS® Workbench*, we employ the following steps:

Step 1: Start an *ANSYS Workbench* Project

Launch *ANSYS Workbench* and save the blank project as "*Woodtruss.wbpj.*"

Step 2: Create a *Static Structural (ANSYS)* Analysis System

Drag the *Static Structural (ANSYS)* icon from the *Analysis Systems Toolbox* window and drop it inside the highlighted green rectangle in the *Project Schematic* window to create a standalone static structural analysis system.

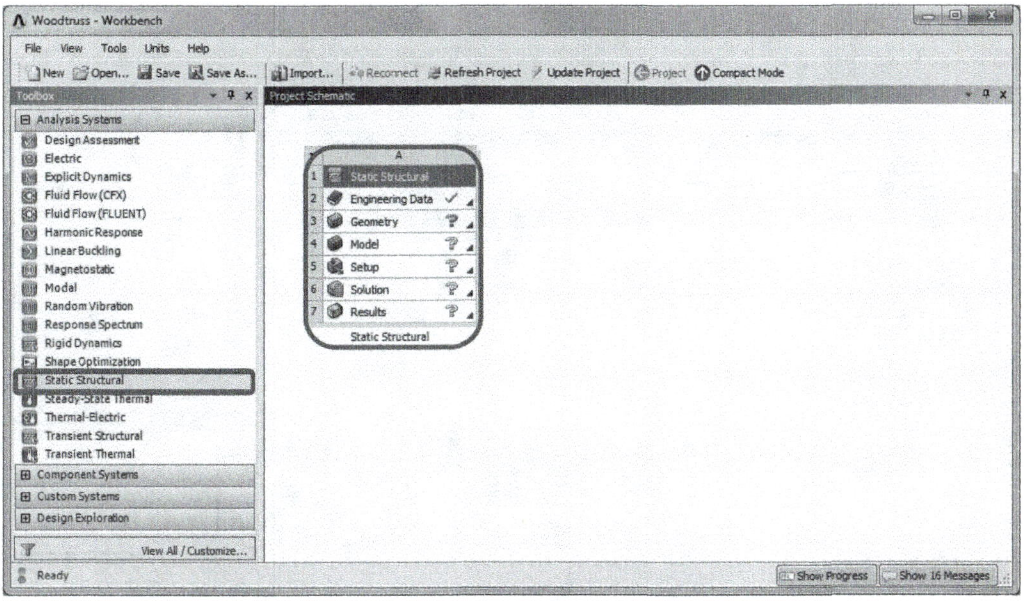

Step 3: Add a New Material

Double-click (or right-click and choose *Edit*) on the *Engineering Data* cell in the above *Project Schematic* to edit or add a material. In the following *Engineering Data* interface which replaces the *Project Schematic*, click the empty box highlighted below and type a name, for example, *"Douglas Fir,"* for the new material.

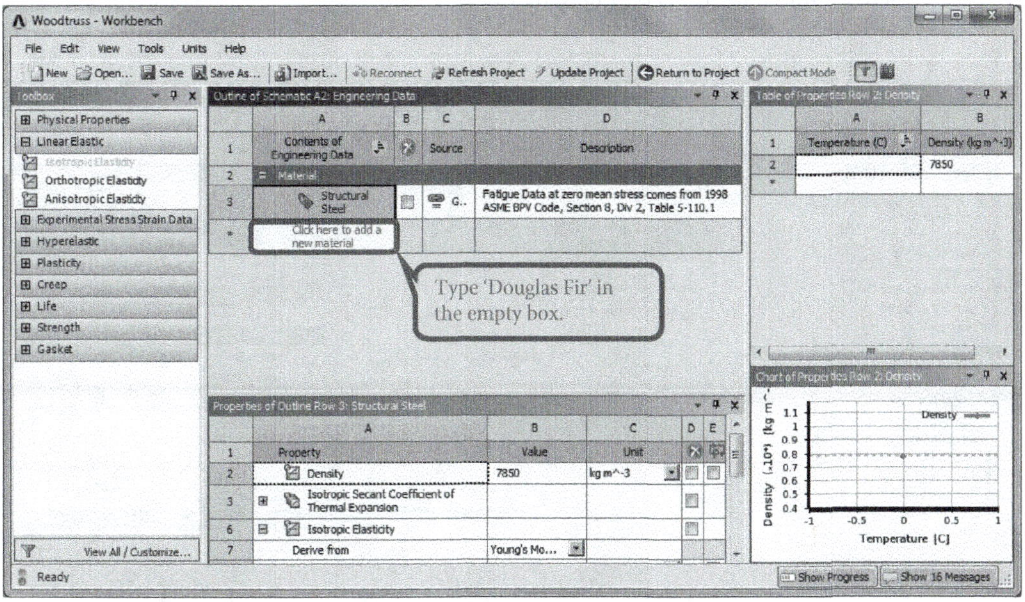

Select *"Douglas Fir"* from the *Outline* window, and double-click *Isotropic Elasticity* under *Linear Elastic* in the leftmost *Toolbox* window. Enter *"1.31E10"* for *Young's Modulus* and *"0.29"* for *Poisson's Ratio* in the bottom center *Properties* window. Click the *Return to Project* button to go back to the *Project Schematic*.

Woodtruss - Workbench

File Edit View Tools Units Help

New Open... Save Save As... Import... Reconnect Refresh Project Update Project Return to Project Compact Mode

Toolbox

Physical Properties

Linear Elastic

Isotropic Elasticity

Orthotropic Elasticity

Anisotropic Elasticity

Experimental Stress Strain Data

Hyperelastic

Plasticity

Creep

Life

Strength

Gasket

View All / Customize...

Outline of Schematic A2: Engineering Data

	A	B	C	D
1	Contents of Engineering Data		Source	Description
2	Material			
3	Structural Steel		G..	Fatigue Data at zero mean stress comes from 1998 ASME BPV Code, Section 8, Div 2, Table 5-110.1
4	Douglas Fir			
*	Click here to add a new material			

Table of Properties Row 6: Isotropic Ela

	A	B
1	Temperature (C)	Bulk Modulus (Pa)
2		1.0397E+10
*		

Properties of Outline Row 4: Douglas Fir

	A	B	C	D	E
1	Property	Value	Unit		
2	Isotropic Elasticity				
3	Derive from	Young's Modul...			
4	Young's Modulus	1.31E+10	Pa		
5	Poisson's Ratio	0.29			
6	Bulk Modulus	1.0397E+10	Pa		

Chart of Properties Row 6: Isotropic Ela

Bulk Modulus

Bulk Modulus [.10^10] [Pa]

1.5

1

-1 -0.5 0 0.5 1

Temperature [C]

Ready

Show Progress Show 16 Messages

Step 4: Launch the *DesignModeler* Program

Ensure *Line Bodies* is checked in the *Properties of Schematic A3: Geometry* window. Double-click the *Geometry* cell to launch *DesignModeler,* and select *"Meter"* as length unit in the *Units* pop-up window.

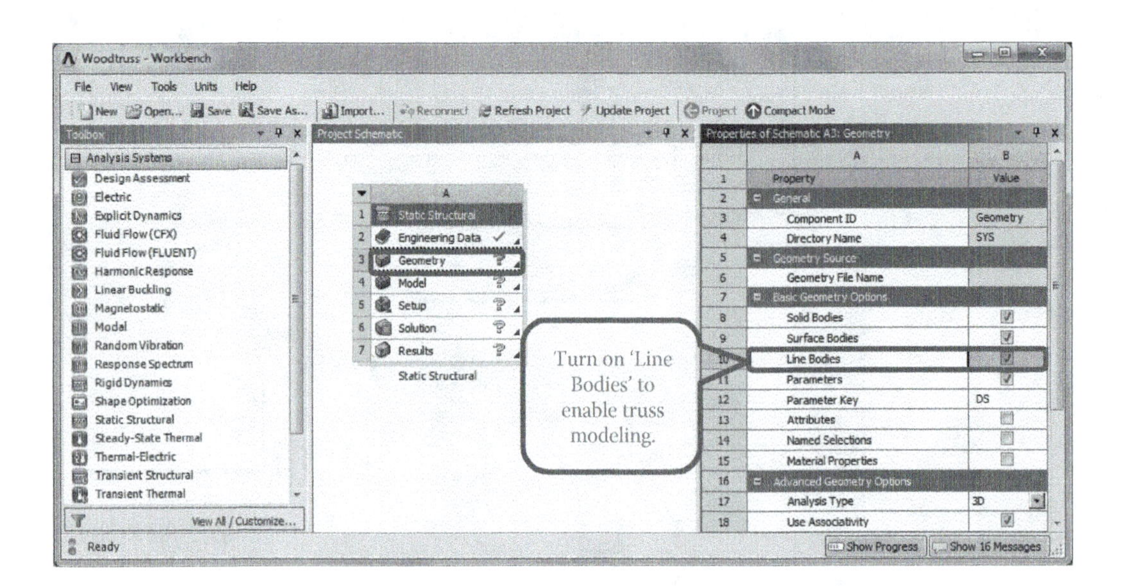

Step 5: Create Line Sketch

Click the *Sketching* tab and select *Settings*. Turn on *Show in 2D* and *Snap* under *Grid* options. Use the default value of "*5 m*" for *Major Grid Spacing* and "*5*" for *Minor-Steps per Major*.

Click a start point and then an end point in the *Graphics* window to draw a line. Draw 13 lines as shown in the sketch below. After completion, click *Generate* to create a line sketch.

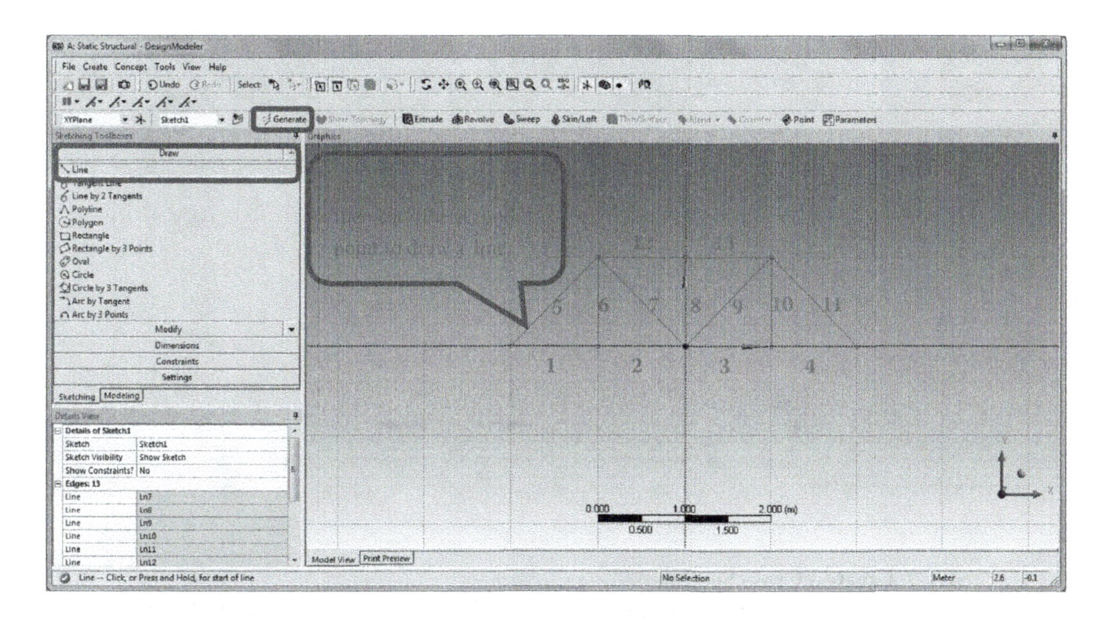

Step 6: Create Line Body from Sketch

Check off the *Grid* options under *Settings* of *Sketching Toolboxes*. Switch to the *Modeling* tab. Note that a new item named *Sketch1* now appears underneath *XYPlane* in the *Tree Outline*.

Select *Lines from Sketches* from the *Concept* drop-down menu.

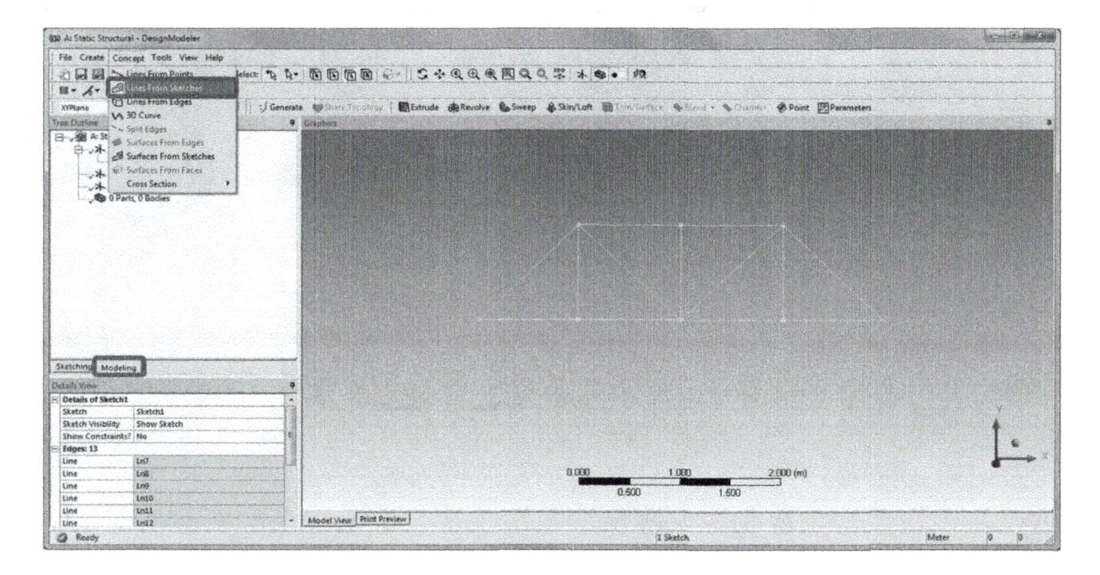

Select *Sketch1* from the *Tree Outline* and click *Apply* to confirm on the *Base Objects* selection in the *Details of Line1*. Click *Generate* to complete the line body creation.

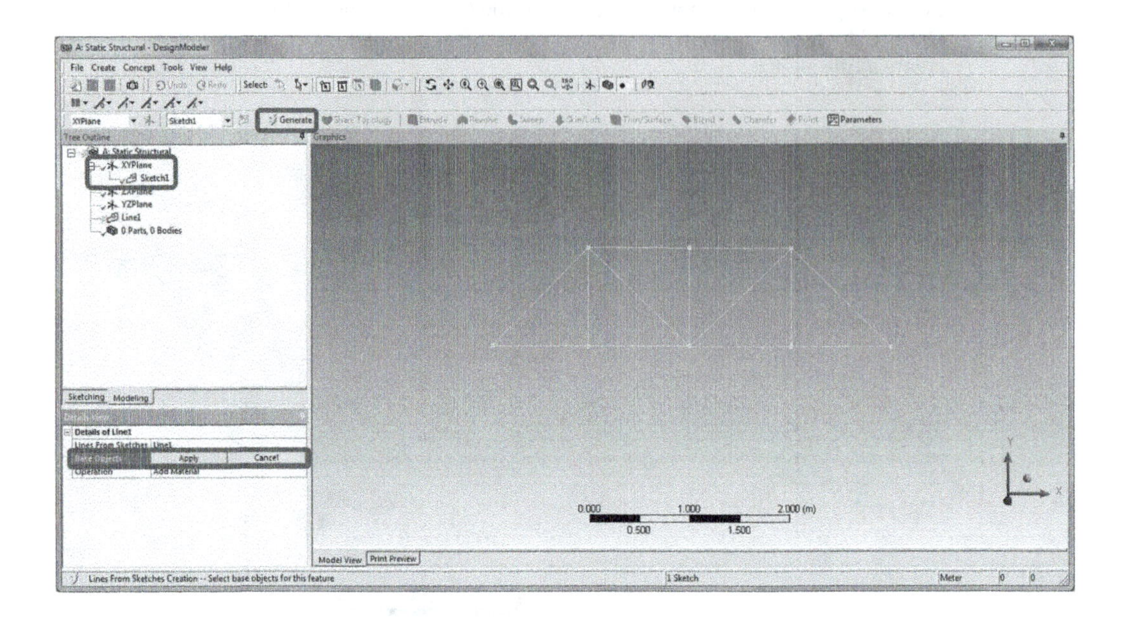

Step 7: Create a Cross Section

Select a *Cross Section* of *Rectangular* from the *Concept* drop-down menu. A new item named *Rect1* is now added underneath the *Cross Section* in the *Tree Outline.*

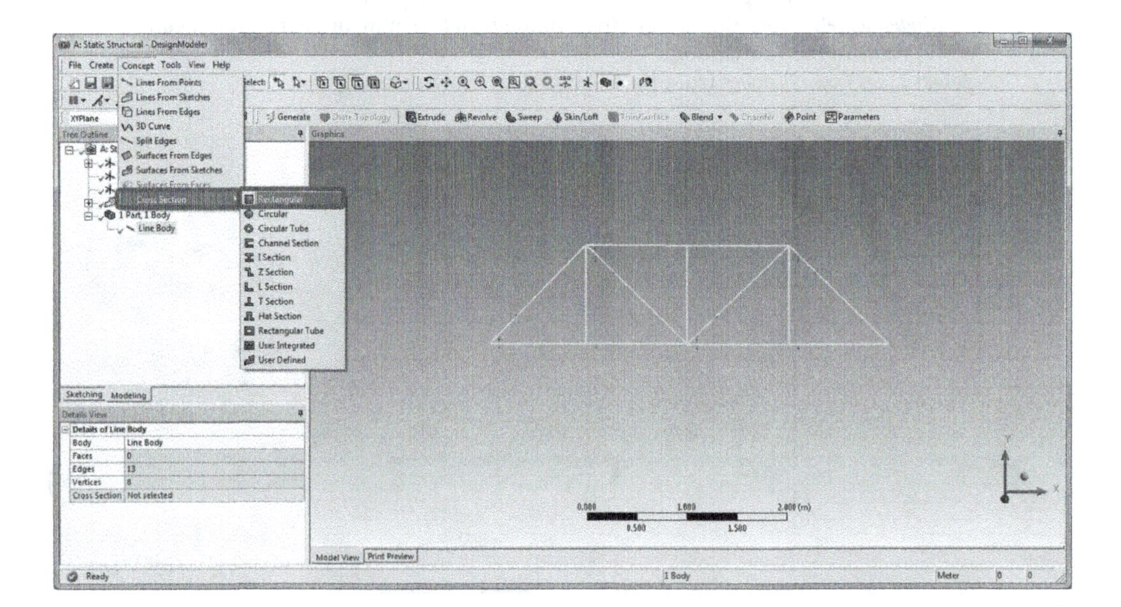

In the *Details of Rect1* under *Dimensions*, enter "*0.06 m*" for both *B* and *H*.

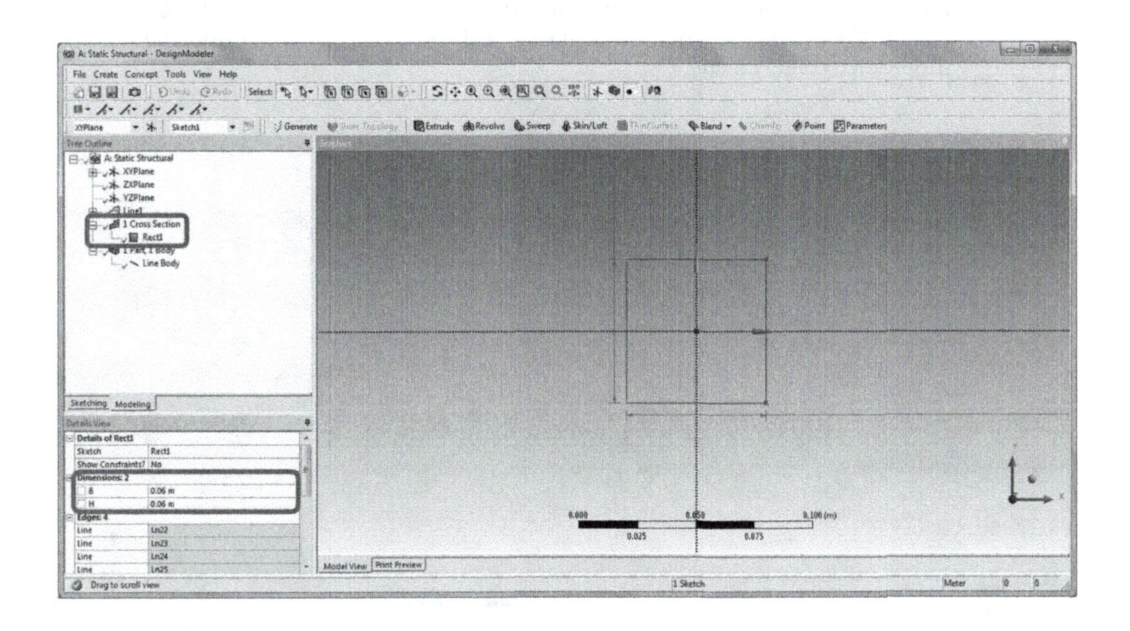

Step 8: Assign Cross Section to Line Body

Select the *Line Body* underneath *1Part, 1 Body* in the *Tree Outline*. In the *Details of Line Body*, assign *Rect1* to the *Cross Section* selection. Click *Close DesignModeler* to exit the program.

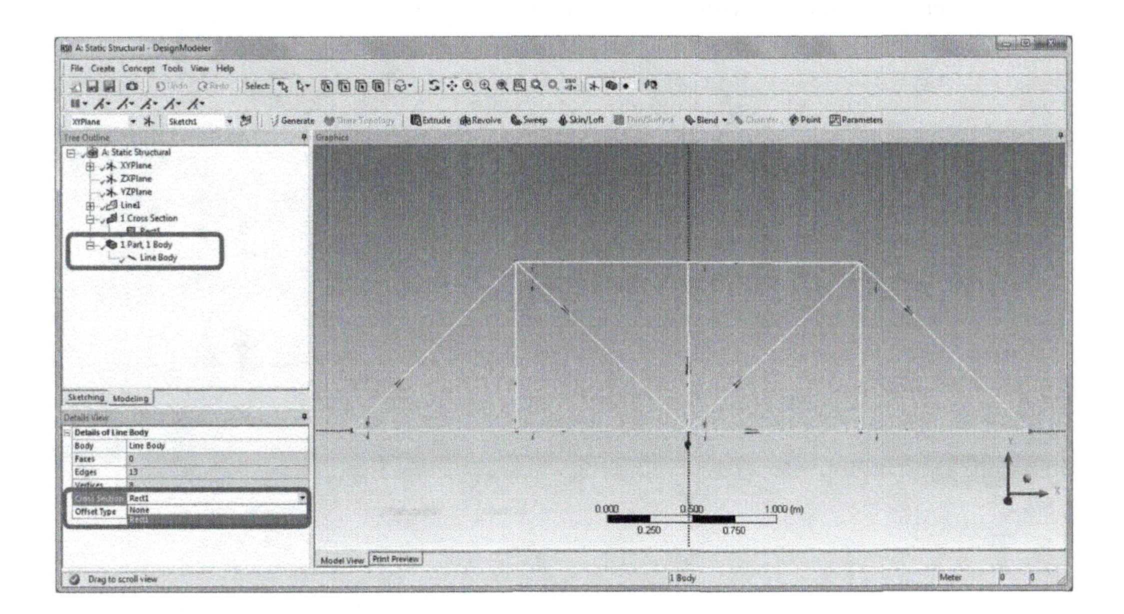

Step 9: Launch the *Static Structural (ANSYS)* Program

Double-click the *Model* cell to launch the *Static Structural (ANSYS)* program. Note that in the *Details of "Line Body"* the material is assigned to *Structural Steel* by default. Click to the right of the *Assignment* field and select *Douglas Fir* from the drop-down context menu.

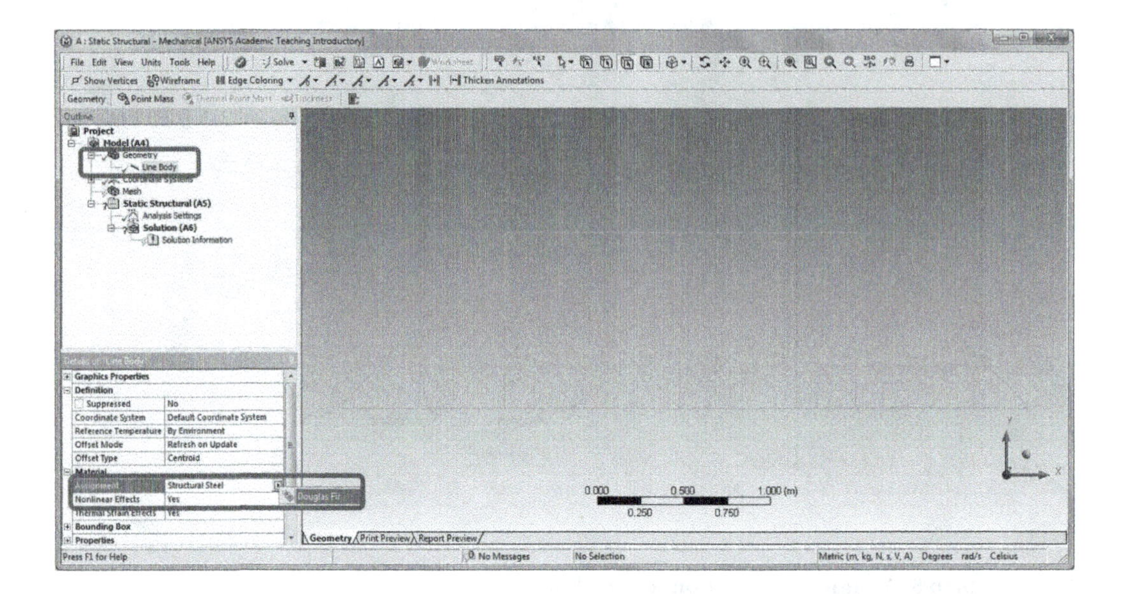

Step 10: Generate Mesh

In the *Details of "Mesh,"* enter a fairly large number, say, *"10 m,"* for the *Element Size*, to ensure each member is meshed with only one element. In the *Outline* of *Project*, right-click on *Mesh* and select *Generate Mesh.*

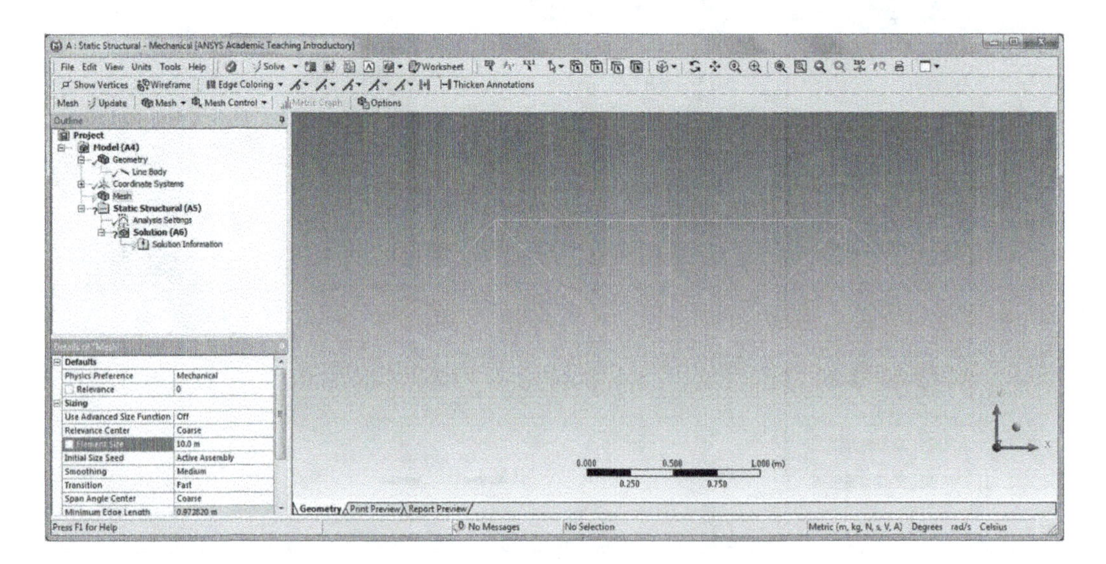

Step 11: Apply Boundary Conditions

In the *Outline* of *Project*, right-click on *Static Structural (A5)* and select *Insert* and then *Fixed Support*. After completion, a *Fixed Support* item is added underneath *Static Structural (A5)* in the project outline tree.

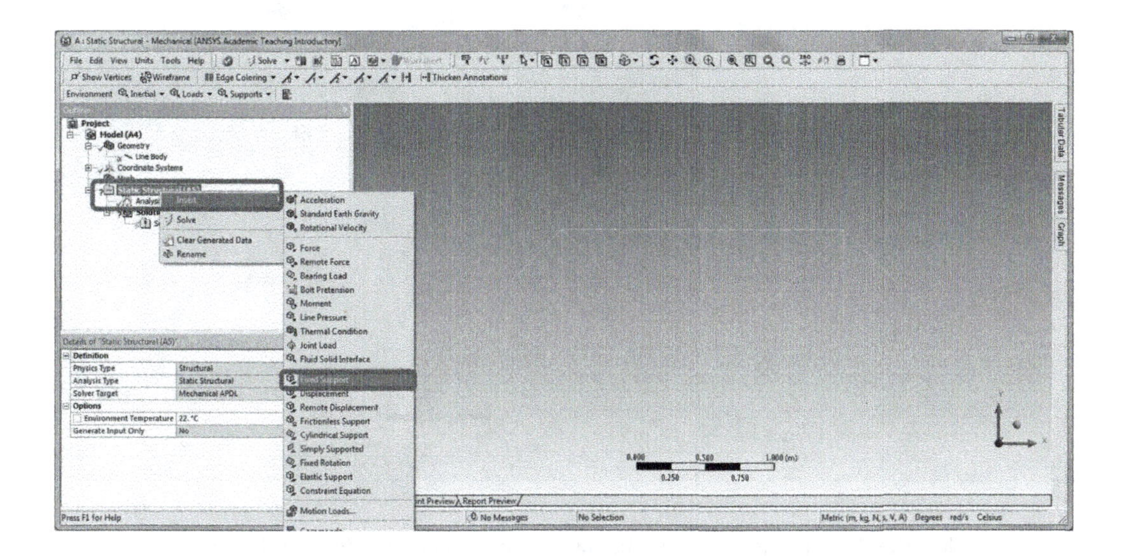

Select the two points as shown below in the *Graphics* window. In the *Details of "Fixed Support,"* click *Apply* to confirm on the *Geometry* selection. After completion, a *Fixed Support* boundary condition will be added to the selected two points.

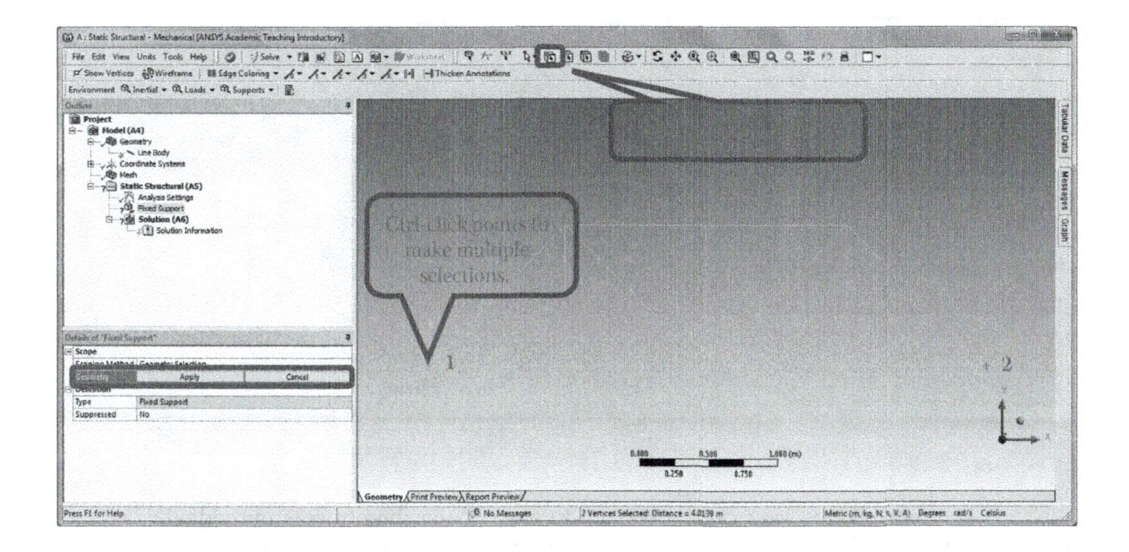

Step 12: Apply Loads

In the *Outline* of *Project*, right-click on *Static Structural (A5)* and select *Insert* and then *Force*.

Select the three points as shown below in the *Graphics* window. In the *Details of "Force,"* click *Apply* to confirm on the Geometry selection. Also underneath the *Details*, change the *Define By* selection to *Components* and enter *"-90000N"* for the *Y Component*. A downward red arrow will appear on the selected three points in the *Graphics* window.

Alternatively, the load can be applied to each of the three points individually by inserting *Force* three times under *Static Structural (A5)*. In this case, enter *"-30000N"* for the *Y Component* of each individual *Force* item.

Step 13: Retrieve Solution
Insert a *Total Deformation* item by right-clicking on *Solution (A6)* in the *Outline* tree.

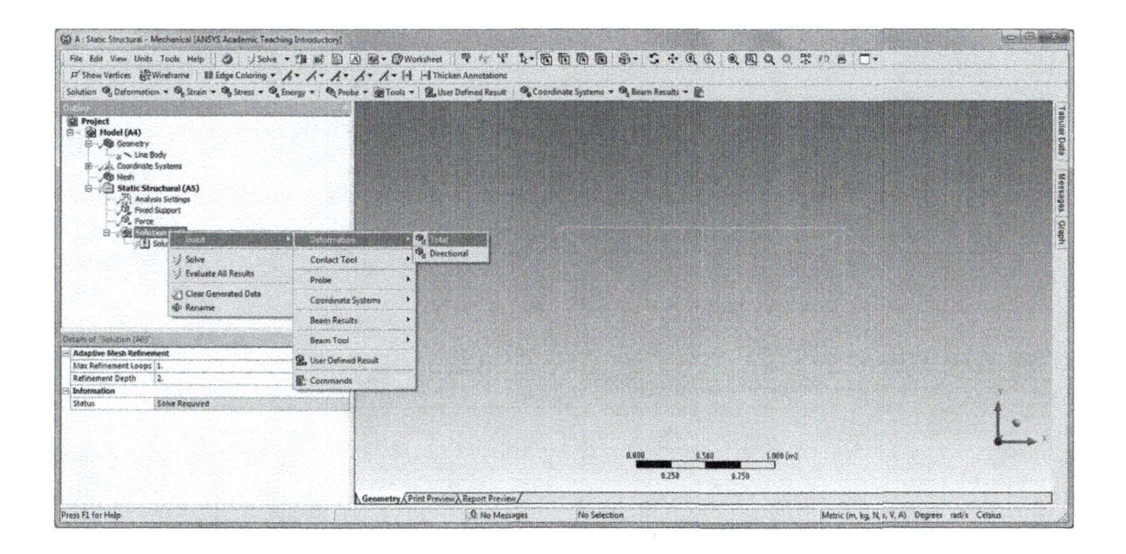

Right-click on *Solution (A6)* in the *Outline* tree and select *Solve*. The program will start to solve the model.

After completion, click *Total Deformation* in the *Outline* to review the total deformation results.

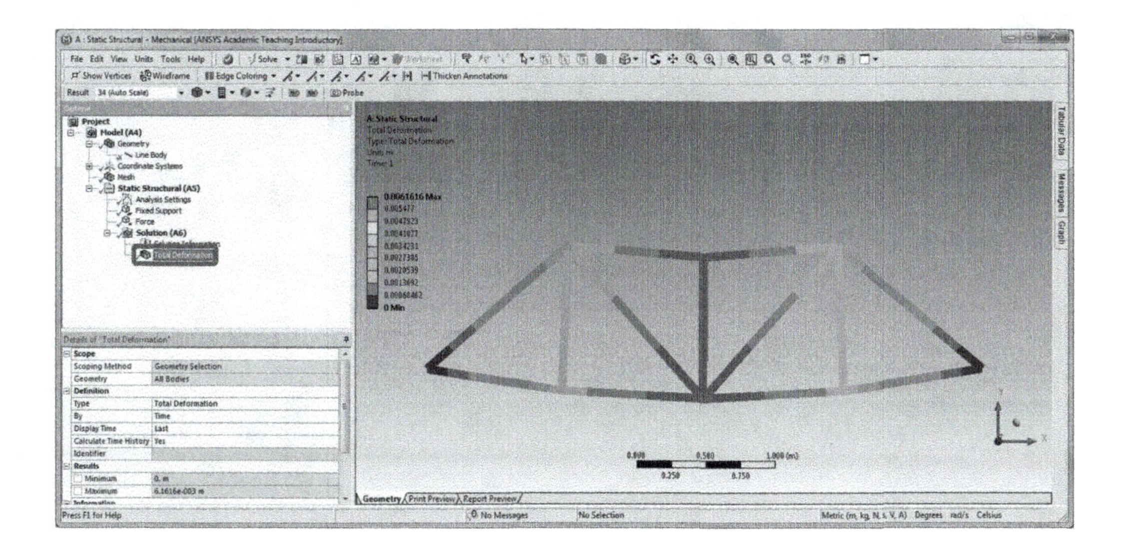

Modeling tips: To get the reaction force, a *Force Reaction* probe can be inserted by right-clicking on *Solution (A6)* in the *Outline* tree as shown below.

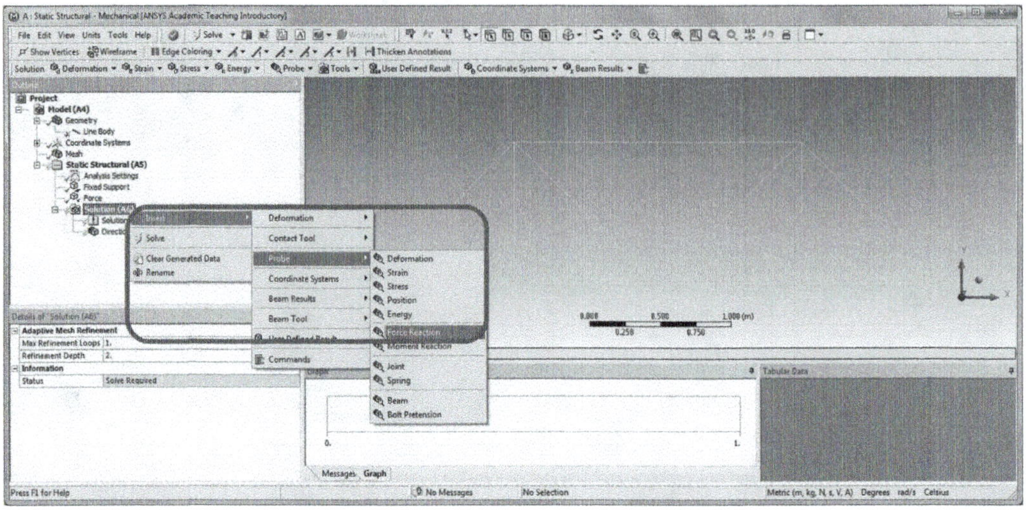

In the *Details of "Force Reaction,"* select the *Fixed Support* as the *Boundary Condition.*

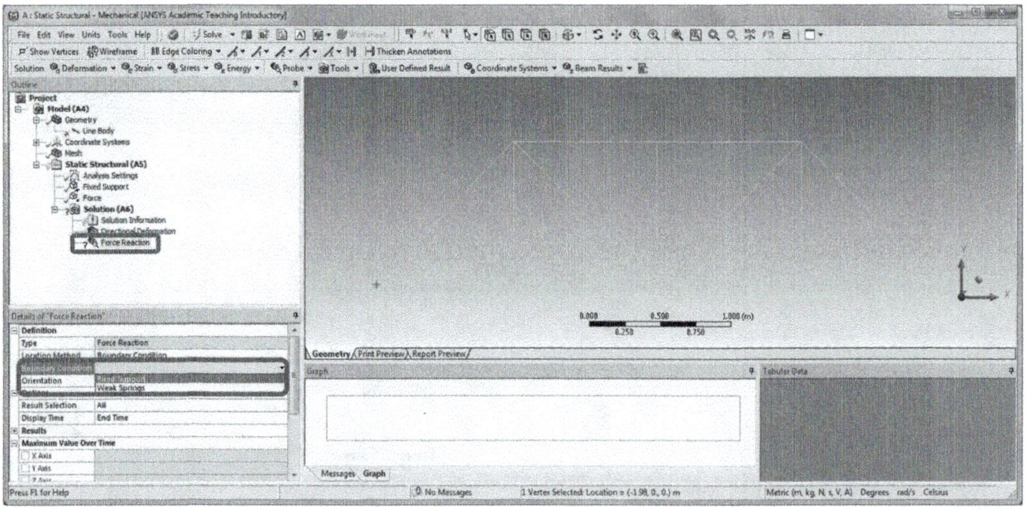

Right-click on *Solution (A6)* in the *Outline* tree and select *Evaluate All Results.*

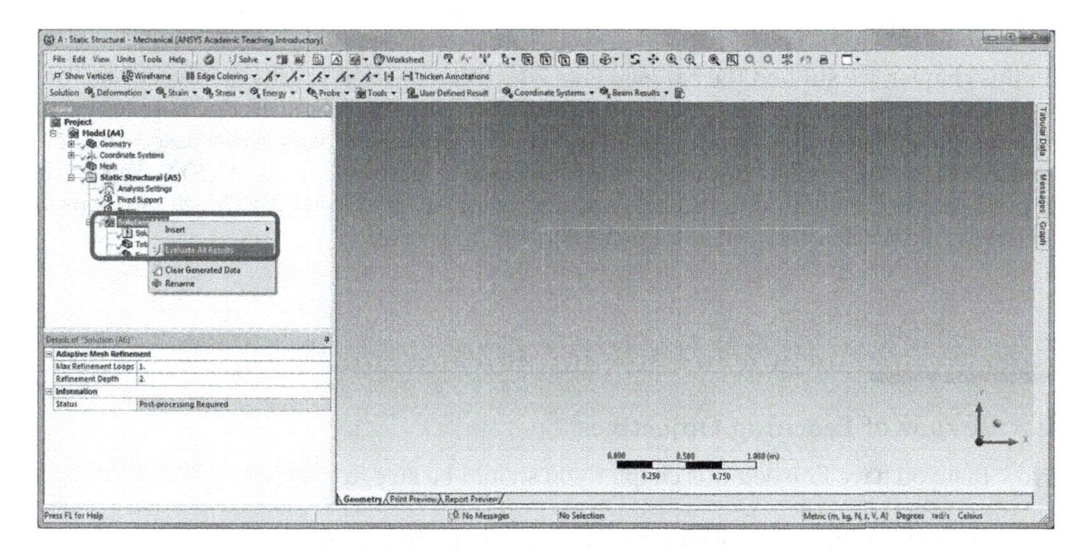

After completion, click *Force Reaction* in the *Outline* to review results.

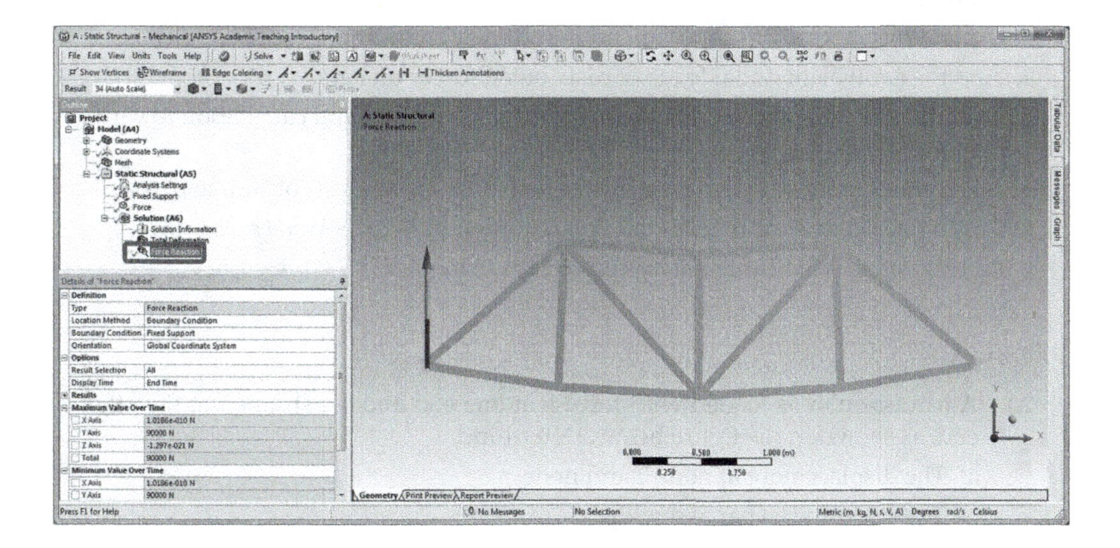

Note here that the reaction force is found to be 90,000 N in the positive Y-direction. This is because a boundary condition has been applied earlier to the two fixed ends in one step (see Step 11). To avoid summing of the force reaction, two fixed conditions can be inserted instead in Step 11, one for each end. The reaction forces at an individual support can then be displayed by selecting the support of interest from the drop-down menu of *Boundary Condition* in the *Details of "Force Reaction."*

2.7 Summary

In this chapter, we studied the bar elements which can be used in truss analysis. The concept of the shape functions is introduced and the derivations of the stiffness matrices using the energy approach are introduced. Treatment of distributed loads is discussed and several examples are studied. A planar truss structure is analyzed using *ANSYS Workbench*. It provides basic modeling techniques and shows step-by-step how *Workbench* can be used to determine the deformation and reaction forces in trusses.

2.8 Review of Learning Objectives

Now that you have finished this chapter you should be able to

1. Set up simplified finite element models for truss structures.
2. Derive the stiffness matrix for plane bar elements using direct and energy approaches.
3. Explain the concept of shape functions and their characteristics for bar elements.
4. Find the equivalent nodal loads of distributed forces on bars.
5. Determine the displacement and stress of a truss using hand calculation to verify the finite element solutions.
6. Apply the general bar element stiffness matrix to the analysis of simple trusses.
7. Create line sketches and new material definition in *ANSYS Workbench*.
8. Perform static structural analyses on trusses using *ANSYS Workbench*.

PROBLEMS

2.1 A bar assembly is loaded with force P at one end and constrained at the other end, as shown in the figure below. Determine:

a. The displacement at node 2 and node 3

b. The stress in the bar assembly

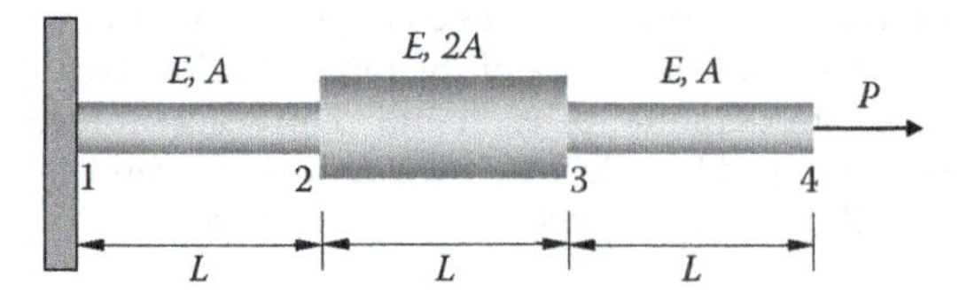

2.2 A simple structure with the profile shown below is loaded with force P, and constrained at one end. Young's modulus E is the same for all the bars. The cross-sectional areas are shown in the figure. Use 1-D bar elements to approximate the deformation and the stresses in the structure.

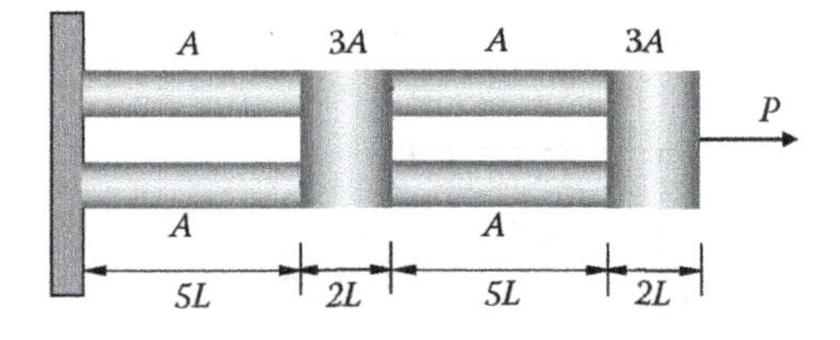

2.3 The plane truss is loaded with force P as shown below. Constants E and A for each bar are as shown in the diagram. Determine:

a. The nodal displacements
b. The reaction forces
c. The stresses in bar elements

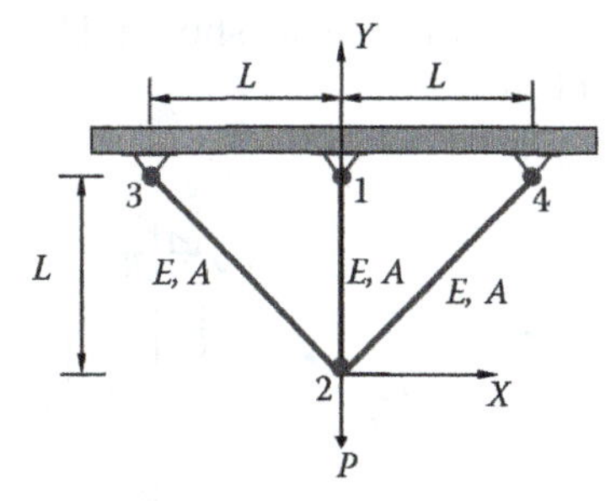

2.4 The plane truss is loaded with force P as shown below. Constants E and A for each bar are as shown in the diagram. Determine:

a. The nodal displacements
b. The reaction forces
c. The stresses in bar elements

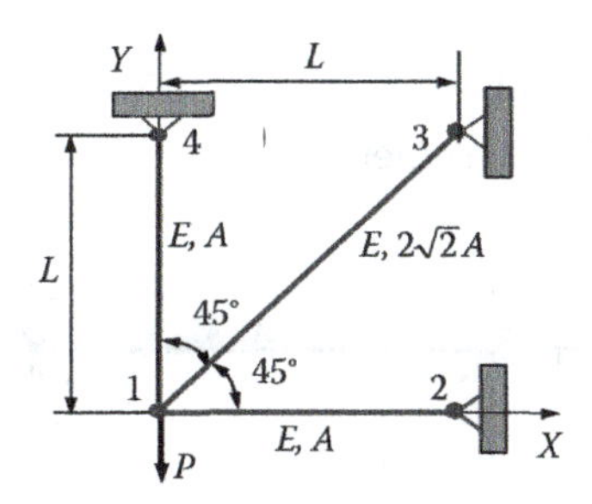

2.5 The plane truss is loaded with force P as shown below. Constants E and A for each bar are as shown in the diagram. Determine:

a. The nodal displacements
b. The stresses in bar elements

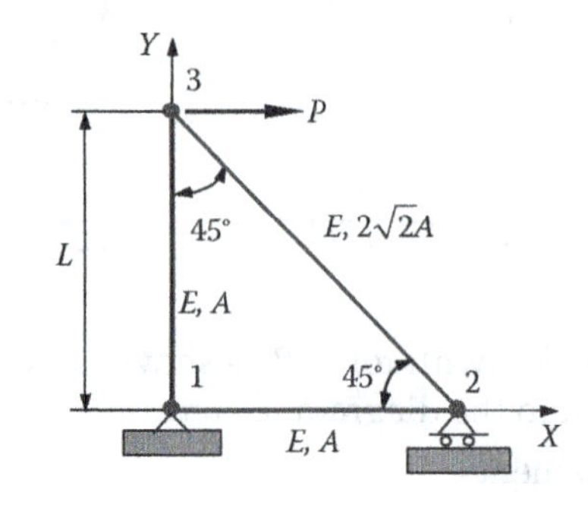

2.6 The plane truss is supported as shown below. Young's modulus E is the same for all the bars. The cross-sectional areas are shown in the figure. Suppose that the node 2 settles by an amount of δ as shown. Determine the stresses in each bar element using the FEM.

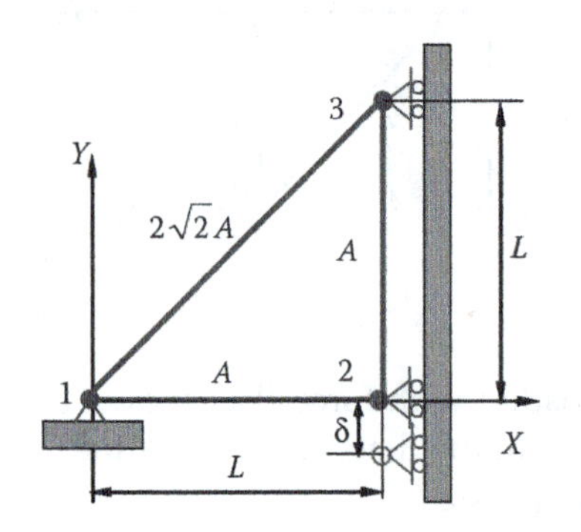

2.7 The plane truss is loaded with force P as shown below. Young's modulus E is the same for all the bars. The cross-sectional areas are shown in the figure. Determine:

a. The nodal displacements
b. The stresses in each bar element

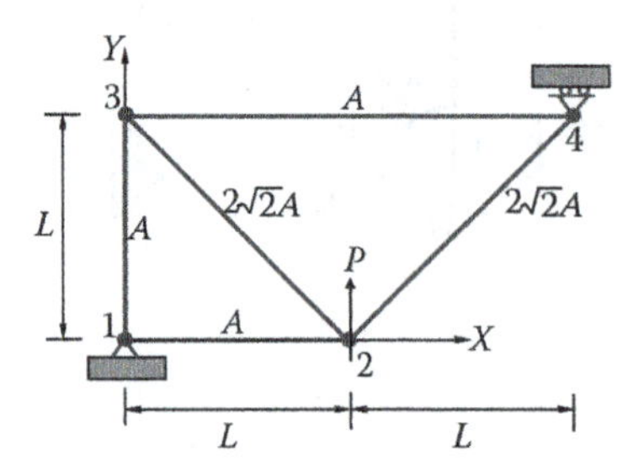

2.8 The plane truss is loaded with force P as shown below. Young's modulus E is the same for all the bars. The cross-sectional areas are shown in the figure. Determine the nodal displacements and the reaction forces.

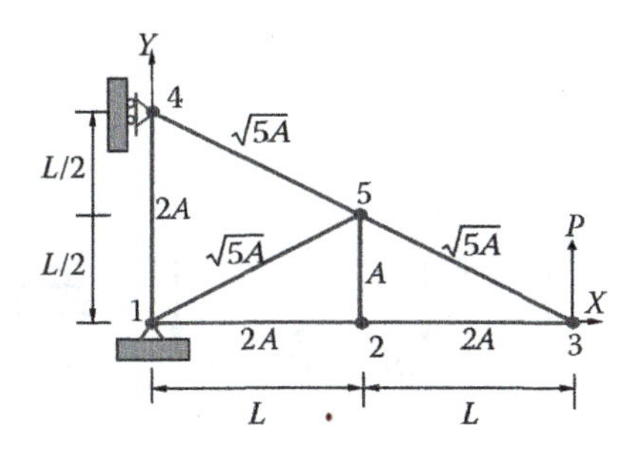

2.9 The roof truss shown below is made of Douglas fir timbers of a 5 mm × 5 mm cross section. Use *ANSYS Workbench* to determine the truss deformation and the support reaction forces.

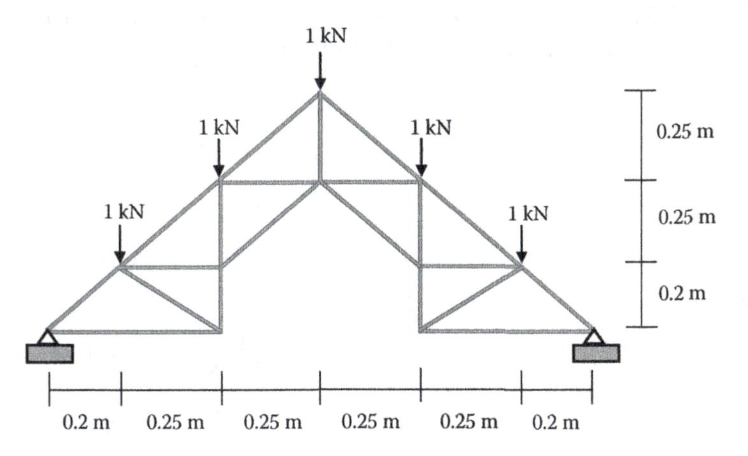

2.10 The truss tower crane shown below is made of structural steel rectangular bars of a 3 mm × 3 mm cross section. Use *ANSYS Workbench* to determine the truss deformation and the support reaction forces.

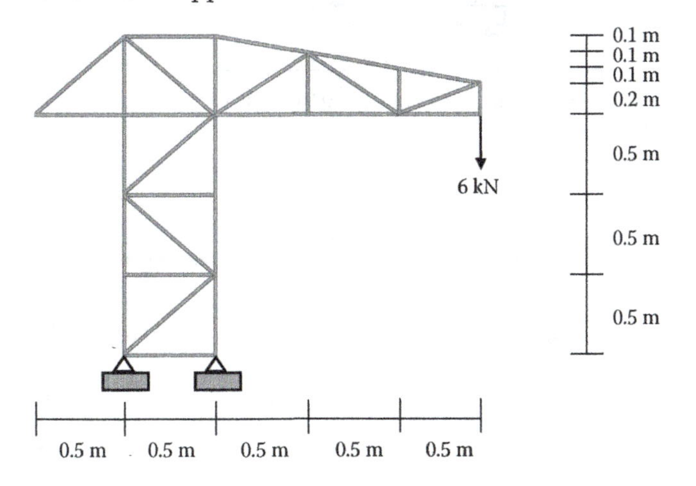

2.11 The truss bridge shown below is made of Douglas fir timbers of a 4 mm × 4 mm cross section. Use *ANSYS Workbench* to determine the truss deformation and the support reactions.

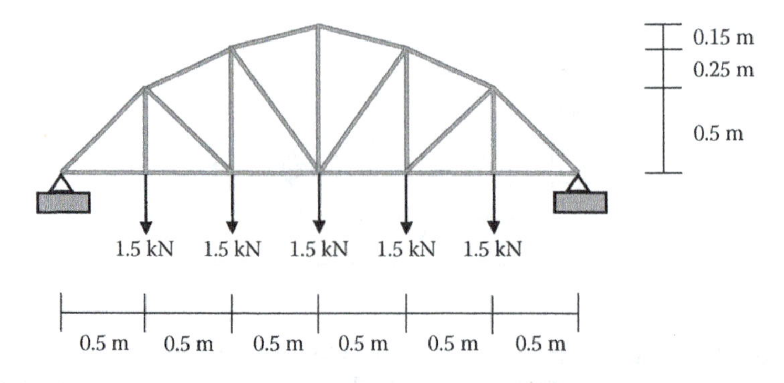

2.12 The truss transmission tower shown below is made of structural steel members of a 2.5 mm × 2.5 mm cross section. Use *ANSYS Workbench* to determine the truss deformation and the support reactions.

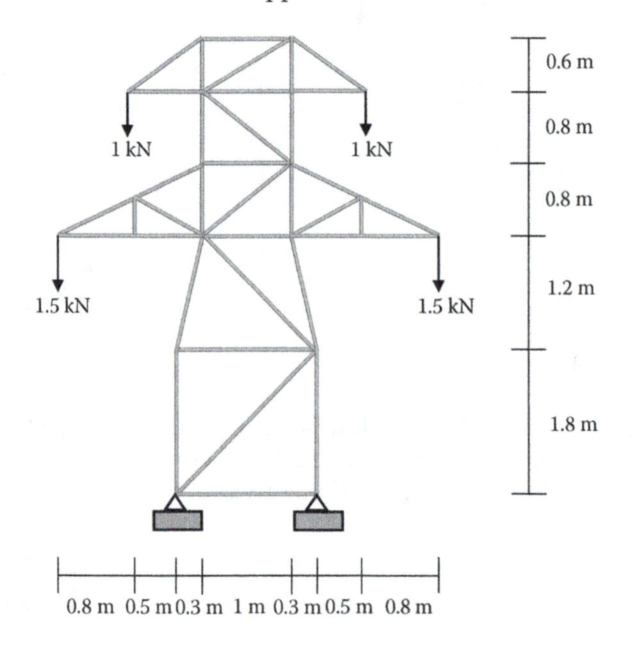

3

Beams and Frames

3.1 Introduction

Beams and frames are frequently used in constructions, in engineering equipment, and in everyday life, for example, in buildings, lifting equipment, vehicles, and exercise machines (Figure 3.1).

Beams are slender structural members subjected primarily to transverse loads. A beam is geometrically similar to a bar in that its longitudinal dimension is significantly larger than the two transverse dimensions. Unlike bars, the deformation in a beam is predominantly bending in transverse directions. Such a bending-dominated deformation is the primary mechanism for a beam to resist transverse loads. In this chapter, we will use the term "general beam" for a beam that is subjected to both bending and axial forces, and the term "simple beam" for a beam subjected to only bending forces. The term "frame" is used for structures constructed of two or more rigidly connected beams.

3.2 Review of the Beam Theory

We start with a brief review of the simple beam theory. Essential features of the two well-known beam models, the Euler–Bernoulli beam and the Timoshenko beam, will be introduced, followed by discussions on the stress, strain, and deflection relations in simple beam theory.

3.2.1 Euler–Bernoulli Beam and Timoshenko Beam

Euler–Bernoulli beam and Timoshenko beam, as shown in Figure 3.2, are two common models that are used in the structural analysis of beams and frames. Both models have at their core the assumption of small deformation and linear elastic isotropic material behavior. They are applicable to beams with uniform cross sections.

For a Euler–Bernoulli beam, it is assumed that the forces on a beam only cause the beam to bend. There is no transverse shear deformation occurring in the beam bending. The neutral axis, an axis that passes through the centroid of each beam cross section, does not change in length after the deformation. A planar cross section perpendicular to the neutral axis remains plane and perpendicular to the neutral axis after deformation. Due to its neglect of shear strain effects, the Euler–Bernoulli model tends to slightly underestimate

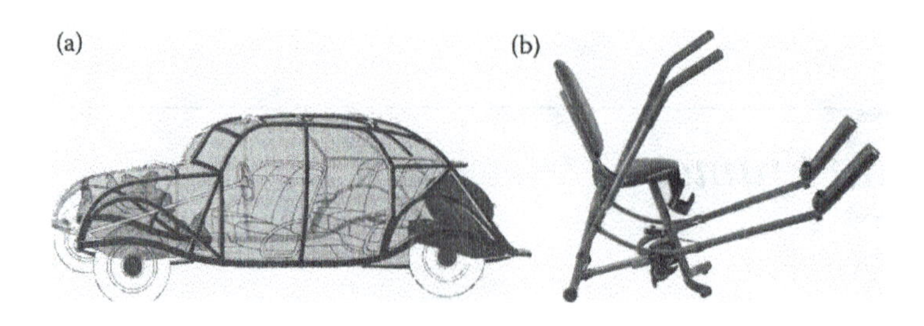

FIGURE 3.1
Examples of beams and frames: (a) a car frame (http://www.carbasics-1950.com/); and (b) an exercise machine. (From http://www.nibbledaily.com/body-by-jake-cardio-cruiser/.)

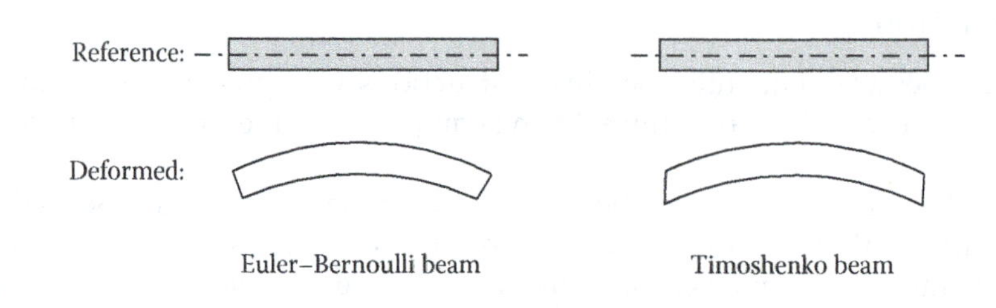

FIGURE 3.2
Two common beam models.

the beam deformation. The model is considered suitable for describing the behavior of thin or long beams.

The Timoshenko beam model accounts for both transverse shear deformation and bending deformation. In this model, a planar cross section remains plane, but does not remain normal to the neutral axis after deformation. By taking into account shear strain effects, the Timoshenko model is physically more realistic for describing the behavior of thick or short beams.

To determine the deflection of a laterally loaded beam of the same material and uniform cross section, we may view the beam as a 1-D model with the transverse load assigned to its neutral axis. As illustrated in Figure 3.3, for a Euler–Bernoulli beam, we need only one variable—the vertical displacement (v), to describe the beam deflection at any point along

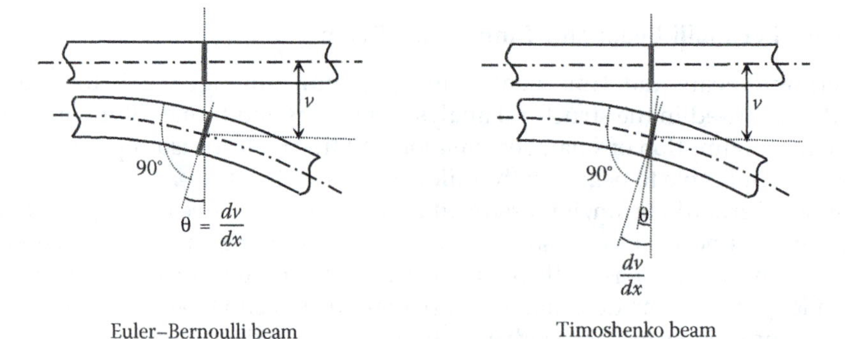

FIGURE 3.3
Deflection of Euler–Bernoulli beam and Timoshenko beam.

the neutral axis. For a Timoshenko beam, both the vertical displacement (v) and the rotation angle of beam cross section (θ) are needed for a complete description of its deformed configuration.

3.2.2 Stress, Strain, Deflection, and Their Relations

Simple beam bending is often analyzed using the Euler–Bernoulli beam theory, which is popular in engineering owing to its simplicity and is referred to as the engineering beam theory by many. To understand the stress state in simple beam bending, imagine that the beam is made of many infinitesimally thin layers deposited on top of each other, as shown in Figure 3.4.

Under the assumption that the neutral axis does not change in length, the material layer located along the axis will stay stress free. However, the beam will bend concave upward to resist a downward transverse load. Bending makes layers above the neutral axis axially compressed and the ones underneath axially stretched. The further away a material layer is from the neutral axis, the more it is strained. The axial strain induced by bending is called the bending strain, which is related to the deflected beam curvature $\kappa(x)$ based on differential geometry as follows:

$$\varepsilon(x) = -y\kappa(x) = -y\frac{d^2v}{dx^2} \tag{3.1}$$

where y is the vertical distance of a thin layer from the neutral axis and v the beam deflection.

The internal resisting bending moment $M(x)$ is a function of the axial bending stress $\sigma(x)$ from the theory of elasticity. Together with Equation 3.1, we have:

$$M(x) = \int -y\sigma(x)dA = \int Ey^2\kappa(x)\,dA = EI\kappa(x) = EI\frac{d^2v}{dx^2} \tag{3.2}$$

where E is the material elastic modulus, I is the area moment of inertia of the beam cross-section with respect to the z-axis, and the product of E and I is called the bending stiffness. Combining Equations 3.1 and 3.2, we arrive at the following flexure formula for the bending stress $\sigma(x)$:

$$\sigma(x) = -\frac{M(x)y}{I} \tag{3.3}$$

The above simple beam theory is analogous to the uniaxial Hooke's law. As shown in Table 3.1, the bending moment is linearly proportional to the deflected beam curvature

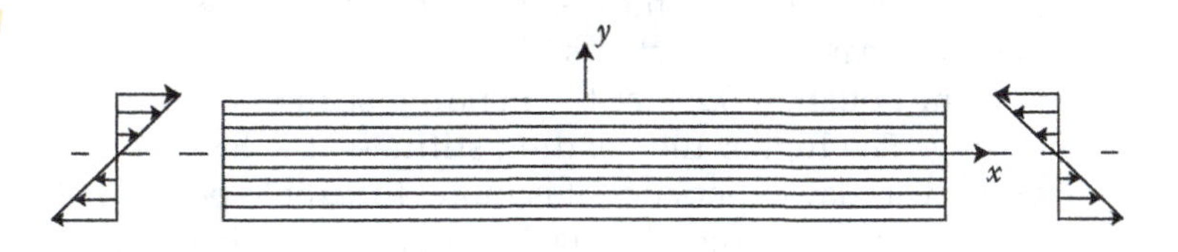

FIGURE 3.4
Bending produces axial stress $\sigma(x)$ along thin material layers in a beam.

TABLE 3.1

Analogy between the Constitutive Equations for Bars and Beams

	Stress Measurement	Strain Measurement	Constitutive Equation
Bar	Axial stress: $\sigma(x)$	Axial strain: $\varepsilon(x)$	$\sigma(x) = E\varepsilon(x)$
Beam	Bending moment: $M(x)$	Curvature: $\frac{d^2v}{dx^2}$	$M(x) = EI\frac{d^2v}{dx^2}$

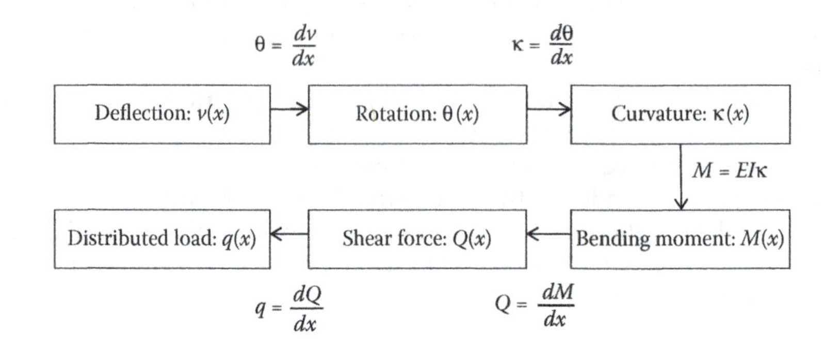

FIGURE 3.5
The governing equations for a simple beam.

through a bending stiffness (EI) constant. This resembles the linear stress–strain relationship described by Hooke's law. The basic equations that govern the problems of simple beam bending are summarized in Figure 3.5. The equations included here will later be used in the formulation of finite element equations for beams.

3.3 Modeling of Beams and Frames

Modeling is an idealization process. Engineers seek to simplify problems and model real physical structures at an adequate level of detail in design and analysis to strike a balance between efficiency and accuracy. Some considerations on cross sections, support conditions, and model simplification of beams and frames are discussed next.

3.3.1 Cross Sections and Strong/Weak Axis

Beams are available in various cross-sectional shapes. There are rectangular hollow tubes, I-beams, C-beams, L-beams, T-beams, and W-beams, to name a few. Figure 3.6 illustrates some common shapes for beam cross sections.

We have learned from Equation 3.2 that the bending stiffness (EI) measures a beam's ability to resist bending. The higher the bending stiffness, the less the beam is likely to bend. Drawing on our everyday experience, it is more difficult to bend a flat ruler with its flat (wide) surface facing forward rather than facing up, as shown in Figure 3.7. It is because the moment of inertia (I) is a cross-sectional property sensitive to the distribution of material with respect to an axis.

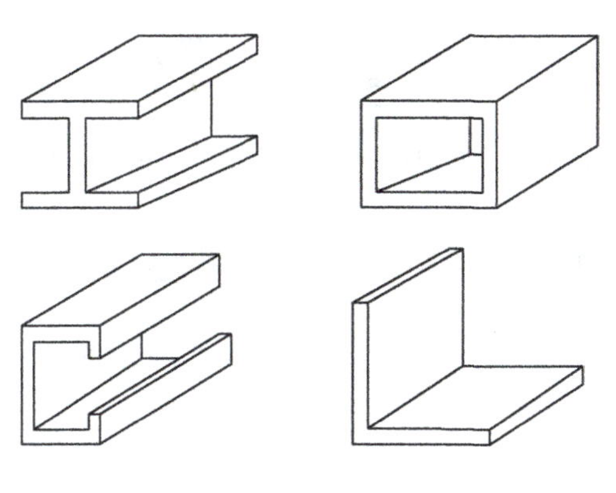

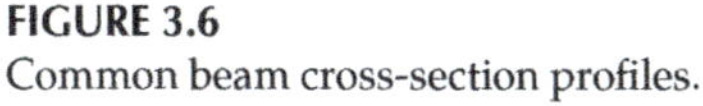

FIGURE 3.6
Common beam cross-section profiles.

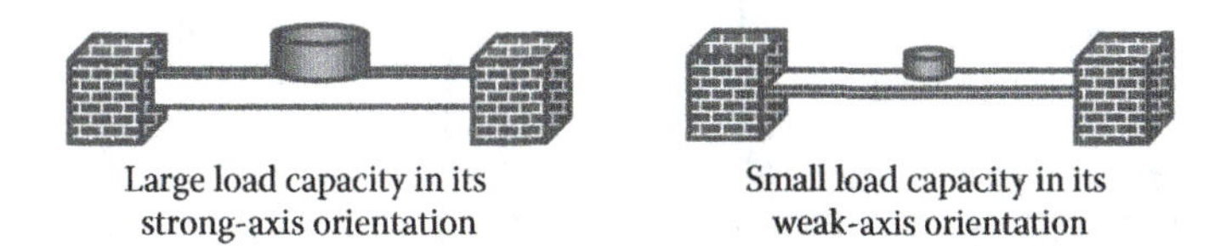

Large load capacity in its
strong-axis orientation
Small load capacity in its
weak-axis orientation

FIGURE 3.7
A beam loaded on its strong and weak axes.

To keep the deflection within acceptable limits, beams need to have sufficiently large bending stiffness to withstand the applied load. It is important that we select beams of efficient cross-section shapes and use beams on their strong-axes of bending, especially in cases when switching to materials of high elastic modulus is deemed costly or infeasible due to other constraints.

3.3.2 Support Conditions

There are many different ways of supporting beams. For example, a beam can be fixed at both ends, or fixed at one end and pinned at the other end. Generally speaking, beams have three types of end support conditions: fixed support, pinned support, and roller support. A fixed support is an anchor condition that has zero translation and zero rotation at the supported end. A pinned support is a hinge condition that prevents any translation but does not prevent rotation about the hinge axis. A roller support allows for both rotation and translation along the surface on which the roller moves but prevents any translation normal to that surface. Figure 3.8 illustrates various types of beams named based on their support conditions. As shown, a simply supported beam is supported by a hinge at one end and by a roller at the other end. A cantilever beam is fixed at one end and free at the other end. A fixed-pinned beam is fixed at one end and pinned at the other end, while a fixed end beam is one that is clamped at both ends.

In general, support conditions are important considerations in determining the mathematical solution of a physical problem, and inappropriate support is a common cause of errors. When we model beams and frames, bear in mind that valid boundary conditions need to reflect the underlying physical construction and be sufficient to prevent the structure from any rigid-body motion (both translation and rotation).

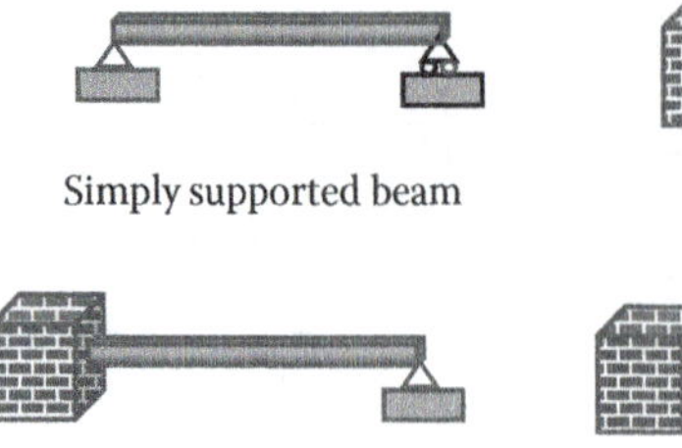

FIGURE 3.8
Beam supports and beam types.

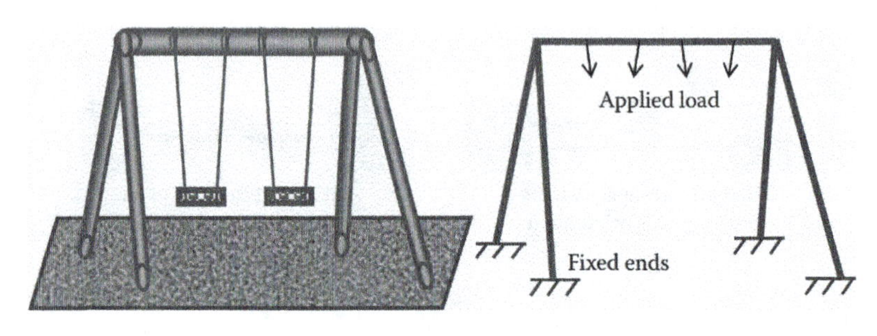

FIGURE 3.9
A swing set and its simplified line model.

3.3.3 Conversion of a Physical Model into a Line Model

Aside from the support idealization, conceptual models are widely adopted in the analyses of beams and frames to achieve modeling efficiency. The idea of model conceptualization stems from the uniform cross-section assumption. Under the assumption, it is apparent that a beam needs only to be modeled at the center axis (neutral axis) of the actual 3-D beam structure, as shown in Figure 3.9 through a swing set example.

For beams and frames, the conceptual model is also known as the line model, which consists of only lines or curves in general. After a line model is created, cross-sectional properties and other data such as material properties, boundary conditions, and loads can be specified for the analysis. Once a problem is fully defined, solutions can be obtained readily after the line model is discretized into a line mesh using beam elements. In contrast to the truss modeling, where each truss member is discretized into a single bar element, many line segments (element divisions) are typically needed in a line mesh made with beam elements in order to obtain more accurate results.

3.4 Formulation of the Beam Element

In this section, we discuss the finite element formulation based on the simple beam theory (Euler–Bernoulli beam). The beam element stiffness matrix will be established using both direct and energy methods.

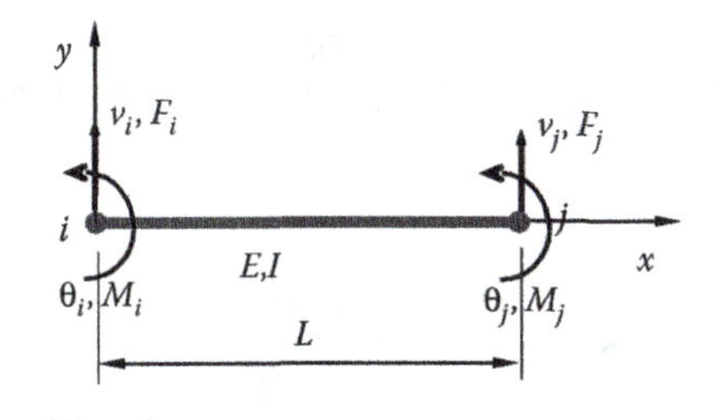

FIGURE 3.10
Notion for a simple beam element in 2-D.

The setup of a simple beam element in 2-D space is shown in Figure 3.10, where the variables are:

L, I, and E = Length, moment of inertia of the cross-sectional area, and elastic modulus of the beam, respectively

$v = v(x)$ deflection (lateral displacement) of the neutral axis of the beam

$\theta = dv/dx$ rotation of the beam about the z-axis

$Q = Q(x)$ (internal) shear force

$M = M(x)$ (internal) bending moment about z-axis

F_i, M_i, F_j, M_j applied (external) lateral forces and moments at node i and j, respectively

For simplicity of presentation, we will restrict our attention to beam element formulation in 2-D space in the following discussion. For a beam element in 3-D space, the element stiffness equation can be formed in the local (2-D) coordinate system first and then transformed into the global (3-D) coordinate system to be assembled.

3.4.1 Element Stiffness Equation: The Direct Approach

We first apply the direct method to establish the beam stiffness matrix using the results from elementary beam theory. The FE equation for a beam takes the form

$$\begin{bmatrix} k_{11} & k_{12} & k_{13} & k_{14} \\ k_{21} & k_{22} & k_{23} & k_{24} \\ k_{31} & k_{32} & k_{33} & k_{34} \\ k_{41} & k_{42} & k_{43} & k_{44} \end{bmatrix} \begin{Bmatrix} v_i \\ \theta_i \\ v_j \\ \theta_j \end{Bmatrix} = \begin{Bmatrix} F_i \\ M_i \\ F_j \\ M_j \end{Bmatrix} \tag{3.4}$$

Recall that each column in the stiffness matrix represents the forces needed to keep the element in a special deformed shape. For example, the first column represents the forces/moments to keep the shape with $v_i = 1$, $\theta_i = v_j = \theta_j = 0$ as shown in Figure 3.11a. Thus, using the results from strength of materials for a cantilever beam with a force k_{11} and moment k_{21} applied at the free end, we have

$$v_i = \frac{k_{11}L^3}{3EI} - \frac{k_{21}L^2}{2EI} = 1 \quad \text{and} \quad \theta_i = -\frac{k_{11}L^2}{2EI} + \frac{k_{21}L}{EI} = 0$$

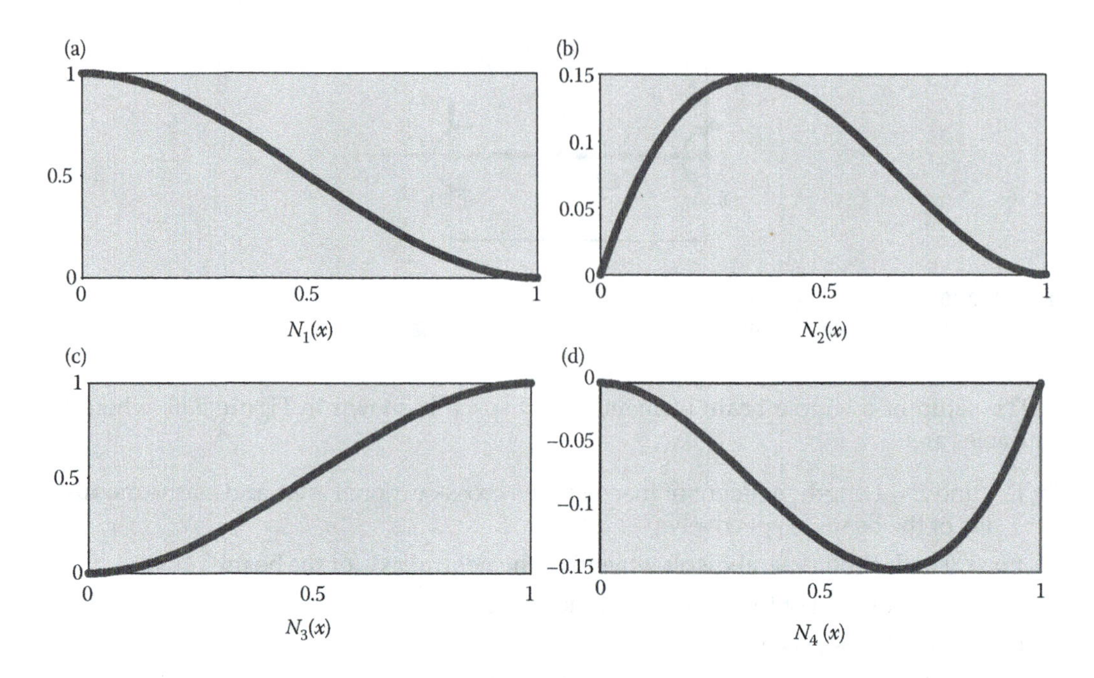

FIGURE 3.11
Four configurations or shapes for the simple beam element.

Solving this system of equations, we obtain k_{11} and k_{21}. Using the equilibrium conditions of the beam, we obtain k_{31} and k_{41}, and thus the first column of the stiffness matrix.

Using the same procedure in calculating columns 2, 3, and 4 of the matrix in Equation 3.4 (corresponding to shapes in Figures 3.11b through d), we obtain the element stiffness equation as:

$$\frac{EI}{L^3}\begin{bmatrix} 12 & 6L & -12 & 6L \\ 6L & 4L^2 & -6L & 2L^2 \\ -12 & -6L & 12 & -6L \\ 6L & 2L^2 & -6L & 4L^2 \end{bmatrix}\begin{Bmatrix} v_i \\ \theta_i \\ v_j \\ \theta_j \end{Bmatrix} = \begin{Bmatrix} F_i \\ M_i \\ F_j \\ M_j \end{Bmatrix} \tag{3.5}$$

3.4.2 Element Stiffness Equation: The Energy Approach

To derive the stiffness matrix in Equation 3.5 using the formal or energy approach, we introduce four shape functions (as shown in Figure 3.11), which can be represented mathematically as

$$\begin{aligned} N_1(x) &= 1 - 3x^2/L^2 + 2x^3/L^3 \\ N_2(x) &= x - 2x^2/L + x^3/L^2 \\ N_3(x) &= 3x^2/L^2 - 2x^3/L^3 \\ N_4(x) &= -x^2/L + x^3/L^2 \end{aligned} \tag{3.6}$$

Then, on each element, we can represent the deflection of the beam (v) using shape functions and the corresponding nodal values as

$$v(x) = \mathbf{N}\mathbf{u} = [N_1(x) \quad N_2(x) \quad N_3(x) \quad N_4(x)]\begin{Bmatrix} v_i \\ \theta_i \\ v_j \\ \theta_j \end{Bmatrix} \tag{3.7}$$

which is a cubic function. Note that,

$$N_1 + N_3 = 1$$
$$N_2 + N_3 L + N_4 = x$$

which implies that the rigid-body motion is represented correctly by the assumed deformed shape of the beam.

To derive the beam element stiffness matrix, we consider the curvature of the beam, which is

$$\frac{d^2 v}{dx^2} = \frac{d^2}{dx^2}\mathbf{N}\mathbf{u} = \mathbf{B}\mathbf{u} \tag{3.8}$$

where the strain–displacement matrix $\mathbf{B}$ is given by

$$\begin{aligned} \mathbf{B} &= \frac{d^2}{dx^2}\mathbf{N} = \begin{bmatrix} N_1''(x) & N_2''(x) & N_3''(x) & N_4''(x) \end{bmatrix} \\ &= \begin{bmatrix} -\frac{6}{L^2} + \frac{12x}{L^3} & -\frac{4}{L} + \frac{6x}{L^2} & \frac{6}{L^2} - \frac{12x}{L^3} & -\frac{2}{L} + \frac{6x}{L^2} \end{bmatrix} \end{aligned} \tag{3.9}$$

Strain energy stored in the beam element is

$$U = \frac{1}{2}\int_V \sigma^T \varepsilon dV$$

Applying the basic equations in the simple beam theory, we have

$$\begin{aligned} U &= \frac{1}{2}\int_0^L \int_A \left(-\frac{My}{I}\right)^T \frac{1}{E}\left(-\frac{My}{I}\right) dA dx = \frac{1}{2}\int_0^L M^T \frac{1}{EI} M dx \\ &= \frac{1}{2}\int_0^L \left(\frac{d^2 v}{dx^2}\right)^T EI \left(\frac{d^2 v}{dx^2}\right) dx = \frac{1}{2}\int_0^L (\mathbf{B}\mathbf{u})^T EI (\mathbf{B}\mathbf{u}) dx \\ &= \frac{1}{2}\mathbf{u}^T \left(\int_0^L \mathbf{B}^T EI \mathbf{B}\, dx\right)\mathbf{u} \end{aligned}$$

We conclude that the stiffness matrix for the simple beam element is

$$\mathbf{k} = \int_0^L \mathbf{B}^T EI \mathbf{B}\, dx \tag{3.10}$$

Applying the result in Equation 3.10 and carrying out the integration, we arrive at the same stiffness matrix as given in Equation 3.5.

3.4.3 Treatment of Distributed Loads

To convert a distributed load into nodal forces and moments (Figure 3.12), we consider again the work done by the distributed load q

$$W_q = \frac{1}{2}\int_0^L v(x)q(x)dx = \frac{1}{2}\int_0^L (\mathbf{N}\mathbf{u})^T q(x)dx = \frac{1}{2}\mathbf{u}^T \int_0^L \mathbf{N}^T q(x)dx$$

The work done by the equivalent nodal forces (and moments) is

$$W_{f_q} = \frac{1}{2}[v_i \quad \theta_i \quad v_j \quad \theta_j]\begin{Bmatrix} F_i^q \\ M_i^q \\ F_j^q \\ M_j^q \end{Bmatrix} = \frac{1}{2}\mathbf{u}^T \mathbf{f}_q$$

By equating $W_q = W_{f_q}$, we obtain the equivalent nodal force vector as

$$\mathbf{f}_q = \int_0^L \mathbf{N}^T q(x)dx \tag{3.11}$$

which is valid for arbitrary distributions of $q(x)$. For constant q, we have the results shown in Figure 3.13. An example of this result is given in Figure 3.14.

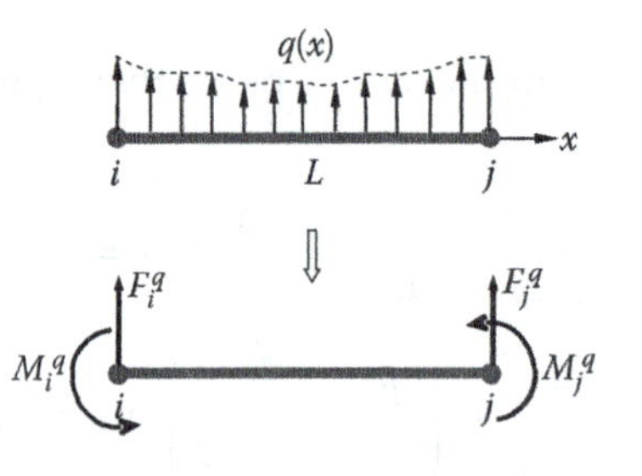

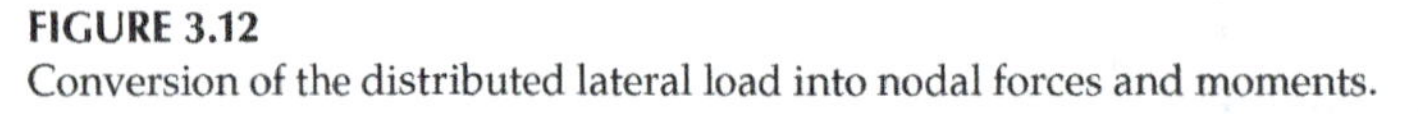
FIGURE 3.12
Conversion of the distributed lateral load into nodal forces and moments.

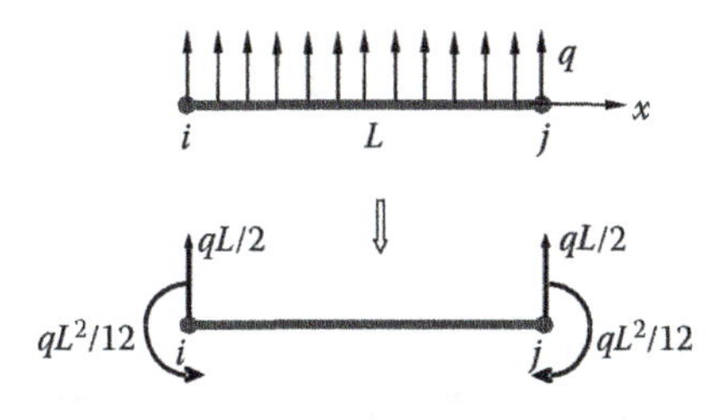

FIGURE 3.13
Conversion of a constant distributed lateral load into nodal forces and moments.

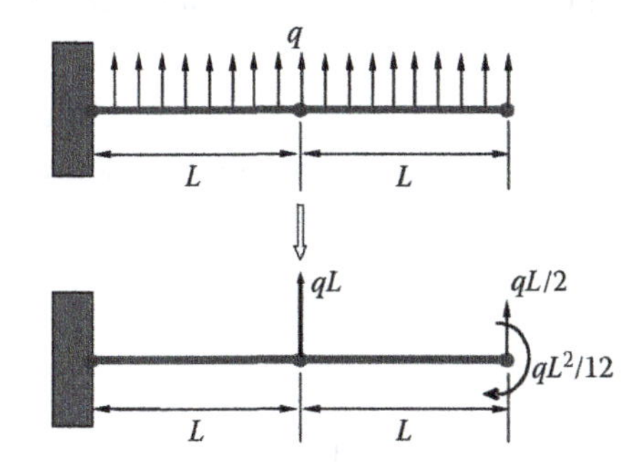

FIGURE 3.14
Conversion of a constant distributed lateral load on two beam elements.

3.4.4 Stiffness Matrix for a General Beam Element

Combining the axial stiffness (from the bar element), we further arrive at the stiffness matrix for a *general 2-D beam element* as

$$\mathbf{k} = \begin{array}{c} \begin{matrix} u_i & v_i & \theta_i & u_j & v_j & \theta_j \end{matrix} \\ \begin{bmatrix} \frac{EA}{L} & 0 & 0 & -\frac{EA}{L} & 0 & 0 \\ 0 & \frac{12EI}{L^3} & \frac{6EI}{L^2} & 0 & -\frac{12EI}{L^3} & \frac{6EI}{L^2} \\ 0 & \frac{6EI}{L^2} & \frac{4EI}{L} & 0 & -\frac{6EI}{L^2} & \frac{2EI}{L} \\ -\frac{EA}{L} & 0 & 0 & \frac{EA}{L} & 0 & 0 \\ 0 & -\frac{12EI}{L^3} & -\frac{6EI}{L^2} & 0 & \frac{12EI}{L^3} & -\frac{6EI}{L^2} \\ 0 & \frac{6EI}{L^2} & \frac{2EI}{L} & 0 & -\frac{6EI}{L^2} & \frac{4EI}{L} \end{bmatrix} \end{array} \quad (3.12)$$

3.5 Examples with Beam Elements

EXAMPLE 3.1

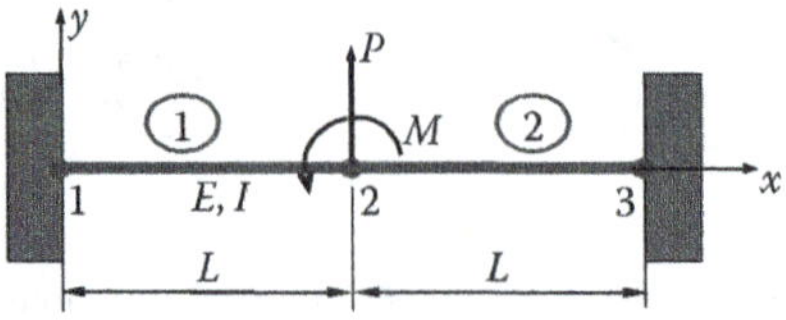

Given:
The beam shown above is clamped at the two ends and acted upon by the force P and moment M in the midspan.

Find:
The deflection and rotation at the center node and the reaction forces and moments at the two ends.

Solution

Element stiffness matrices are

$$\mathbf{k}_1 = \frac{EI}{L^3}\begin{array}{c}\begin{matrix} \;\;v_1 & \;\;\theta_1 & \;\;v_2 & \;\;\theta_2 \end{matrix}\\ \begin{bmatrix} 12 & 6L & -12 & 6L \\ 6L & 4L^2 & -6L & 2L^2 \\ -12 & -6L & 12 & -6L \\ 6L & 2L^2 & -6L & 4L^2 \end{bmatrix}\end{array}$$

$$\mathbf{k}_2 = \frac{EI}{L^3}\begin{array}{c}\begin{matrix} \;\;v_2 & \;\;\theta_2 & \;\;v_3 & \;\;\theta_3 \end{matrix}\\ \begin{bmatrix} 12 & 6L & -12 & 6L \\ 6L & 4L^2 & -6L & 2L^2 \\ -12 & -6L & 12 & -6L \\ 6L & 2L^2 & -6L & 4L^2 \end{bmatrix}\end{array}$$

Global FE equation is

$$\frac{EI}{L^3}\begin{array}{c}\begin{matrix} \;\;v_1 & \;\;\theta_1 & \;\;v_2 & \;\;\theta_2 & \;\;v_3 & \;\;\theta_3 \end{matrix}\\ \begin{bmatrix} 12 & 6L & -12 & 6L & 0 & 0 \\ 6L & 4L^2 & -6L & 2L^2 & 0 & 0 \\ -12 & -6L & 24 & 0 & -12 & 6L \\ 6L & 2L^2 & 0 & 8L^2 & -6L & 2L^2 \\ 0 & 0 & -12 & -6L & 12 & -6L \\ 0 & 0 & 6L & 2L^2 & -6L & 4L^2 \end{bmatrix}\end{array}\begin{Bmatrix} v_1 \\ \theta_1 \\ v_2 \\ \theta_2 \\ v_3 \\ \theta_3 \end{Bmatrix} = \begin{Bmatrix} F_{1Y} \\ M_1 \\ F_{2Y} \\ M_2 \\ F_{3Y} \\ M_3 \end{Bmatrix}$$

Loads and constraints (BCs) are

$$F_{2Y} = -P, \qquad M_2 = M, \qquad v_1 = v_3 = \theta_1 = \theta_3 = 0$$

Reduced FE equation

$$\frac{EI}{L^3}\begin{bmatrix} 24 & 0 \\ 0 & 8L^2 \end{bmatrix}\begin{Bmatrix} v_2 \\ \theta_2 \end{Bmatrix} = \begin{Bmatrix} -P \\ M \end{Bmatrix}$$

Solving this, we obtain

$$\begin{Bmatrix} v_2 \\ \theta_2 \end{Bmatrix} = \frac{L}{24EI}\begin{Bmatrix} -PL^2 \\ 3M \end{Bmatrix}$$

From the global FE equation, we obtain the reaction forces and moments

$$\begin{Bmatrix} F_{1Y} \\ M_1 \\ F_{3Y} \\ M_3 \end{Bmatrix} = \frac{EI}{L^3}\begin{bmatrix} -12 & 6L \\ -6L & 2L^2 \\ -12 & -6L \\ 6L & 2L^2 \end{bmatrix}\begin{Bmatrix} v_2 \\ \theta_2 \end{Bmatrix} = \frac{1}{4}\begin{Bmatrix} 2P + 3M/L \\ PL + M \\ 2P - 3M/L \\ -PL + M \end{Bmatrix}$$

Stresses in the beam at the two ends can be calculated using the formula

$$\sigma = \sigma_x = -\frac{My}{I}$$

Note that the FE solution is exact for this problem according to the simple beam theory, since no distributed load is present between the nodes. Recall that

$$EI\frac{d^4v}{dx^4} = q(x)$$

If $q(x) = 0$, then exact solution for the deflection v is a cubic function of x, which is exactly what described by the shape functions given in Equation 3.6.

EXAMPLE 3.2

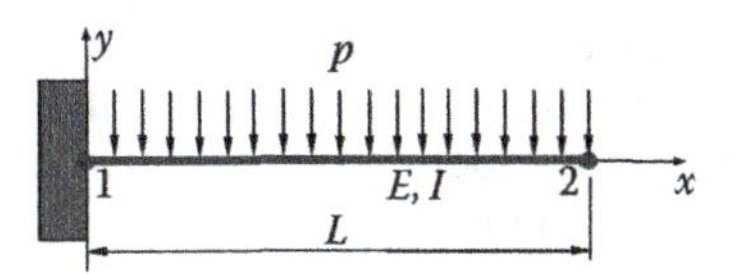

Given:
A cantilever beam with distributed lateral load p as shown above.

Find:
The deflection and rotation at the right end, the reaction force and moment at the left end.

Solution

The work-equivalent nodal loads are shown below,

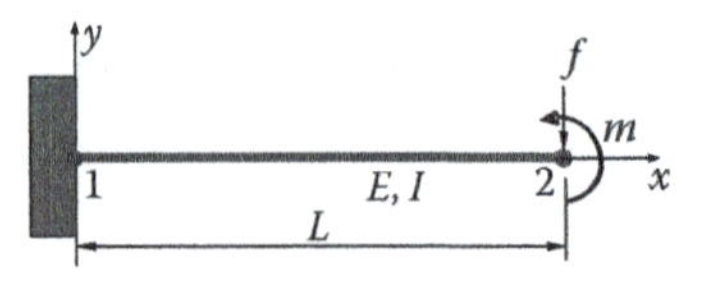

where

$$f = pL/2, \qquad m = pL^2/12$$

Applying the FE equation, we have

$$\frac{EI}{L^3}\begin{bmatrix} 12 & 6L & -12 & 6L \\ 6L & 4L^2 & -6L & 2L^2 \\ -12 & -6L & 12 & -6L \\ 6L & 2L^2 & -6L & 4L^2 \end{bmatrix}\begin{Bmatrix} v_1 \\ \theta_1 \\ v_2 \\ \theta_2 \end{Bmatrix} = \begin{Bmatrix} F_{1Y} \\ M_1 \\ F_{2Y} \\ M_2 \end{Bmatrix}$$

Load and constraints (BCs) are

$$F_{2Y} = -f, \qquad M_2 = m$$
$$v_1 = \theta_1 = 0$$

Reduced equation is

$$\frac{EI}{L^3}\begin{bmatrix} 12 & -6L \\ -6L & 4L^2 \end{bmatrix}\begin{Bmatrix} v_2 \\ \theta_2 \end{Bmatrix} = \begin{Bmatrix} -f \\ m \end{Bmatrix}$$

Solving this, we obtain

$$\begin{Bmatrix} v_2 \\ \theta_2 \end{Bmatrix} = \frac{L}{6EI}\begin{Bmatrix} -2L^2 f + 3Lm \\ -3Lf + 6m \end{Bmatrix} = \begin{Bmatrix} -pL^4/8EI \\ -pL^3/6EI \end{Bmatrix} \qquad \text{(A)}$$

These nodal values are the same as the exact solution. Note that the deflection $v(x)$ (for $0 < x < L$) in the beam by the FEM is, however, different from that by the exact solution. The exact solution by the simple beam theory is a fourth-order polynomial of x, while the FE solution of v is only a third-order polynomial of x.

If the equivalent moment m is ignored, we have

$$\begin{Bmatrix} v_2 \\ \theta_2 \end{Bmatrix} = \frac{L}{6EI}\begin{Bmatrix} -2L^2 f \\ -3Lf \end{Bmatrix} = \begin{Bmatrix} -pL^4/6EI \\ -pL^3/4EI \end{Bmatrix} \qquad \text{(B)}$$

The errors in (B) will decrease if more elements are used. The equivalent moment m is often ignored in the FEM applications. The FE solutions still converge as more elements are applied.

From the FE equation, we can calculate the reaction force and moment as

$$\begin{Bmatrix} F_{1Y} \\ M_1 \end{Bmatrix} = \frac{EI}{L^3}\begin{bmatrix} -12 & 6L \\ -6L & 2L^2 \end{bmatrix}\begin{Bmatrix} v_2 \\ \theta_2 \end{Bmatrix} = \begin{Bmatrix} pL/2 \\ 5pL^2/12 \end{Bmatrix}$$

where the result in (A) has been used. This force vector gives the total *effective nodal forces*, which include the equivalent nodal forces for the distributed lateral load p given by

$$\begin{Bmatrix} -pL/2 \\ -pL^2/12 \end{Bmatrix}$$

The correct *reaction forces* can be obtained as follows:

$$\begin{Bmatrix} F_{1Y} \\ M_1 \end{Bmatrix} = \begin{Bmatrix} pL/2 \\ 5pL^2/12 \end{Bmatrix} - \begin{Bmatrix} -pL/2 \\ -pL^2/12 \end{Bmatrix} = \begin{Bmatrix} pL \\ pL^2/2 \end{Bmatrix}$$

Check the results: Draw the FBD for the FE model (with the equivalent nodal force vector) and check the equilibrium condition.

EXAMPLE 3.3

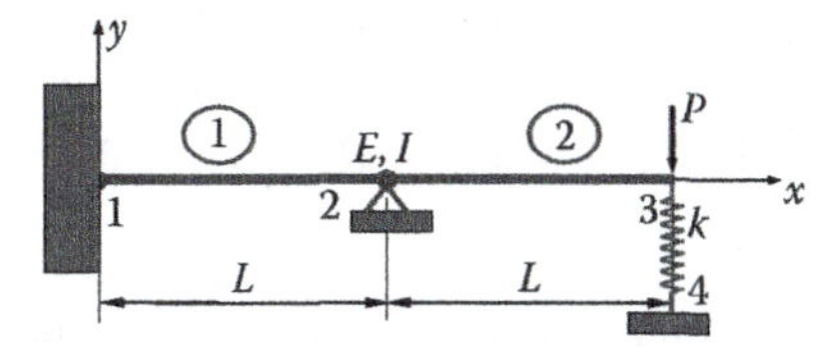

Given:
$P = 50$ kN, $k = 200$ kN/m, $L = 3$ m, $E = 210$ GPa, $I = 2 \times 10^{-4}$ m^4.

Find:
Deflections, rotations, and reaction forces.

Solution

The beam has a roller (or hinge) support at node 2 and a spring support at node 3. We use two beam elements and one spring element to solve this problem.

The spring stiffness matrix is given by

$$\mathbf{k}_s = \begin{bmatrix} k & -k \\ -k & k \end{bmatrix} \quad (v_3, v_4)$$

Adding this stiffness matrix to the global FE equation (see Example 3.1), we have

$$
\frac{EI}{L^3}\overset{\begin{matrix} v_1 & \theta_1 & v_2 & \theta_2 & v_3 & \theta_3 & v_4 \end{matrix}}{\begin{bmatrix} 12 & 6L & -12 & 6L & 0 & 0 & 0 \\ & 4L^2 & -6L & 2L^2 & 0 & 0 & 0 \\ & & 24 & 0 & -12 & 6L & 0 \\ & & & 8L^2 & -6L & 2L^2 & 0 \\ & & & & 12+k' & -6L & -k' \\ & & & & & 4L^2 & 0 \\ & \textit{Symmetry} & & & & & k' \end{bmatrix}}\begin{Bmatrix} v_1 \\ \theta_1 \\ v_2 \\ \theta_2 \\ v_3 \\ \theta_3 \\ v_4 \end{Bmatrix} = \begin{Bmatrix} F_{1Y} \\ M_1 \\ F_{2Y} \\ M_2 \\ F_{3Y} \\ M_3 \\ F_{4Y} \end{Bmatrix}
$$

in which

$$
k' = \frac{L^3}{EI}k
$$

is used to simplify the notation.

We now apply the boundary conditions

$$
\begin{aligned}
&v_1 = \theta_1 = v_2 = v_4 = 0, \\
&M_2 = M_3 = 0, \qquad\qquad F_{3Y} = -P
\end{aligned}
$$

"Deleting" the first three and seventh equations (rows and columns), we have the following reduced equation:

$$
\frac{EI}{L^3}\begin{bmatrix} 8L^2 & -6L & 2L^2 \\ -6L & 12+k' & -6L \\ 2L^2 & -6L & 4L^2 \end{bmatrix}\begin{Bmatrix} \theta_2 \\ v_3 \\ \theta_3 \end{Bmatrix} = \begin{Bmatrix} 0 \\ -P \\ 0 \end{Bmatrix}
$$

Solving this equation, we obtain the deflection and rotations at nodes 2 and 3,

$$
\begin{Bmatrix} \theta_2 \\ v_3 \\ \theta_3 \end{Bmatrix} = -\frac{PL^2}{EI(12+7k')}\begin{Bmatrix} 3 \\ 7L \\ 9 \end{Bmatrix}
$$

The influence of the spring k is easily seen from this result. Plugging in the given numbers, we can calculate

$$
\begin{Bmatrix} \theta_2 \\ v_3 \\ \theta_3 \end{Bmatrix} = \begin{Bmatrix} -0.002492\ \text{rad} \\ -0.01744\ \text{m} \\ -0.007475\ \text{rad} \end{Bmatrix}
$$

From the global FE equation, we obtain the nodal reaction forces as,

$$\begin{Bmatrix} F_{1Y} \\ M_1 \\ F_{2Y} \\ F_{4Y} \end{Bmatrix} = \begin{Bmatrix} -69.78 \text{ kN} \\ -69.78 \text{ kN} \cdot \text{m} \\ 116.2 \text{ kN} \\ 3.488 \text{ kN} \end{Bmatrix}$$

Checking the results: Draw *free-body diagram* of the beam

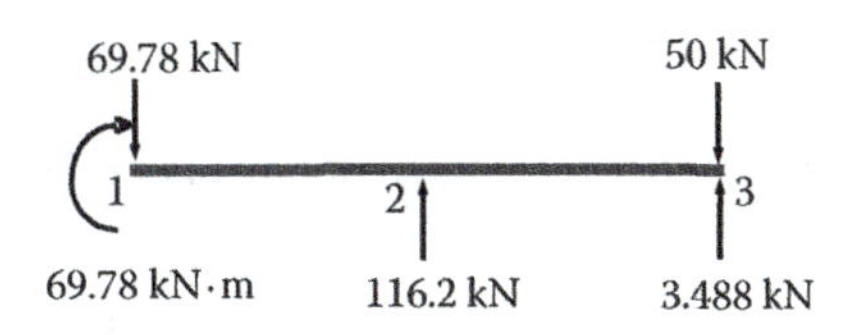

Sum the forces and moments to verify that equilibrium of the beam is satisfied.

We use the following example to show how to model frames using the general beam elements. This example can be used to verify the FEA results when using a software package.

EXAMPLE 3.4

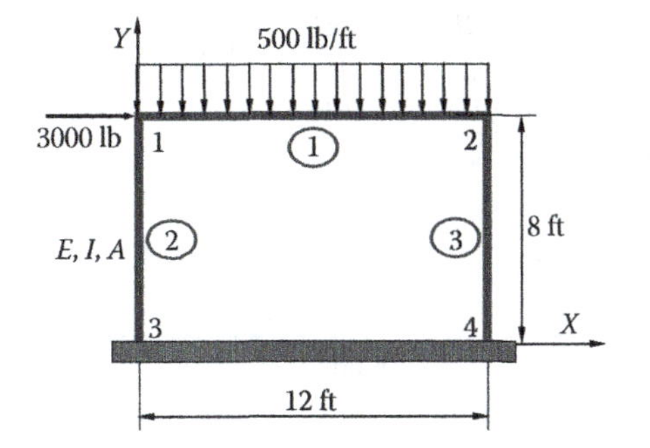

Given: $E = 30 \times 10^6 \text{psi}, \ I = 65 \text{ in.}^4, A = 6.8 \text{ in.}^2$

Find:
Displacements and rotations of the two joints 1 and 2.

Solution

This is a problem of analyzing a frame. Members in a frame are considered to be rigidly connected (e.g., welded together). Both forces and moments can be transmitted through their joints. We need the *general beam element* (combinations of bar and simple beam elements) to model frames.

For this example, we first convert the distributed load into its equivalent nodal loads to obtain the following FE mode.

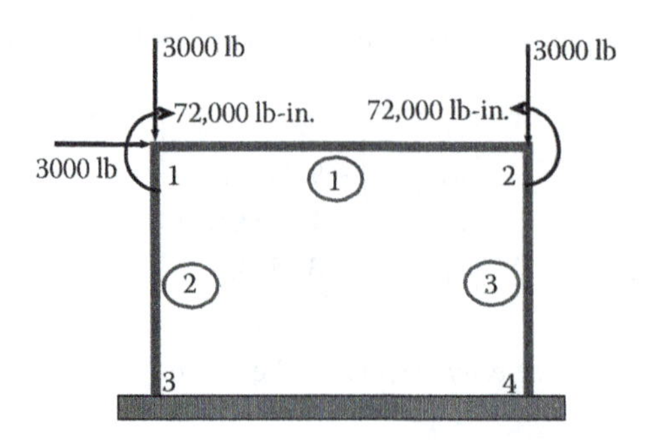

In *local coordinate system,* the stiffness matrix for a general 2-D beam element is

$$\mathbf{k} = \begin{array}{c} \begin{matrix} u_i & v_i & \theta_i & u_j & v_j & \theta_j \end{matrix} \\ \begin{bmatrix} \frac{EA}{L} & 0 & 0 & -\frac{EA}{L} & 0 & 0 \\ 0 & \frac{12EI}{L^3} & \frac{6EI}{L^2} & 0 & -\frac{12EI}{L^3} & \frac{6EI}{L^2} \\ 0 & \frac{6EI}{L^2} & \frac{4EI}{L} & 0 & -\frac{6EI}{L^2} & \frac{2EI}{L} \\ -\frac{EA}{L} & 0 & 0 & \frac{EA}{L} & 0 & 0 \\ 0 & -\frac{12EI}{L^3} & -\frac{6EI}{L^2} & 0 & \frac{12EI}{L^3} & -\frac{6EI}{L^2} \\ 0 & \frac{6EI}{L^2} & \frac{2EI}{L} & 0 & -\frac{6EI}{L^2} & \frac{4EI}{L} \end{bmatrix} \end{array}$$

Element Connectivity Table

Element	Node *i* (1)	Node *j* (2)
1	1	2
2	3	1
3	4	2

For element 1, we have

$$\mathbf{k}_1 = \mathbf{k}'_1 = 10^4 \times \begin{array}{c} \begin{matrix} u_1 & v_1 & \theta_1 & u_2 & v_2 & \theta_2 \end{matrix} \\ \begin{bmatrix} 141.7 & 0 & 0 & -141.7 & 0 & 0 \\ 0 & 0.784 & 56.4 & 0 & -0.784 & 56.4 \\ 0 & 56.4 & 5417 & 0 & -56.4 & 2708 \\ -141.7 & 0 & 0 & 141.7 & 0 & 0 \\ 0 & -0.784 & -56.4 & 0 & 0.784 & -56.4 \\ 0 & 56.4 & 2708 & 0 & -56.4 & 5417 \end{bmatrix} \end{array}$$

For elements 2 and 3, the stiffness matrix in *local system* is

$$
\mathbf{k}_2' = \mathbf{k}_3' = 10^4 \times
\begin{array}{c}
\begin{matrix} u_i' & v_i' & \theta_i' & u_j' & v_j' & \theta_j' \end{matrix} \\
\begin{bmatrix}
212.5 & 0 & 0 & -212.5 & 0 & 0 \\
0 & 2.65 & 127 & 0 & -2.65 & 127 \\
0 & 127 & 8125 & 0 & -127 & 4063 \\
-212.5 & 0 & 0 & 212.5 & 0 & 0 \\
0 & -2.65 & -127 & 0 & 2.65 & -127 \\
0 & 127 & 4063 & 0 & -127 & 8125
\end{bmatrix}
\end{array}
$$

where $i = 3, j = 1$ for element 2, and $i = 4, j = 2$ for element 3.

The transformation matrix $\mathbf{T}$ is

$$
\mathbf{T} = \begin{bmatrix}
l & m & 0 & 0 & 0 & 0 \\
-m & l & 0 & 0 & 0 & 0 \\
0 & 0 & 1 & 0 & 0 & 0 \\
0 & 0 & 0 & l & m & 0 \\
0 & 0 & 0 & -m & l & 0 \\
0 & 0 & 0 & 0 & 0 & 1
\end{bmatrix}
$$

We have $l = 0, m = 1$ for both elements 2 and 3. Thus,

$$
\mathbf{T} = \begin{bmatrix}
0 & 1 & 0 & 0 & 0 & 0 \\
-1 & 0 & 0 & 0 & 0 & 0 \\
0 & 0 & 1 & 0 & 0 & 0 \\
0 & 0 & 0 & 0 & 1 & 0 \\
0 & 0 & 0 & -1 & 0 & 0 \\
0 & 0 & 0 & 0 & 0 & 1
\end{bmatrix}
$$

Using the transformation relation

$$
\mathbf{k} = \mathbf{T}^T \mathbf{k}' \mathbf{T}
$$

we obtain the stiffness matrices in the *global coordinate system* for elements 2 and 3

$$
\mathbf{k}_2 = 10^4 \times
\begin{array}{c}
\begin{matrix} u_3 & v_3 & \theta_3 & u_1 & v_1 & \theta_1 \end{matrix} \\
\begin{bmatrix}
2.65 & 0 & -127 & -2.65 & 0 & -127 \\
0 & 212.5 & 0 & 0 & -212.5 & 0 \\
-127 & 0 & 8125 & 127 & 0 & 4063 \\
-2.65 & 0 & 127 & 2.65 & 0 & 127 \\
0 & -212.5 & 0 & 0 & 212.5 & 0 \\
-127 & 0 & 4063 & 127 & 0 & 8125
\end{bmatrix}
\end{array}
$$

and

$$\mathbf{k}_3 = 10^4 \times \begin{array}{c} \begin{matrix} u_4 & v_4 & \theta_4 & u_2 & v_2 & \theta_2 \end{matrix} \\ \begin{bmatrix} 2.65 & 0 & -127 & -2.65 & 0 & -127 \\ 0 & 212.5 & 0 & 0 & -212.5 & 0 \\ -127 & 0 & 8125 & 127 & 0 & 4063 \\ -2.65 & 0 & 127 & 2.65 & 0 & 127 \\ 0 & -212.5 & 0 & 0 & 212.5 & 0 \\ -127 & 0 & 4063 & 127 & 0 & 8125 \end{bmatrix} \end{array}$$

Assembling the global FE equation and noticing the following boundary conditions

$$\begin{aligned} &u_3 = v_3 = \theta_3 = u_4 = v_4 = \theta_4 = 0 \\ &F_{1X} = 3000\ \text{lb}, \quad F_{2X} = 0, \quad F_{1Y} = F_{2Y} = -3000\ \text{lb}, \\ &M_1 = -72{,}000\ \text{lb} \cdot \text{in.}, \quad M_2 = 72{,}000\ \text{lb} \cdot \text{in.} \end{aligned}$$

we obtain the condensed FE equation

$$10^4 \times \begin{bmatrix} 144.3 & 0 & 127 & -141.7 & 0 & 0 \\ 0 & 213.3 & 56.4 & 0 & -0.784 & 56.4 \\ 127 & 56.4 & 13{,}542 & 0 & -56.4 & 2708 \\ -141.7 & 0 & 0 & 144.3 & 0 & 127 \\ 0 & -0.784 & -56.4 & 0 & 213.3 & -56.4 \\ 0 & 56.4 & 2708 & 127 & -56.4 & 13{,}542 \end{bmatrix} \begin{Bmatrix} u_1 \\ v_1 \\ \theta_1 \\ u_2 \\ v_2 \\ \theta_2 \end{Bmatrix} = \begin{Bmatrix} 3000 \\ -3000 \\ -72{,}000 \\ 0 \\ -3000 \\ 72{,}000 \end{Bmatrix}$$

Solving this, we obtain

$$\begin{Bmatrix} u_1 \\ v_1 \\ \theta_1 \\ u_2 \\ v_2 \\ \theta_2 \end{Bmatrix} = \begin{Bmatrix} 0.092\ \text{in.} \\ -0.00104\ \text{in.} \\ -0.00139\ \text{rad} \\ 0.0901\ \text{in.} \\ -0.0018\ \text{in.} \\ -3.88 \times 10^{-5}\ \text{rad} \end{Bmatrix}$$

To calculate the reaction forces and moments at the two ends, we employ the element FE equations for elements 2 and 3 with known nodal displacement vectors. We obtain

$$\begin{Bmatrix} F_{3X} \\ F_{3Y} \\ M_3 \end{Bmatrix} = \begin{Bmatrix} -672.7\ \text{lb} \\ 2210\ \text{lb} \\ 60{,}364\ \text{lb} \cdot \text{in.} \end{Bmatrix}$$

and

$$\begin{Bmatrix} F_{4X} \\ F_{4Y} \\ M_4 \end{Bmatrix} = \begin{Bmatrix} -2338 \text{ lb} \\ 3825 \text{ lb} \\ 112{,}641 \text{ lb} \cdot \text{in.} \end{Bmatrix}$$

Check the results: Draw the free-body diagram of the frame as shown below. Equilibrium is maintained with the calculated forces and moments. Recall that the problem we solved is the one with the equivalent loads, not the one with the distributed load. Thus, the corresponding FBD for the FE model should be applied for verifying the results.

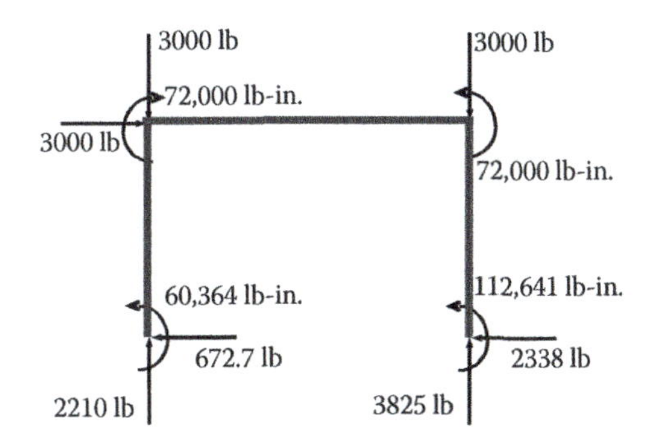

3.6 Case Study with *ANSYS Workbench*

Problem Description: Steel framing systems provide cost-effective solutions for low-rise buildings. They have high strength-to-weight ratios, and can be prefabricated and custom-designed. Consider the following two-storey building constructed with structural steel I-beams. Determine the deformations and the stresses in the frame when a uniform load of 50 kN/m is applied on the second floor as shown below.

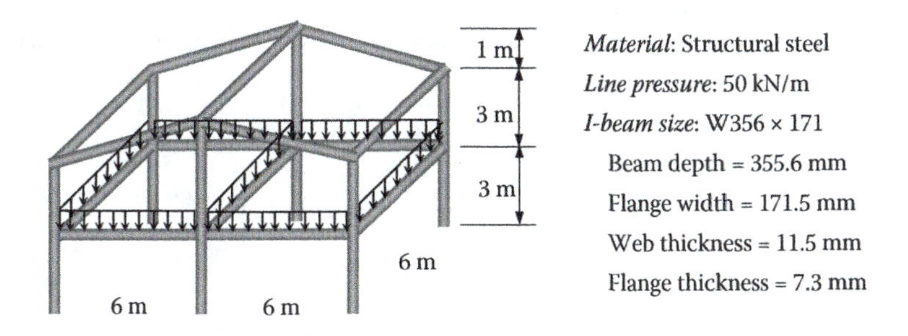

Material: Structural steel
Line pressure: 50 kN/m
I-beam size: W356 × 171
Beam depth = 355.6 mm
Flange width = 171.5 mm
Web thickness = 11.5 mm
Flange thickness = 7.3 mm

Solution

To solve the problem with *ANSYS® Workbench*, we employ the following steps:

Step 1: Start an *ANSYS Workbench* Project
Launch *ANSYS Workbench* and save the blank project as *"Steelframe.wbpj."*

Step 2: Create a *Static Structural (ANSYS)* Analysis System

Drag the *Static Structural (ANSYS)* icon from the *Analysis Systems Toolbox* window and drop it inside the highlighted green rectangle in the *Project Schematic* window to create a standalone static structural analysis system.

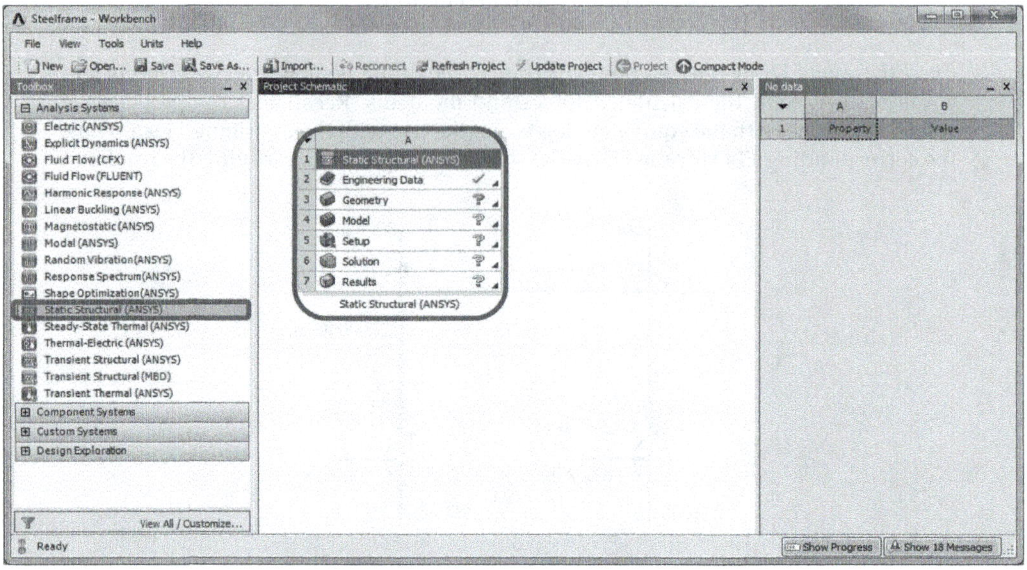

Steelframe - Workbench
File View Tools Units Help
New Open... Save Save As... Import... Reconnect Refresh Project Update Project Project Compact Mode
Toolbox
Analysis Systems
Electric (ANSYS)
Explicit Dynamics (ANSYS)
Fluid Flow (CFX)
Fluid Flow (FLUENT)
Harmonic Response (ANSYS)
Linear Buckling (ANSYS)
Magnetostatic (ANSYS)
Modal (ANSYS)
Random Vibration (ANSYS)
Response Spectrum (ANSYS)
Shape Optimization (ANSYS)
Static Structural (ANSYS)
Steady-State Thermal (ANSYS)
Thermal-Electric (ANSYS)
Transient Structural (ANSYS)
Transient Structural (MBD)
Transient Thermal (ANSYS)
Component Systems
Custom Systems
Design Exploration
View All / Customize...
Ready
Project Schematic

	A
1	Static Structural (ANSYS)
2	Engineering Data
3	Geometry
4	Model
5	Setup
6	Solution
7	Results

Static Structural (ANSYS)

No data

	A	B
1	Property	Value

Show Progress Show 18 Messages

Step 3: Launch the *DesignModeler* Program

Double-click the *Geometry* cell to launch *DesignModeler,* and select *"Meter"* as length unit in the *Units* pop-up window. Ensure *Line Bodies* is selected in the *Properties of Schematic A3: Geometry* window.

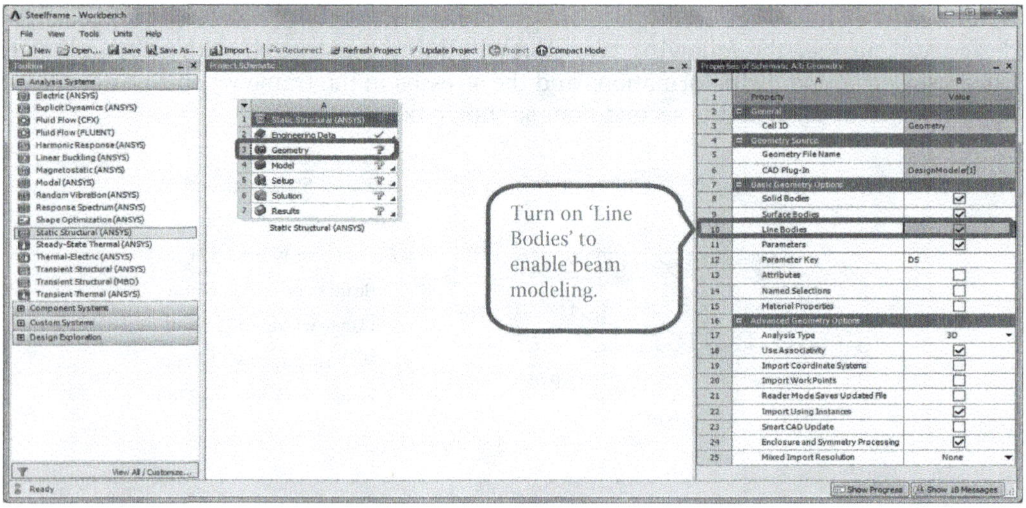

Steelframe - Workbench
File View Tools Units Help
New Open... Save Save As... Import... Reconnect Refresh Project Update Project Project Compact Mode
Toolbox
Analysis Systems
Electric (ANSYS)
Explicit Dynamics (ANSYS)
Fluid Flow (CFX)
Fluid Flow (FLUENT)
Harmonic Response (ANSYS)
Linear Buckling (ANSYS)
Magnetostatic (ANSYS)
Modal (ANSYS)
Random Vibration (ANSYS)
Response Spectrum (ANSYS)
Shape Optimization (ANSYS)
Static Structural (ANSYS)
Steady-State Thermal (ANSYS)
Thermal-Electric (ANSYS)
Transient Structural (ANSYS)
Transient Structural (MBD)
Transient Thermal (ANSYS)
Component Systems
Custom Systems
Design Exploration
View All / Customize...
Ready
Project Schematic

	A
1	Static Structural (ANSYS)
2	Engineering Data
3	Geometry
4	Model
5	Setup
6	Solution
7	Results

Static Structural (ANSYS)

Turn on 'Line Bodies' to enable beam modeling.

Properties of Schematic A3: Geometry

	A	B
1	Property	Value
2	General	
3	Cell ID	Geometry
4	Geometry Source	
5	Geometry File Name	
6	CAD Plug-In	DesignModeler[1]
7	Basic Geometry Options	
8	Solid Bodies	☑
9	Surface Bodies	☑
10	Line Bodies	☑
11	Parameters	☑
12	Parameter Key	DS
13	Attributes	☐
14	Named Selections	☐
15	Material Properties	☐
16	Advanced Geometry Options	
17	Analysis Type	3D
18	Use Associativity	☑
19	Import Coordinate Systems	☐
20	Import Work Points	☐
21	Reader Mode Saves Updated File	☐
22	Import Using Instances	☑
23	Smart CAD Update	☐
24	Enclosure and Symmetry Processing	☑
25	Mixed Import Resolution	None

Show Progress Show 18 Messages

Step 4: Create Line Sketch

Click the *Sketching* tab and select *Settings*. Turn on *Show in 2D* and *Snap* under *Grid* options. Use the default value of *"5 m"* for *Major Grid Spacing* and *"5"* for *Minor-Steps per Major.*

Click a start point and then an end point in the *Graphics* window to draw a line. Draw 10 lines as shown in the sketch below. After completion, click *Generate* to create a line sketch.

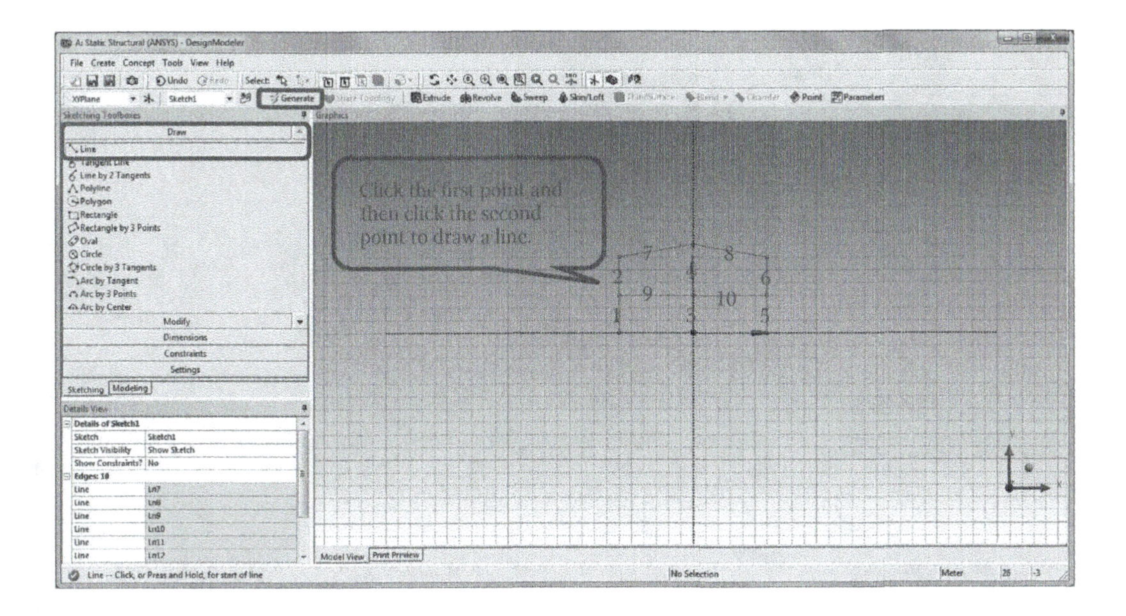

Step 5: Create Line Body from Sketch

Check off the *Grid* options under *Settings* of *Sketching Toolboxes*. Switch to the *Modeling* tab. Note that a new item named *Sketch1* now appears underneath *XYPlane* in the *Tree Outline*.

Select *Lines from Sketches* from the *Concept* drop-down menu. Click *Zoom to Fit*.

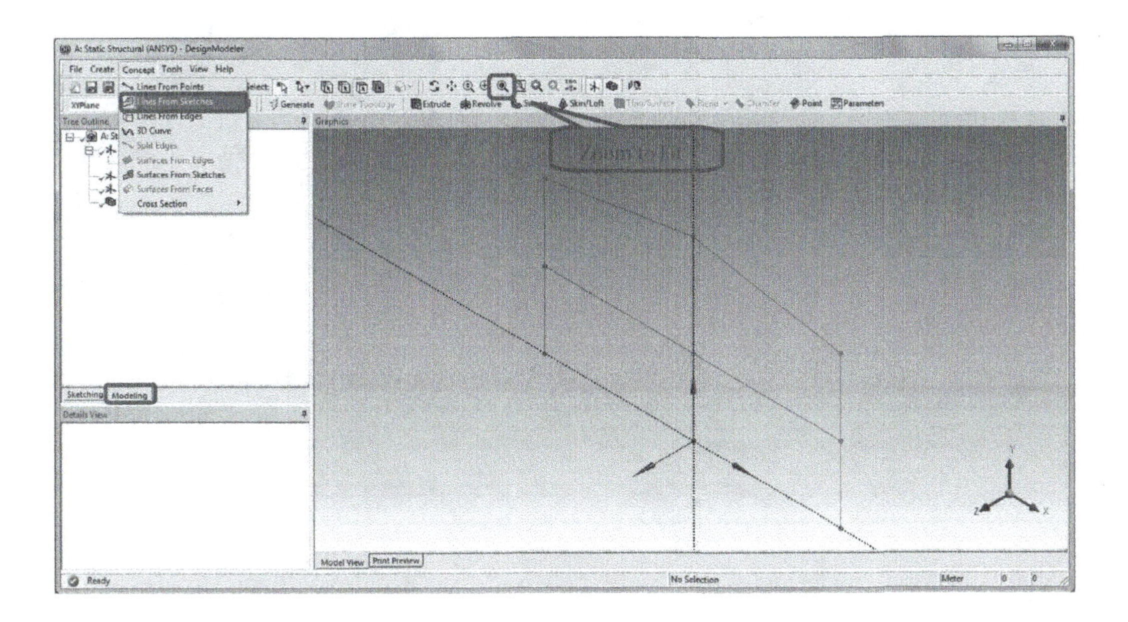

Select *Sketch1* from the *Tree Outline* and click *Apply* to confirm on the *Base Objects* selection in the *Details of Line1*. Click *Generate* to complete the line body creation.

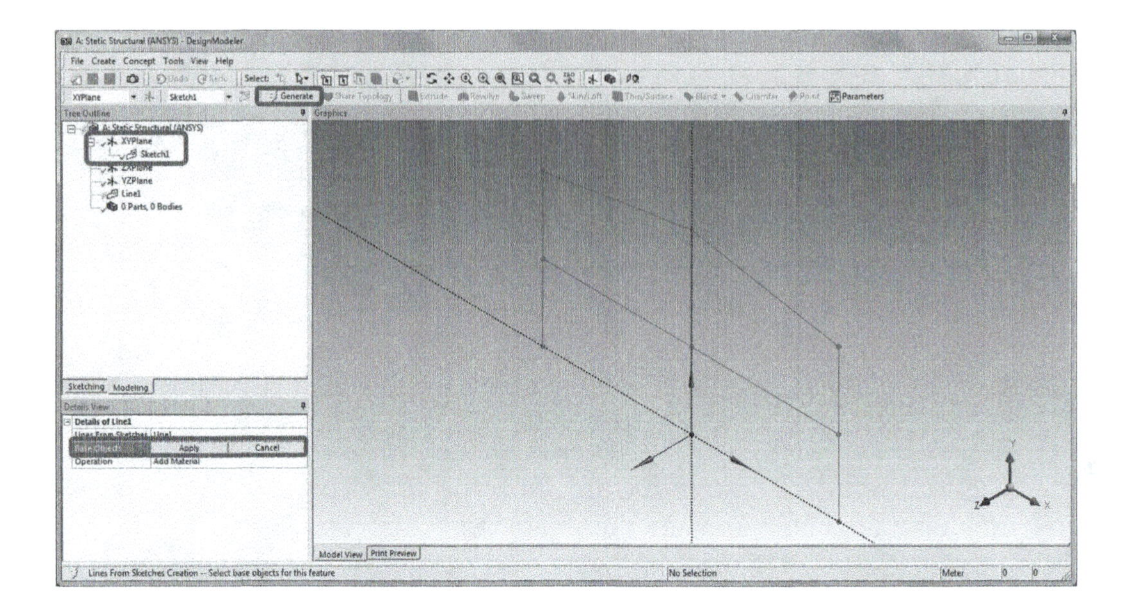

Step 6: Create Line Body through Translation

Select *Body Operation* from the *Create* drop-down menu. A new item named *BodyOp1* is now added to the *Tree Outline.*

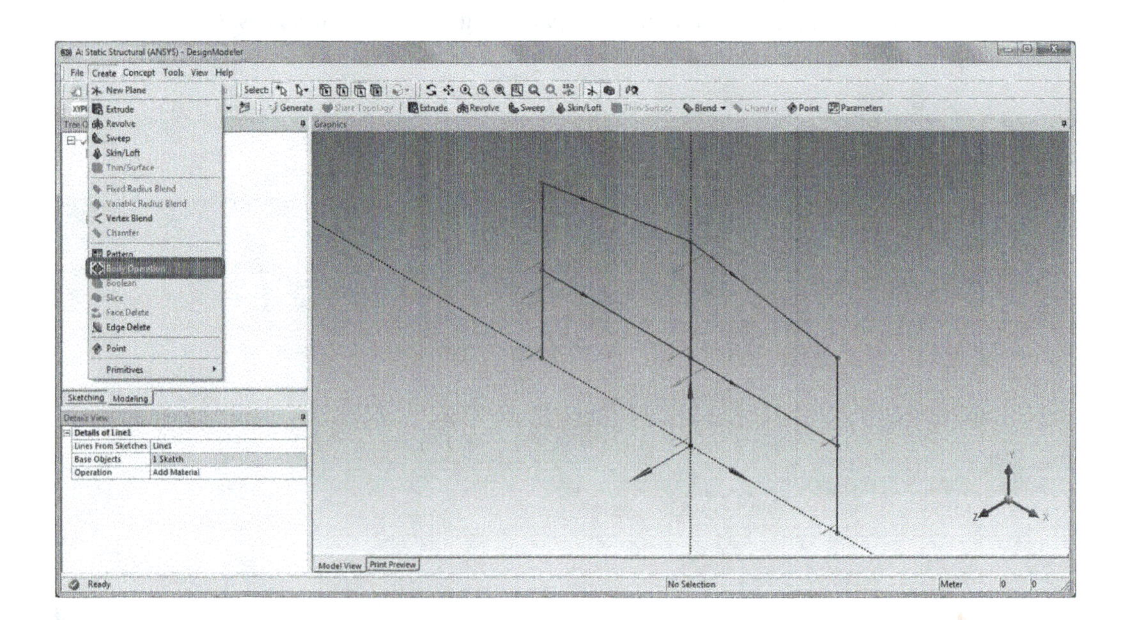

In the *Details of BodyOp1*, click anywhere on the *Type* cell and select *Translate* from the drop-down menu.

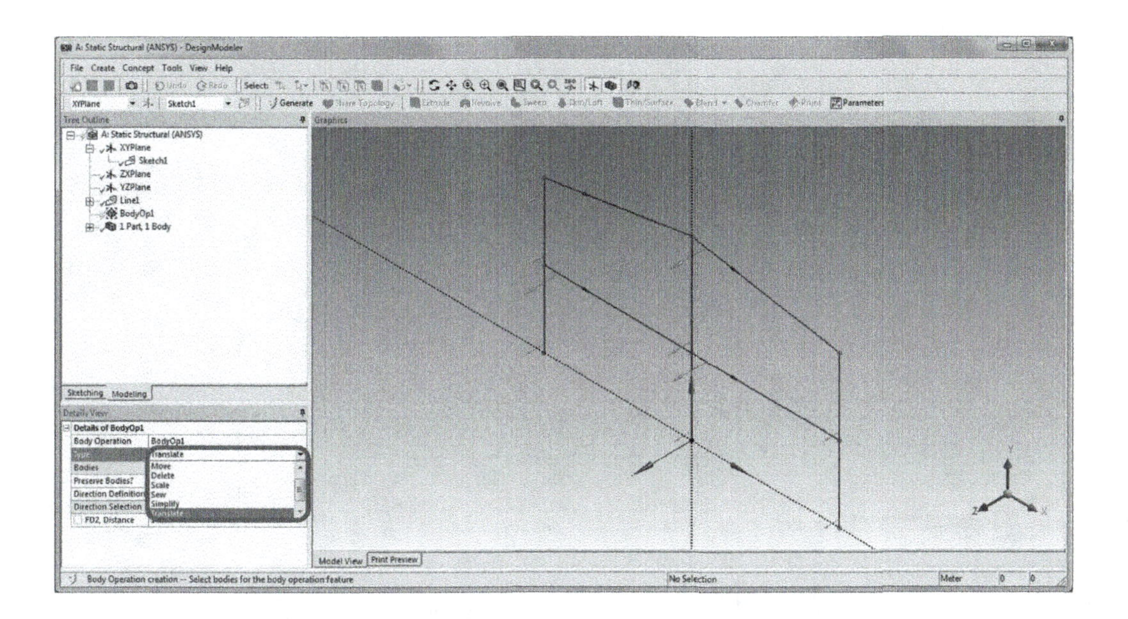

Select the line body from the *Graphics* window and then click *Apply* to confirm on the *Bodies* selection in the *Details of BodyOp1*. After completion, change the *Preserve Bodies?* selection to *Yes*. This will help preserve a copy of the selected line body at the current location while translating it to a new location.

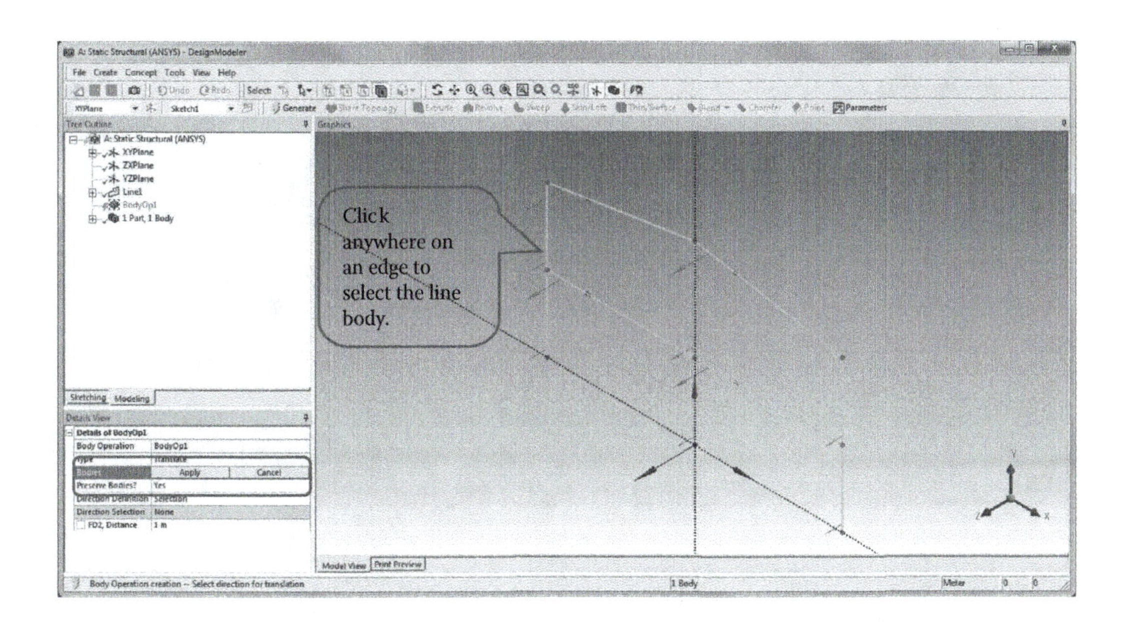

In the *Details of BodyOp1*, change the *Direction Definition* to *Coordinates*, and enter "**–6**" for the *Z Offset*. Click *Generate*. After completion, the line body will be copied backward by 6 m.

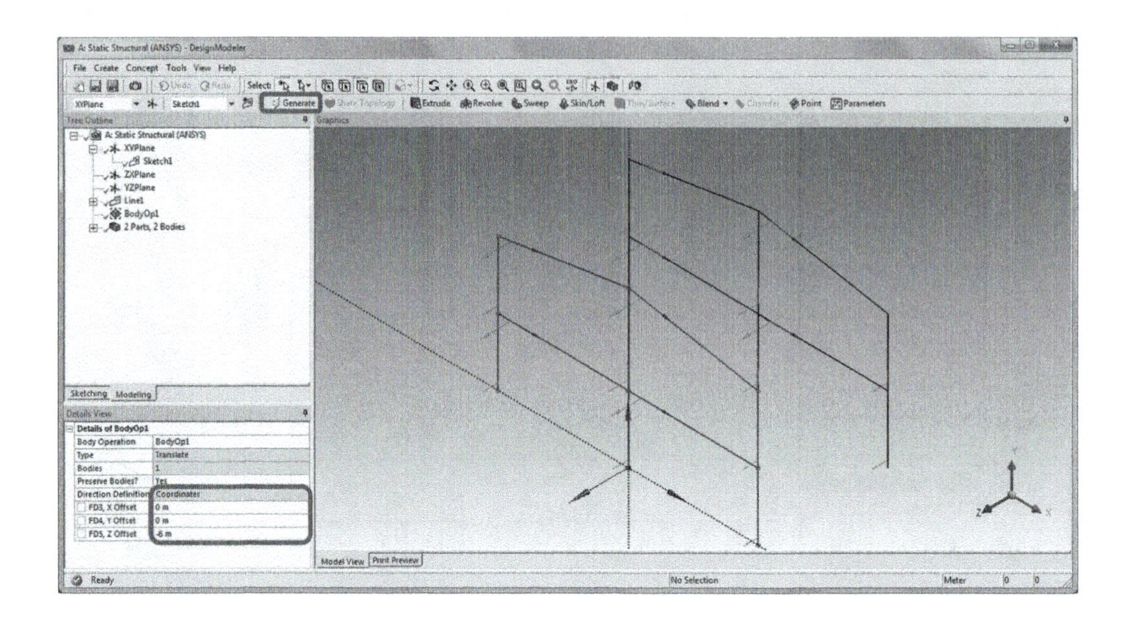

Step 7: Create Line Body from Points

Select *Lines From Points* from the *Concept* drop-down menu. After completion, a new item named *Line2* is added to the *Tree Outline*.

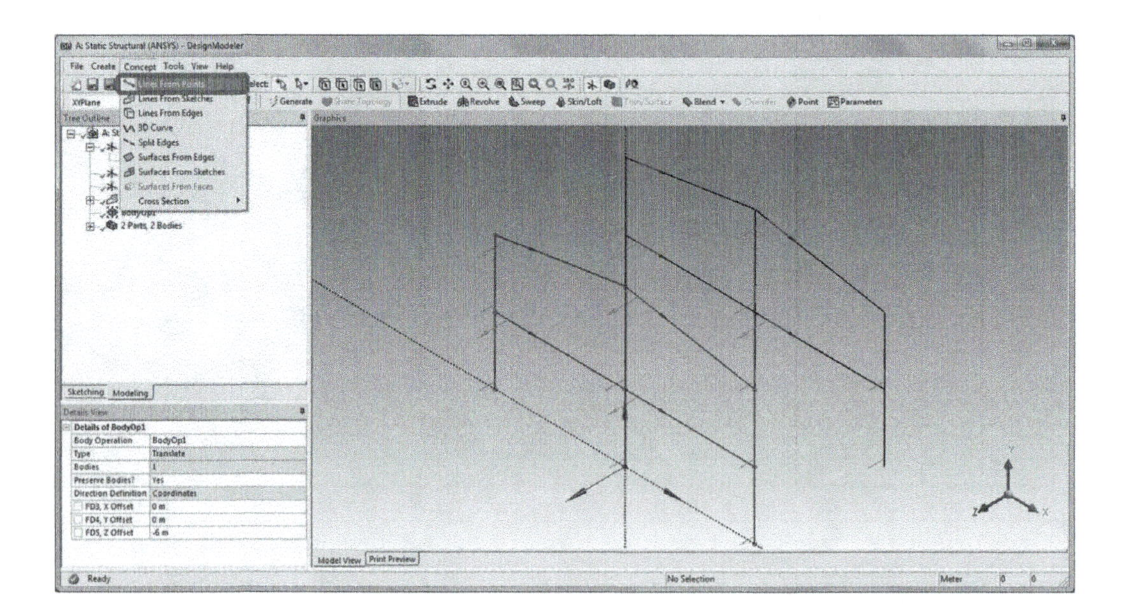

In the *Graphics* window, select a start point and Ctrl-select an end point to draw a line. Draw six new lines connecting the two planar frames as shown below. Click *Apply* to confirm on the *Point Segments* selection in the *Details of Line2*. Click *Generate* to complete the line creation.

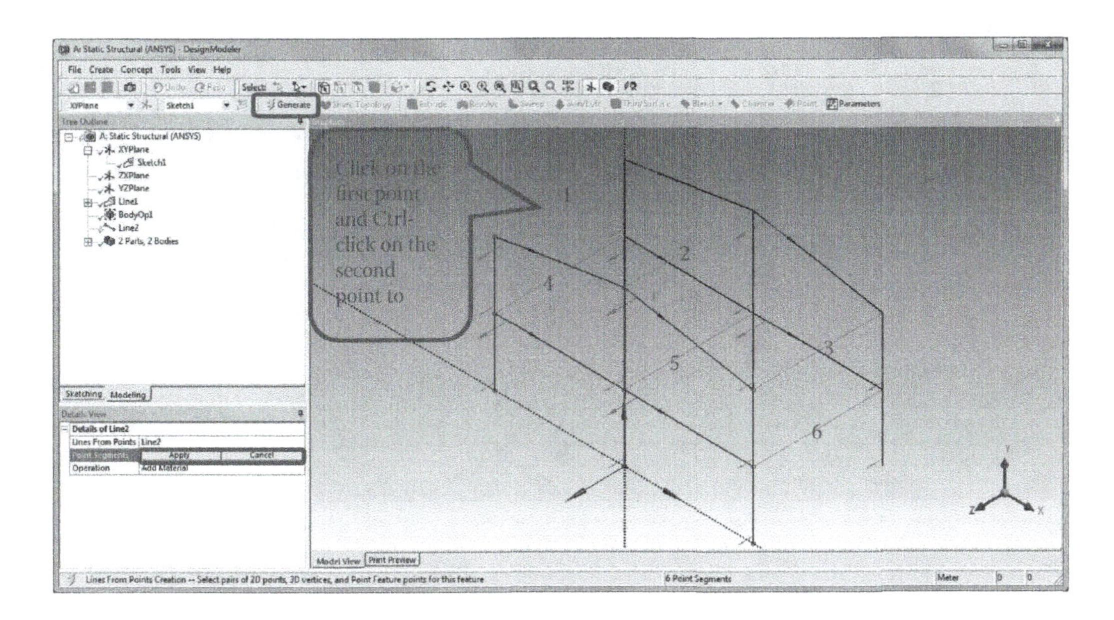

Note that the line bodies created in all previous steps now merge into a single line body. Check off the *Cross Section Alignments* from the *View* drop-down menu to switch-off the display of local coordinate systems.

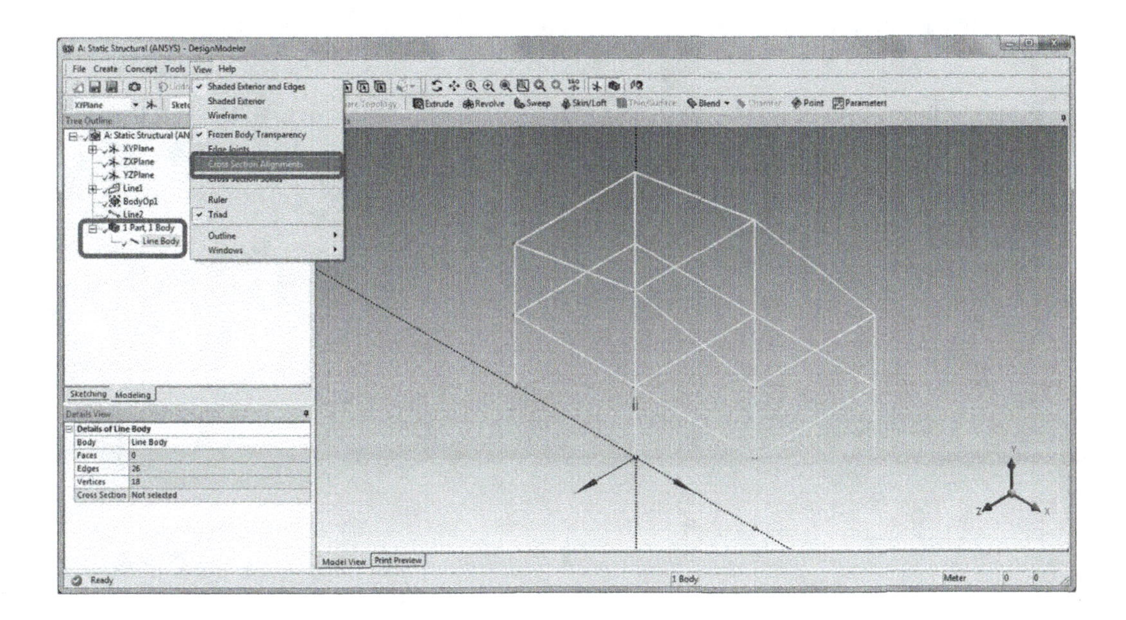

Step 8: Create a Cross Section

Select a *Cross Section* of *I Section* from the *Concept* drop-down menu. A new item named *I1* is now added underneath the *Cross Section* in the *Tree Outline*.

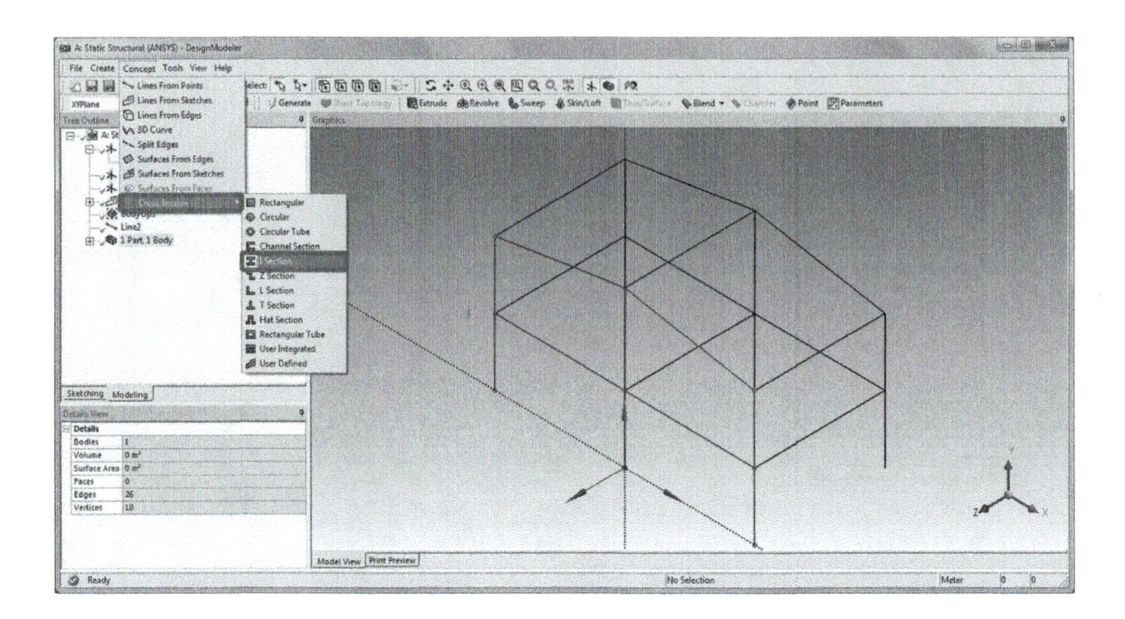

In the *Details of I1* under *Dimensions*, enter "*0.1715*" for *W1* and *W2*, "0.3556" for *W3*, "0.0073" for *t1* and *t2*, and "*0.0115*" for *t3*.

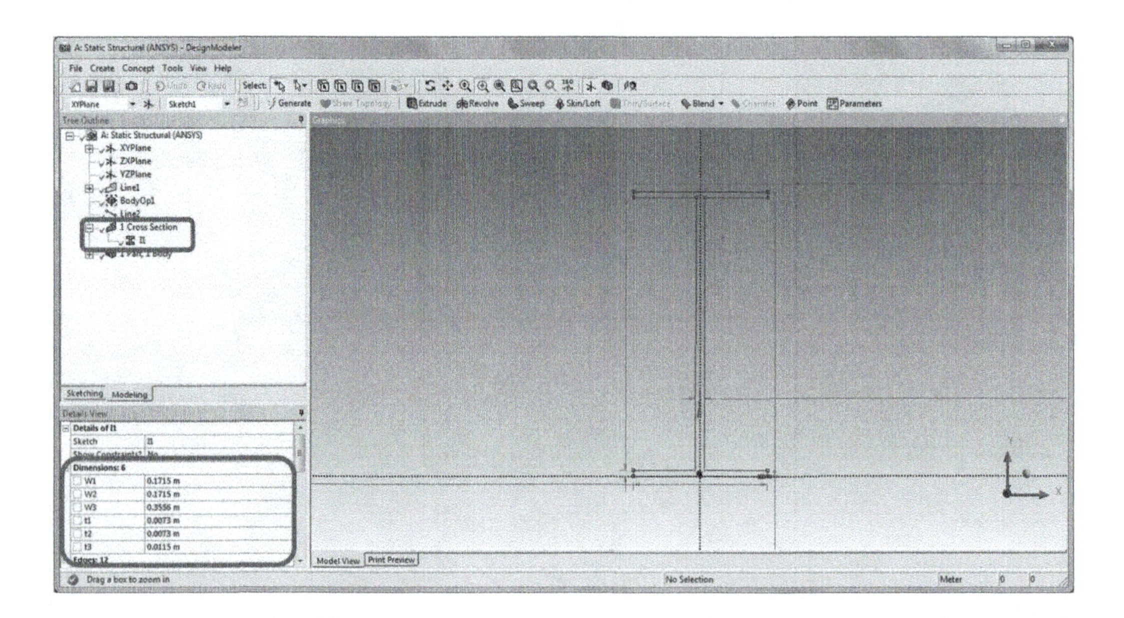

Step 9: Assign Cross Section to Line Body

Select the *Line Body* underneath *1Part, 1 Body* in the *Tree Outline*. In the *Details of Line Body*, assign *I1* to the *Cross Section* selection.

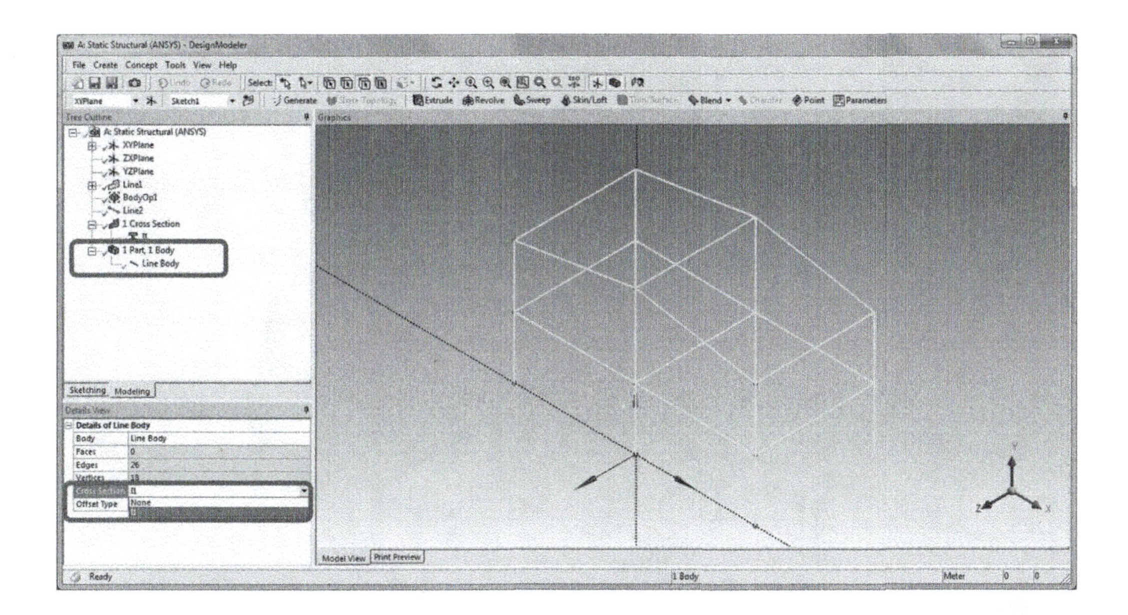

Turn on the *Cross Section Solids* from the *View* drop-down menu to view the frame as a solid structure.

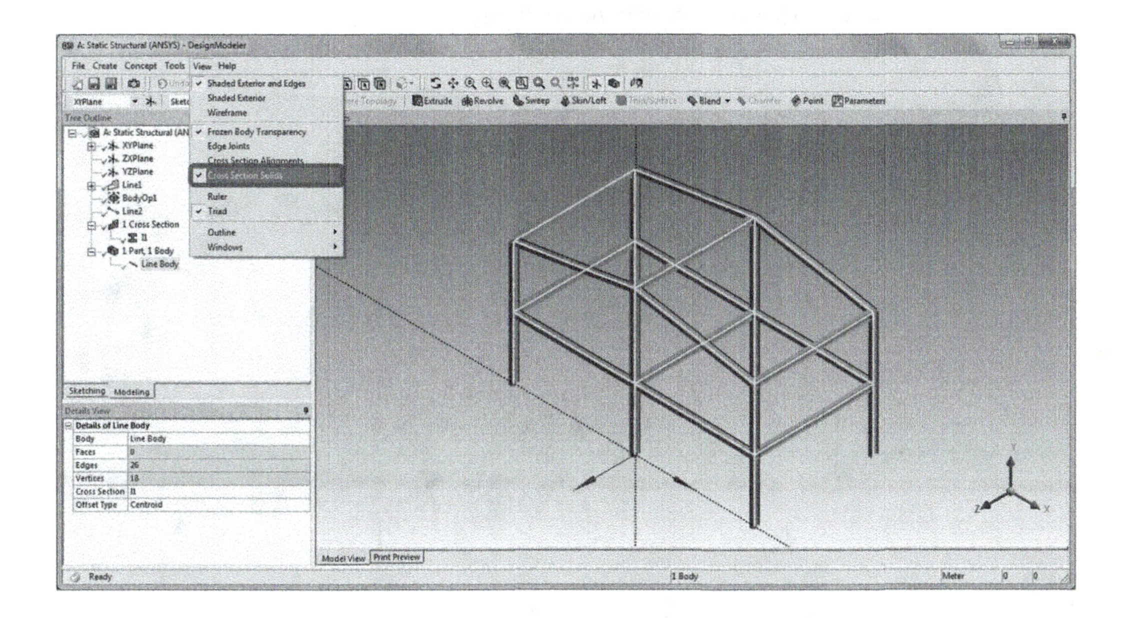

Note that some I-beams in the above structure are used as H-beams. To fix the misaligned cross sections, turn on the *Edge Selection Filter* and select the eight line edges shown below from the *Graphics* window. In the *Details of Line-Body Edges,* enter "*90*" for *Rotate* to turn the beams 90° about their neutral axes.

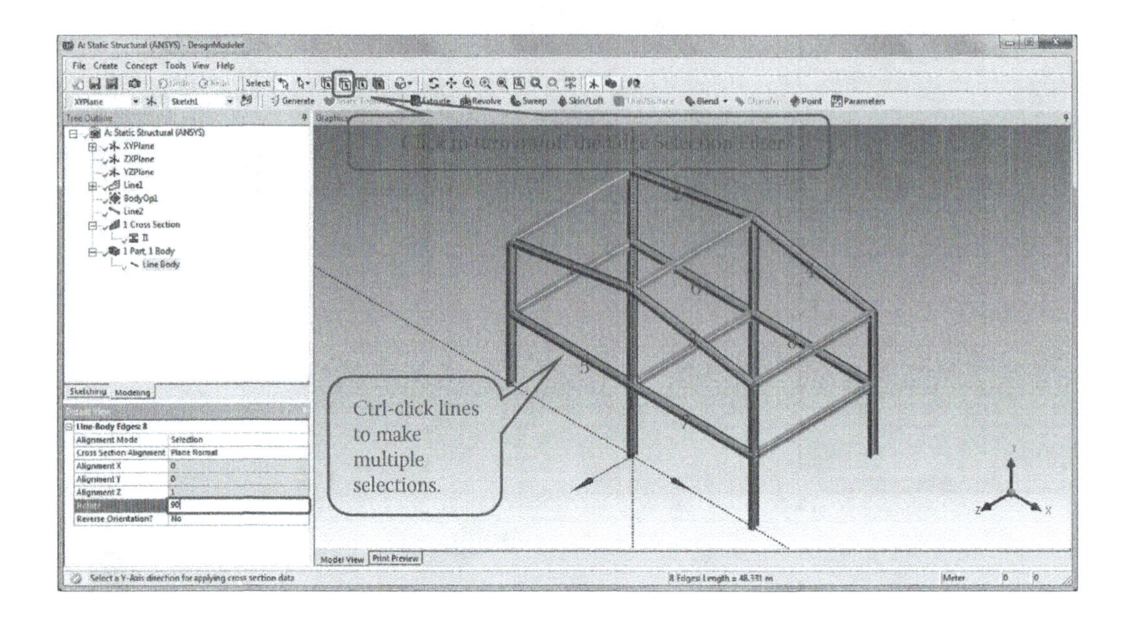

The adjusted frame shown below now has all the I-beams oriented in the strong-axis configuration. This completes the geometry creation of a frame structure. Click *Close DesignModeler* to exit the program.

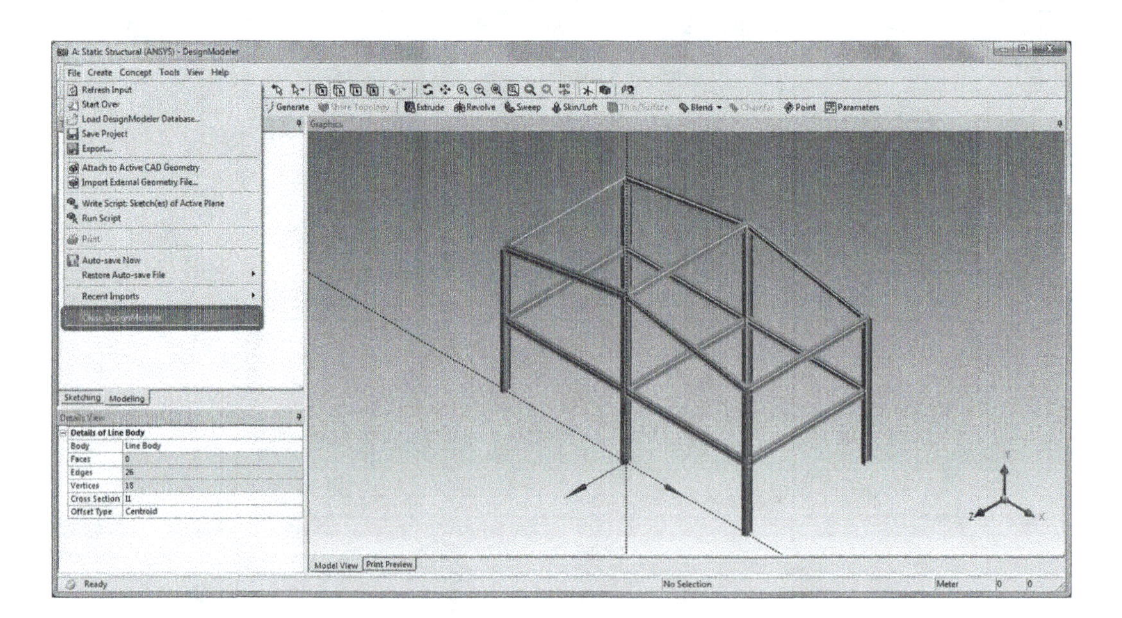

Step 10: Launch the *Static Structural (ANSYS)* Program

Double-click the *Model* cell to launch the *Static Structural (ANSYS)* program. From its *Units* drop-down menu, select *Metric (m, Kg, N, s, V, A)*. Note that in the *Details of "Line Body"* the material is assigned to *Structural Steel* by default.

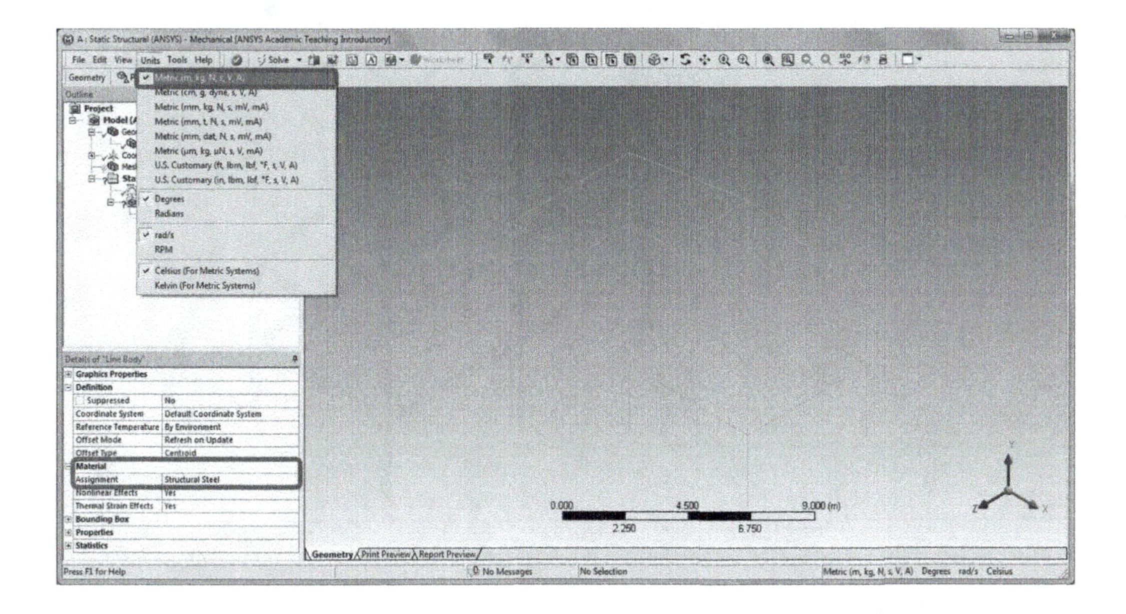

Step 11: Generate Mesh

In the *Details of "Mesh,"* enter *"0.2 m"* for the *Element Size*. In the *Outline* of *Project*, right-click on *Mesh* and select *Generate Mesh*.

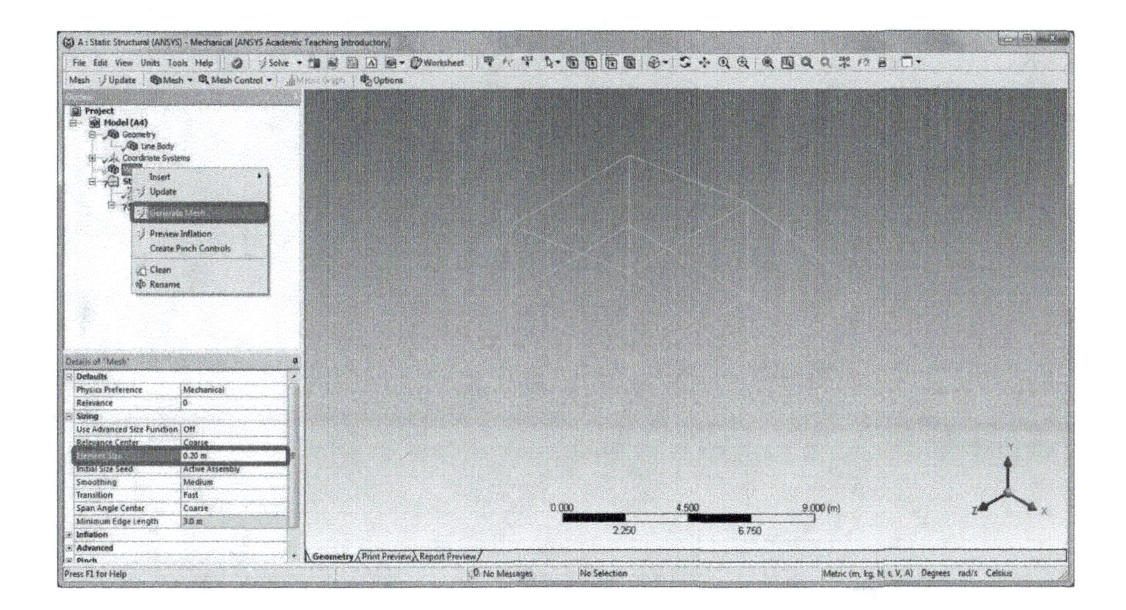

After completion, the meshed structure appears in the *Graphics* window. You may deselect the *Ruler* from the *View* drop-down menu to turn off the ruler display in the *Graphics* window.

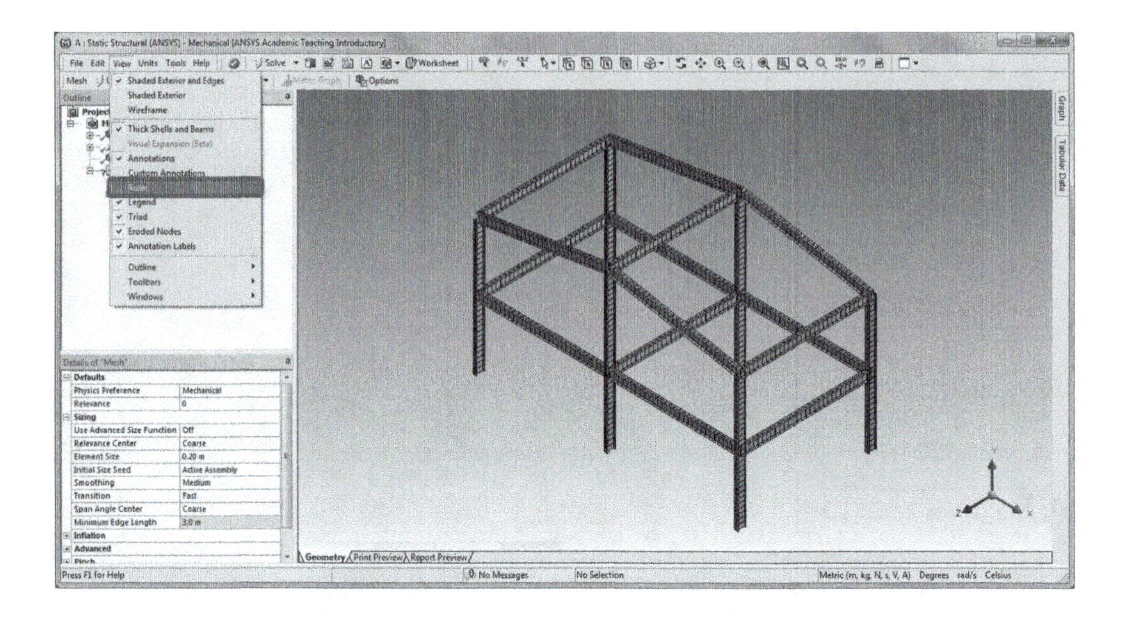

Step 12: Apply Boundary Conditions

In the *Outline* of *Project*, right-click on *Static Structural (A5)* and select *Insert* and then *Fixed Support*. After completion, a *Fixed Support* item is added underneath *Static Structural (A5)* in the project outline tree.

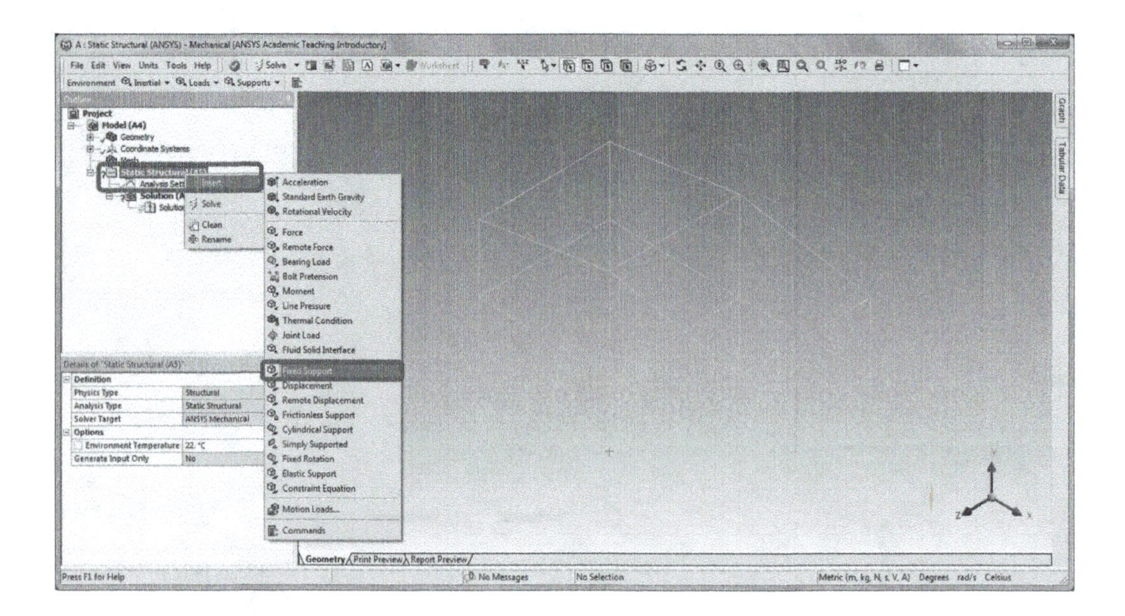

Select the six points as shown below in the *Graphics* window. In the *Details of "Fixed Support,"* click *Apply* to confirm on the Geometry selection. After completion, a *Fixed Support* boundary condition will be added to the selected six points.

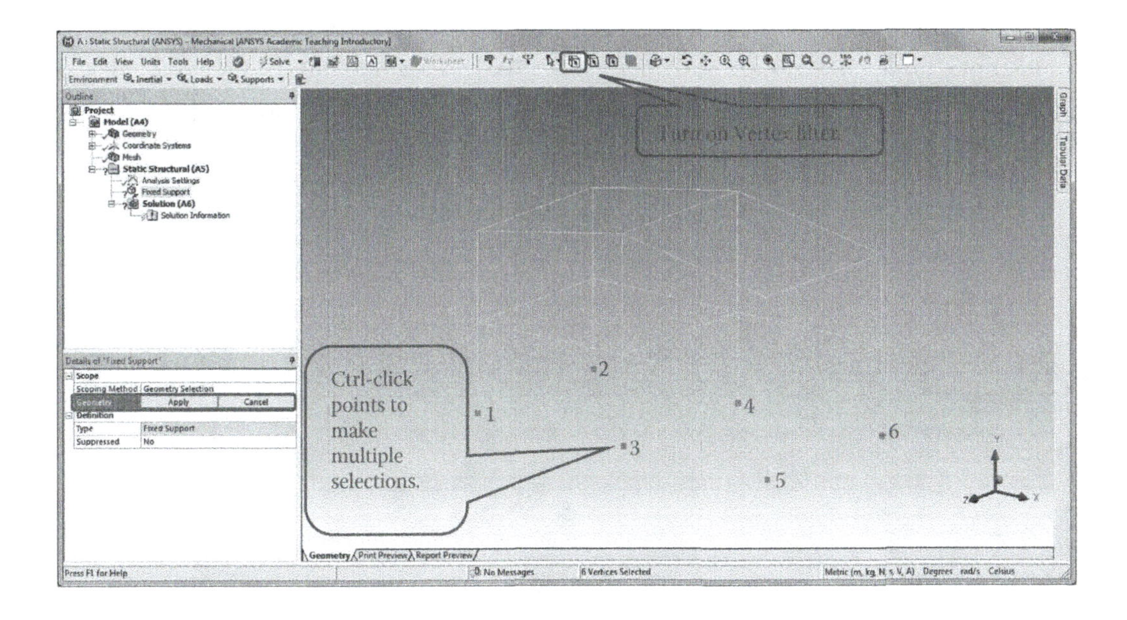

Step 13: Apply Loads

In the *Outline* of *Project*, right-click on *Static Structural (A5)* and select *Insert* and then *Line Pressure*.

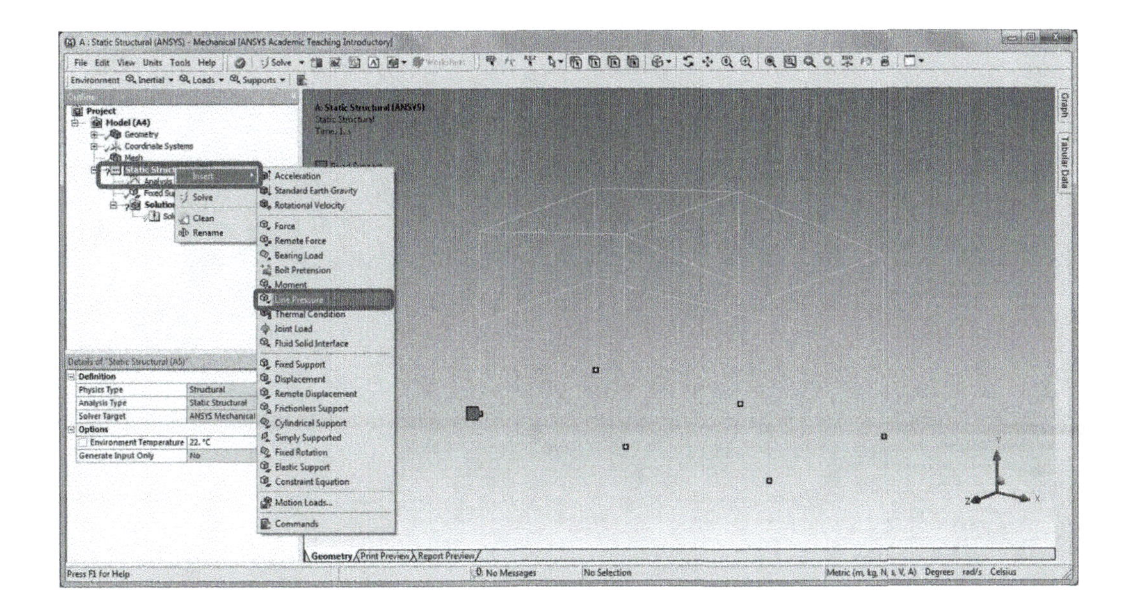

Select the line as shown below in the *Graphics* window. In the *Details of "Line Pressure,"* click *Apply* to confirm on the *Geometry* selection.

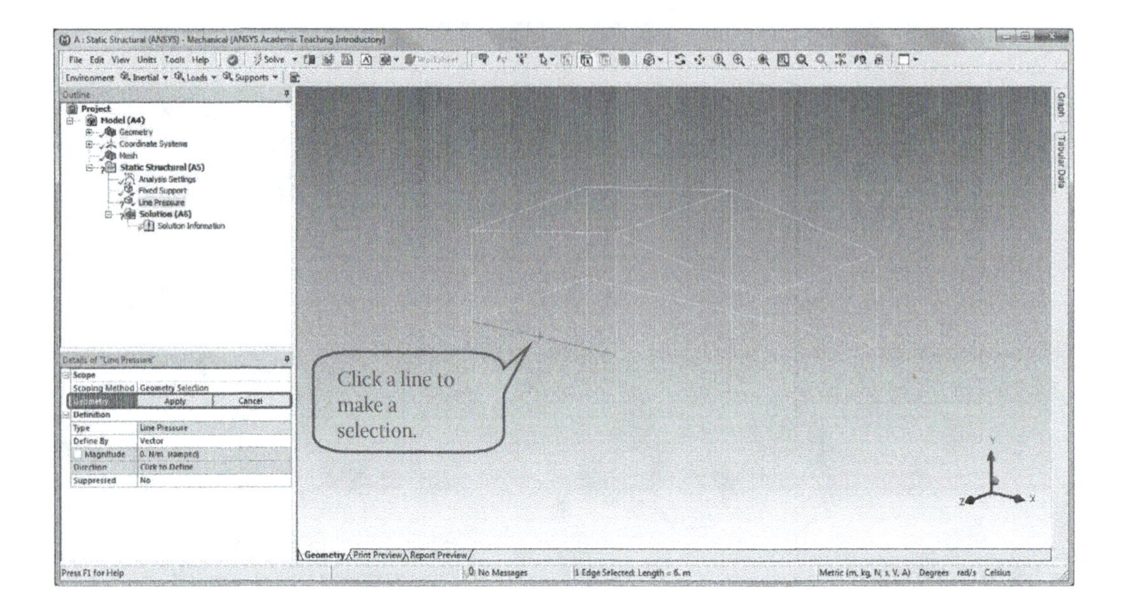

In the *Details of "Line Pressure,"* change the *Define By* selection to *Components* and enter "-50000" for the *Y Component*. A downward red arrow will appear on the selected line in the *Graphics* window.

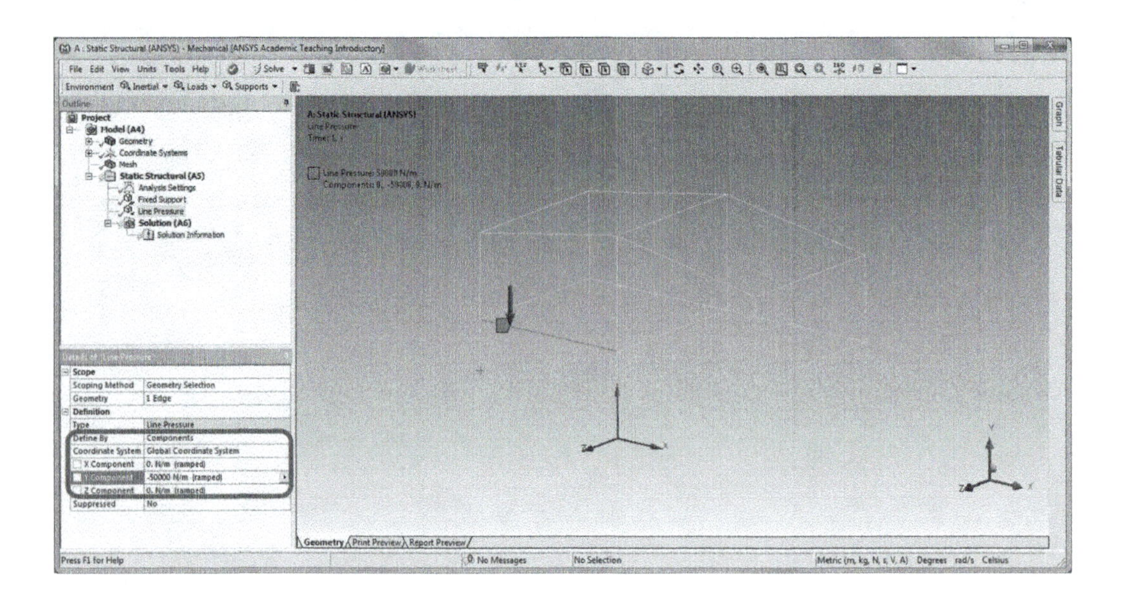

Repeat the steps of adding a line pressure, and insert the second *Line Pressure* item underneath *Static Structural (A5)* in the *Project Outline* tree. Apply the same exact load to the selected line highlighted in the following figure.

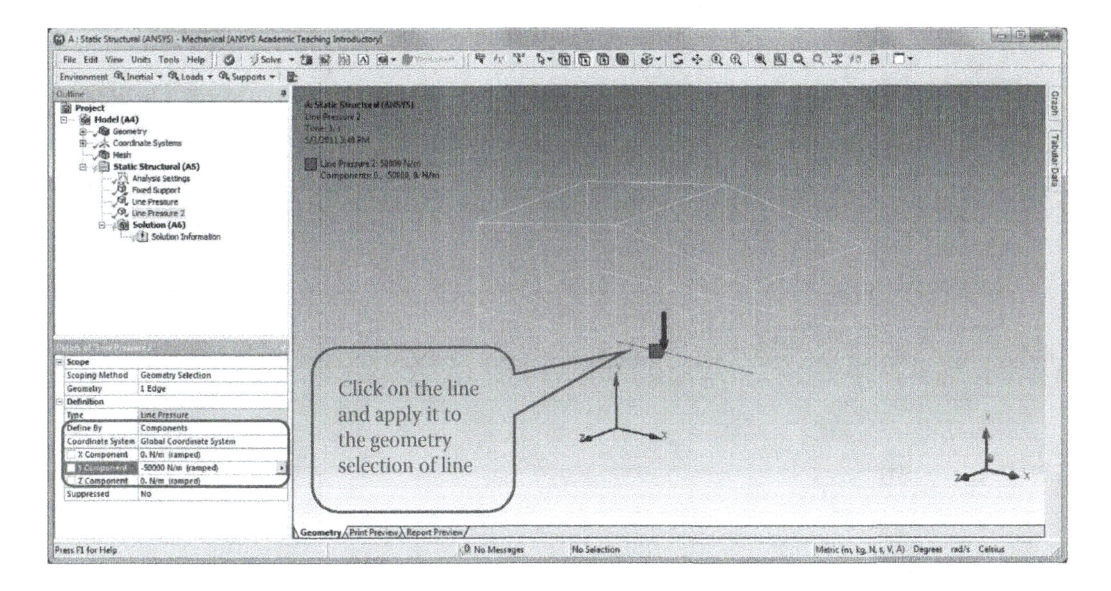

Repeat the steps until the same load is applied to all seven edges highlighted in the figure below.

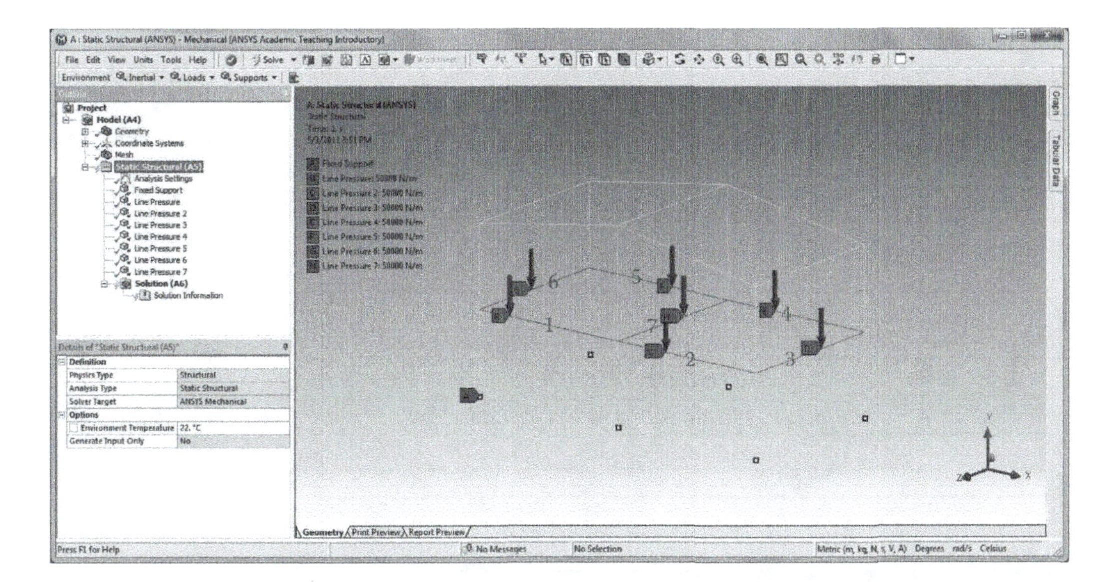

Alternative Procedure: Note that the *Line Pressure* item in the *Outline* can be copied and pasted under *Static Structural (A5)* for repeated use. To make a copy, right-click on *Line Pressure* and select *Copy* from the menu. To paste, right-click on the *Static Structural (A5)* and select *Paste*. Remember to apply each newly pasted *Line Pressure* to a different line edge on the *Geometry* selection in the *Details of "Line Pressure"* until the same load is applied to all seven edges.

Step 14: Retrieve Solution
Insert a *Total Deformation* item by right-clicking on *Solution (A6)* in the *Outline* tree.

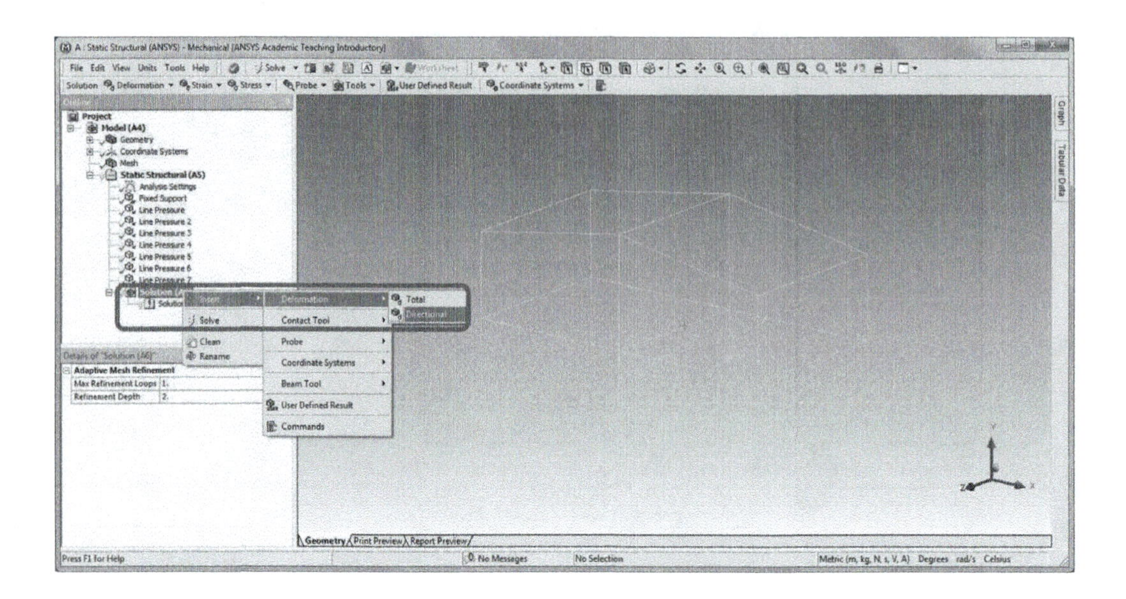

Insert a *Beam Tool* item by right-clicking on *Solution (A6)* in the *Outline* tree.

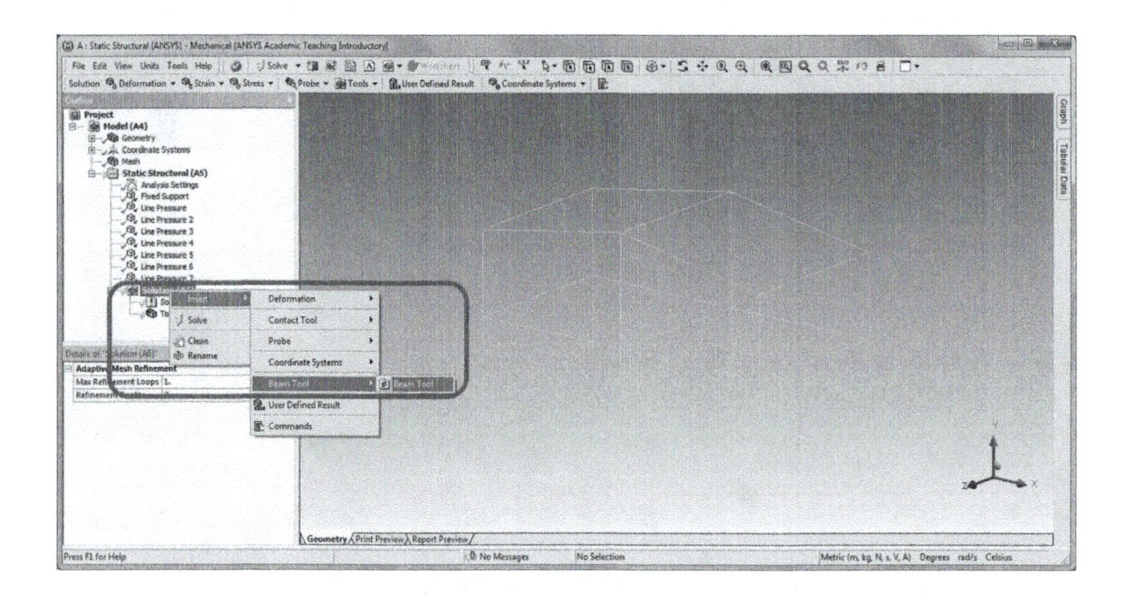

Right-click on *Solution (A6)* in the *Outline* tree and select *Solve.* The program will start to solve the model.

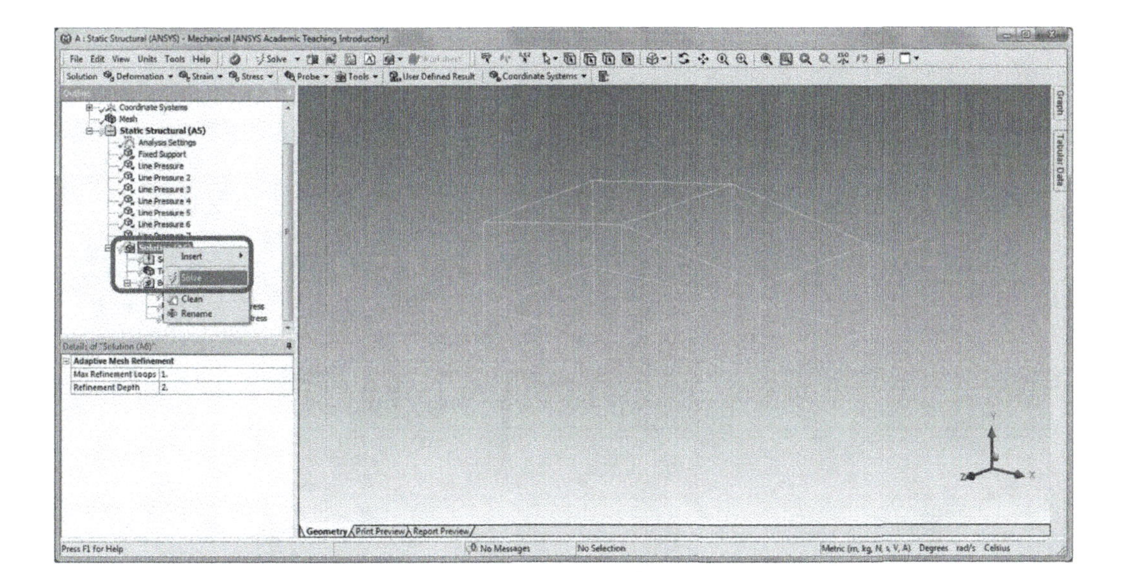

After completion, click *Total Deformation* in the *Outline* to review the total deformation results.

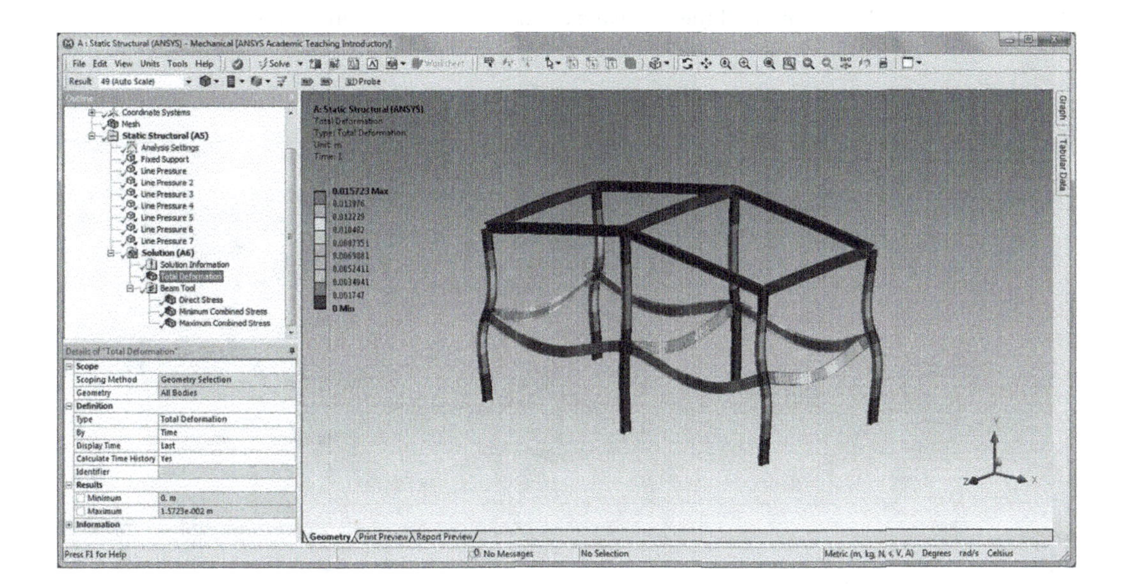

Click *Direct Stress* under *Beam Tool* in the *Outline* to review the axial stress results in beams.

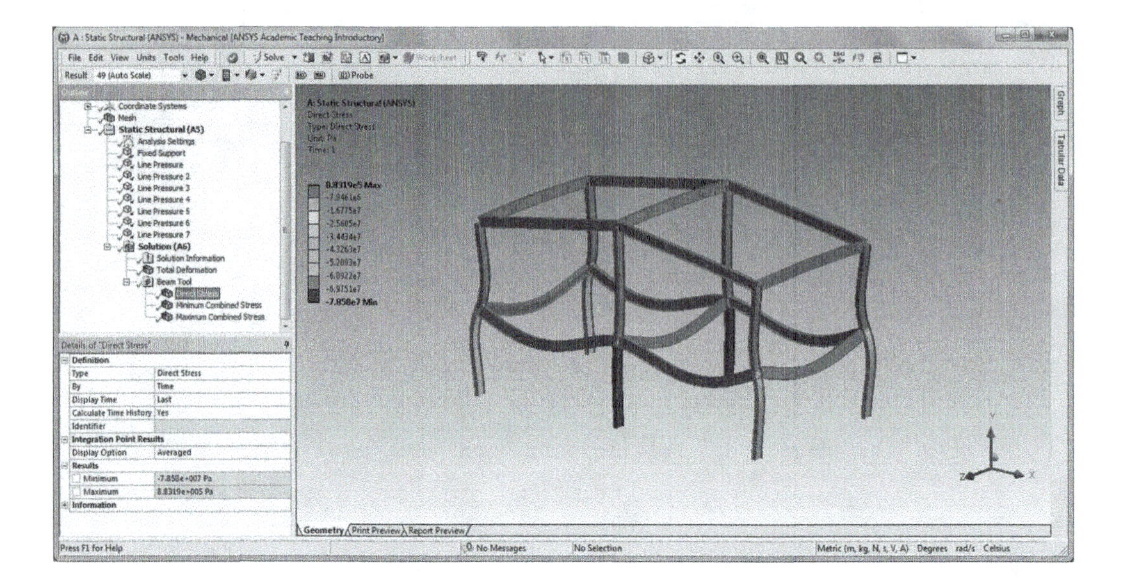

Click *Minimum Combined Stress* under *Beam Tool* to retrieve the linear combination of the *Direct Stress* and the *Minimum Bending Stress* results in beams.

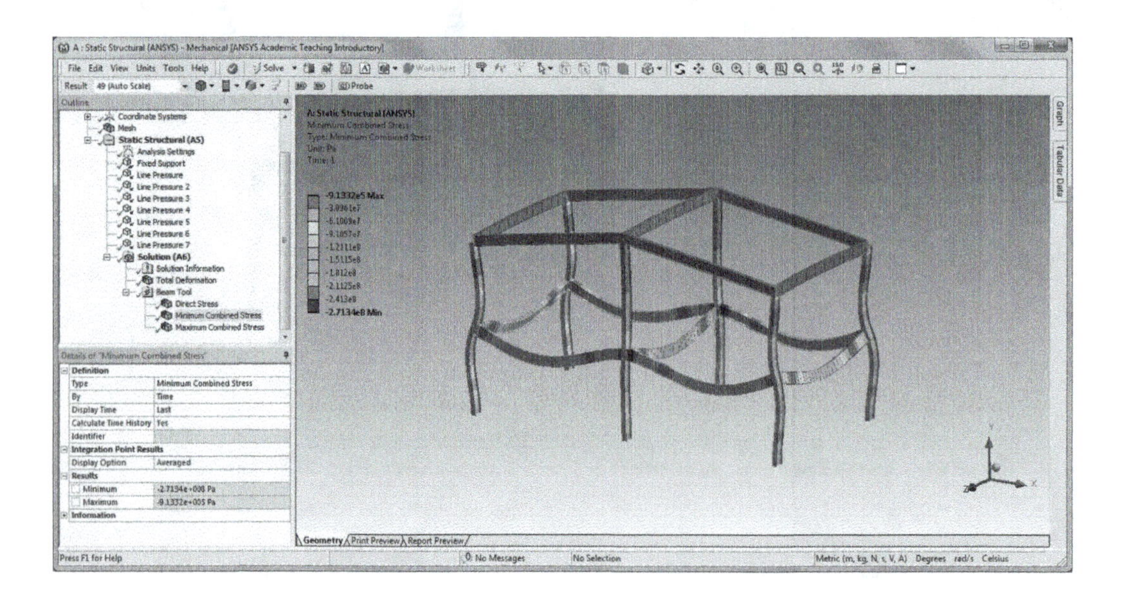

Click *Maximum Combined Stress* under *Beam Tool* to retrieve the linear combination of the *Direct Stress* and the *Maximum Bending Stress* results in beams.

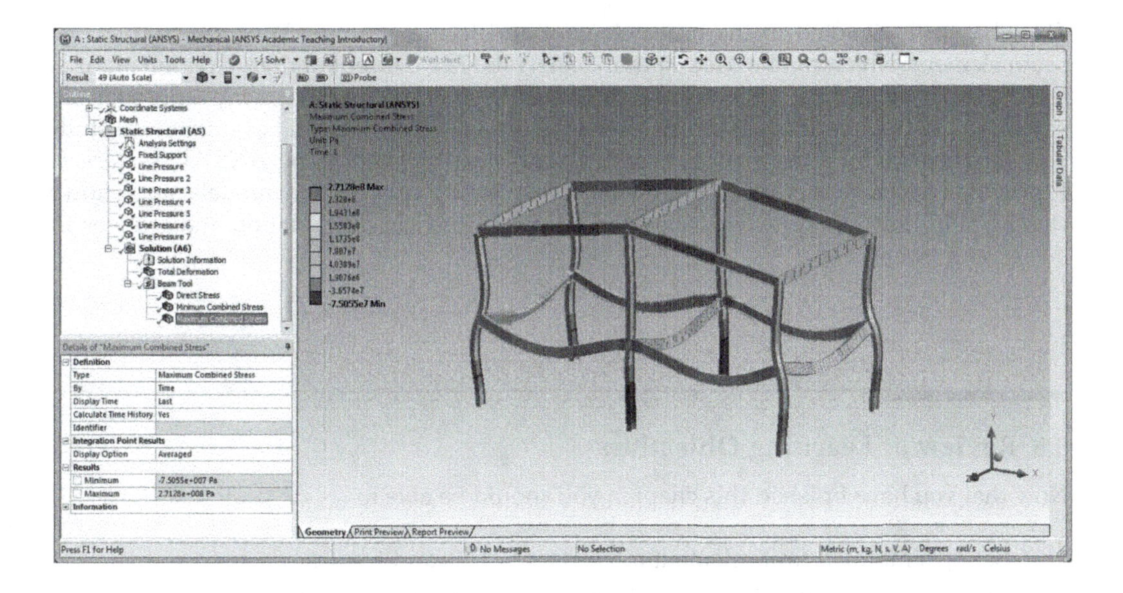

Close the *Static Structural (ANSYS)* program. Save project and exit *Workbench.*

Modeling tips: A group of points can be constructed in *DesignModeler* through a text file that contains the following five columns: group number, point number, x-coordinate, *y*-coordinate, and *z*-coordinate. The data columns can be separated by spaces or Tabs.

For example, to create five points with their coordinates given as *x*(*i*), *y*(i), and *z*(*i*), type the data in the following form in *Notepad* or *WordPad*, and save as a text file, for example, "points.txt," in local disk:

1	1	$x(1)$	$y(1)$	$z(1)$
1	2	$x(2)$	$y(2)$	$z(2)$
1	3	$x(3)$	$y(3)$	$z(3)$
1	4	$x(4)$	$y(4)$	$z(4)$
1	5	$x(5)$	$y(5)$	$z(5)$

Then, follow the steps described in the following figure:

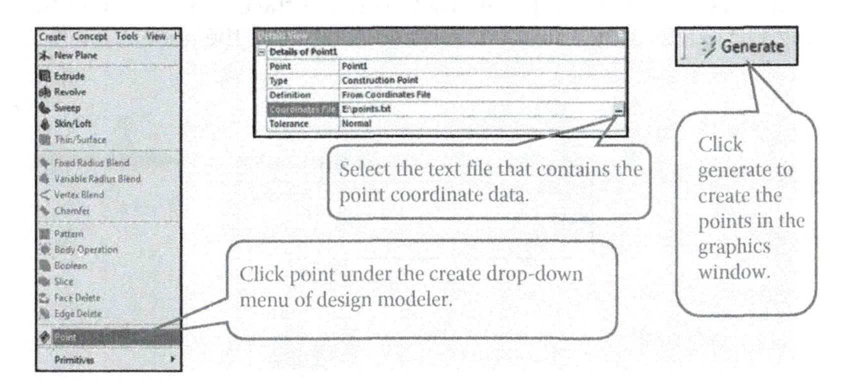

3.7 Summary

This chapter is devoted to the beam element and frame analysis. We have studied the beam element, which can be used in frame analysis. The concept of the shape functions is further explored and the derivations of the stiffness matrix using the energy approach are emphasized. Treatment of distributed loads is discussed and several examples are studied. A two-story building structure with I-beams is analyzed using *ANSYS Workbench*. It provides useful modeling techniques in constructing concept line models, and shows step-by-step how *Workbench* can be used to determine the deformation and stresses in beams and frames.

3.8 Review of Learning Objectives

Now that you have finished this chapter, you should be able to

1. Set up simplified finite element models for beams and frames.
2. Derive the element stiffness matrix for plane beams using direct/energy approach.
3. Explain the concept of shape functions and their characteristics for beam elements.
4. Find the equivalent nodal forces of distributed loads on beams.
5. Determine the deflection and rotation at a point of a beam using hand calculation to verify the finite element solutions.
6. Apply the general beam element stiffness matrix to the analysis of simple frames.
7. Create line models from concept points, sketches, or by body translation in *Workbench*.
8. Perform static structural analyses on beams and frames using *Workbench*.

PROBLEMS

3.1 Using Equation 3.11, derive the results of the equivalent nodal forces and moments for a beam element with uniformly distributed lateral load.

3.2 The cantilever beam is supported by a spring at the end as shown in the figure. Using FEM, determine the deflection and rotation at the node 2.

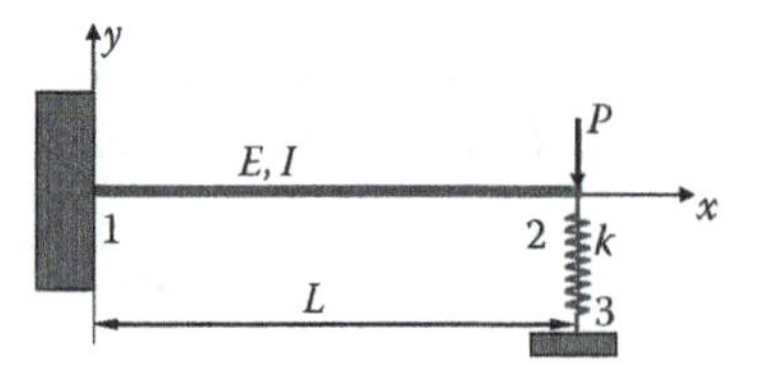

3.3 Using FEM, determine the nodal displacement, rotations, and reaction forces for the propped cantilever beam shown below. The beam is assumed to have constant EI and length $2L$. It is supported by a roller at mid length and is built in at the right end.

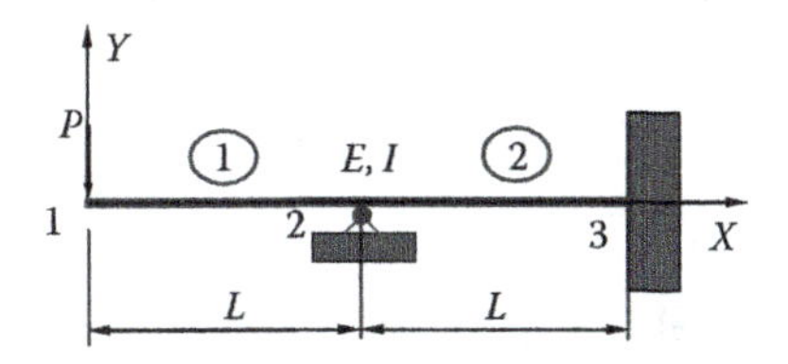

3.4 The 2-D frame is supported as shown in the figure. Constants E, A, I of the beam and the length L are given. Using FEM, determine the displacement and rotation at node 2.

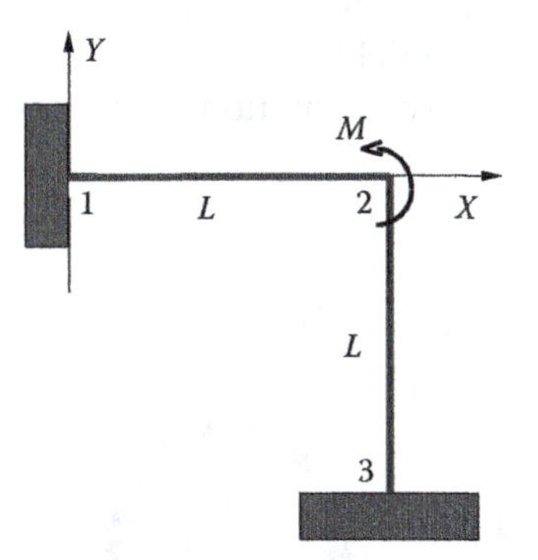

3.5 The plane frame is subjected to the uniformly distributed load and is fixed at the ends as shown in the figure. Assume $E = 30 \times 10^6$ psi, $A = 100$ in.^{2}, and $I = 1000$ in.^{4} for both elements of the frame. Using FEM, find the displacement and rotation of node 2.

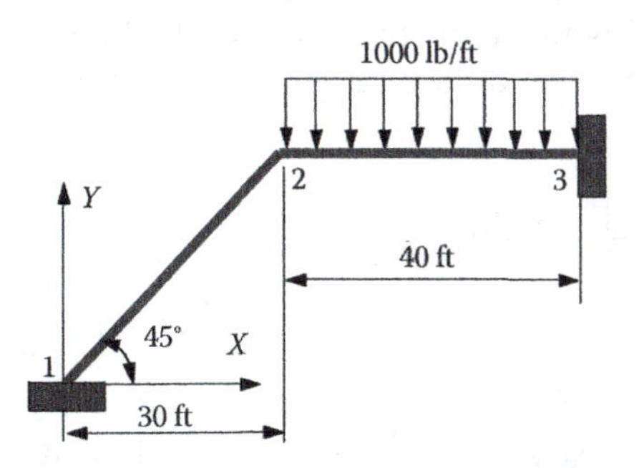

3.6 Using *ANSYS Workbench*, solve the frame problem in Problem 3.5, and determine the reaction forces and moments at the two fixed ends.

3.7 The physical construction of two representative beam connections is shown in the figure below, where an I-beam is connected to a floor slab through a slotted bolt hole in (a), and is connected to a column through an angle bracket in (b). What simplified support conditions would you use to represent the physical construction in the two cases and explain your reasons?

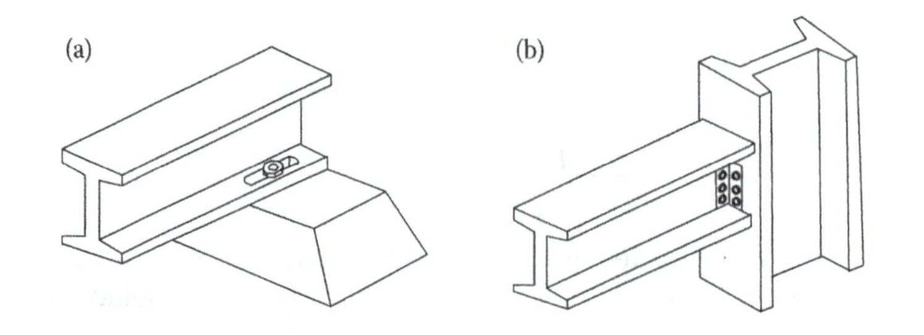

3.8 The bike frame shown in the figure below has hollow circular tubes (24 mm outer diameter and 2 mm thick) and is made of aluminum alloy. Use *ANSYS Workbench* to determine the deformation and stresses of the frame members.

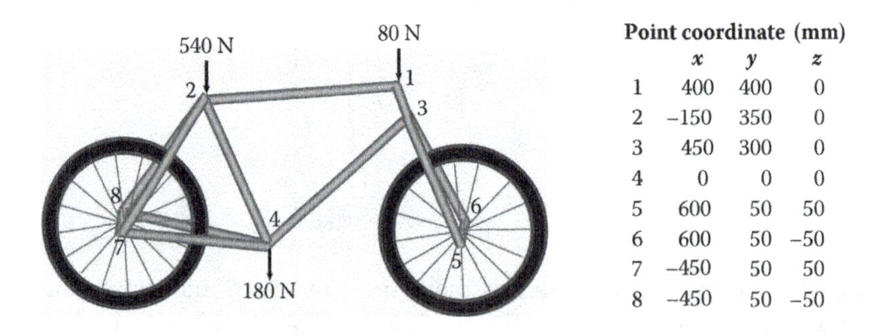

Point coordinate (mm)

	x	y	z
1	400	400	0
2	−150	350	0
3	450	300	0
4	0	0	0
5	600	50	50
6	600	50	−50
7	−450	50	50
8	−450	50	−50

3.9 The hoist frame shown in the following figure has 76 mm wide and 3 mm thick square tubes and is made of structural steel. Use *ANSYS Workbench* to determine the deformation and the stresses in the frame.

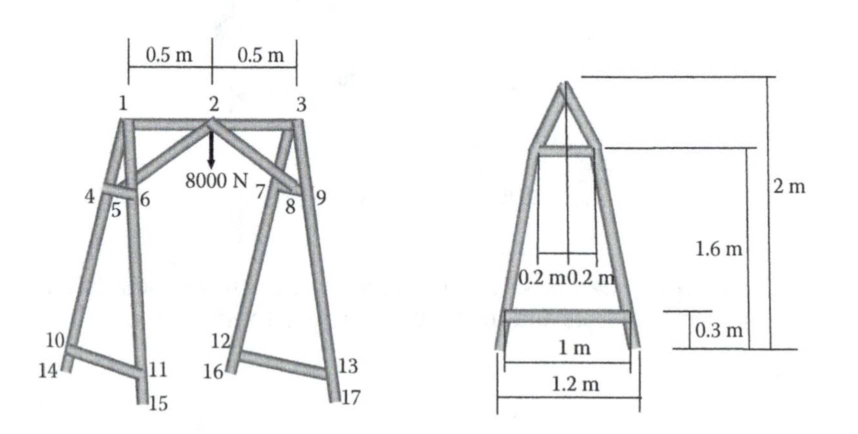

3.10 The frame structure is constructed with structural steel I-beams of given cross-section dimensions and is used to support a uniform load of 6 kN/m as shown below. Use *ANSYS Workbench* to determine the deformation and stresses in the frame members.

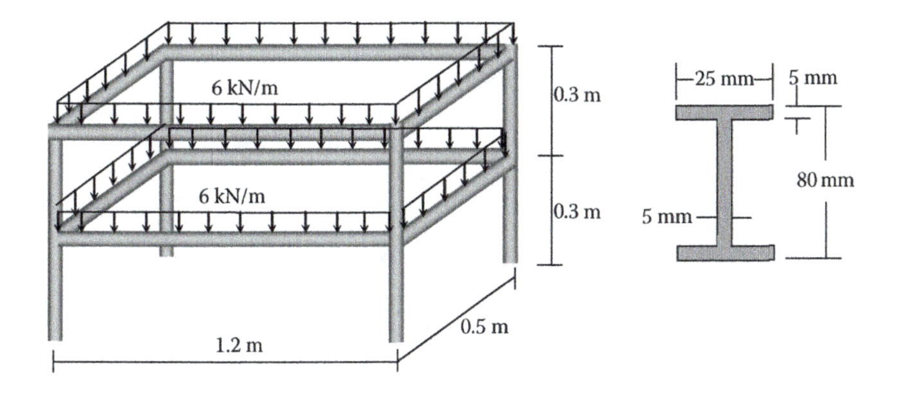

3.11 The following bridge structure is used to carry a uniform load of 50 kN/m on the bridge surface. Use *ANSYS Workbench* to determine the deformation and stresses in the bridge if structural steel I-beams of the given cross-section are used.

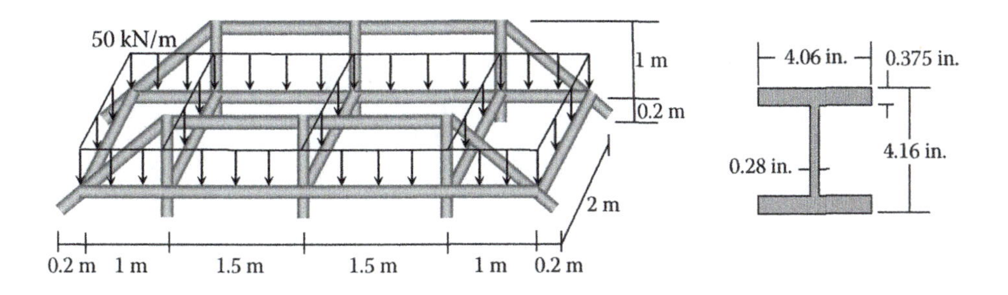

4

Two-Dimensional Elasticity

4.1 Introduction

Many structures that are three-dimensional can be satisfactorily treated as two-dimensional problems (Figure 4.1). This chapter introduces you to the use of the finite element method for deformation and stress analyses of two-dimensional elasticity problems. First, the basic equations in plane elasticity theory are reviewed. Then several types of 2-D finite elements for plane elasticity analysis are presented. Applications of these elements are demonstrated and their accuracies and efficiencies are discussed. This presentation is followed by an example illustrating analysis of a 2-D elasticity problem using *ANSYS® Workbench*.

4.2 Review of 2-D Elasticity Theory

In general, the stresses and strains at any point in a structure consist of six independent components, that is (Figure 4.2),

$$\sigma_x, \sigma_y, \sigma_z, \tau_{xy}, \tau_{yz}, \tau_{zx}$$

for stresses, and

$$\varepsilon_x, \varepsilon_y, \varepsilon_z, \gamma_{xy}, \gamma_{yz}, \gamma_{zx}$$

for strains.

Under certain conditions, the state of stresses and strains can be simplified. A 3-D stress analysis can, therefore, be reduced to a 2-D analysis. There are two general types of models involved in this 2-D analysis: plane stress and plane strain.

4.2.1 Plane Stress

In the plane stress case, any stress component related to the z direction is zero, that is,

$$\sigma_z = \tau_{yz} = \tau_{zx} = 0 \qquad (\varepsilon_z \neq 0) \tag{4.1}$$

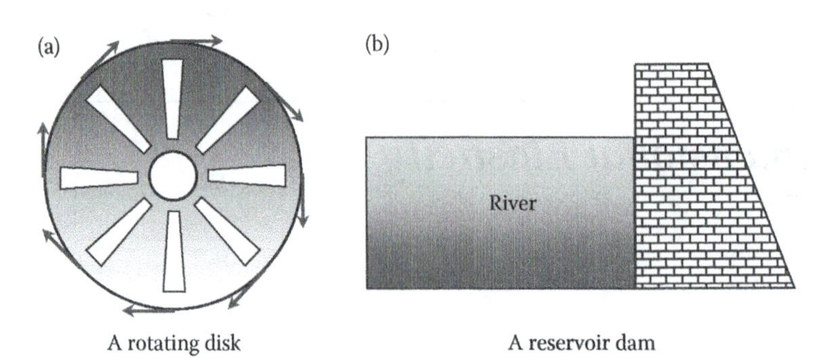

FIGURE 4.1
Examples of 2-D elasticity problems.

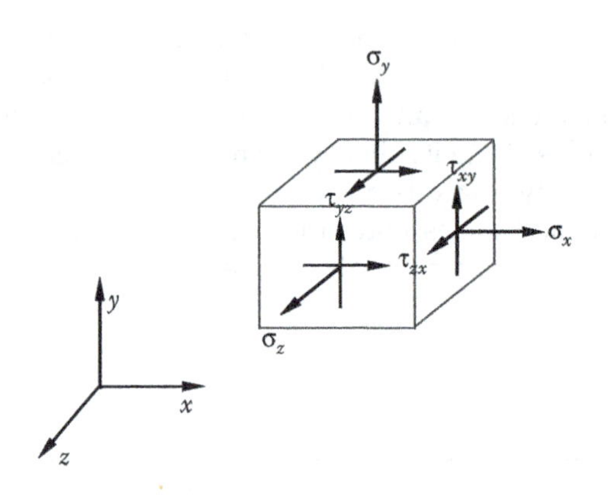

FIGURE 4.2
Stress components at a point in a structure.

A thin planar structure with constant thickness and loading within the plane of the structure (xy-plane) can be regarded as a plane stress case (Figure 4.3).

4.2.2 Plane Strain

In the plane strain case, any strain component related to the z direction is zero, that is,

$$\varepsilon_z = \gamma_{yz} = \gamma_{zx} = 0 \qquad (\sigma_z \neq 0) \tag{4.2}$$

A long structure with a uniform cross section and transverse loading along its thickness (z-direction), such as a tunnel or a dam, can be regarded a plane strain case (Figure 4.4).

As illustrated in the above figures, structures in plane stress or plane strain conditions can be modeled as 2-D boundary value problems. At an arbitrary point, the in-plane displacement field may be defined by two components (u and v), the in-plane strain field by three components ($\varepsilon_x, \varepsilon_y, \gamma_{xy}$), and the in-plane stress field by three components ($\sigma_x, \sigma_y, \tau_{xy}$). The following is a review of equations describing how the field variables are related.

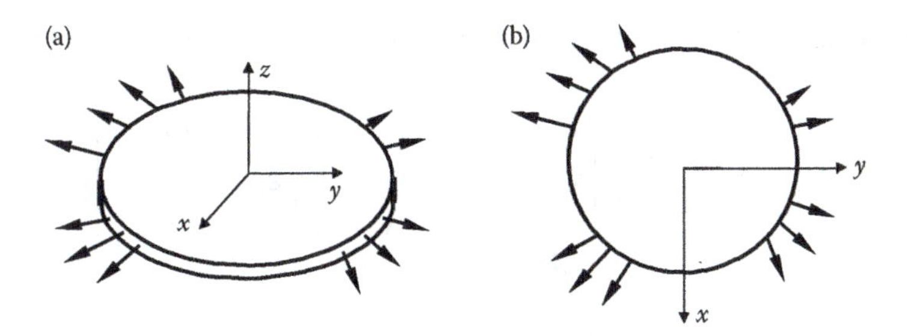

FIGURE 4.3
Plane stress condition: (a) structure with in-plane loading; and (b) 2-D model.

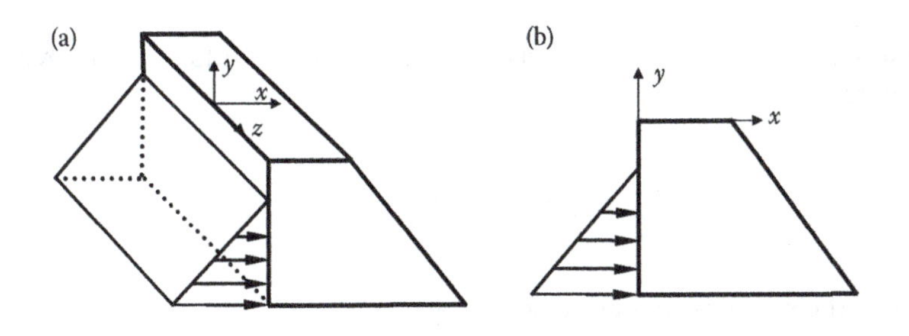

FIGURE 4.4
Plane strain condition: (a) structure with uniform transverse load; and (b) 2-D model.

4.2.3 Stress–Strain (Constitutive) Equations

For linear elastic isotropic materials, we have the following strain–stress relation for the plane stress case:

$$\begin{Bmatrix} \varepsilon_x \\ \varepsilon_y \\ \gamma_{xy} \end{Bmatrix} = \begin{bmatrix} 1/E & -\nu/E & 0 \\ -\nu/E & 1/E & 0 \\ 0 & 0 & 1/G \end{bmatrix} \begin{Bmatrix} \sigma_x \\ \sigma_y \\ \tau_{xy} \end{Bmatrix} + \begin{Bmatrix} \varepsilon_{x0} \\ \varepsilon_{y0} \\ \gamma_{xy0} \end{Bmatrix} \tag{4.3}$$

or,

$$\varepsilon = \mathbf{E}^{-1}\sigma + \varepsilon_0$$

where ε_0 is the initial strain (e.g., due to a temperature change), E Young's modulus, ν Poisson's ratio, and G the shear modulus. Note that

$$G = \frac{E}{2(1+\nu)} \tag{4.4}$$

for *homogeneous* and *isotropic* materials.

We can also express stresses in terms of strains by solving the above equation

$$\begin{Bmatrix} \sigma_x \\ \sigma_y \\ \tau_{xy} \end{Bmatrix} = \frac{E}{1-\nu^2} \begin{bmatrix} 1 & \nu & 0 \\ \nu & 1 & 0 \\ 0 & 0 & (1-\nu)/2 \end{bmatrix} \left(\begin{Bmatrix} \varepsilon_x \\ \varepsilon_y \\ \gamma_{xy} \end{Bmatrix} - \begin{Bmatrix} \varepsilon_{x0} \\ \varepsilon_{y0} \\ \gamma_{xy0} \end{Bmatrix} \right) \tag{4.5}$$

or,

$$\boldsymbol{\sigma} = \mathbf{E}\boldsymbol{\varepsilon} + \boldsymbol{\sigma}_0$$

where $\boldsymbol{\sigma}_0 = -\mathbf{E}\boldsymbol{\varepsilon}_0$ is the initial stress.

For *plane strain* case, we need to replace the material constants in the above equations in the following fashion:

$$E \to \frac{E}{1-\nu^2}; \qquad \nu \to \frac{\nu}{1-\nu}; \qquad G \to G \tag{4.6}$$

For example, the stress is related to strain by

$$\begin{Bmatrix} \sigma_x \\ \sigma_y \\ \tau_{xy} \end{Bmatrix} = \frac{E}{(1+\nu)(1-2\nu)} \begin{bmatrix} 1-\nu & \nu & 0 \\ \nu & 1-\nu & 0 \\ 0 & 0 & (1-2\nu)/2 \end{bmatrix} \left(\begin{Bmatrix} \varepsilon_x \\ \varepsilon_y \\ \gamma_{xy} \end{Bmatrix} - \begin{Bmatrix} \varepsilon_{x0} \\ \varepsilon_{y0} \\ \gamma_{xy0} \end{Bmatrix} \right)$$

in the *plane strain* case.

Initial strain due to a *temperature change* (thermal loading) is given by the following for the plane stress case

$$\begin{Bmatrix} \varepsilon_{x0} \\ \varepsilon_{y0} \\ \gamma_{xy0} \end{Bmatrix} = \begin{Bmatrix} \alpha\Delta T \\ \alpha\Delta T \\ 0 \end{Bmatrix} \tag{4.7}$$

where α is the coefficient of thermal expansion, ΔT the change of temperature. For the plane strain case, α should be replaced by $(1+\nu)\alpha$ in Equation 4.7. Note that if the structure is free to deform under thermal loading, there will be no (elastic) stresses in the structure.

4.2.4 Strain and Displacement Relations

For small strains and small rotations, we have,

$$\varepsilon_x = \frac{\partial u}{\partial x}, \qquad \varepsilon_y = \frac{\partial v}{\partial y}, \qquad \gamma_{xy} = \frac{\partial u}{\partial y} + \frac{\partial v}{\partial x}$$

In matrix form, we write

$$\begin{Bmatrix} \varepsilon_x \\ \varepsilon_y \\ \gamma_{xy} \end{Bmatrix} = \begin{bmatrix} \partial/\partial x & 0 \\ 0 & \partial/\partial y \\ \partial/\partial y & \partial/\partial x \end{bmatrix} \begin{Bmatrix} u \\ v \end{Bmatrix}, \quad \text{or } \varepsilon = \mathbf{D}\mathbf{u} \tag{4.8}$$

From this relation, we know that, if the displacements are represented by polynomials, the strains (and thus stresses) will be polynomials of an order that is one order lower than the displacements.

4.2.5 Equilibrium Equations

In plane elasticity, the stresses in the structure must satisfy the following equilibrium equations:

$$\begin{aligned} \frac{\partial \sigma_x}{\partial x} + \frac{\partial \tau_{xy}}{\partial y} + f_x &= 0 \\ \frac{\partial \tau_{xy}}{\partial x} + \frac{\partial \sigma_y}{\partial y} + f_y &= 0 \end{aligned} \tag{4.9}$$

where f_x and f_y are body forces (forces per unit volume, such as gravity forces). In the FEM, these equilibrium conditions are satisfied in an approximate sense.

4.2.6 Boundary Conditions

The boundary S of the 2-D region can be divided into two parts, S_u and S_t (Figure 4.5). The boundary conditions (BCs) can be described as

$$\begin{aligned} u = \bar{u}, \quad v = \bar{v}, \qquad & \text{on } S_u \\ t_x = \bar{t}_x, \quad t_y = \bar{t}_y, \qquad & \text{on } S_t \end{aligned} \tag{4.10}$$

in which t_x and t_y are tractions (stresses on the boundary) and the barred quantities are those with known values.

4.2.7 Exact Elasticity Solution

The exact solution (displacements, strains, and stresses) of a given problem must satisfy the constitutive relations, equilibrium equations, and compatibility conditions (structures

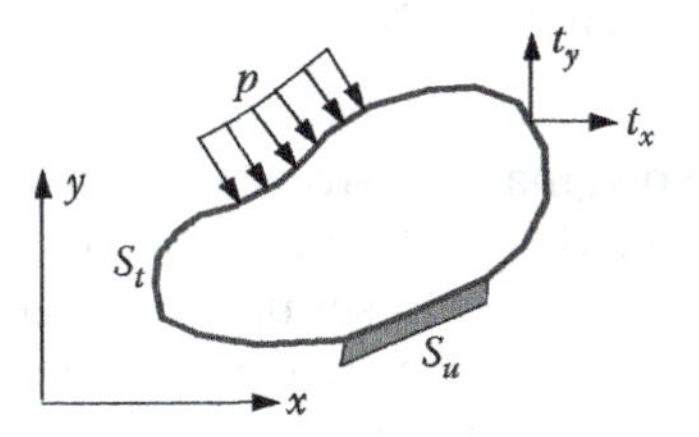

FIGURE 4.5
Boundary conditions for a structure.

should deform in a continuous manner, with no cracks or overlaps in the obtained displacement fields).

EXAMPLE 4.1

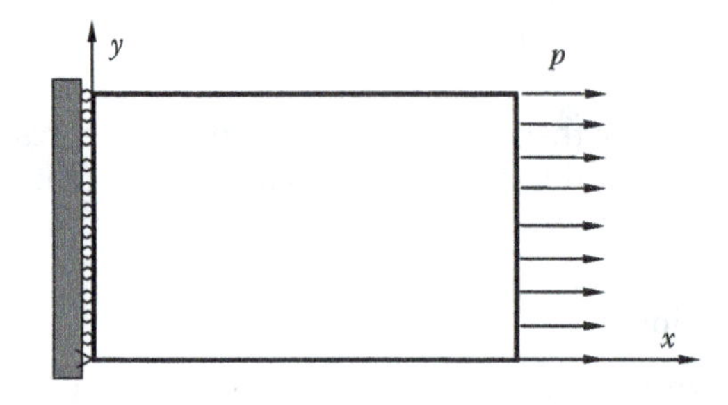

A plate is supported and loaded with distributed force p as shown in the figure. The material constants E and ν are known.

The exact solution for this simple problem is found to be as follows:

Displacement:

$$u = \frac{p}{E}x, \quad v = -\nu\frac{p}{E}y$$

Strain:

$$\varepsilon_x = \frac{p}{E}, \quad \varepsilon_y = -\nu\frac{p}{E}, \quad \gamma_{xy} = 0$$

Stress:

$$\sigma_x = p, \quad \sigma_y = 0, \quad \tau_{xy} = 0$$

Exact (or analytical) solutions for simple problems are numbered (suppose there is a hole in the plate or the roller supports are replaced by fixed ones). That is why we need the FEM for solutions of 2-D elasticity problems in general.

4.3 Modeling of 2-D Elasticity Problems

For a plane stress or plane strain analysis, we model only a 2-D region, that is, a planar surface or cross section of the original 3-D structure. The region then needs to be divided into an element discretization made of triangles, quadrilaterals, or a mixture of both. The element behaviors need to be specified to set up the problem type as either plane stress or plane strain. The discretization can be structured (mapped mesh on a three-sided or four-sided surface region with equal numbers of element divisions for the opposite sides) or unstructured (free mesh), as shown in Figure 4.6. A structured surface mesh has regular

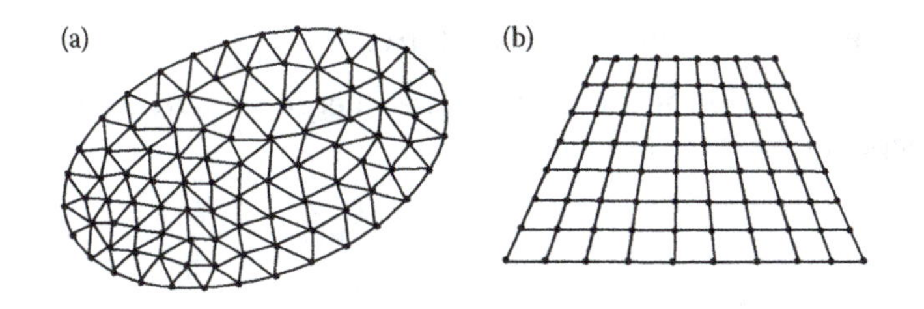

FIGURE 4.6
Finite element discretization: (a) unstructured (free) mesh; and (b) structured (mapped) mesh.

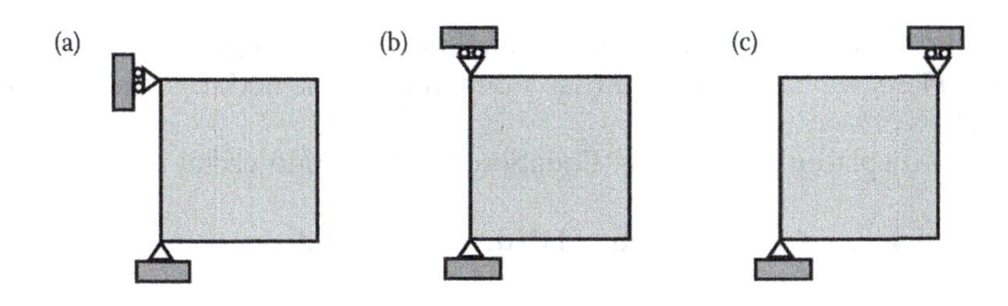

FIGURE 4.7
Support conditions in 2-D: (a) and (c) effective constraints; (b) ineffective constraints.

connectivity that can be described as a 2-D array, and it generally has better solution reliability compared to a free mesh.

Mesh properties such as the mesh density and the element shape quality are important factors that affect the solution accuracy and efficiency. Finer mesh is usually needed in areas where a high stress/strain is expected. The type of finite element employed also plays a vital role in the analysis, as will be discussed in the next section.

After meshing is complete, the surface discretization is represented in terms of nodal coordinates as well as element connectivity showing how different elements are connected together through nodes. All types of loads (distributed forces, concentrated forces and moments, and so on) are converted into point forces acting at individual nodes. Boundary conditions are no exception, and will be converted into nodal DOF constraints. Remember that support conditions need to suppress rigid-body motion, and the minimum conditions in 2-D include restraining the body with displacement in x, y directions and rotation about the z-axis. Figure 4.7 gives examples of effective and ineffective support conditions in 2-D. Explain why a case is effective or not.

4.4 Formulation of the Plane Stress/Strain Element

In this section, we formulate the element equations based on the plane elasticity theory. A general formula for the stiffness matrix is derived first, followed by presentation of several common element types for 2-D elasticity problems.

4.4.1 A General Formula for the Stiffness Matrix

Displacements (u, v) in a plane element can be interpolated from nodal displacements (u_i, v_i) using shape functions N_i as follows:

$$\begin{Bmatrix} u \\ v \end{Bmatrix} = \begin{bmatrix} N_1 & 0 & N_2 & 0 & \cdots \\ 0 & N_1 & 0 & N_2 & \cdots \end{bmatrix} \begin{Bmatrix} u_1 \\ v_1 \\ u_2 \\ v_2 \\ \vdots \end{Bmatrix} \quad \text{or} \quad \mathbf{u} = \mathbf{Nd} \tag{4.11}$$

where **N** is the *shape function matrix*, **u** the displacement vector, and **d** the *nodal* displacement vector. Here, we have assumed that u depends on the nodal values of u only, and v on nodal values of v only.

From strain–displacement relation (Equation 4.8), the strain vector is

$$\boldsymbol{\varepsilon} = \mathbf{Du} = \mathbf{DNd}, \quad \text{or} \quad \boldsymbol{\varepsilon} = \mathbf{Bd} \tag{4.12}$$

where $\mathbf{B} = \mathbf{DN}$ is the strain–displacement matrix.

Consider the strain energy stored in an element,

$$\begin{aligned} U &= \frac{1}{2}\int_V \sigma^T \varepsilon \, dV = \frac{1}{2}\int_V \left(\sigma_x \varepsilon_x + \sigma_y \varepsilon_y + \tau_{xy}\gamma_{xy}\right) dV \\ &= \frac{1}{2}\int_V (\mathbf{E}\varepsilon)^T \varepsilon \, dV = \frac{1}{2}\int_V \varepsilon^T \mathbf{E}\varepsilon \, dV = \frac{1}{2}\mathbf{d}^T \int_V \mathbf{B}^T\mathbf{E}\mathbf{B} \, dV \mathbf{d} \\ &= \frac{1}{2}\mathbf{d}^T\mathbf{k}\mathbf{d} \end{aligned}$$

From this, we obtain the general formula for the *element stiffness matrix*

$$\mathbf{k} = \int_V \mathbf{B}^T\mathbf{E}\mathbf{B} \, dV \tag{4.13}$$

Note that unlike the 1-D cases, **E** here is a *matrix* which is given by the stress–strain relation (e.g., Equation 4.5 for plane stress).

The stiffness matrix **k** defined by Equation 4.13 is symmetric since **E** is symmetric. Also note that given the material property, the behavior of **k** depends on the **B** matrix only, which in turn depends on the shape functions. Thus, the quality of finite elements in representing the behavior of a structure is mainly determined by the choice of shape functions. Most commonly employed 2-D elements are linear or quadratic triangles and quadrilaterals.

4.4.2 Constant Strain Triangle (CST or T3)

This is the simplest 2-D element (Figure 4.8), which is also called *linear triangular element*.

For this element, we have three nodes at the vertices of the triangle, which are numbered around the element in the counterclockwise direction. Each node has two DOFs (can move

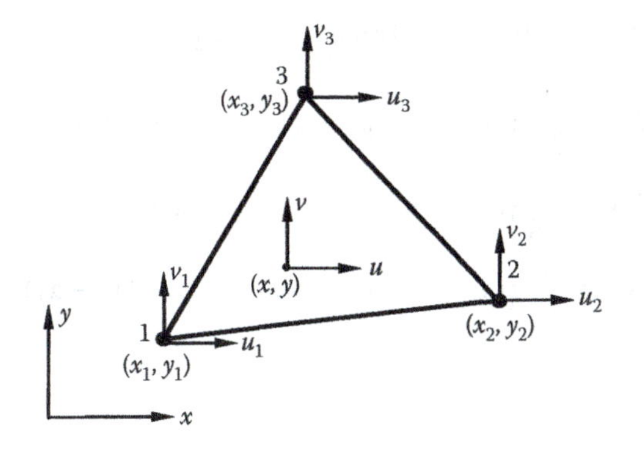

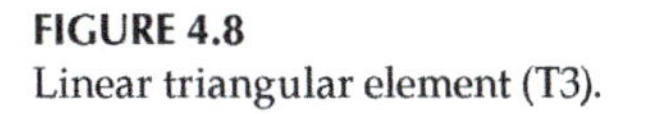
FIGURE 4.8
Linear triangular element (T3).

in the x and y directions). The displacements u and v are assumed to be linear functions within the element, that is

$$u = b_1 + b_2x + b_3y, \quad v = b_4 + b_5x + b_6y \tag{4.14}$$

where b_i $(i = 1, 2, \ldots, 6)$ are constants. From these, the strains are found to be

$$\varepsilon_x = b_2, \quad \varepsilon_y = b_6, \quad \gamma_{xy} = b_3 + b_5 \tag{4.15}$$

which are constant throughout the element. Thus, we have the name "constant strain triangle" (CST).

Displacements given by Equation 4.14 should satisfy the following six equations:

$$\begin{aligned} u_1 &= b_1 + b_2x_1 + b_3y_1 \\ u_2 &= b_1 + b_2x_2 + b_3y_2 \\ &\vdots \\ v_3 &= b_4 + b_5x_3 + b_6y_3 \end{aligned}$$

Solving these equations, we can find the coefficients b_1, b_2, …, and b_6 in terms of nodal displacements and coordinates. Substituting these coefficients into Equation 4.14 and rearranging the terms, we obtain

$$\begin{Bmatrix} u \\ v \end{Bmatrix} = \begin{bmatrix} N_1 & 0 & N_2 & 0 & N_3 & 0 \\ 0 & N_1 & 0 & N_2 & 0 & N_3 \end{bmatrix} \begin{Bmatrix} u_1 \\ v_1 \\ u_2 \\ v_2 \\ u_3 \\ v_3 \end{Bmatrix} \tag{4.16}$$

where the shape functions (linear functions in x and y) are

$$
\begin{aligned}
N_1 &= \frac{1}{2A}\{(x_2y_3 - x_3y_2) + (y_2 - y_3)x + (x_3 - x_2)y\} \\
N_2 &= \frac{1}{2A}\{(x_3y_1 - x_1y_3) + (y_3 - y_1)x + (x_1 - x_3)y\} \\
N_3 &= \frac{1}{2A}\{(x_1y_2 - x_2y_1) + (y_1 - y_2)x + (x_2 - x_1)y\}
\end{aligned}
\tag{4.17}
$$

and

$$
A = \frac{1}{2}\det\begin{bmatrix} 1 & x_1 & y_1 \\ 1 & x_2 & y_2 \\ 1 & x_3 & y_3 \end{bmatrix} \tag{4.18}
$$

is the area of the triangle.

Using the strain–displacement relation (4.8), results (4.16) and (4.17), we have

$$
\begin{Bmatrix} \varepsilon_x \\ \varepsilon_y \\ \gamma_{xy} \end{Bmatrix} = \mathbf{Bd} = \frac{1}{2A}\begin{bmatrix} y_{23} & 0 & y_{31} & 0 & y_{12} & 0 \\ 0 & x_{32} & 0 & x_{13} & 0 & x_{21} \\ x_{32} & y_{23} & x_{13} & y_{31} & x_{21} & y_{12} \end{bmatrix}\begin{Bmatrix} u_1 \\ v_1 \\ u_2 \\ v_2 \\ u_3 \\ v_3 \end{Bmatrix} \tag{4.19}
$$

where $x_{ij} = x_i - x_j$ and $y_{ij} = y_i - y_j$ $(i, j = 1, 2, 3)$. Again, we see constant strains within the element. From stress–strain relation (e.g., Equation 4.5), we see that stresses obtained using the CST element are also constant.

Applying the formula in Equation 4.13, we obtain the element stiffness matrix for the CST element

$$
\mathbf{k} = \int_V \mathbf{B}^T\mathbf{E}\mathbf{B}\,dV = tA(\mathbf{B}^T\mathbf{E}\mathbf{B}) \tag{4.20}
$$

in which t is the thickness of the element. Notice that $\mathbf{k}$ for CST is a 6×6 *symmetric* matrix.

Both the expressions of the shape functions in Equation 4.17 and their derivations are lengthy and offer little insight into the behavior of the element.

We introduce the *natural coordinates* (ξ, η) on the triangle where $\xi = A_1/A$, $\eta = A_2/A$ with A_1 and A_2 being the areas of the smaller triangles indicated in Figure 4.9. Then *the shape functions* can be represented simply by

$$
N_1 = \xi, \quad N_2 = \eta, \quad N_3 = 1 - \xi - \eta \tag{4.21}
$$

Notice that,

$$
N_1 + N_2 + N_3 = 1 \tag{4.22}
$$

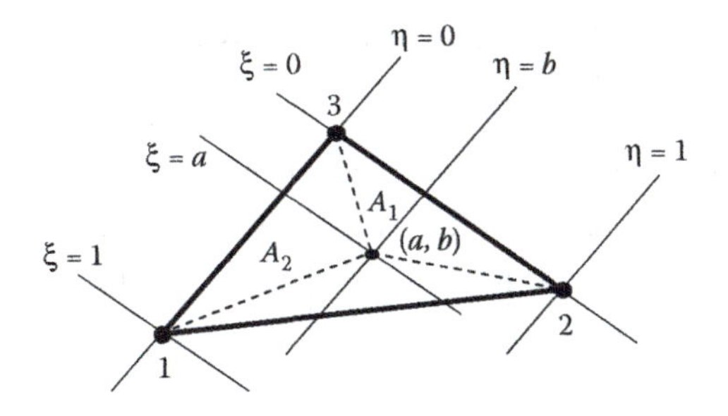

FIGURE 4.9
The natural coordinate system defined on the triangle.

which ensures that the rigid-body translation is represented by the chosen shape functions. Also, as in the 1-D case,

$$N_i = \begin{cases} 1, & \text{at node } i; \\ 0, & \text{at other nodes} \end{cases} \tag{4.23}$$

and varies linearly within the element. The plot for shape function N_1 is shown in Figure 4.10. N_2 and N_3 have similar features.

We have two coordinate systems for the element: the global coordinates (x, y) and the natural (local) coordinates (ξ, η). The relation between the two is given by

$$\begin{aligned} x &= N_1x_1 + N_2x_2 + N_3x_3 \\ y &= N_1y_1 + N_2y_2 + N_3y_3 \end{aligned} \tag{4.24}$$

or,

$$\begin{aligned} x &= x_{13}\xi + x_{23}\eta + x_3 \\ y &= y_{13}\xi + y_{23}\eta + y_3 \end{aligned} \tag{4.25}$$

where $x_{ij} = x_i - x_j$ and $y_{ij} = y_i - y_j$ $(i, j = 1, 2, 3)$ as defined earlier.

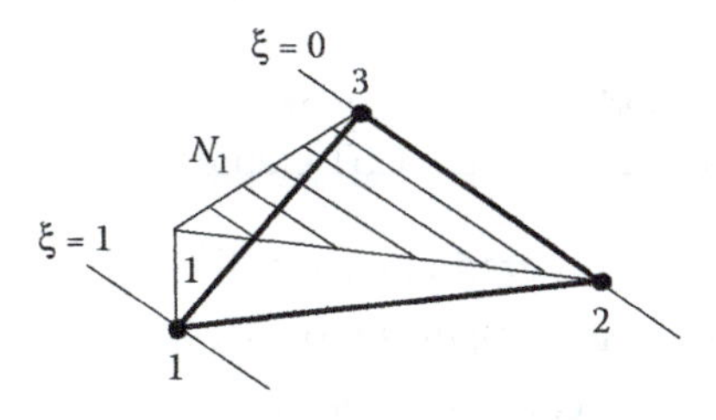

FIGURE 4.10
Plot of the shape function N_1 for T3 element.

Displacement u or v on the element can be viewed as functions of (x, y) or (ξ, η). Using the chain rule for derivatives, we have

$$\begin{Bmatrix} \dfrac{\partial u}{\partial \xi} \\ \dfrac{\partial u}{\partial \eta} \end{Bmatrix} = \begin{bmatrix} \dfrac{\partial x}{\partial \xi} & \dfrac{\partial y}{\partial \xi} \\ \dfrac{\partial x}{\partial \eta} & \dfrac{\partial y}{\partial \eta} \end{bmatrix} \begin{Bmatrix} \dfrac{\partial u}{\partial x} \\ \dfrac{\partial u}{\partial y} \end{Bmatrix} = \mathbf{J} \begin{Bmatrix} \dfrac{\partial u}{\partial x} \\ \dfrac{\partial u}{\partial y} \end{Bmatrix} \tag{4.26}$$

where **J** is called the *Jacobian matrix* of the transformation.

From Equation 4.25, we calculate

$$\mathbf{J} = \begin{bmatrix} x_{13} & y_{13} \\ x_{23} & y_{23} \end{bmatrix}, \quad \mathbf{J}^{-1} = \frac{1}{2A} \begin{bmatrix} y_{23} & -y_{13} \\ -x_{23} & x_{13} \end{bmatrix} \tag{4.27}$$

where $\det \mathbf{J} = x_{13}y_{23} - x_{23}y_{13} = 2A$ has been used (A is the area of the triangle).

From Equations 4.26, 4.27, 4.16, and 4.21, we have

$$\begin{Bmatrix} \dfrac{\partial u}{\partial x} \\ \dfrac{\partial u}{\partial y} \end{Bmatrix} = \frac{1}{2A} \begin{bmatrix} y_{23} & -y_{13} \\ -x_{23} & x_{13} \end{bmatrix} \begin{Bmatrix} \dfrac{\partial u}{\partial \xi} \\ \dfrac{\partial u}{\partial \eta} \end{Bmatrix} = \frac{1}{2A} \begin{bmatrix} y_{23} & -y_{13} \\ -x_{23} & x_{13} \end{bmatrix} \begin{Bmatrix} u_1 - u_3 \\ u_2 - u_3 \end{Bmatrix} \tag{4.28}$$

Similarly,

$$\begin{Bmatrix} \dfrac{\partial v}{\partial x} \\ \dfrac{\partial v}{\partial y} \end{Bmatrix} = \frac{1}{2A} \begin{bmatrix} y_{23} & -y_{13} \\ -x_{23} & x_{13} \end{bmatrix} \begin{Bmatrix} v_1 - v_3 \\ v_2 - v_3 \end{Bmatrix} \tag{4.29}$$

Using the results in Equations 4.28 and 4.29, and the relations $\varepsilon = \mathbf{Du} = \mathbf{DNd} = \mathbf{Bd}$, we obtain the strain–displacement matrix,

$$\mathbf{B} = \frac{1}{2A} \begin{bmatrix} y_{23} & 0 & y_{31} & 0 & y_{12} & 0 \\ 0 & x_{32} & 0 & x_{13} & 0 & x_{21} \\ x_{32} & y_{23} & x_{13} & y_{31} & x_{21} & y_{12} \end{bmatrix} \tag{4.30}$$

which is the same as we derived earlier in Equation 4.19.

We should note the following about the CST element:

- Use in areas where the strain gradient is small.
- Use in mesh transition areas (fine mesh to coarse mesh).
- Avoid using CST in stress concentration or other crucial areas in the structure, such as edges of holes and corners.
- Recommended only for quick and preliminary FE analysis of 2-D problems.

4.4.3 Quadratic Triangular Element (LST or T6)

This type of element (Figure 4.11) is also called *quadratic triangular element*. There are six nodes on this element: three corner nodes and three midside nodes. Each node has two DOFs as before. The displacements (u, v) are assumed to be quadratic functions of (x, y),

$$\begin{aligned} u &= b_1 + b_2x + b_3y + b_4x^2 + b_5xy + b_6y^2 \\ v &= b_7 + b_8x + b_9y + b_{10}x^2 + b_{11}xy + b_{12}y^2 \end{aligned} \tag{4.31}$$

where b_i $(i = 1, 2, \ldots, 12)$ are constants. From these, the strains are found to be

$$\begin{aligned} \varepsilon_x &= b_2 + 2b_4x + b_5y \\ \varepsilon_y &= b_9 + b_{11}x + 2b_{12}y \\ \gamma_{xy} &= (b_3 + b_8) + (b_5 + 2b_{10})x + (2b_6 + b_{11})y \end{aligned} \tag{4.32}$$

which are linear functions. Thus, we have the "linear strain triangle" (LST), which provides better results than the CST.

In the natural coordinate system, the six shape functions for the LST element are

$$\begin{aligned} N_1 &= \xi(2\xi - 1) \\ N_2 &= \eta(2\eta - 1) \\ N_3 &= \zeta(2\zeta - 1) \\ N_4 &= 4\xi\eta \\ N_5 &= 4\eta\zeta \\ N_6 &= 4\zeta\xi \end{aligned} \tag{4.33}$$

in which $\zeta = 1 - \xi - \eta$. Each of these six shape functions represents a quadratic form on the element as shown in Figure 4.12.

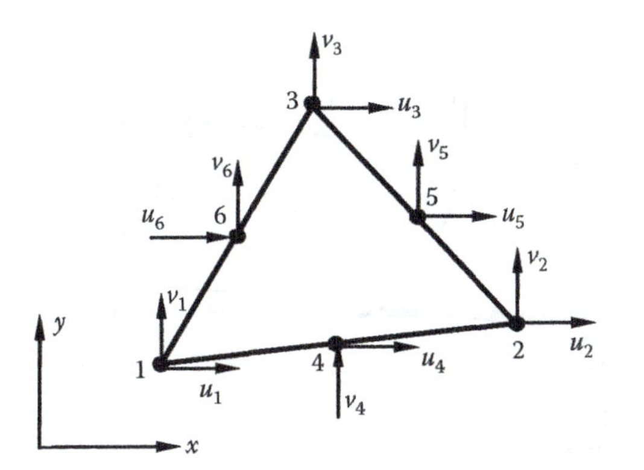

FIGURE 4.11
Quadratic triangular element (T6).

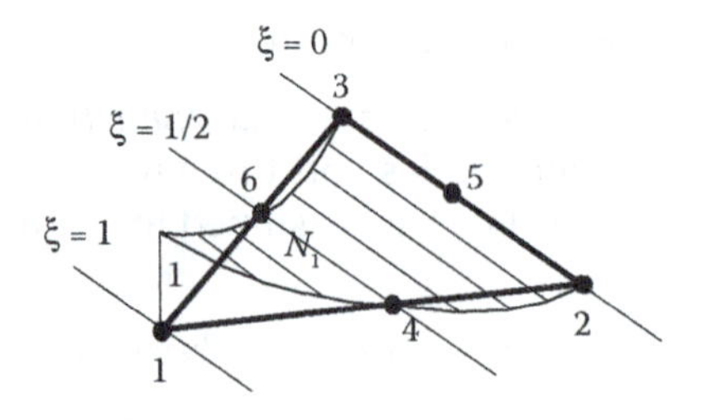

FIGURE 4.12
Plot of the shape function N_1 for T6 element.

Displacements can be written as

$$u = \sum_{i=1}^{6} N_i u_i, \qquad v = \sum_{i=1}^{6} N_i v_i \tag{4.34}$$

The element stiffness matrix is still given by $\mathbf{k} = \int_V \mathbf{B}^T\mathbf{E}\mathbf{B}\,dV$, but here $\mathbf{B}^T\mathbf{E}\mathbf{B}$ is quadratic in x and y. In general, the integral has to be computed numerically.

4.4.4 Linear Quadrilateral Element (Q4)

There are four nodes at the corners of the quadrilateral element (Figure 4.13). In the natural coordinate system (ξ, η), the four shape functions are

$$\begin{aligned}
N_1 &= \frac{1}{4}(1-\xi)(1-\eta) \\
N_2 &= \frac{1}{4}(1+\xi)(1-\eta) \\
N_3 &= \frac{1}{4}(1+\xi)(1+\eta) \\
N_4 &= \frac{1}{4}(1-\xi)(1+\eta)
\end{aligned} \tag{4.35}$$

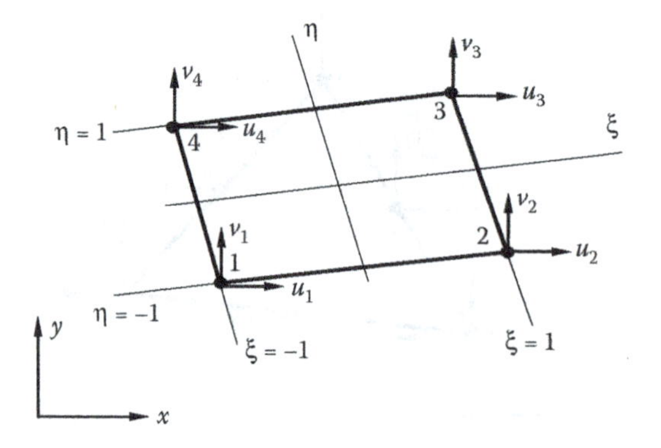

FIGURE 4.13
Linear quadrilateral element (Q4).

Note that $\sum_{i=1}^{4} N_i = 1$ at any point inside the element, as expected.

The displacement field is given by

$$u = \sum_{i=1}^{4} N_i u_i, \qquad v = \sum_{i=1}^{4} N_i v_i \tag{4.36}$$

which are bilinear functions over the element. The stress and strain fields are constant on this type of elements.

4.4.5 Quadratic Quadrilateral Element (Q8)

This is the most widely used element for 2-D problems due to its high accuracy in analysis and flexibility in modeling.

There are eight nodes for this element (Figure 4.14), four corners nodes and four midside nodes. In the natural coordinate system (ξ, η), the eight shape functions are

$$\begin{aligned}
N_1 &= \frac{1}{4}(1-\xi)(\eta-1)(\xi+\eta+1)\\
N_2 &= \frac{1}{4}(1+\xi)(\eta-1)(\eta-\xi+1)\\
N_3 &= \frac{1}{4}(1+\xi)(1+\eta)(\xi+\eta-1)\\
N_4 &= \frac{1}{4}(\xi-1)(\eta+1)(\xi-\eta+1)\\
N_5 &= \frac{1}{2}(1-\eta)(1-\xi^2)\\
N_6 &= \frac{1}{2}(1+\xi)(1-\eta^2)\\
N_7 &= \frac{1}{2}(1+\eta)(1-\xi^2)\\
N_8 &= \frac{1}{2}(1-\xi)(1-\eta^2)
\end{aligned} \tag{4.37}$$

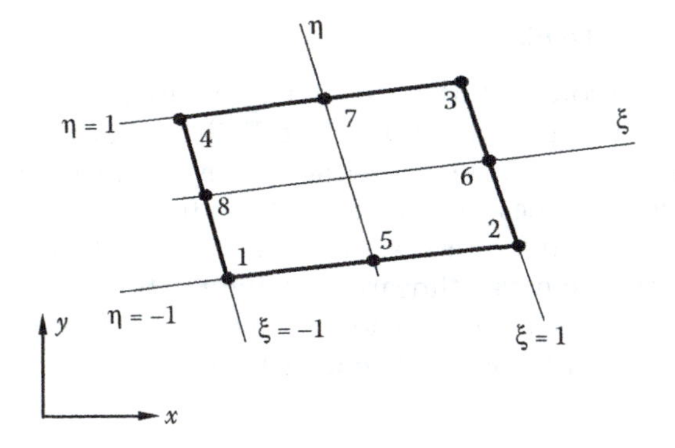

FIGURE 4.14
Quadratic quadrilateral element (Q8).

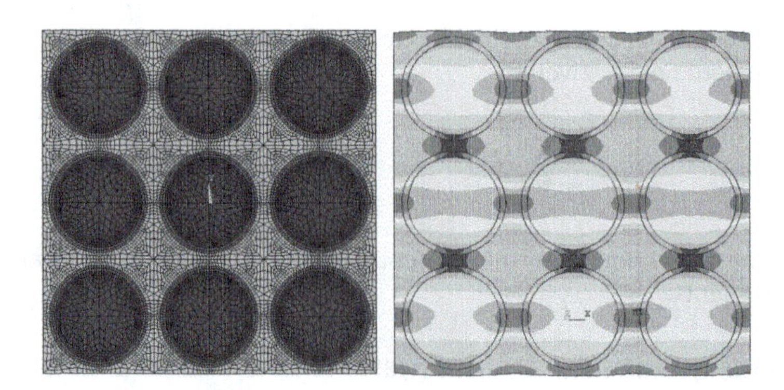

FIGURE 4.15
Analysis of composite materials (mesh and contour stress plots).

Again, we have $\sum_{i=1}^{8} N_i = 1$ at any point inside the element.

The displacement field is given by

$$u = \sum_{i=1}^{8} N_i u_i, \qquad v = \sum_{i=1}^{8} N_i v_i \tag{4.38}$$

which are quadratic functions over the element. Strains and stresses over a quadratic quadrilateral element are linear functions, which are better representations. A model of fiber-reinforced composite materials using the Q8 elements is shown in Figure 4.15.

We need to note the following when applying the 2-D elements:

- Q4 and T3 are usually used together in a mesh with linear elements.
- Q8 and T6 are usually applied in a mesh composed of quadratic elements.
- Linear elements are good for deformation analysis, that is, when global responses need to be determined.
- Quadratic elements are preferred for stress analysis, because of their high accuracy and the flexibility in modeling complex geometry, such as curved boundaries.

4.4.6 Transformation of Loads

Concentrated load (point forces), surface traction (pressure loads), and body force (weight) are the main types of loads applied to a structure. Both traction and body forces need to be converted into nodal forces in the FE model, since they cannot be applied to the FE model directly. The conversions of these loads are based on the same idea (the equivalent-work concept), which we have used for the cases of bar and beam elements.

Suppose, for example, we have a linearly varying traction q on a Q4 element edge, as shown in the Figure 4.16. The traction is normal to the boundary. Using the local (tangential) coordinate s, we can write the work done by the traction q as

$$W_q = \frac{1}{2} t \int_0^L u_n(s) q(s) ds$$

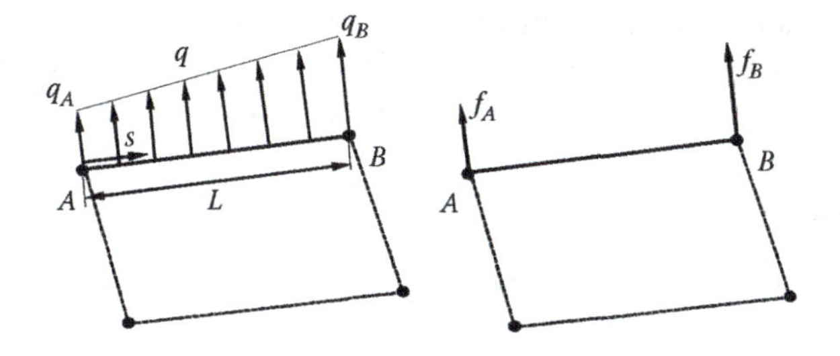

FIGURE 4.16
Traction applied on the edge of a Q4 element.

where t is the thickness, L the side length, and u_n the component of displacement normal to the edge AB.

For the Q4 element (linear displacement field), we have

$$u_n(s) = (1 - s/L)u_{nA} + (s/L)u_{nB}$$

The traction $q(s)$, which is also linear, is given in a similar way

$$q(s) = (1 - s/L)q_A + (s/L)q_B$$

Thus, we have

$$\begin{aligned}
W_q &= \frac{1}{2}t\int_0^L \left(\begin{bmatrix} u_{nA} & u_{nB} \end{bmatrix} \begin{bmatrix} 1 - s/L \\ s/L \end{bmatrix} \right) \left(\begin{bmatrix} 1 - s/L & s/L \end{bmatrix} \begin{bmatrix} q_A \\ q_B \end{bmatrix} \right) ds \\
&= \frac{1}{2}\begin{bmatrix} u_{nA} & u_{nB} \end{bmatrix} t\int_0^L \begin{bmatrix} (1 - s/L)^2 & (s/L)(1 - s/L) \\ (s/L)(1 - s/L) & (s/L)^2 \end{bmatrix} ds \begin{bmatrix} q_A \\ q_B \end{bmatrix} \\
&= \frac{1}{2}\begin{bmatrix} u_{nA} & u_{nB} \end{bmatrix} \frac{tL}{6}\begin{bmatrix} 2 & 1 \\ 1 & 2 \end{bmatrix} \begin{bmatrix} q_A \\ q_B \end{bmatrix} \\
&= \frac{1}{2}\begin{bmatrix} u_{nA} & u_{nB} \end{bmatrix} \begin{Bmatrix} f_A \\ f_B \end{Bmatrix}
\end{aligned}$$

and hence the equivalent nodal force vector is

$$\begin{Bmatrix} f_A \\ f_B \end{Bmatrix} = \frac{tL}{6}\begin{bmatrix} 2 & 1 \\ 1 & 2 \end{bmatrix} \begin{Bmatrix} q_A \\ q_B \end{Bmatrix}$$

Note, for constant q, we have

$$\begin{Bmatrix} f_A \\ f_B \end{Bmatrix} = \frac{qtL}{2}\begin{Bmatrix} 1 \\ 1 \end{Bmatrix}$$

For quadratic elements (either triangular or quadrilateral), the traction is converted into forces at three nodes along the edge, instead of two nodes. Traction tangent to the boundary, as well as body forces, are converted into nodal forces in a similar way.

4.4.7 Stress Calculation

The stress in an element is determined by the following relation:

$$\begin{Bmatrix} \sigma_x \\ \sigma_y \\ \tau_{xy} \end{Bmatrix} = \mathbf{E} \begin{Bmatrix} \varepsilon_x \\ \varepsilon_y \\ \gamma_{xy} \end{Bmatrix} = \mathbf{EBd} \tag{4.39}$$

where **B** is the strain-nodal displacement matrix and **d** is the nodal displacement vector, which is known for each element once the global FE equation has been solved.

Stresses can be evaluated at any point inside the element (such as the center) or at the nodes. Contour plots are usually used in FEA software packages (during postprocess) for users to visually inspect the stress results.

4.4.7.1 The von Mises Stress

The von Mises stress is the *effective* or *equivalent* stress for 2-D and 3-D stress analysis. For a ductile material, the stress level is considered to be safe, if

$$\sigma_e \leq \sigma_Y$$

where σ_e is the von Mises stress and σ_Y the yield stress of the material. This is a generalization of the 1-D (experimental) result to 2-D and 3-D situations.

The von Mises stress is defined by

$$\sigma_e = \frac{1}{\sqrt{2}} \sqrt{(\sigma_1 - \sigma_2)^2 + (\sigma_2 - \sigma_3)^2 + (\sigma_3 - \sigma_1)^2} \tag{4.40}$$

in which σ_1, σ_2 and σ_3 are the three principle stresses at the considered point in a structure.

For 2-D problems, the two principle stresses in the plane are determined by

$$\begin{aligned} \sigma_1^P &= \frac{\sigma_x + \sigma_y}{2} + \sqrt{\left(\frac{\sigma_x - \sigma_y}{2}\right)^2 + \tau_{xy}^2} \\ \sigma_2^P &= \frac{\sigma_x + \sigma_y}{2} - \sqrt{\left(\frac{\sigma_x - \sigma_y}{2}\right)^2 + \tau_{xy}^2} \end{aligned} \tag{4.41}$$

Thus, we can also express the von Mises stress in terms of the stress components in the xy coordinate system. For plane stress conditions, we have

$$\sigma_e = \sqrt{(\sigma_x + \sigma_y)^2 - 3(\sigma_x\sigma_y - \tau_{xy}^2)} \tag{4.42}$$

4.4.7.2 Averaged Stresses

Stresses are usually averaged at nodes in FEA software packages to provide more accurate stress values. This option should be turned off at nodes between two materials or other geometry discontinuity locations where stress discontinuity does exist.

EXAMPLE 4.2

A square plate with a hole at the center is under a tension load p in x direction as shown in the figure.

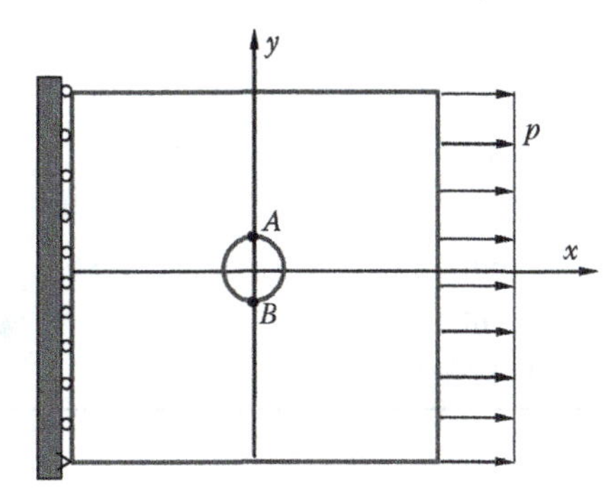

The dimension of the plate is 10 in. × 10 in., thickness is 0.1 in., and radius of the hole is 1 in. Assume $E = 10 \times 10^6$ psi, $\nu = 0.3$, and $p = 100$ psi. Find the maximum stress in the plate.

FE Analysis

This is a plane stress case. From the knowledge of stress concentrations, we should expect the maximum stresses occur at points A and B on the edge of the hole. Value of this stress should be around $3p$ (=300 psi), which is the exact solution for an infinitely large plate with a circular hole.

We use the *ANSYS* to do the modeling (meshing) and analysis, using quadratic triangular (T6), linear quadrilateral (Q4), and quadratic quadrilateral (Q8) elements. The FEM results by using the three different elements are compared and their accuracies and efficiencies are discussed. One mesh plot and one stress contour plot are shown below.

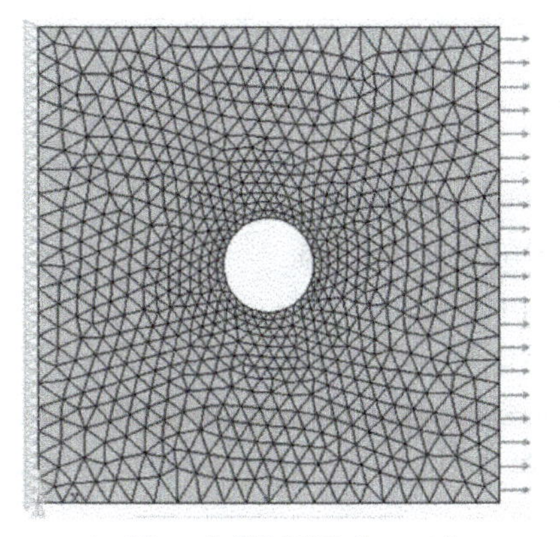

An FE mesh (T6, 1518 elements)

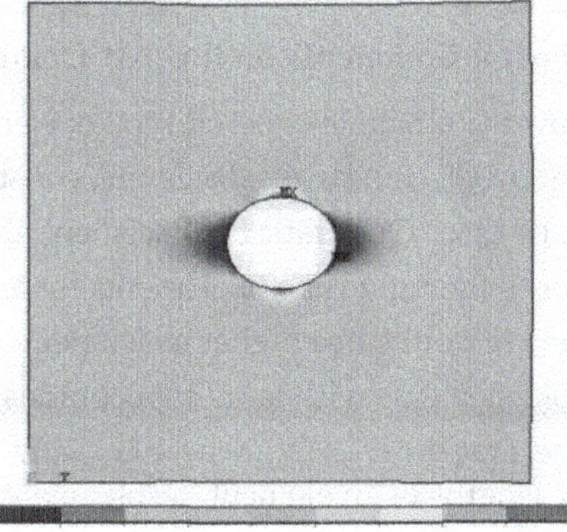

-8.006 64.925 137.856 210.787 283.719
28.46 101.391 174.322 247.253 320.184

FE stress plot and deformed shape
(T6, 1518 elements)

TABLE 4.1

FEA Stress Results

Elem. Type	No. of Elem.	Total DOFs	Max. σ (psi)
Q4	506	1102	312.42
Q4	3352	7014	322.64
Q4	31,349	64,106	322.38
...	...	...	...
T6	1518	6254	320.18
T6	2562	10,494	321.23
T6	24,516	100,702	322.24
...	...	...	...
Q8	501	3188	320.58
Q8	2167	13,376	321.70
Q8	14,333	88,636	322.24

The stress calculations with several meshes are listed in Table 4.1, along with the number of elements and DOFs used.

The converged results are obtained with all three types of elements with the differences in the maximum stress values less than 0.05%. However, Q8 and T6 elements are more efficient and converge much faster than the Q4 elements which are a linear representation and cannot model curved boundaries accurately. If the required accuracy is set at 1%, then the mesh with 501 Q8 elements should be sufficient. Note also that we need to check the deformed shape of the plate for each model to make sure the BCs are applied correctly. Fewer elements should be enough to achieve the same accuracy with a better or "smarter" mesh (mapped mesh). We will redo this example in the next chapter employing the symmetry features of the problem.

4.4.8 General Comments on the 2-D Elements

a. Know the attributes of each type of elements:

   T3 and Q4: Linear displacement, constant strain and stress

   T6 and Q8: Quadratic displacement, linear strain and stress

b. Choose the right type of elements for a given problem:

   When in doubt, use higher order elements (T6 or Q8) or a finer mesh.

c. Avoid elements with large aspect ratios and corner angles (Figure 4.17):

   Aspect ratio = L_{max}/L_{min} where L_{max} and L_{min} are the largest and smallest characteristic lengths of an element, respectively.

d. Make sure the elements are connected properly:

   Do not leave unintended gaps or free elements in FE models (Figure 4.18).

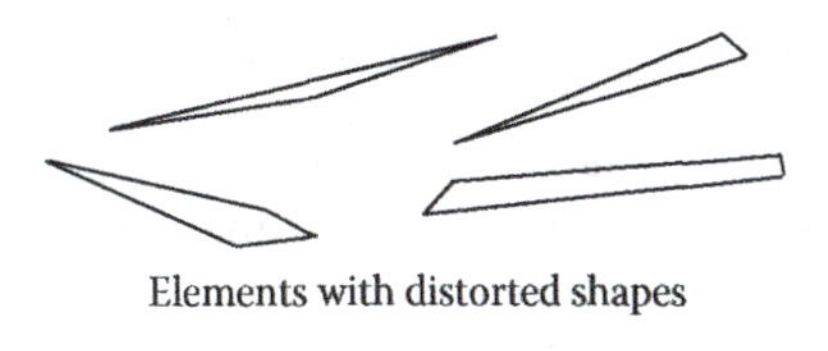

FIGURE 4.17
Elements with distorted (irregular) and normal (regular) shapes.

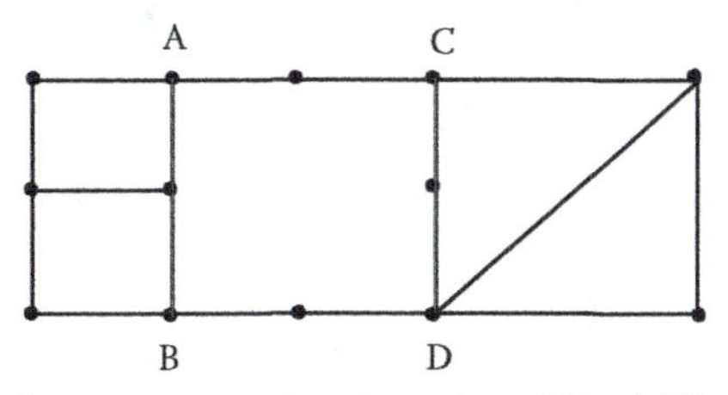

FIGURE 4.18
Unintended gaps in the FE mesh.

4.5 Case Study with *ANSYS Workbench*

Problem Description: A combination wrench is a convenient tool that is used to apply torque to loosen or tighten a fastener. The wrench shown below is made of stainless steel and has a thickness of 3 mm. Determine the maximum deformation and the distribution of von Mises stresses under the given distributed load and boundary conditions.

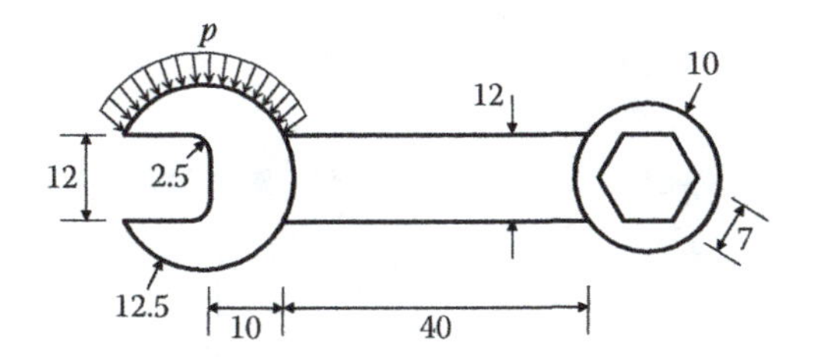

Material: Stainless steel
E = 193 GPa
ν = 0.27

Boundary conditions:
Hexagon on the right end is fixed on all sides.
Pressure p = 2 MPa
(All units are in millimeters.)

Solution

To solve the problem with *ANSYS Workbench*, we employ the following steps:

Step 1: Start an *ANSYS Workbench* Project
Launch *ANSYS Workbench* and save the blank project as "*Wrench.wbpj*."

Step 2: Create a *Static Structural (ANSYS)* Analysis System

Drag the *Static Structural (ANSYS)* icon from the *Analysis Systems Toolbox* window and drop it inside the highlighted green rectangle in the *Project Schematic* window to create a standalone static structural analysis system.

Step 3: Add a New Material

Double-click (or right-click and choose *Edit*) on the *Engineering Data* cell in the above *Project Schematic* to edit or add a material. In the following *Engineering Data* interface which replaces the *Project Schematic*, click the empty box highlighted below and type *"Stainless Steel"* as name for the new material.

Select *"Stainless Steel"* from the *Outline* window, and double-click *Isotropic Elasticity* under *Linear Elastic* in the leftmost *Toolbox* window. Enter *"193E9"* for *Young's Modulus* and *"0.27"* for *Poisson's Ratio* in the bottom center

Properties window. Click the *Return to Project* button to go back to the *Project Schematic*.

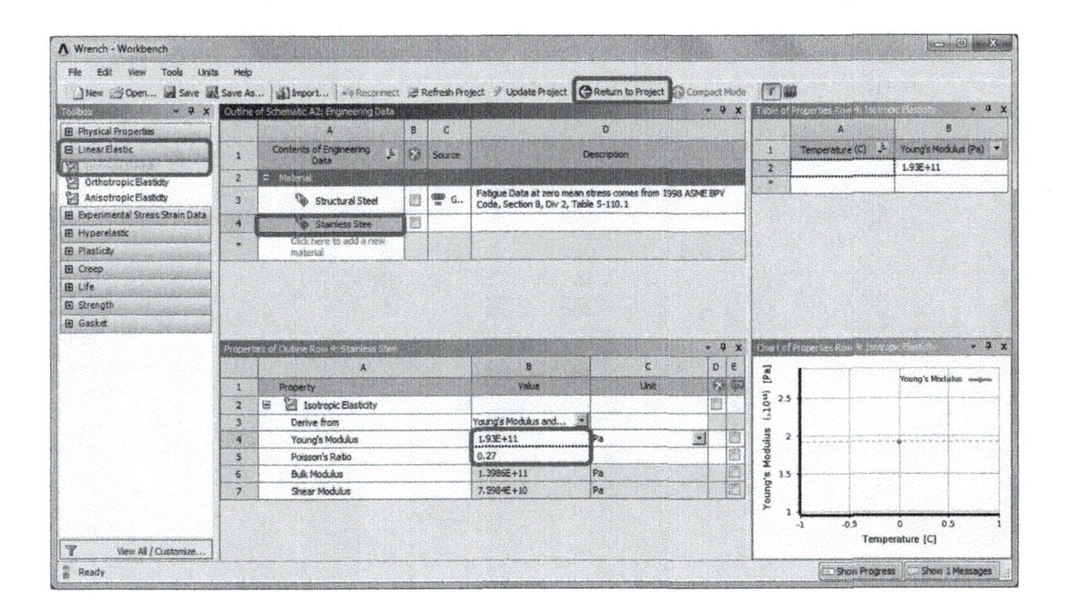

Step 4: Launch the *DesignModeler* Program

Ensure *Surface Bodies* is checked in the *Properties of Schematic A3: Geometry* window (select *Properties* from the *View* drop-down menu to enable display of this window). Select *2D* for *Analysis Type* in this *Properties* window. Double-click the *Geometry* cell to launch *DesignModeler*, and select *"Millimeter"* in the *Units* pop-up window.

Step 5: Create Surface Sketch

Click on the *Sketching* tab. Select the *Draw* toolbox and then *Rectangle.* Draw a rectangle in the *XY*-plane in the *Graphics* window by first clicking on the *Y*-axis and then dragging anywhere to the right. This makes the left side of the rectangle coincide with the *Y*-axis. Select the *Dimensions* toolbox to specify one horizontal (H1) and two vertical dimensions (V2 and V3) as shown below.

Select the *Draw* toolbox and then *Circle.* Draw a circle by first clicking on the *X*-axis for the center of the circle and then at a corner of the rectangle for a point on the circle, as shown below. The letter *C* indicates the center has a coincident relation with *X*-axis, and the letter *P* indicates the cursor is on the corner point. Select *Dimensions* and specify a radius of *10 mm* for the circle.

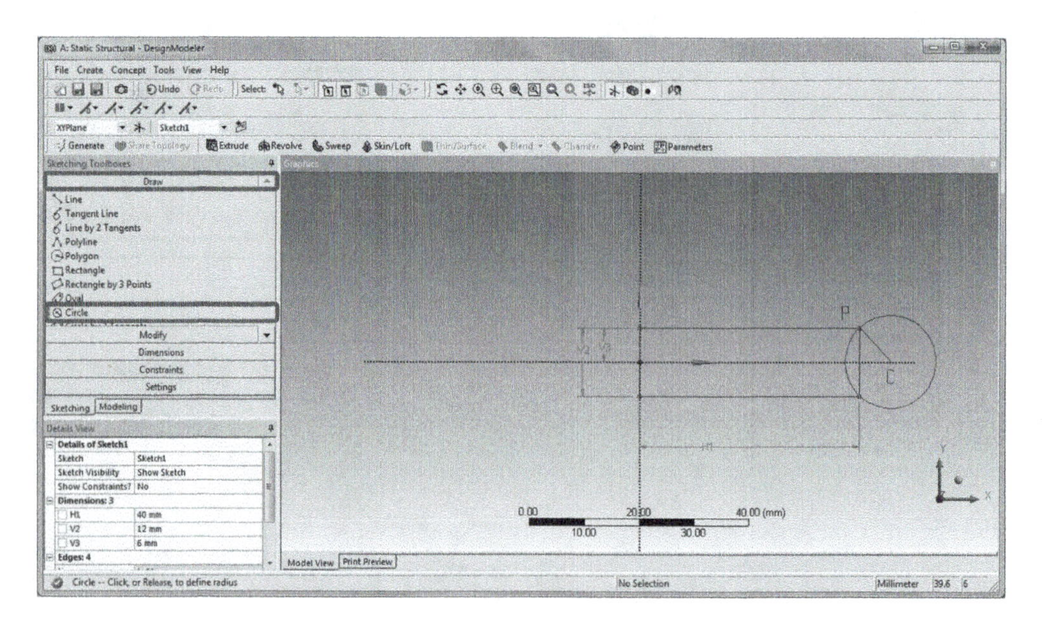

Draw another circle on the left by first clicking on the *X*-axis and then at the upper-left corner of the rectangle. Make sure letters *C* and *P* show up as you draw. Specify a radius dimension of *12.5 mm*.

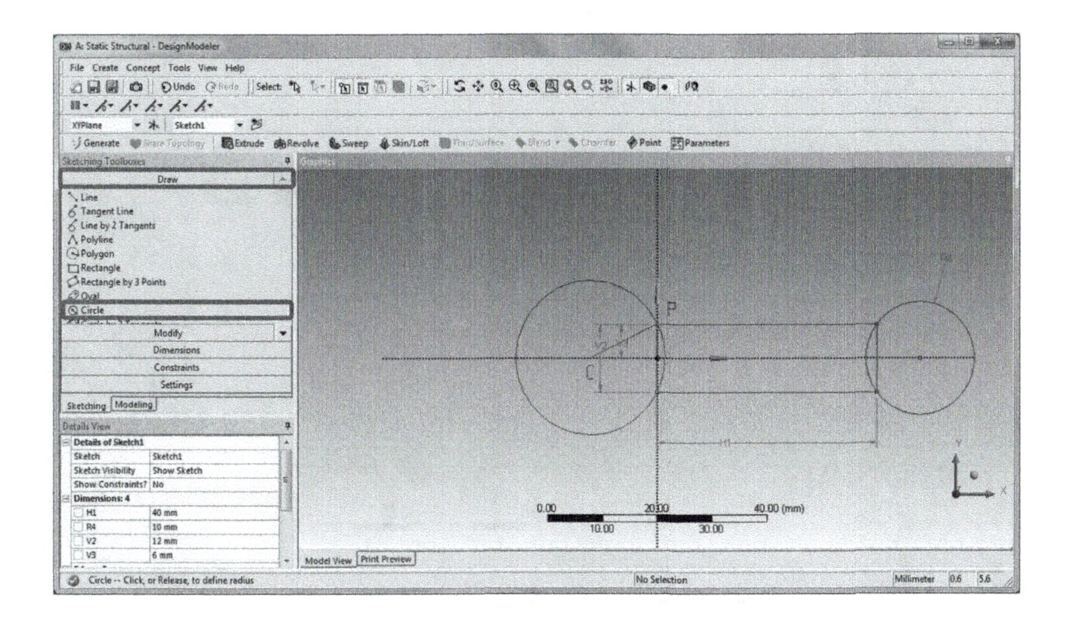

Select the *Draw* toolbox and then *Polygon*. Draw a six-sided polygon (set $n = 6$) in the *Graphics* window by first clicking at the center of the right circle. Then, drag the cursor anywhere inside the circle and click when there is a letter *H* showing above the top edge of the polygon. The letter *H* indicates the edge is a horizontal line. Select the *Dimensions* toolbox and specify the polygon side length as *7 mm*.

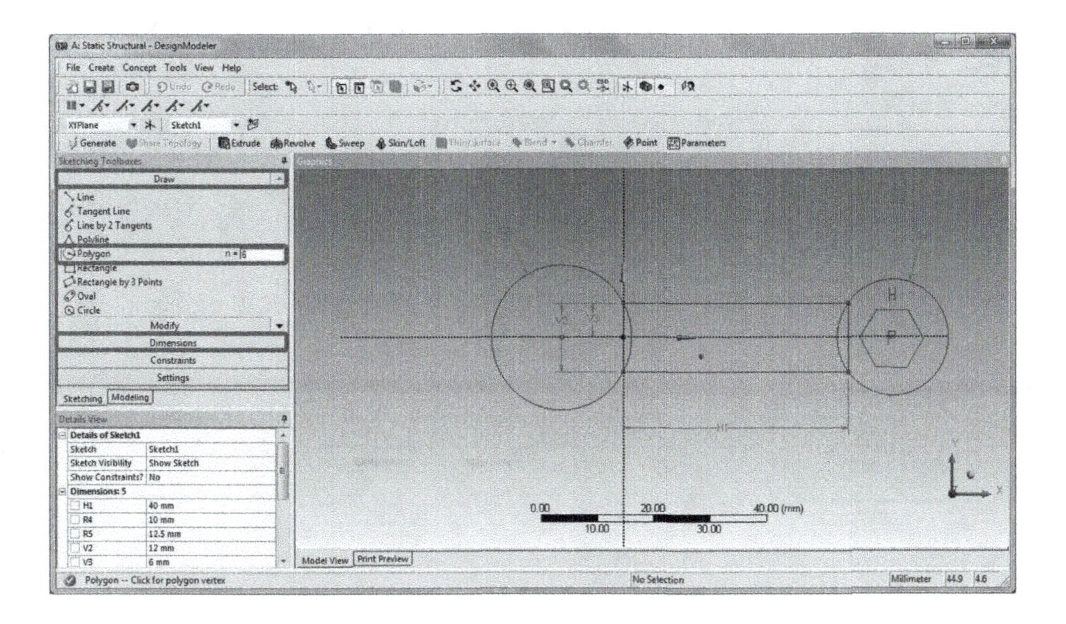

Select the *Modify* toolbox and then *Trim*. Click on unwanted line segments and arcs (a total of four lines and four arcs) in the *Graphics* window to trim them away. The trimmed sketch looks like the following.

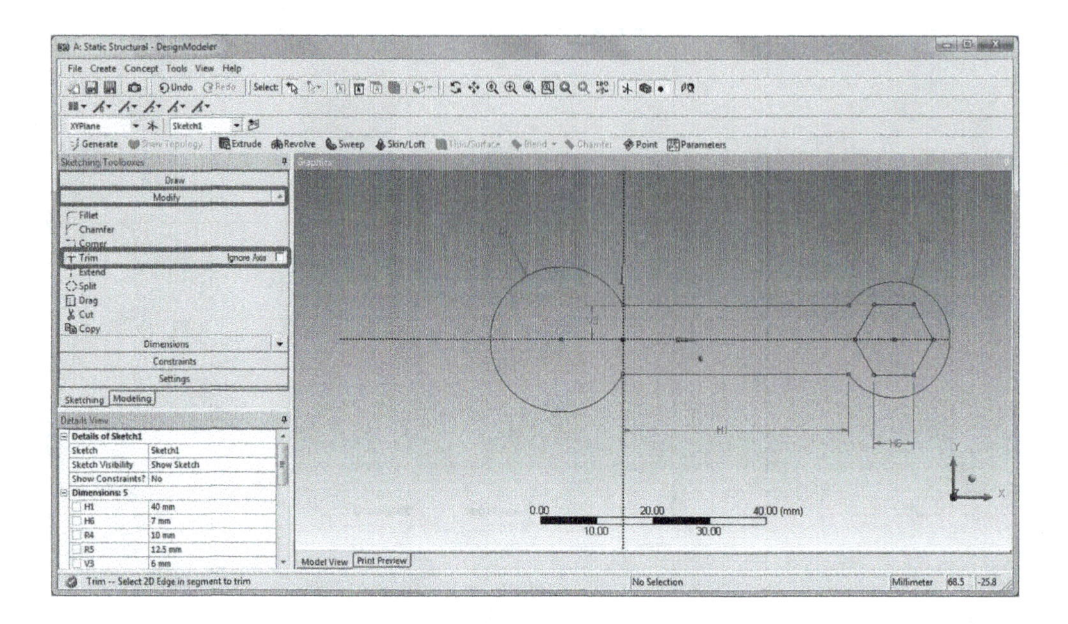

Select the *Draw* toolbox and then *Rectangle*. Draw a rectangle on the left as shown below. Select the *Dimensions* toolbox and specify one horizontal dimension (H7) and two vertical dimensions (V8 and V9) as given below.

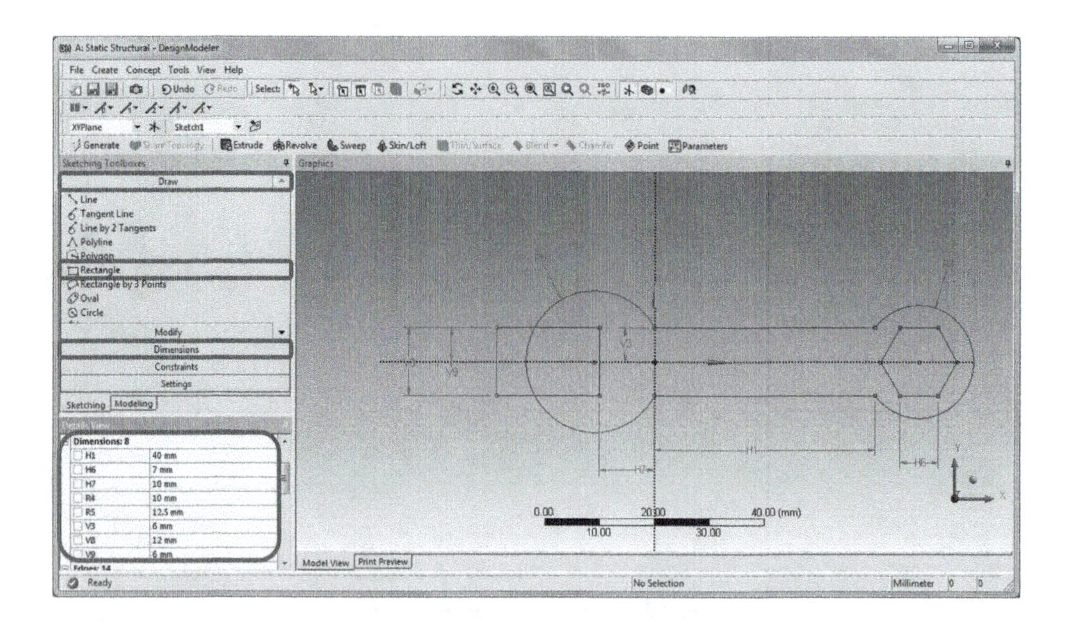

Select the *Modify* toolbox and then *Trim*. Trim away unwanted line segments and arcs (a total of two lines and two arcs) by clicking on them. The trimmed sketch looks like the following.

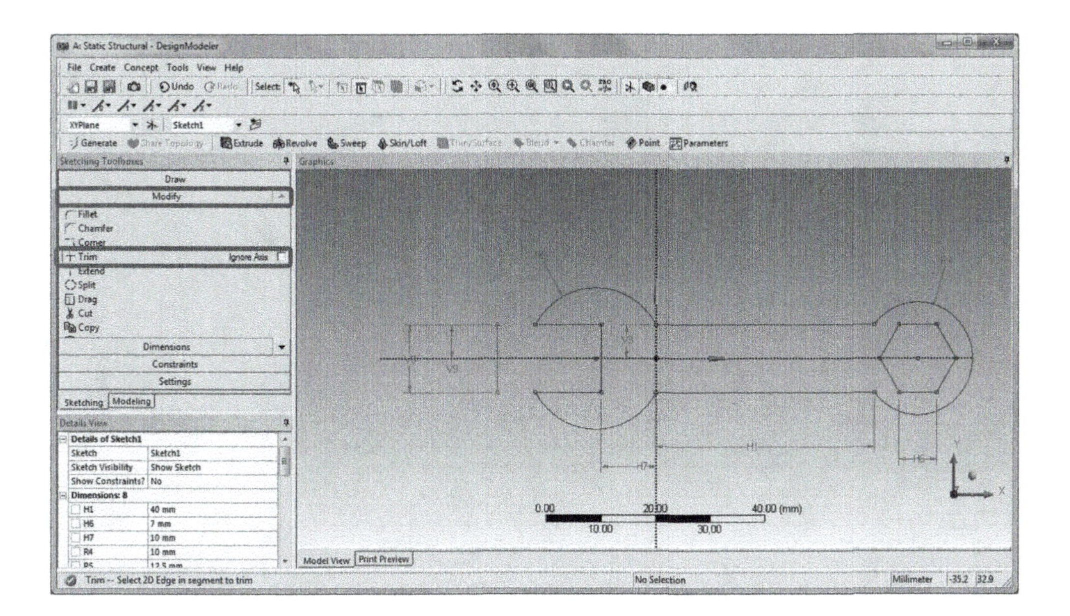

Select the *Modify* toolbox and then *Fillet*. Set the radius of fillet as *2.5 mm*. Click on two intersecting edges to define a fillet. Create two fillets as shown below at the corners of the rectangle.

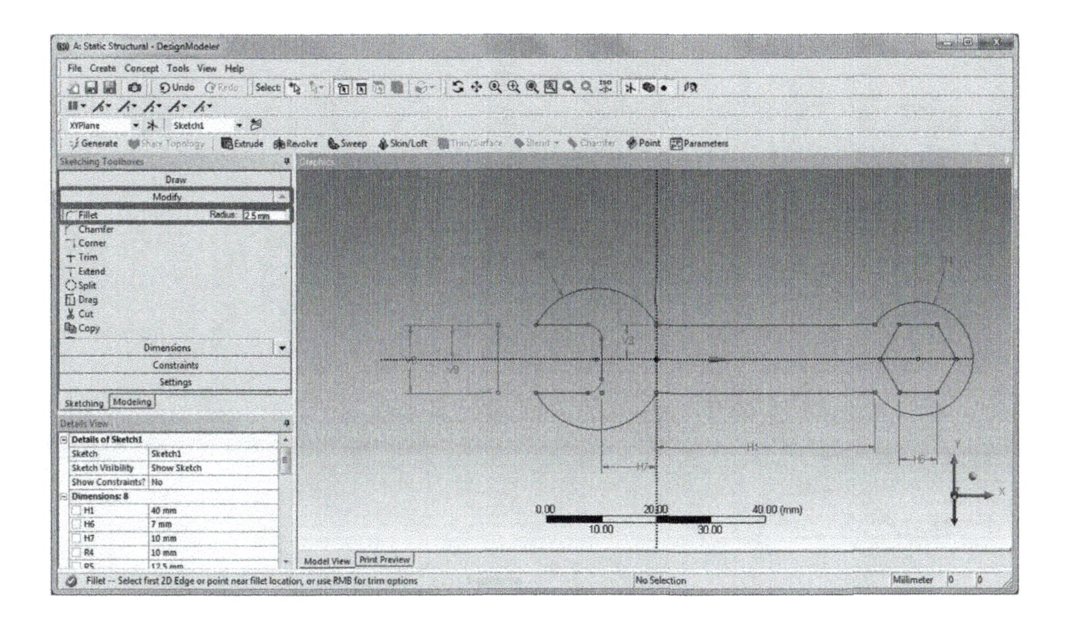

Step 6: Create Surface Body

Switch to the *Modeling* tab and select *Surfaces from Sketches* from the drop-down menu of *Concept.*

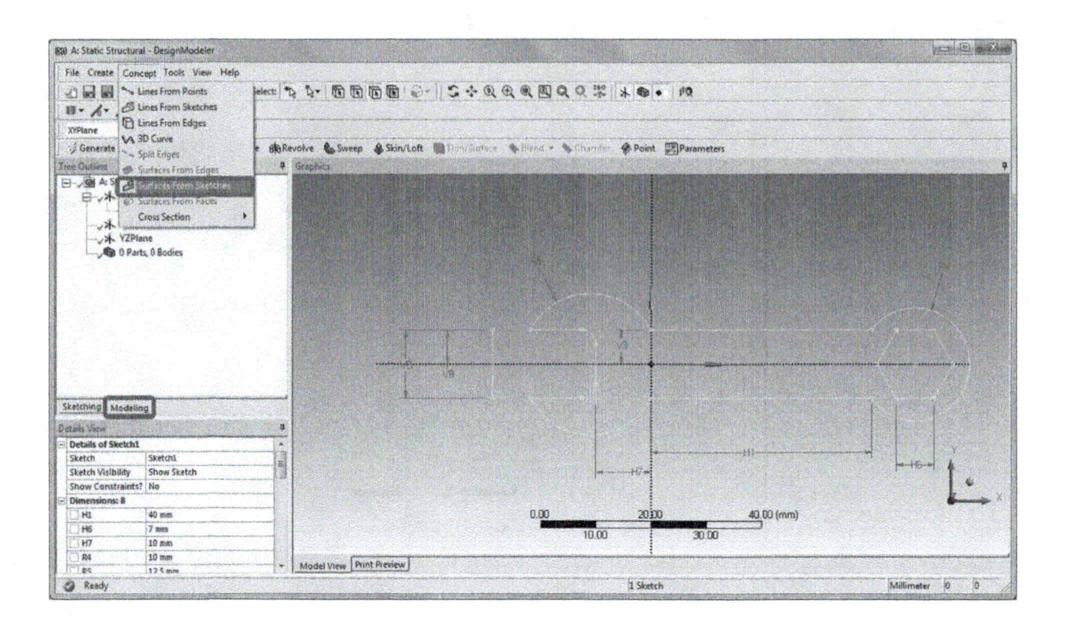

Select *Sketch1* from the *Tree Outline* and click *Apply* on the *Base Objects* selection in the *Details of SurfaceSK1.* Then click *Generate.*

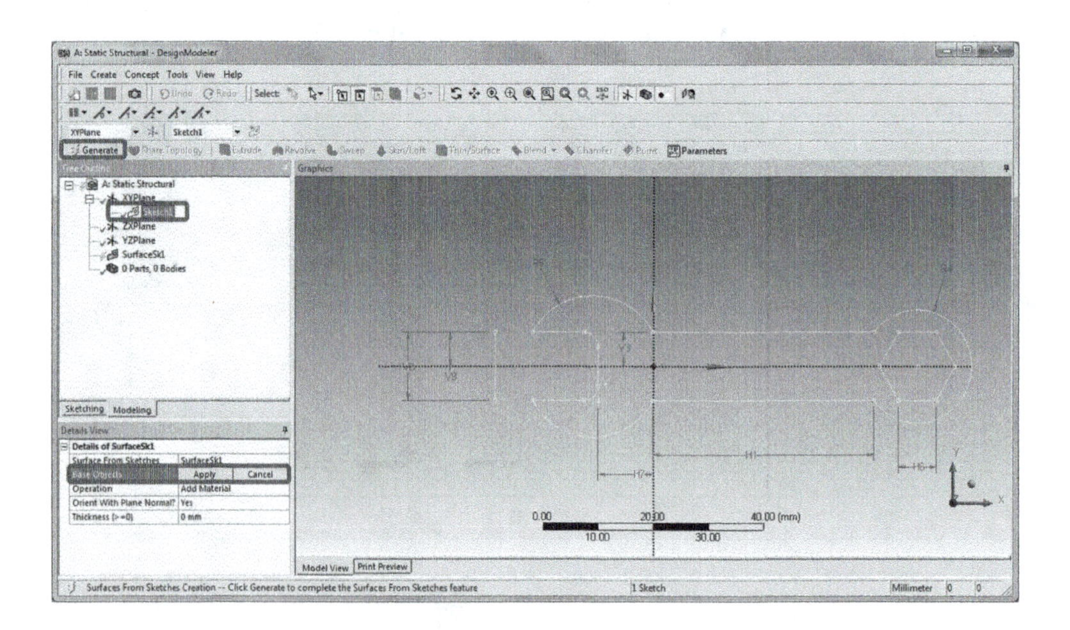

A *Surface Body* is now created from the surface sketch. Set the *Thickness* to *3 mm* in the *Details of Surface Body*. Exit the *DesignModeler.*

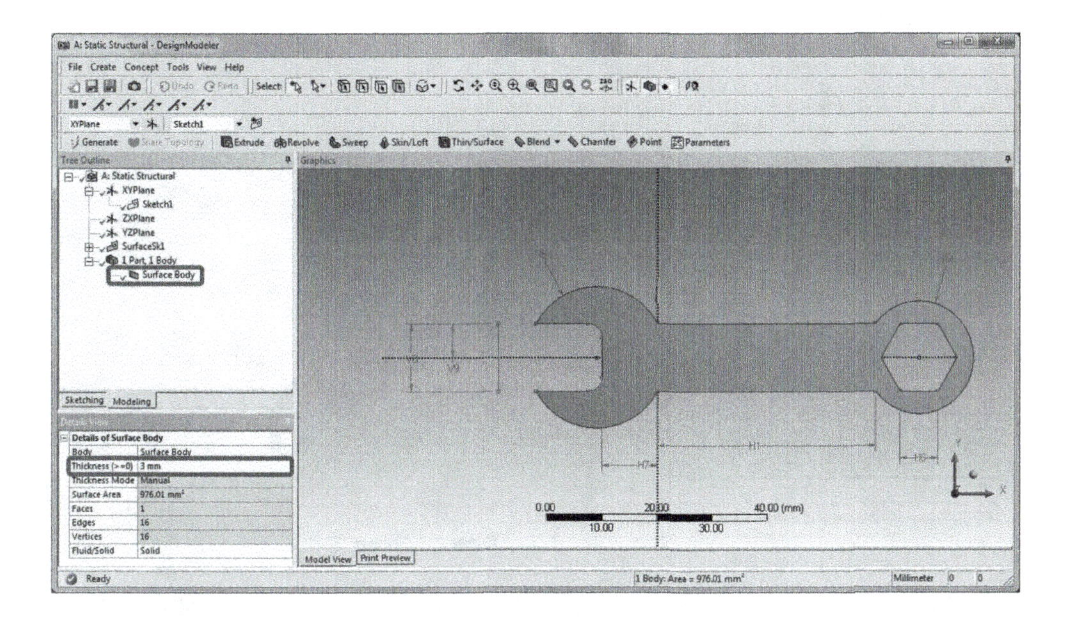

Step 7: Launch the *Static Structural* Program

Double-click on the *Model* cell to launch the *Static Structural* program. Click on *Geometry* in the *Outline.* In the *Details of "Geometry,"* the following options are available for *2D Behavior*: plane stress, axisymmetric, plane strain, generalized plane strain (assuming a finite length in the z direction, as opposed to the infinite value assumed for the standard plane stain option), and by body (allowing you to set 2D behavior options for individual bodies that appear under Geometry in the outline). Choose *Plan Stress* as the desired *2D Behavior.*

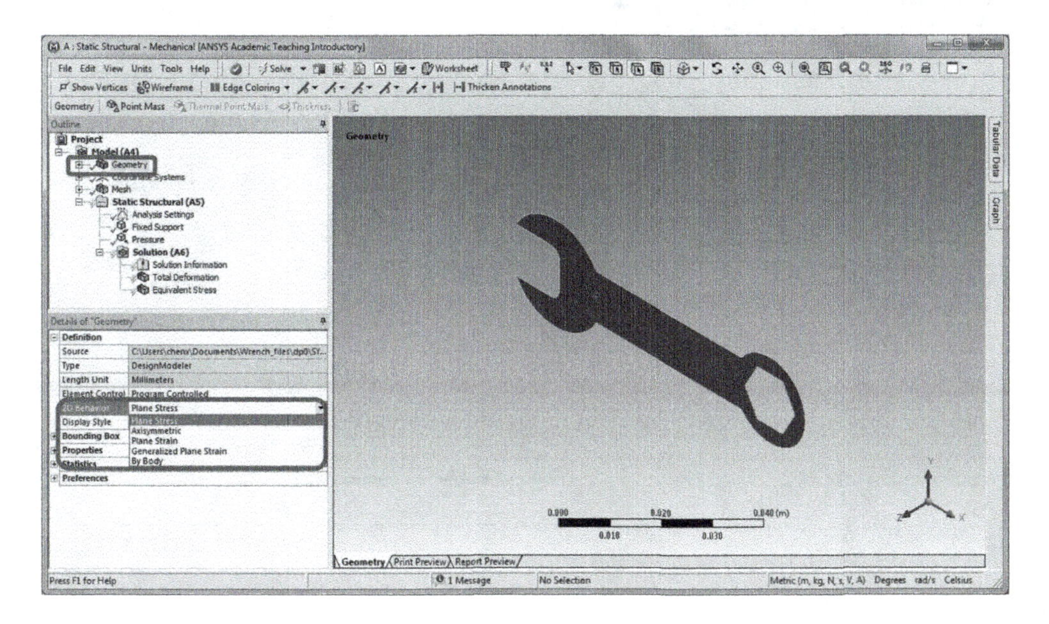

In the *Details of "Surface Body,"* click to the right of the *Material Assignment* field and select *Stainless Steel* from the drop-down context menu.

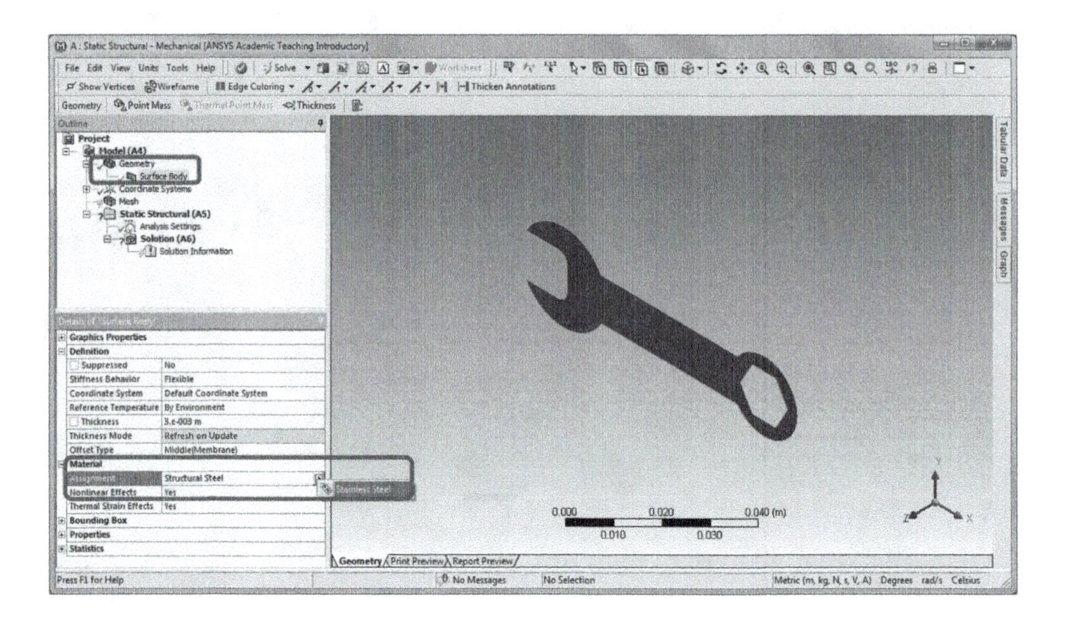

Step 8: Generate Mesh

Right click on *Mesh* in the *Project Outline* tree. Select *Insert* and then *Method* from the context menu. Click on the surface body in the *Graphics* window, and apply it to the *Geometry* selection in the *Details of "Automatic Method."*

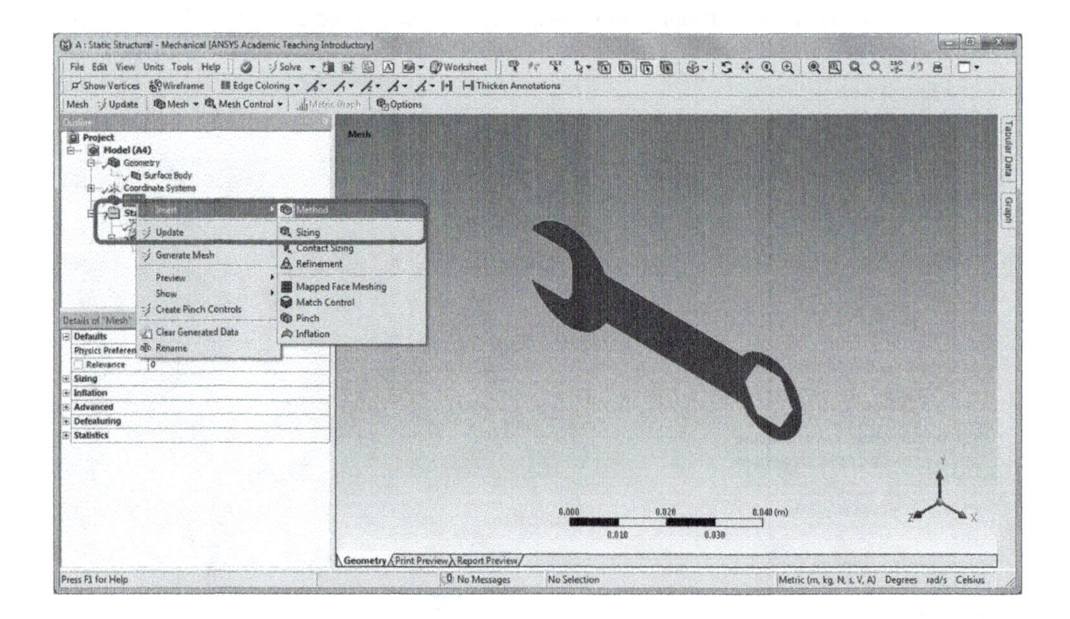

Right click on *Mesh* in the *Project Outline*. Select *Insert* and then *Sizing* from the context menu.

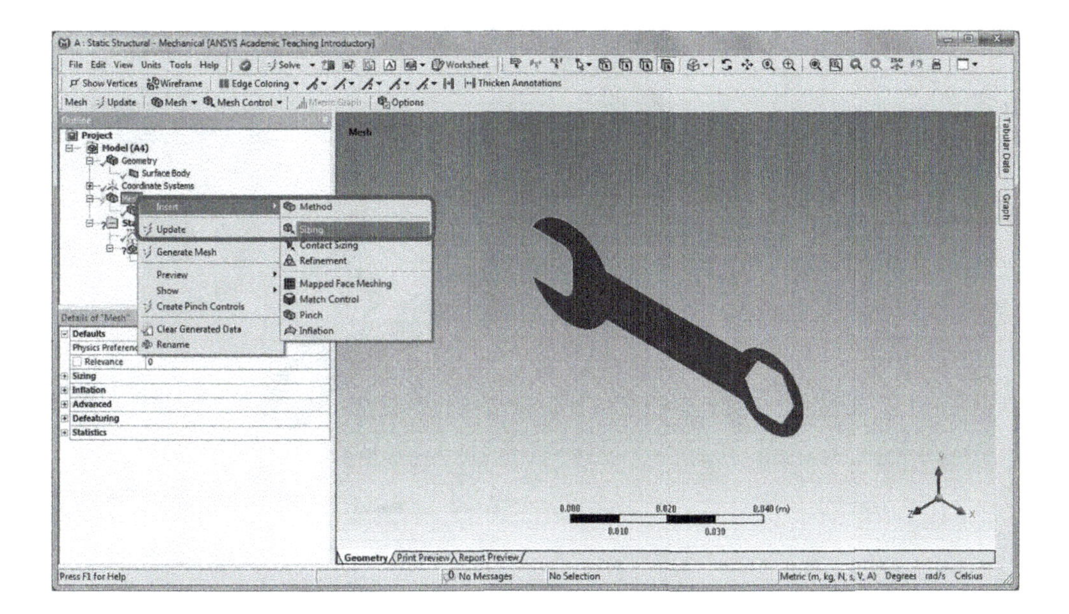

In the *Details of "Body Sizing,"* enter *"'1.5e-3 m"* for the *Element Size*. Click on the surface body in the *Graphics* window and apply it to the *Geometry* selection. In the *Outline* of *Project*, right-click on the *Mesh* and select *Generate Mesh*.

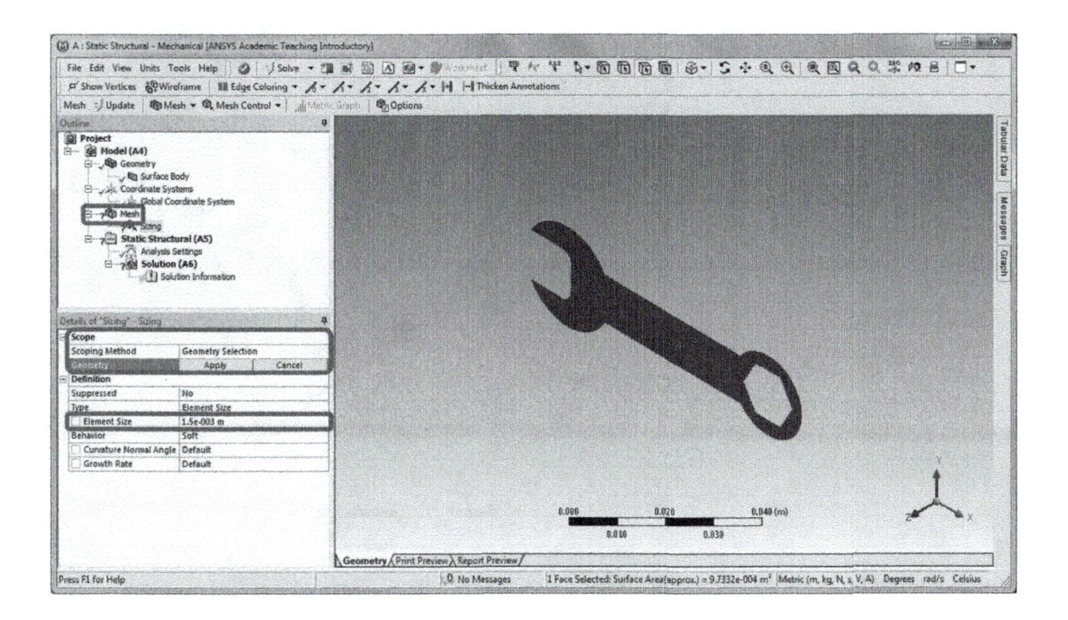

The following mesh plot will show up.

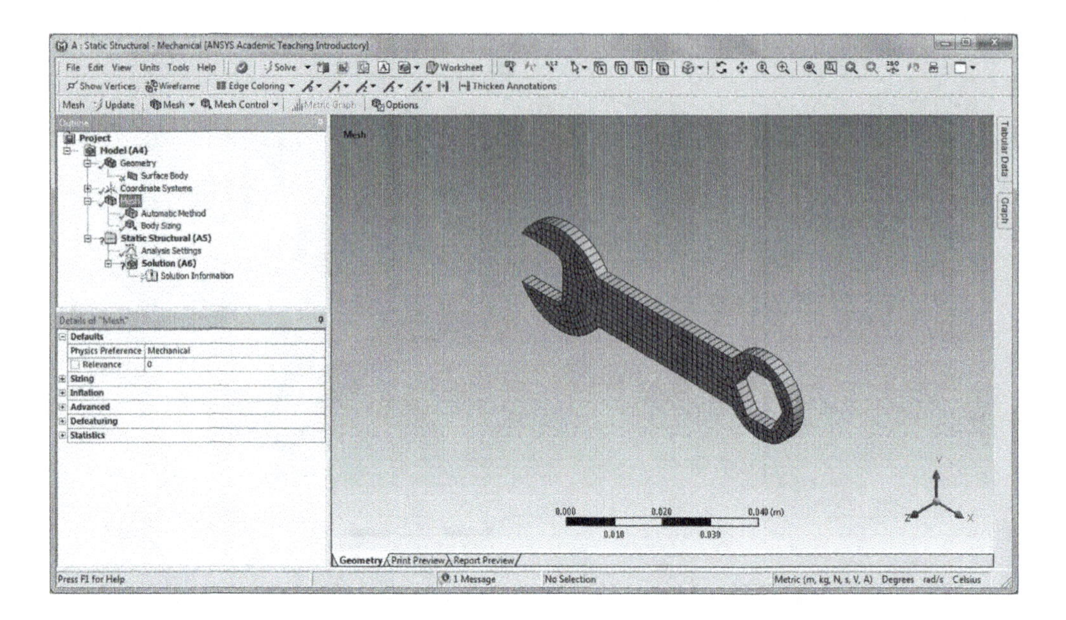

Step 9: Apply Boundary Conditions

Right-click on *Static Structural (A5)* and select *Insert* and then *Fixed Support* from the context menu.

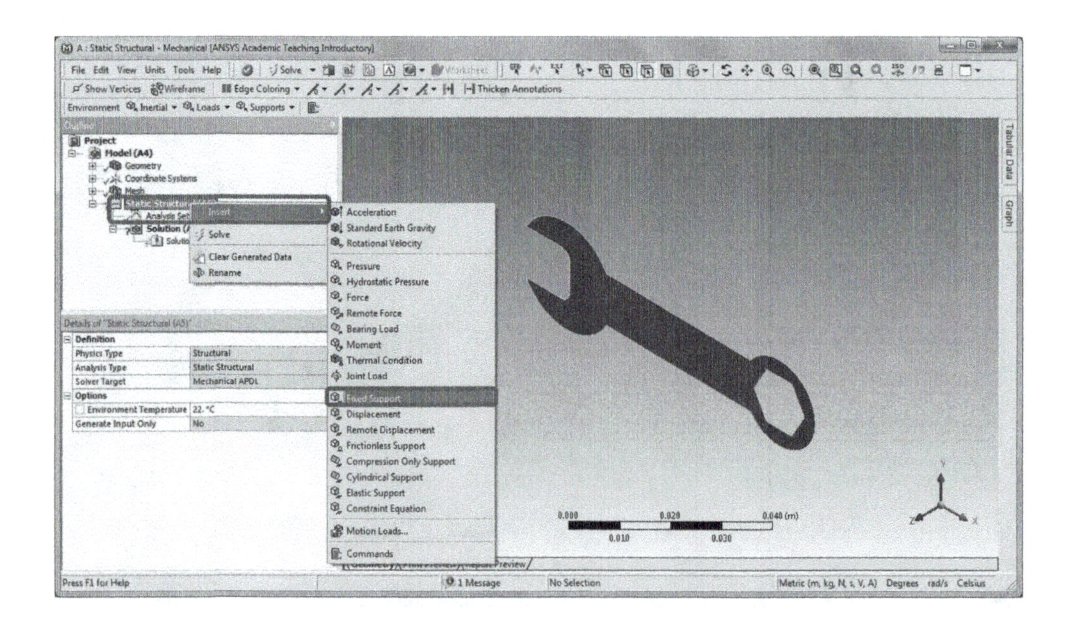

Ctrl-Click all six sides of the polygon as highlighted in the *Graphics* window. You must turn on the *Edge Selection filter* to select a line. Click *Apply* on the *Geometry* selection field in the *Details of "Fixed Support."*

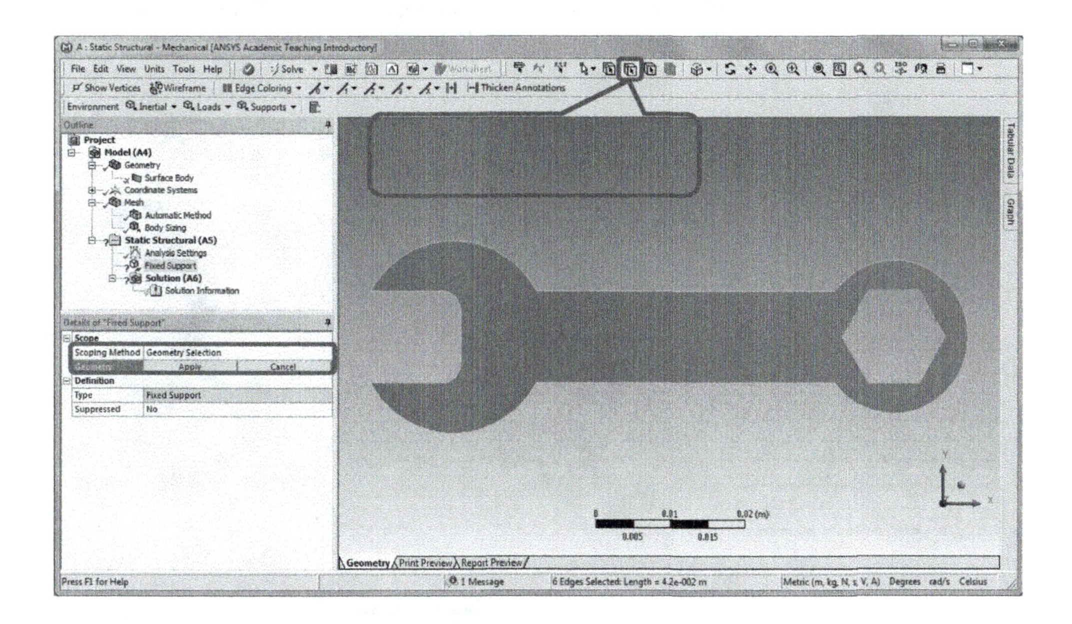

Step 10: Apply Loads

In the *Outline*, right-click on *Static Structural (A5)*. Choose *Insert* and then *Pressure*.

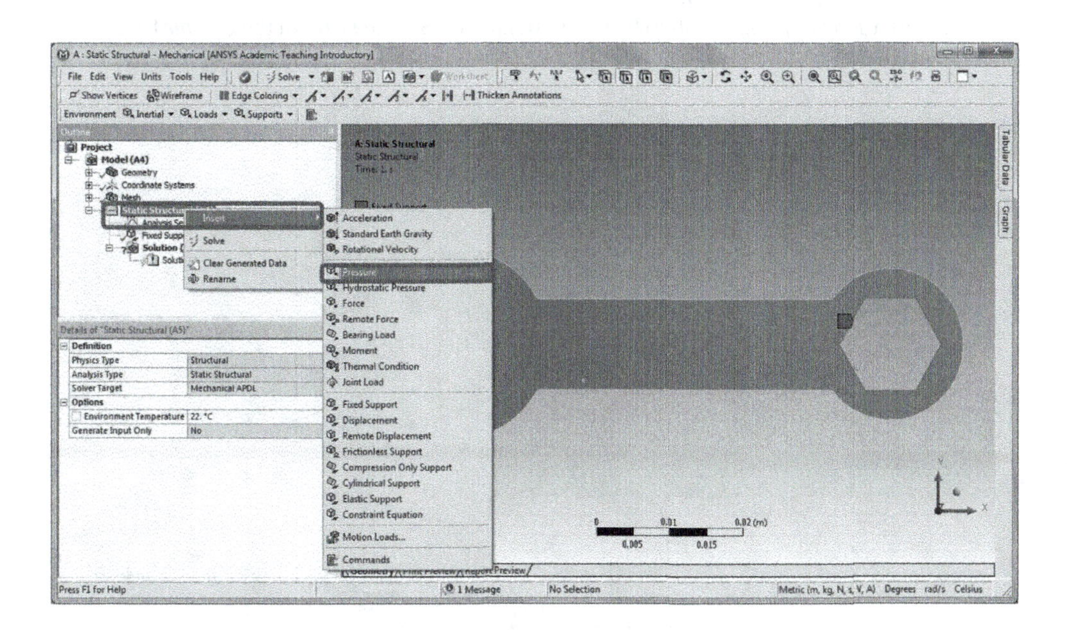

In the *Details of "Pressure"*, click on the arc as shown below and apply it to the *Geometry selection*. The *Edge Selection filter* must be turned on to allow this selection. Choose *Vector* for the *Define By* field, and enter "2e6" for *Magnitude*. Click on the vertical line as shown below and click *Apply* to confirm on the *Direction* selection. Make sure the vector is pointing downward. If not, click the left or right arrow at the bottom left corner to toggle direction.

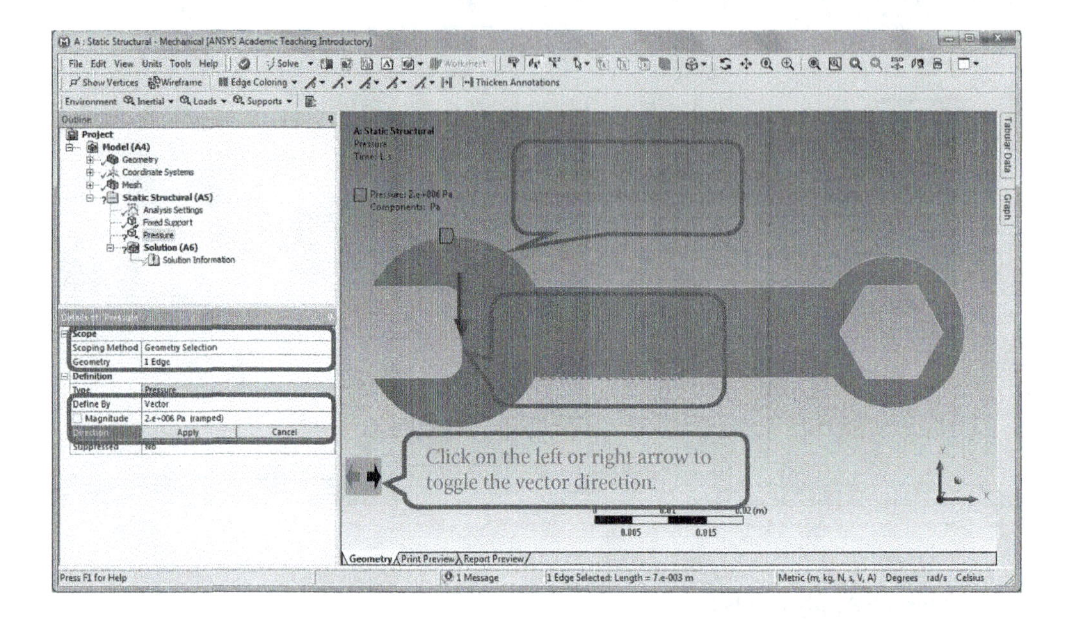

Step 11: Retrieve Solution

Insert a *Total Deformation* item by right-clicking on *Solution (A6)* in the *Outline* tree.

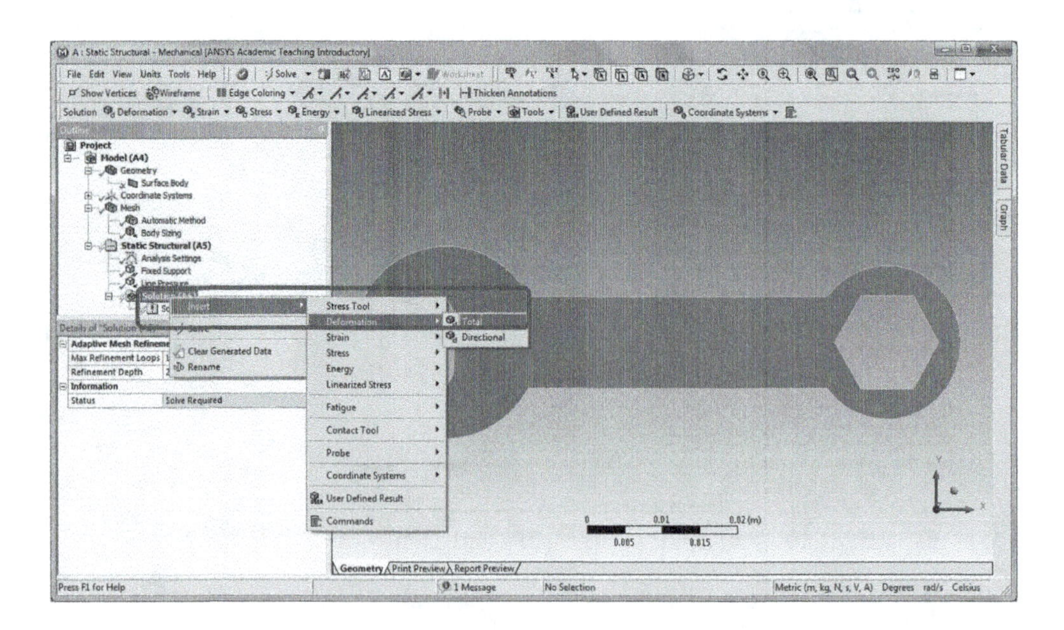

Insert an *Equivalent Stress* item by right-clicking on *Solution (A6)* in the *Outline* tree.

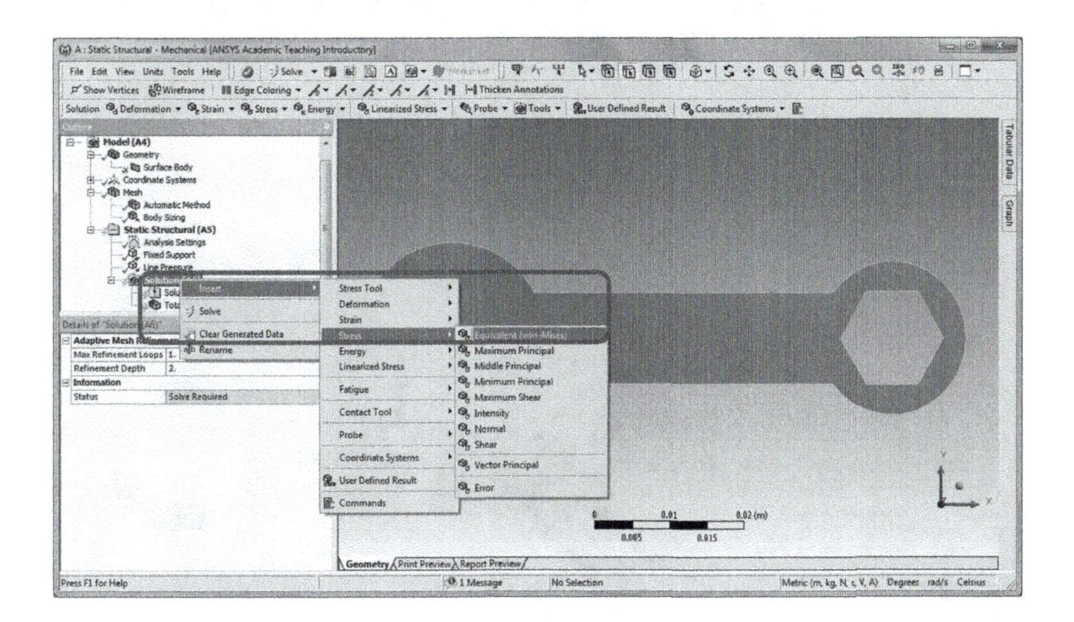

Right-click on *Solution (A6)* in the *Outline* tree and select *Solve*. The program will start to solve the model. After completion, click *Total Deformation* in the *Outline* to review the total deformation results. As shown below, the maximum deformation occurs at the two end tips on the left side of the wrench with a magnitude of *0.138 mm* under the assigned pressure load.

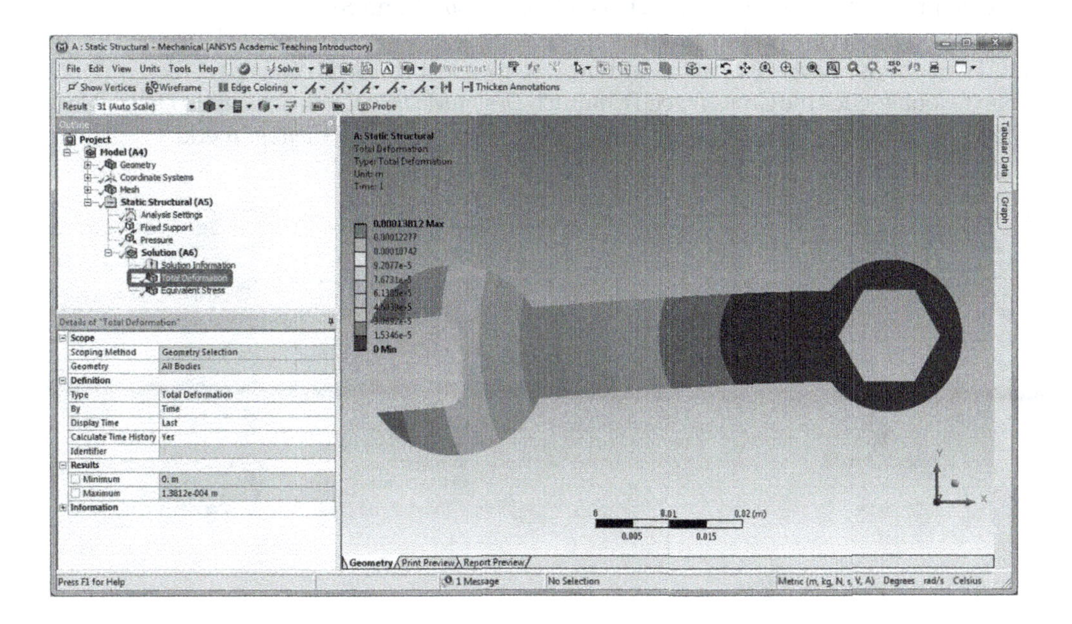

Select *Equivalent Stress* in the *Outline* to review the von Mises stress distribution. The following figure shows that the maximum von Mises stress is located at the two entrant corners on the right side of the wrench with a magnitude of *165.68 MPa,* which is below *207 MPa,* the yield strength of stainless steel. Decrease the element size in the *Details of "Body Sizing"* and solve again to see if there is any increase in the solution accuracy.

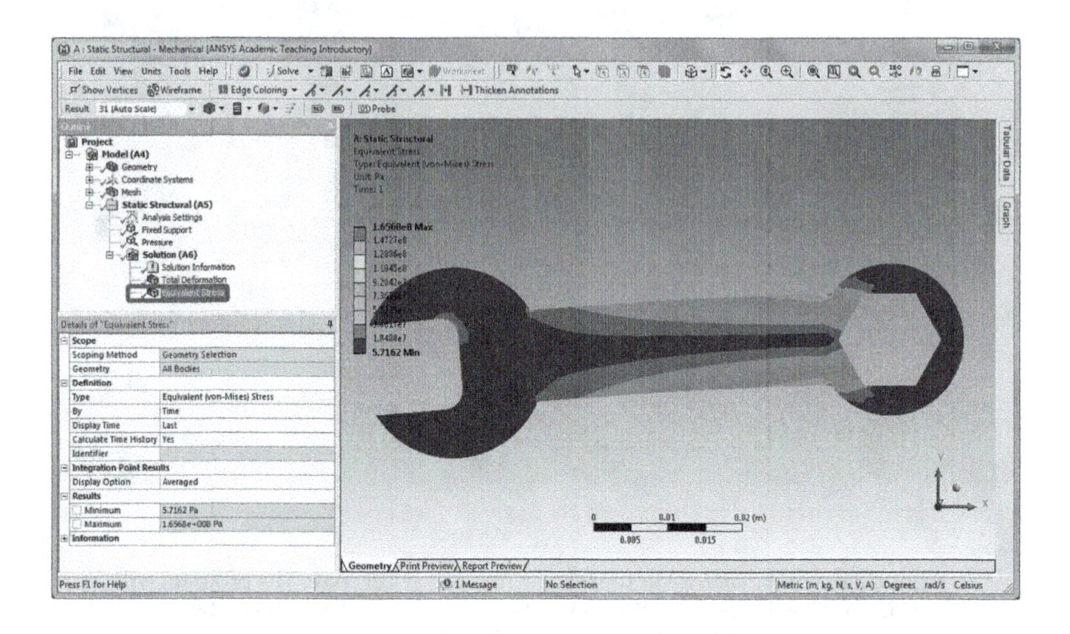

Modeling tips: *ANSYS Workbench* can work with native geometric models created in other CAD systems such as *SolidWorks, CATIA,* or *Pro/E.* It also allows geometry to be imported in neutral file formats such as *IGES, Parasolid,* and *SAT.*

To import a part created in *SolidWorks,* for example, you can right-click on the *Geometry* cell. Then choose *Import Geometry* and browse to select the *SolidWorks* part file.

Next, double-click the *Geometry* cell to launch *DesignModeler.* Click *Generate* in *DesignModeler* to load the geometry. Then, choose *Surfaces From Faces* from the *Concept* menu.

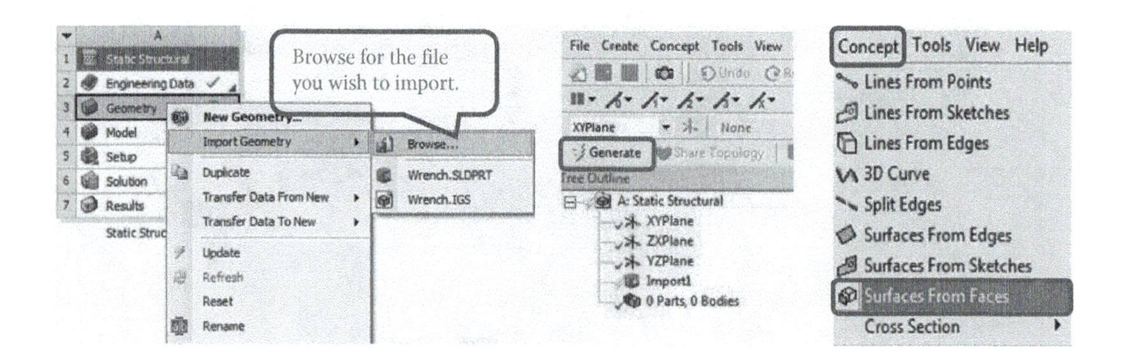

Click on the front wrench face and apply it to the *Faces* selection in the *Details of SurfFromFace1*. Click *Generate* to complete the surface creation.

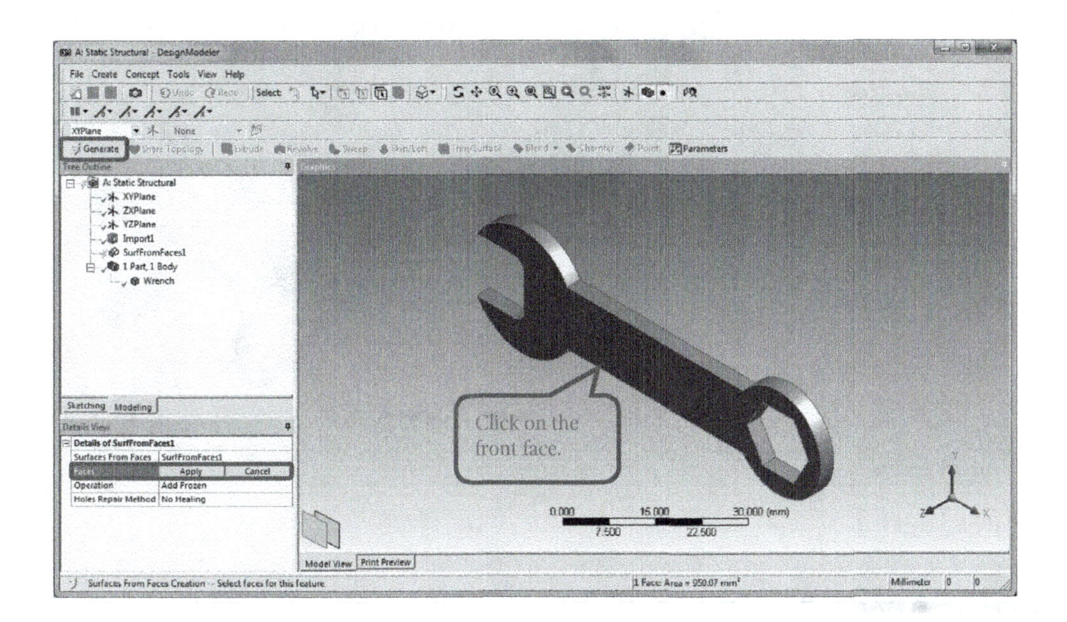

In the *Tree Outline,* a new *Wrench* item is now added underneath *2 Parts, 2 Bodies.* Right-click on the first *Wrench* item in the *Outline,* and choose *Suppress Body* in the context menu. This will hide the solid wrench and display only the new surface model. Click on the new *Wrench* item, and enter a *"3 mm"* for the *Thickness* in the *Details of Surface Body* and exit *DesignModeler.* Double-click on the *Model* cell in the *Project Schematic* to launch the *Static. Structural* program. A concept wrench surface is now ready for downstream analysis.

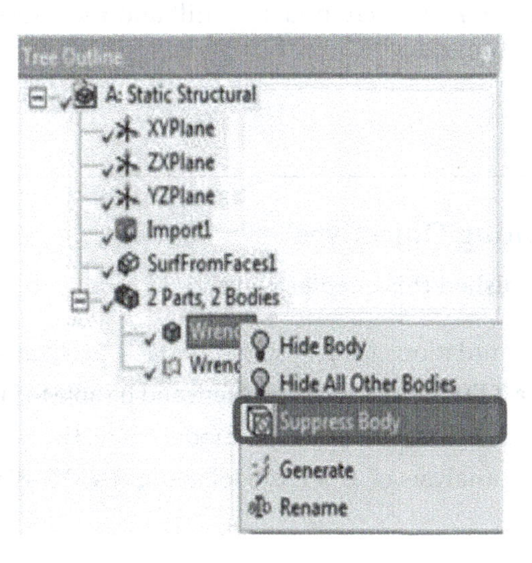

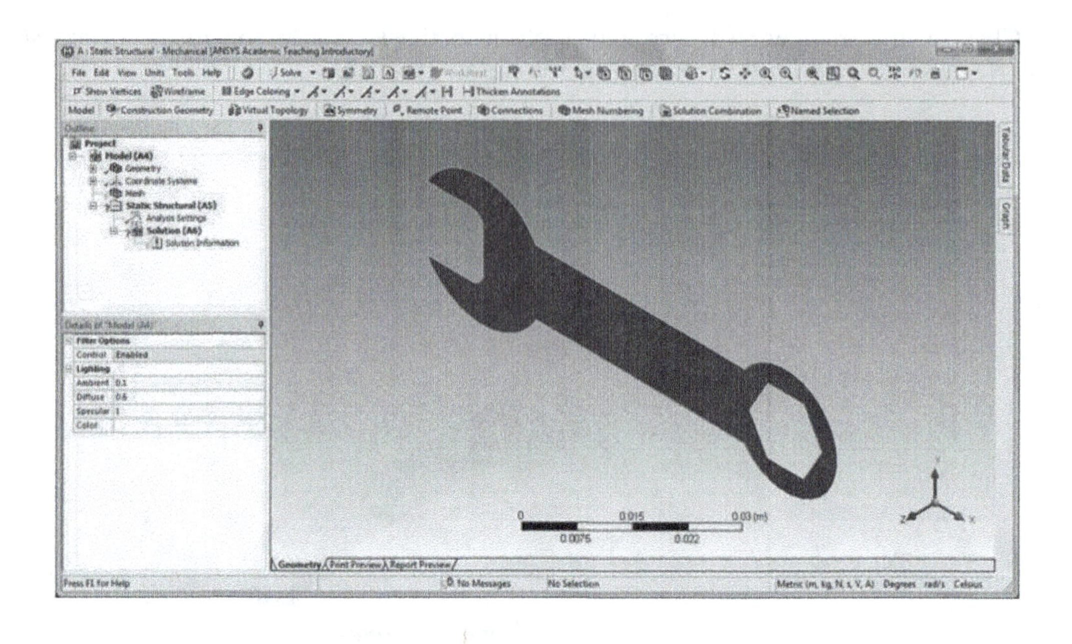

4.6 Summary

The 2-D elasticity equations are reviewed in this chapter and 2-D elements for analyzing plane stress and plane strain problems are discussed. FE formulations for 2-D stress analysis are introduced. It is emphasized that linear triangular (T3) and linear quadrilateral (Q4) elements are good for deformation analysis and not accurate for stress analysis. For stress analysis, quadratic triangular (T6) and quadratic quadrilateral (Q8) elements are recommended. Bad-shaped elements with large aspect ratios and large or small angles should be avoided in an FE mesh. A wrench model is built and the stress is analyzed in *ANSYS Workbench* to show how to conduct a 2-D FEA.

4.7 Review of Learning Objectives

Now that you have finished this chapter you should be able to

1. Develop the FE formulations for 2-D stress analysis problems.
2. Create a mesh for a 2-D stress analysis problem and be able to judge the mesh quality.
3. Apply the load and constraints to the FE model correctly.
4. Conduct 2-D stress analysis of an FE model using *ANSYS Workbench*.

PROBLEMS

4.1 List the boundary conditions in Example 4.1.

4.2 The plate shown below is constrained at the left end and loaded with a linearly varying pressure load at the right end. Constants E, ν, and thickness t are given.

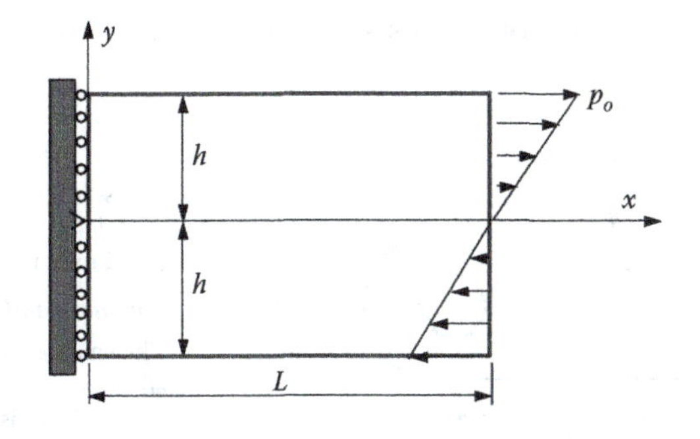

Suppose we have found the displacement field as follows:

$$u = \frac{p_0}{Eh}xy,$$

$$v = -\frac{p_0}{2Eh}(x^2 + \nu y^2)$$

Find:

a. strains in the plate;
b. stresses in the plate;
c. check if or not the equilibrium equations are satisfied by the stresses;
d. check if or not the boundary conditions are satisfied by the solution.

Optional:

Assume $E = 10 \times 10^6$ psi, $\nu = 0.3$, $p_o = 100$ psi, $L = 12$ in., $h = 4$ in., and thickness $t = 0.1$ in. Use *ANSYS Workbench* to check your results.

4.3 Derive the shape functions in Equation 4.17 for T3 elements and prove (Equation 4.18).

4.4 From Equation 4.27, prove $\det \mathbf{J} = x_{13}y_{23} - x_{23}y_{13} = 2A$ and discuss why "bad-shaped" elements can cause numerical errors in the FEM.

4.5 The torque arm shown below is a 5 mm thick automotive component made of structural steel with a Young's modulus of 200 GPa and a Poisson's ratio of 0.3. Using *ANSYS Workbench*, determine the deformation and von Mises stress distributions under the given load and boundary conditions.

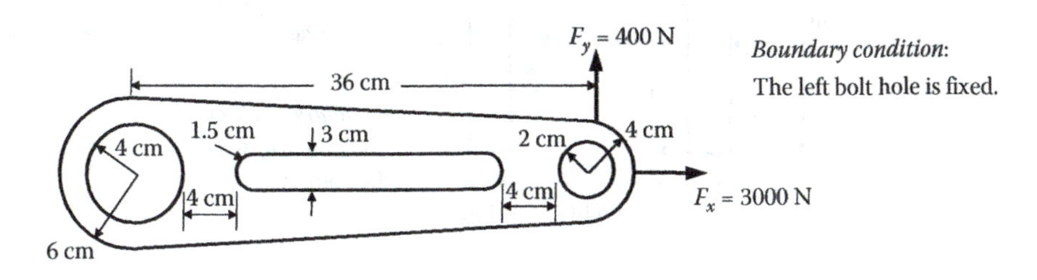

4.6 A bicycle wrench shown below is made of stainless steel with a Young's modulus of 193 GPa and a Poisson's ratio of 0.27. The wrench is 2 mm thick. Using *ANSYS Workbench,* determine the location and magnitude of the maximum deformation and the maximum von Mises stress under the given load and boundary conditions.

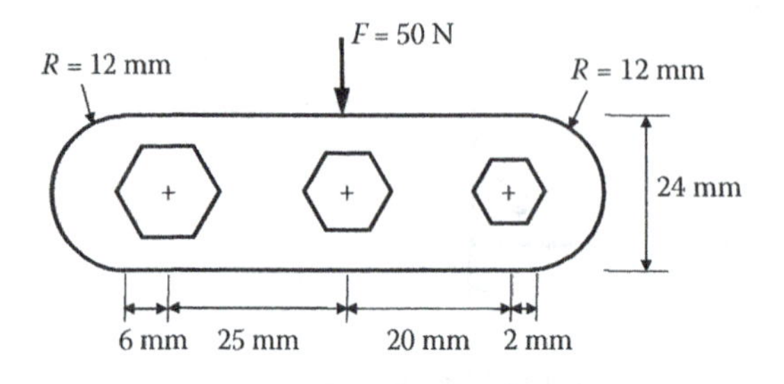

The side lengths of the hexagons from left to right are 9, 7, and 5 mm, respectively.

Boundary and load condition:
The smallest hexagon is fixed on all sides.
A 50 N force is applied on the top edge of the wrench.

4.7 Consider a straight and long hexagonal pipe under internal pressure as shown below. The pipe is made of stainless steel with a Young's modulus of 193 GPa and a Poisson's ratio of 0.27. Using *ANSYS Workbench,* determine the maximum in-plane deformation and the maximum von Mises stress under the given load and boundary conditions.

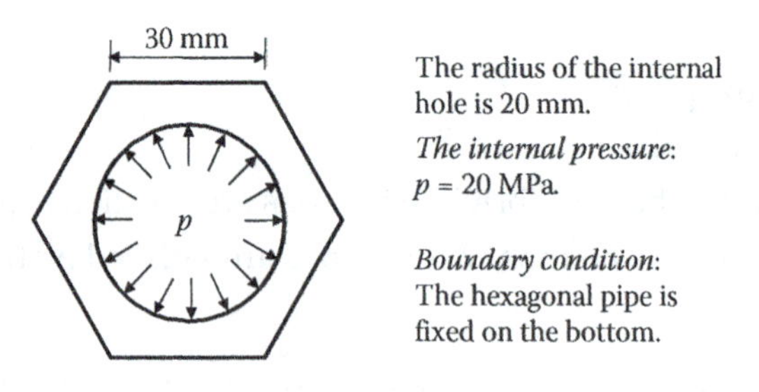

30 mm
p
The radius of the internal hole is 20 mm.
The internal pressure:
p = 20 MPa.
Boundary condition:
The hexagonal pipe is fixed on the bottom.

4.8 Consider a straight and deep tunnel under external pressure. The tunnel is made of concrete with a Young's modulus of 29 GPa and a Poisson's ratio of 0.15. Using *ANSYS Workbench,* determine the maximum in-plane deformation and the maximum in-plane von Mises stress under the given load and boundary conditions.

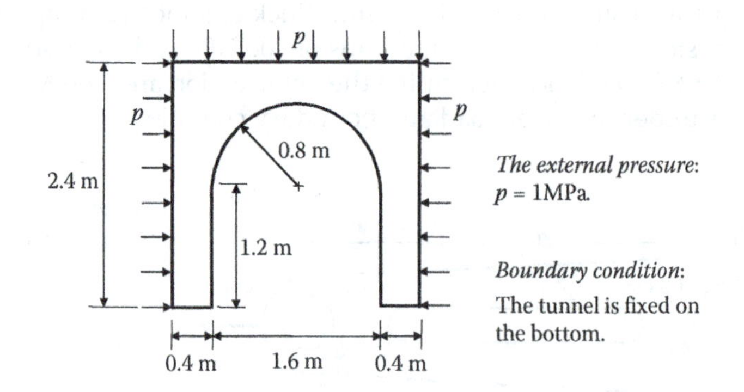

4.9 Using *ANSYS Workbench,* design a steel shelf bracket. Some dimensions of the bracket are fixed as shown in the figure. The three holes have the same diameter, which can be changed in increment of 0.2 cm in the design. The goal is using as less material as possible for the bracket, while supporting the given distributed load p.

a. For steel, use $E = 200$ GPa, $\nu = 0.3$, and yield stress $\sigma_Y = 250$ MPa.

b. Use a factor of safety = 2.5 for the design.

c. Report the configuration and dimensions of the bracket of your final design.

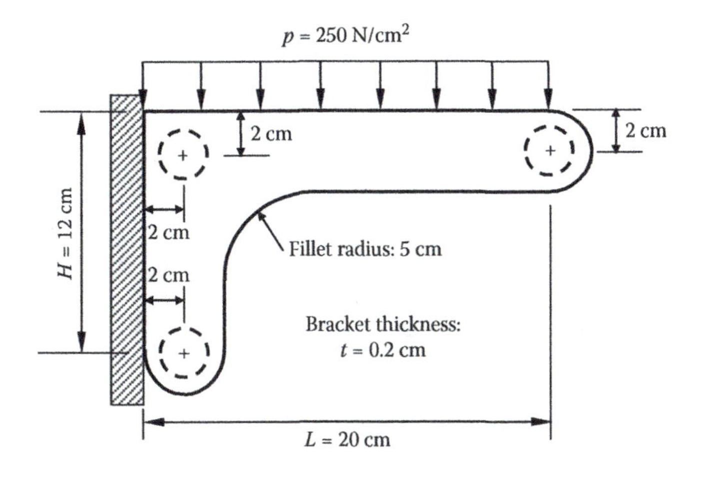

4.10 Similar to the previous problem, design a steel bracket. Some dimensions of the bracket are fixed as shown in the figure, while others can be changed. The shape and topology of the bracket can also be changed. The goal of this design is using least material for the bracket, while supporting the given loads.

a. For steel, use $E = 200$ GPa, $\nu = 0.32$, and yield stress $\sigma_Y = 250$ MPa.

b. Use a safety factor of 3 for the design.

c. Report the configuration and dimensions of the bracket of your final design.

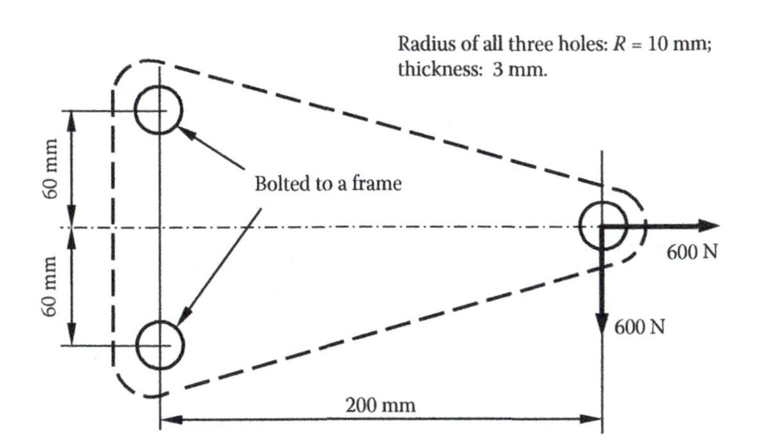

5

Modeling and Solution Techniques

5.1 Introduction

In this chapter, we discuss several techniques in the modeling and solution options of using the FEM. Applying these techniques can greatly improve the efficiencies and accuracies of the FEA.

5.2 Symmetry

Symmetry feature of a structure is the first thing one should look into and explore in the FE modeling and analysis. The model size can be cut almost in half and the solution efficiencies can be improved by several times. A structure possesses *symmetry* if its components are arranged in a periodic or reflective manner. Types of symmetries are (Figure 5.1):

- Reflective (mirror, bilateral) symmetry
- Axisymmetry
- Rotational (cyclic) symmetry
- Translational symmetry
- Others (or combinations of the above)

In the FEM, symmetry properties can be applied to

- Reduce the size of the problems and thus save CPU time, disk space, postprocessing efforts, and so on
- Simplify the modeling task
- Check the FEM results (make sure the results are symmetrical if the geometry and loading of the structure are symmetrical)

Symmetry properties of a structure should be fully exploited and retained in the FE model to ensure the efficiency and quality of FE solutions.

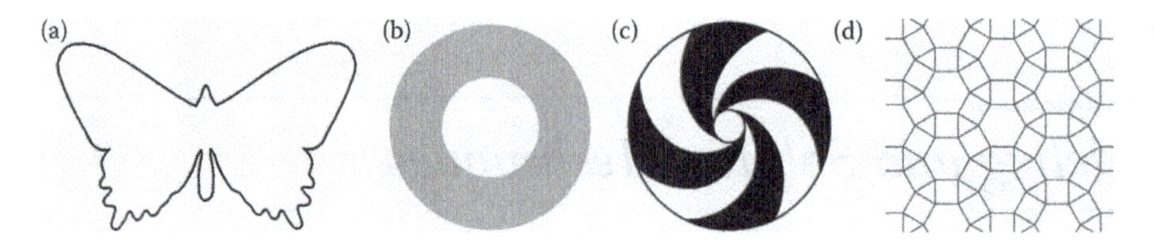

FIGURE 5.1
Some examples of symmetry: (a) reflective symmetry (From http://www.thinkingfountain.org/s/symmetry/butterflypattern.gif); (b) axisymmetry; (c) rotational symmetry (From http://csdt.rpi.edu/na/pnwb/symmetry2a.html); and (d) translational symmetry (From http://library.thinkquest.org/16661/background/symmetry.1.html).

5.2.1 An Example

For the problem of a plate with a center hole as discussed in Example 4.2 of the previous chapter, we redo the FE mesh using the symmetry features of the plate. To do this, we first model just one-quarter of the plate using *mapped mesh*, and then reflect the model (with the mesh) twice to obtain the model and mesh for the entire plate, as shown in Figure 5.2. Only 896 Q8 elements are used in this symmetrical model and the results are comparable to those in Chapter 4 using more elements with the *free mesh*. The quarter model can also be applied in the analysis, if the boundary conditions are also symmetrical about the *xz* and *yz* planes.

In vibration or buckling analysis, however, the symmetry concept should not be used in the FEA solutions (it is still applicable in the modeling stage), since symmetric structures often have antisymmetric vibration or buckling modes.

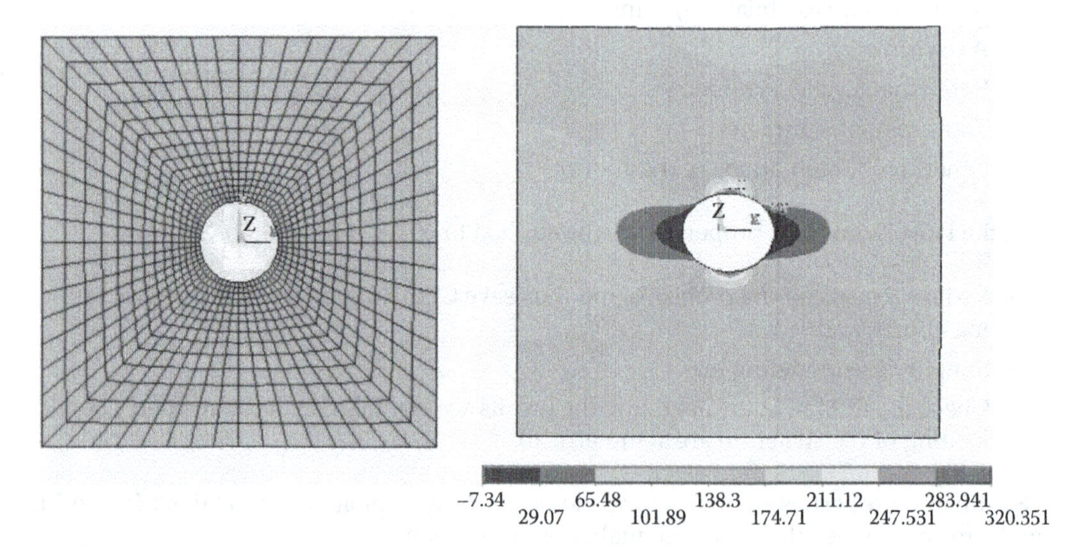

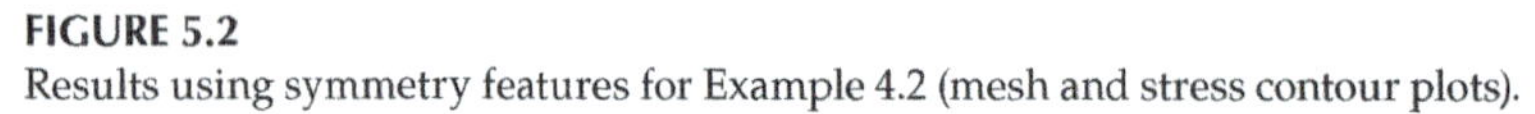

FIGURE 5.2
Results using symmetry features for Example 4.2 (mesh and stress contour plots).

5.3 Substructures (Superelements)

Another very useful technique for analyzing very large FEA models of mechanical systems is to apply the concept of substructures or superelements (SEs). Substructuring is a process of analyzing a large structure as a collection of (natural) components. The FEA models for these components are called *substructures* or *superelements* (SEs). The physical meaning of a substructure is simply a finite element model of a portion of the structure. Mathematically, it presents a boundary matrix which is condensed by eliminating the *interior* points and keeping only the *exterior* or boundary points of the portion of the structure. In other words, instead of solving the FEA system of equations once, one can use partitions of the matrix so that larger models can be solved on relatively smaller computers. More details of the theory and implementations of the substructures or SEs can be found in the documentation of the FEA software packages.

Figure 5.3 shows an FEA model of a truck used to conduct the full vehicle static or dynamic analysis. The entire model can have several millions of DOFs that can be beyond the capabilities of some computers. Using the substructuring technique, one can build the FEA model for each subsystem first (such as the cab, chassis, steering system, suspension system, payload, and so on) and then condense the FEA equations to smaller ones relating only to DOFs on the interfaces between the subsystems and residing on a residual structure (e.g., the chassis). The condensed system is much smaller than the original system and can be solved readily.

The *advantages* of using the substructuring technique are

- Good for large problems (which will otherwise exceed your computer capabilities)
- Less CPU time per run once the SEs have been processed (i.e., matrices have been condensed and saved)
- Components may be modeled by different groups
- Partial redesign requires only partial reanalysis (reduced cost)
- Efficient for problems with local nonlinearities (such as confined plastic deformations) which can be placed in one SE (residual structure)
- Exact for static deformation and stress analysis

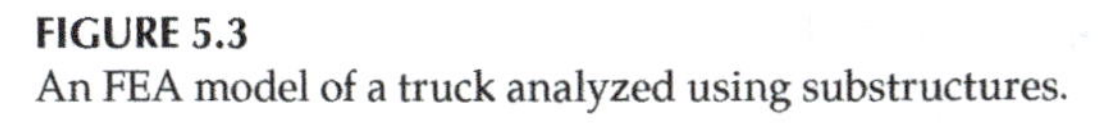

FIGURE 5.3
An FEA model of a truck analyzed using substructures.

The *disadvantages* of using the substructuring technique are

- Increased overhead for file management
- Increased initial time for setting up the system
- Matrix condensations for dynamic problems introduce new approximations

5.4 Equation Solving

There are two types of solvers used in the FEA for solving the linear systems of algebraic equations, mainly, the *direct* methods and *iterative* methods.

5.4.1 Direct Methods (Gauss Elimination)

- Solution time proportional to NB^2 (with N being the dimension of the matrix and B the bandwidth of the FEA systems)
- Suitable for small to medium problems (with DOFs in the 100,000 range), or slender structures (small bandwidth)
- Easy to handle multiple load cases

5.4.2 Iterative Methods

- Solution time is unknown beforehand
- Reduced storage requirement
- Suitable for large problems, or bulky structures (large bandwidth, converge faster)
- Need to solve the system again for different load cases

5.4.3 An Example: Gauss Elimination

Solve the following given system of equations:

$$\begin{bmatrix} 8 & -2 & 0 \\ -2 & 4 & -3 \\ 0 & -3 & 3 \end{bmatrix} \begin{Bmatrix} x_1 \\ x_2 \\ x_3 \end{Bmatrix} = \begin{Bmatrix} 2 \\ -1 \\ 3 \end{Bmatrix} \quad \text{or } \mathbf{Ax} = \mathbf{b} \tag{5.1}$$

Forward elimination:
Form

$$\begin{matrix} (1) \\ (2) \\ (3) \end{matrix} \left[\begin{array}{ccc|c} 8 & -2 & 0 & 2 \\ -2 & 4 & -3 & -1 \\ 0 & -3 & 3 & 3 \end{array}\right] \tag{5.2}$$

(1) + 4x(2) ⇒ (2):

$$\begin{matrix}(1)\\(2)\\(3)\end{matrix}\left[\begin{array}{ccc|c} 8 & -2 & 0 & 2 \\ 0 & 14 & -12 & -2 \\ 0 & -3 & 3 & 3 \end{array}\right] \tag{5.3}$$

$(2) + \frac{14}{3}(3) \Rightarrow (3)$

$$\begin{matrix}(1)\\(2)\\(3)\end{matrix}\left[\begin{array}{ccc|c} 8 & -2 & 0 & 2 \\ 0 & 14 & -12 & -2 \\ 0 & 0 & 2 & 12 \end{array}\right] \tag{5.4}$$

Back substitutions (to obtain the solution):

$$x_3 = 12/2 = 6$$

$$x_2 = (-2 + 12x_3)/14 = 5 \quad \text{or} \quad \mathbf{x} = \begin{Bmatrix} 1.5 \\ 5 \\ 6 \end{Bmatrix} \tag{5.5}$$

$$x_1 = (2 + 2x_2)/8 = 1.5$$

5.4.4 An Example: Iterative Method

The Gauss–Seidel method (as an example):

$$\mathbf{Ax} = \mathbf{b} \quad (\mathbf{A} \text{ is symmetric}) \tag{5.6}$$

or

$$\sum_{j=1}^{N} a_{ij}x_j = b_i, \quad i = 1,2,\ldots,N$$

Start with an estimate $\mathbf{x}^{(0)}$ of the solution vector and then iterate using the following:

$$x_i^{(k+1)} = \frac{1}{a_{ii}}\left[b_i - \sum_{j=1}^{i-1} a_{ij}x_j^{(k+1)} - \sum_{j=i+1}^{N} a_{ij}x_j^{(k)}\right], \quad \text{for } i = 1,2,\ldots,N \tag{5.7}$$

In vector form,

$$\mathbf{x}^{(k+1)} = \mathbf{A}_D^{-1}[\mathbf{b} - \mathbf{A}_L\mathbf{x}^{(k+1)} - \mathbf{A}_L^T\mathbf{x}^{(k)}] \tag{5.8}$$

where

$\mathbf{A}_D = \langle a_{ii} \rangle$ is the diagonal matrix of **A**,

$\mathbf{A}_L$ is the *lower triangular matrix* of **A**,

such that

$$\mathbf{A} = \mathbf{A}_D + \mathbf{A}_L + \mathbf{A}_L{}^T \tag{5.9}$$

Iterations continue until solution $\mathbf{x}$ converges, that is

$$\frac{\left\| \mathbf{x}^{(k+1)} - \mathbf{x}^{(k)} \right\|}{\left\| \mathbf{x}^{(k)} \right\|} \leq \varepsilon \tag{5.10}$$

where ε is the tolerance for convergence control.

Iterative solvers with moderate selections of the tolerance are usually much faster than direct solvers in solving large-scale models. However, for ill-conditioned systems, direct solvers should be applied to ensure the accuracy of the solutions.

5.5 Nature of Finite Element Solutions

FEA model is a mathematical model of the real structure, based on many approximations. An observation is made on the FEA models and solutions:

- Real structure—infinite number of nodes (physical points or particles), thus infinite number of DOFs
- FEA model—finite number of nodes, thus finite number of DOFs

In particular, one can argue that the displacement field is controlled (or constrained) by the values at a limited number of nodes (Figure 5.4).

Therefore, we have the so-called stiffening effect:

- FEA model is stiffer than the real structure.
- In general, displacement results are smaller in magnitudes than the exact values.

Hence, the FEM solution of displacement is a *lower bound* of the exact solution.

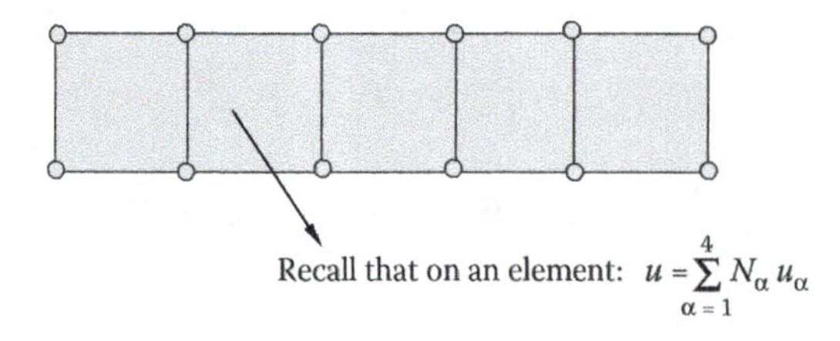
Recall that on an element: $u = \sum_{\alpha=1}^{4} N_\alpha u_\alpha$

FIGURE 5.4
Elements in an FEA model.

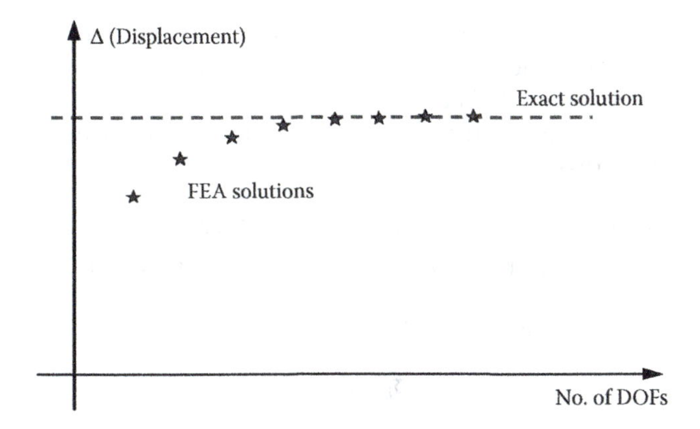

FIGURE 5.5
Convergence of FEM solutions with exact solution.

That is, FEA displacement solutions approach the exact solution from below (Figure 5.5), which can be used to monitor the FEA solutions. However, this is true only for the displacement-based FEA (which is the FEA formulation discussed in this book).

5.6 Convergence of FEA Solutions

As the mesh in an FEA model is "refined" (with smaller and smaller elements) repeatedly, the FEA solution will converge to the exact solution of the mathematical model of the problem (the model based on beam, plane stress/strain, plate, shell, or 3-D elasticity theories or assumptions). Several types of refinements have been devised in the FEA, which include:

h-refinement: Reduce the size of the element ("*h*" refers to the typical size of the elements).

p-refinement: Increase the order of the polynomials on an element (linear to quadratic, and so on; "*p*" refers to the highest order in a polynomial).

r-refinement: Rearrange the nodes in the mesh.

hp-refinement: Combination of the *h*- and *p*-refinements (to achieve better results).

With any of the above type of refinements, the FEA solutions will converge to the analytical solutions of the mathematical models. Some FEA software can automate the process of refinements in the FEA solutions to achieve the so-called adaptive solutions.

5.7 Adaptivity (*h*-, *p*-, and *hp*-Methods)

Adaptive FEA represents the future of the FEA applications. With proper error control, automatic refinements of an FEA mesh can be generated by the program until the converged FEA solutions are obtained. With the adaptive FEA capability, users' interactions are reduced, in the sense that a user only needs to provide a good initial mesh for the model (even this step can be done by the software automatically).

Error estimates are crucial in the adaptive FEA. Interested readers can refer to Reference [5] for more details. In the following, we introduce one type of the error estimates.

We first define two stress fields:

σ—element by element stress field (discontinuous across elements)

σ^*—averaged or smoothed stress field (continuous across elements)

Then, the error stress field can be defined as

$$\sigma_E = \sigma - \sigma^* \tag{5.11}$$

Compute strain energies,

$$U = \sum_{i=1}^{M} U_i, \quad U_i = \int_{V_i} \frac{1}{2} \sigma^T \mathbf{E}^{-1} \sigma dV \tag{5.12}$$

$$U^* = \sum_{i=1}^{M} U_i^*, \quad U_i^* = \int_{V_i} \frac{1}{2} \sigma^{*T} \mathbf{E}^{-1} \sigma^* dV \tag{5.13}$$

$$U_E = \sum_{i=1}^{M} U_{Ei}, \quad U_{Ei} = \int_{V_i} \frac{1}{2} \sigma_E^T \mathbf{E}^{-1} \sigma_E dV \tag{5.14}$$

where M is the total number of elements and V_i is the volume of the element i.

One error indicator—the *relative energy error* is defined as

$$\eta = \left[\frac{U_E}{U + U_E} \right]^{1/2}. \quad (0 \le \eta \le 1) \tag{5.15}$$

The indicator η is computed after each FEA solution. Refinement of the FEA model continues until, say

$$\eta \le 0.05$$

When this condition is satisfied, we conclude that the converged FE solution is obtained.

Some examples of using different error estimates in the FEA solutions can be found in Reference [5].

5.8 Case Study with *ANSYS Workbench*

Problem Description: Garden fountains are popular amenities that are often found at theme parks and hotels. As a fountain structure is usually an axisymmetric geometry with axisymmetric loads and support, only a 2-D model, sliced through the 3-D geometry, is needed to correctly predict the deformation of or stress in the structure. The figure below gives the cross section of an axisymmetric model of a two-tier garden fountain made of concrete. Determine the maximum deformation and von Mises stress under the given hydrostatic pressure. Use adaptive meshing to improve solution convergence.

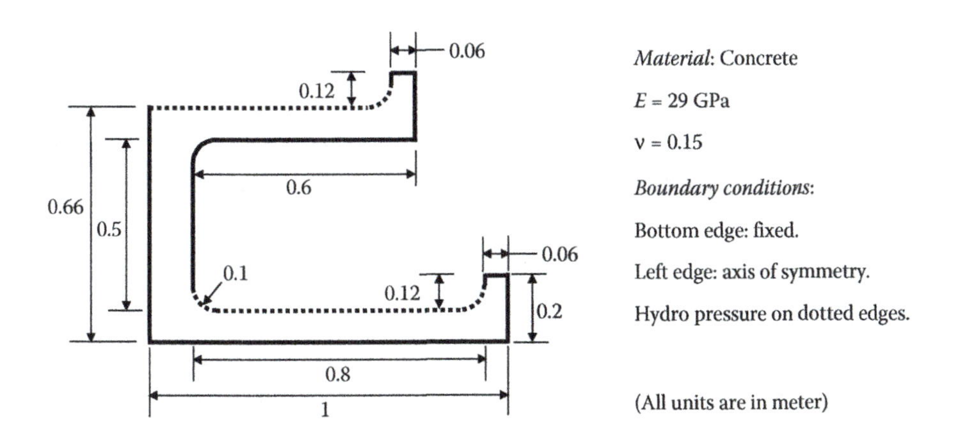

Solution

To solve the problem with *ANSYS® Workbench*, we employ the following steps:

Step 1: Start an *ANSYS Workbench* Project

Launch *ANSYS Workbench* and save the blank project as *"Fountain.wbpj"*

Step 2: Create a *Static Structural (ANSYS)* Analysis System

Drag the *Static Structural (ANSYS)* icon from the *Analysis Systems Toolbox* window and drop it inside the highlighted green rectangle in the *Project Schematic* window to create a standalone static structural analysis system.

Step 3: Add a New Material

Double-click on the *Engineering Data* cell to add a new material. In the following *Engineering Data* interface which replaces the *Project Schematic*, type *"Concrete"* as the name for the new material, and double-click *Isotropic Elasticity* under *Linear Elastic* in the leftmost *Toolbox* window. Enter *"29E9"* for *Young's Modulus* and *"0.15"* for *Poisson's Ratio* in the *Properties* window. Click the *Return to Project* button to go back to *Project Schematic*.

Step 4: Launch the *DesignModeler* Program

Ensure *Surface Bodies* is checked in the *Properties of Schematic A3: Geometry* window (select *Properties* from the *View* drop-down menu to enable display of this window). Select *2D* for *Analysis Type* in this *Properties* window. Double-click the *Geometry* cell to launch *DesignModeler*, and select *"Meter"* in the *Units* pop-up window.

Step 5: Create Surface Sketch

Click on the *Sketching* tab. Select the *Draw* toolbox and then *Line*. Draw a closed-loop line profile as dimensioned below. Make sure a horizontal constraint (H) and a vertical constraint (V) appear when drawing a horizontal and a vertical line, respectively. Use the *Fillet* tool in the *Draw* toolbox to create line fillets with a radius of 0.1 m as shown below.

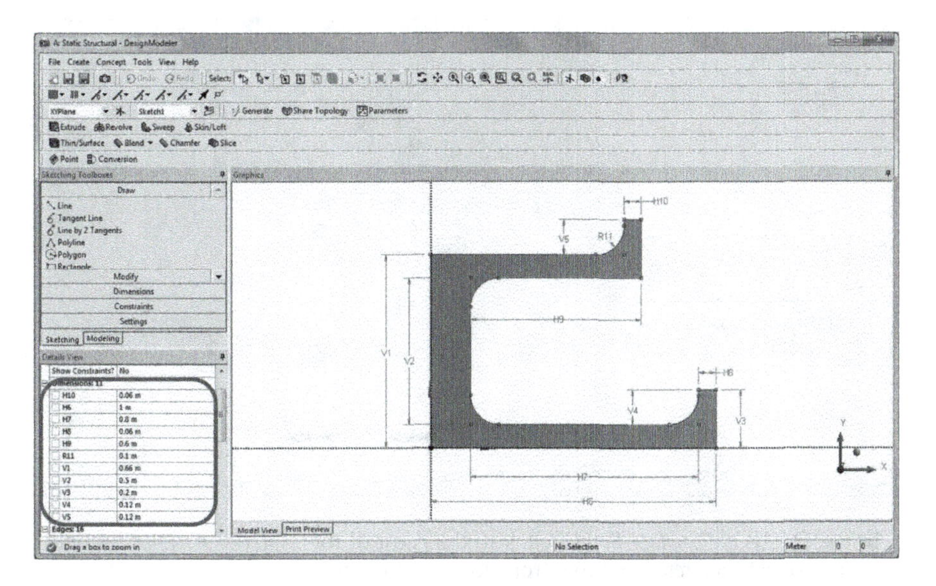

Step 6: Create Surface Body

Switch to the *Modeling* tab and choose *Surfaces from Sketches* from the *Concept* menu.

Select *Sketch1* from the *Tree Outline* shown below, and apply it to the *Base Objects* selection in the *Details of SurfaceSK1*. Then click *Generate*.

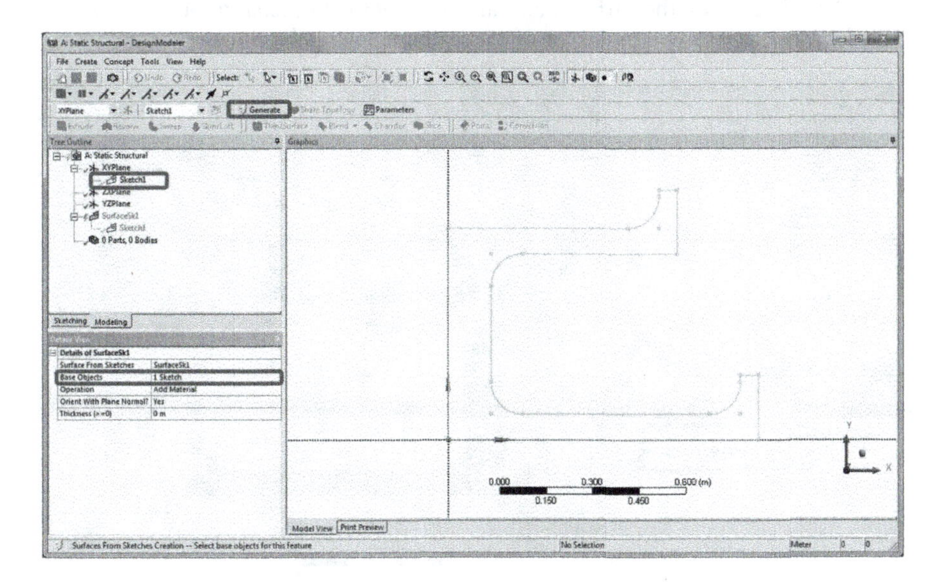

A *Surface Body* will be created from the surface sketch. Exit the *DesignModeler*.

Step 7: Launch the *Static Structural* Program

Double-click on the *Model* cell to launch the *Static Structural* program. Click on *Geometry* in the *Outline*. In the *Details of "Geometry,"* choose *axisymmetric* for *2D Behavior*.

In the *Details of "Surface Body,"* click to the right of the *Material Assignment* field and select *Concrete* from the drop-down context menu.

Step 8: Generate Mesh

Right click on *Mesh* in the *Outline*. Select *Insert* and then *Sizing* from the context menu. In the *Details of "Body Sizing,"* enter *"0.05 m"* for *Element Size*. Click on the surface body in the *Graphics* window and apply it to the *Geometry* selection.

Right click on *Mesh*. Select *Insert* and then *Method*. In the *Details of "Automatic Method,"* click on the surface body, and apply it to the *Geometry*. Select *Triangles* for *Method*. This will make use of triangular elements for the mesh generation.

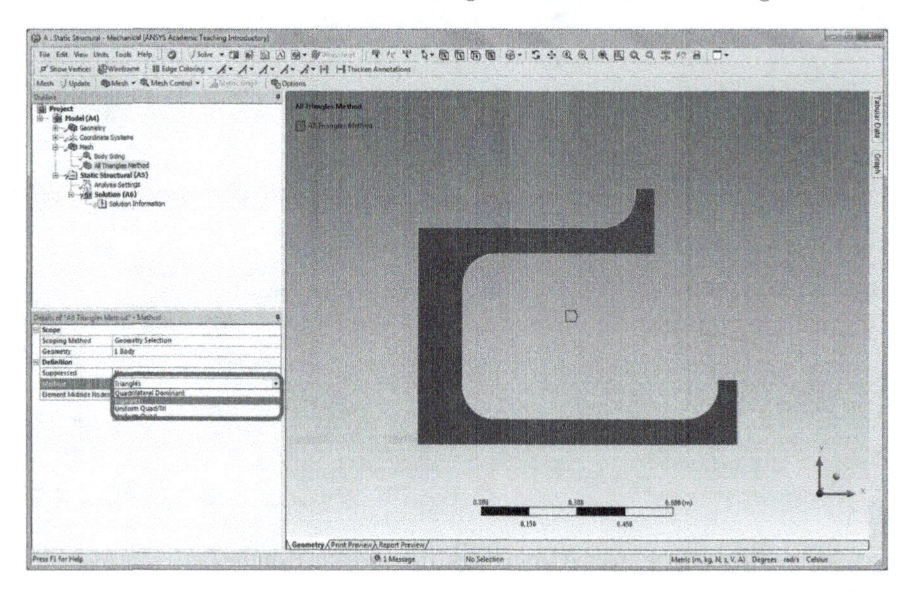

In the *Details of "Mesh,"* choose *Dropped* for the *Element Midside Nodes* field. This will specify the use of linear elements in the mesh. Note that linear triangular elements are employed here to show the convergence of linear FEA approximate solutions; they are in general not recommended for stress analysis. Right-click on *Mesh* and select *Generate Mesh.*

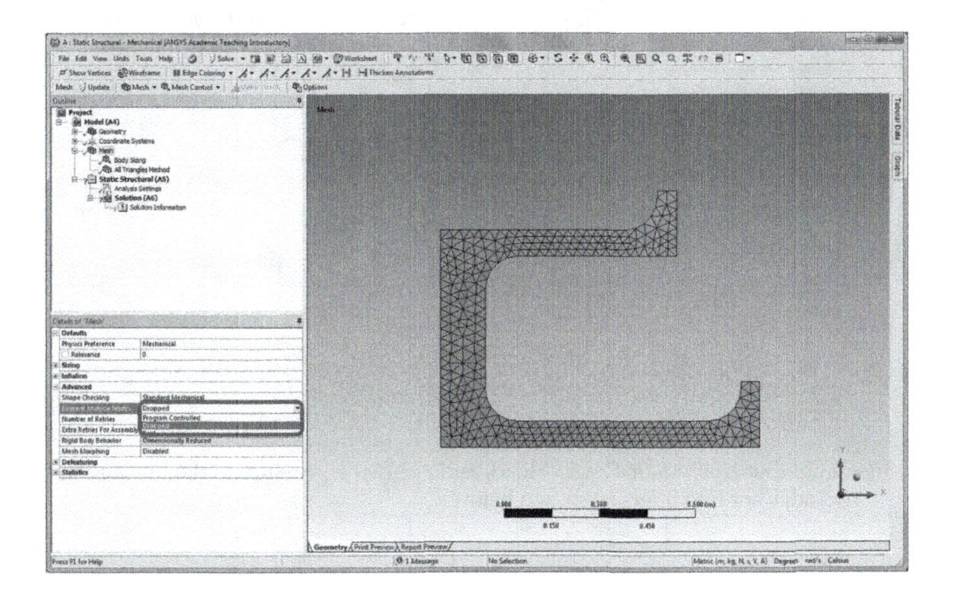

Step 9: Apply Boundary Conditions

Right-click on *Static Structural (A5).* Choose *Insert* and then *Fixed Support* from the context menu. Apply this support to the horizontal line at the bottom.

Right-click on *Static Structural (A5).* Choose *Insert* and then *Frictionless Support* from the context menu. Apply this support to the leftmost vertical line (center line of the fountain). The frictionless support prevents the line from moving or deforming in the normal direction, and thus is equivalent to a symmetry condition.

Step 10: Apply Loads

In the *Project Outline,* right-click on *Static Structural (A5).* Choose *Insert* and then *Hydrostatic Pressure.* The hydrostatic load simulates pressure due to fluid weight.

In the *Details of "Hydrostatic Pressure,"* ctrl-click the horizontal line and the adjacent line fillet shown below, and apply the two edges to the *Geometry* selection. Enter *1000* kg/m^3 for *Fluid Density.* Change the *Define By* selection to *Components,* and enter *9.8* m/s^2 for the *Y Component* of *Hydrostatic Acceleration.* Enter *0.68* m for the *X Coordinate* and *0.76* m for the *Y Coordinate* for the *Free Surface Location.* The location corresponds to the upper endpoint of the line fillet.

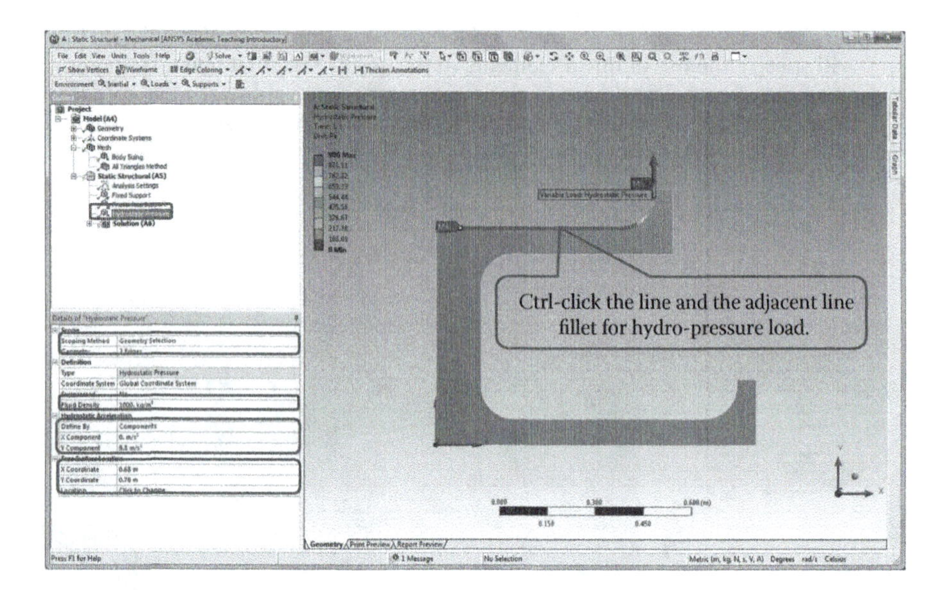

Insert another *Hydrostatic Pressure* load, and apply the pressure to the line and its two adjacent line fillets as shown below.

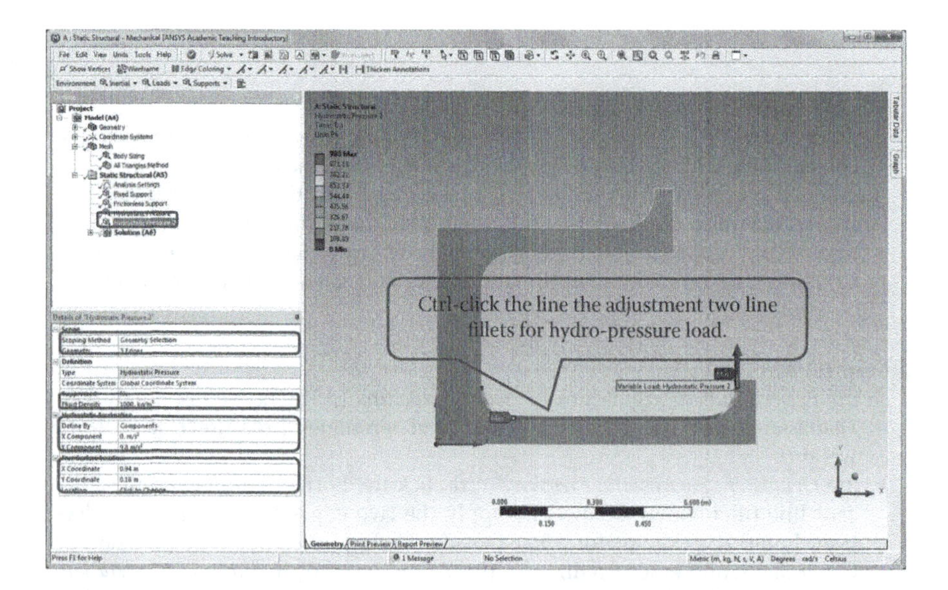

Step 11: Retrieve Solution

Insert a *Total Deformation* item by right-clicking on *Solution (A6)* in the *Outline* tree.

Right-click on the *Total Deformation* in the *Outline* tree, and choose *Insert* then *Convergence.*

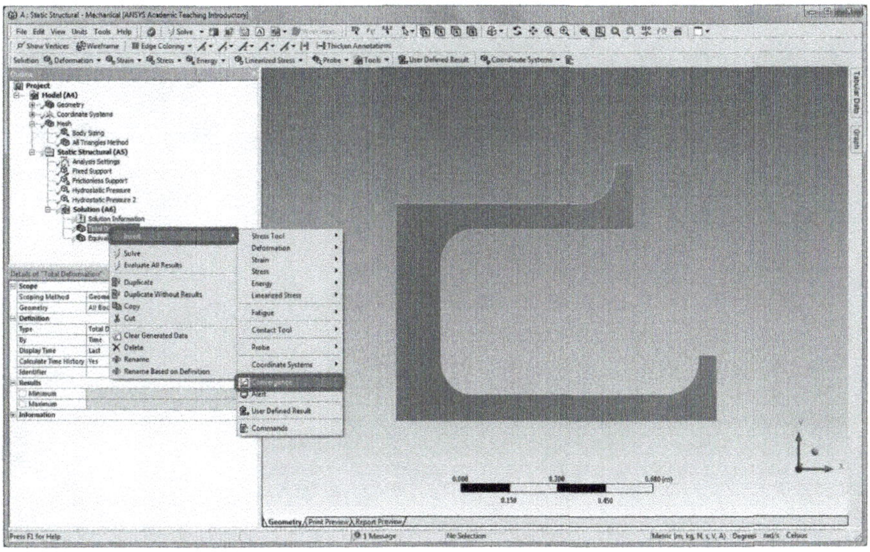

In the *Details of "Convergence,"* enter *1%* for the *Allowable Change* field.

In the *Details of "Solution (A6),"* set *Max Refinement Loops* as *10,* and *Refinement Depth* as 1. The refinement depth controls the aggressiveness of the mesh refinement; it has a range from 0 to 3 with a larger number indicating more aggressive refinement.

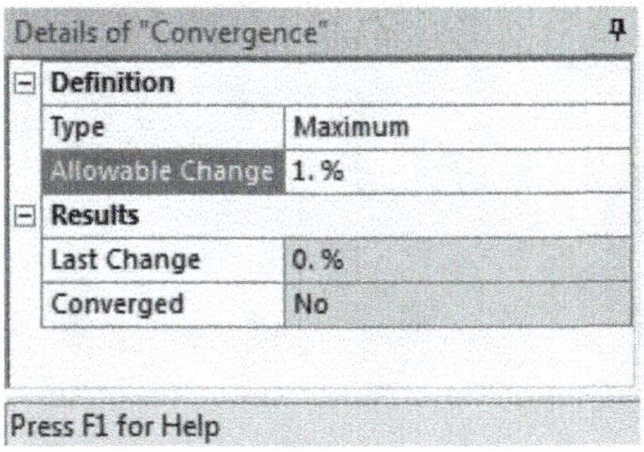

Details of "Convergence"

Definition	
Type	Maximum
Allowable Change	1. %
Results	
Last Change	0. %
Converged	No

Press F1 for Help

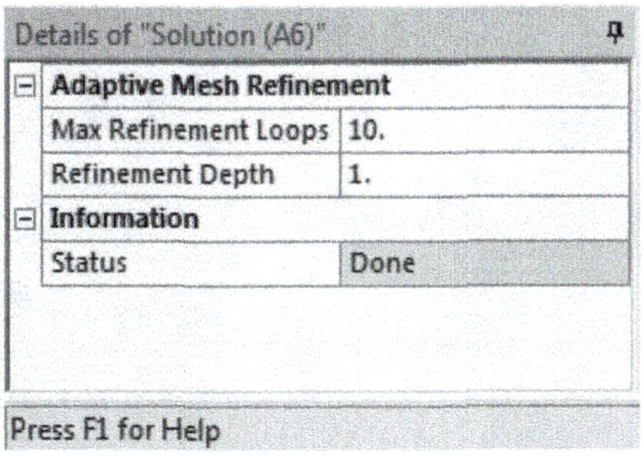

Details of "Solution (A6)"

Adaptive Mesh Refinement	
Max Refinement Loops	10.
Refinement Depth	1.
Information	
Status	Done

Press F1 for Help

Insert an *Equivalent Stress* item by right-clicking on *Solution (A6)* in the *Outline* tree.

Right-click on *Solution (A6)* in the *Outline* tree and choose *Solve.* The program will start to iterate the solution until the difference between two consecutive iterations is less than 1% or the maximum number of mesh refinement loops reaches 10. After completion, click on *Convergence* in the *Outline* to review the convergence curve. The resulting maximum deformations at different mesh iterations are also recorded in the table below the curve.

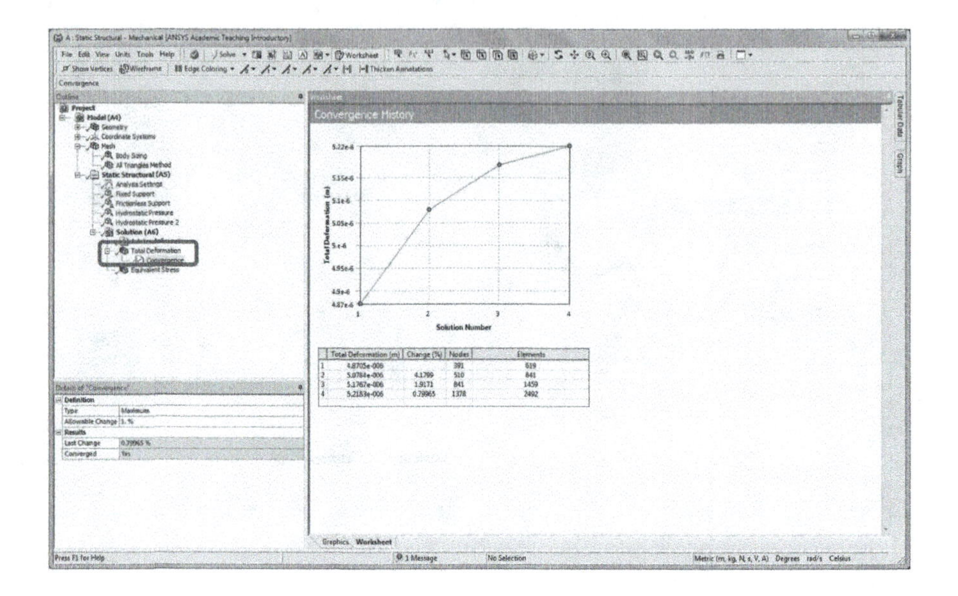

Click on *Total Deformation* in the *Outline* to review the converged deformation results.

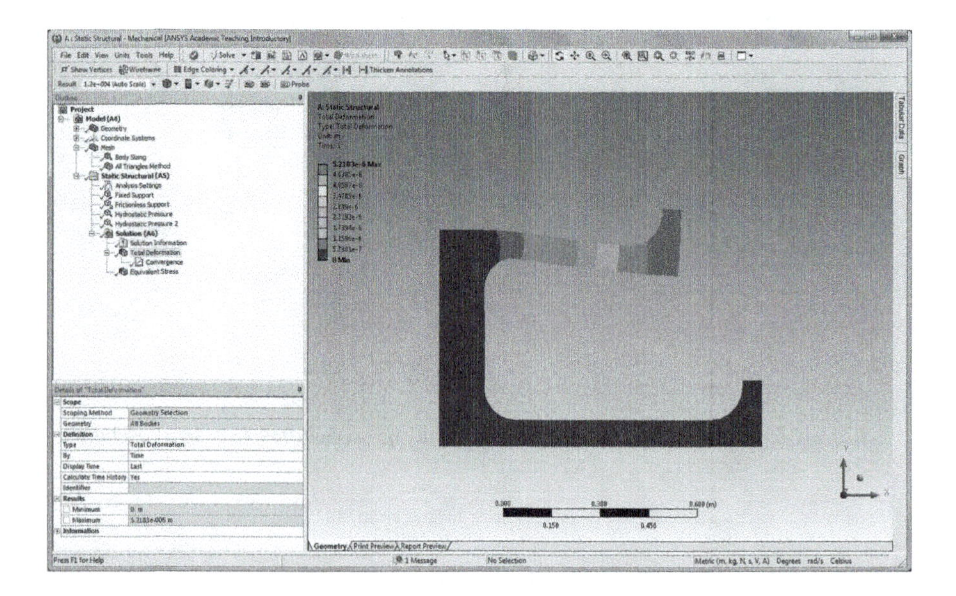

Click on *Equivalent Stress* in the *Outline* to review the stress results.

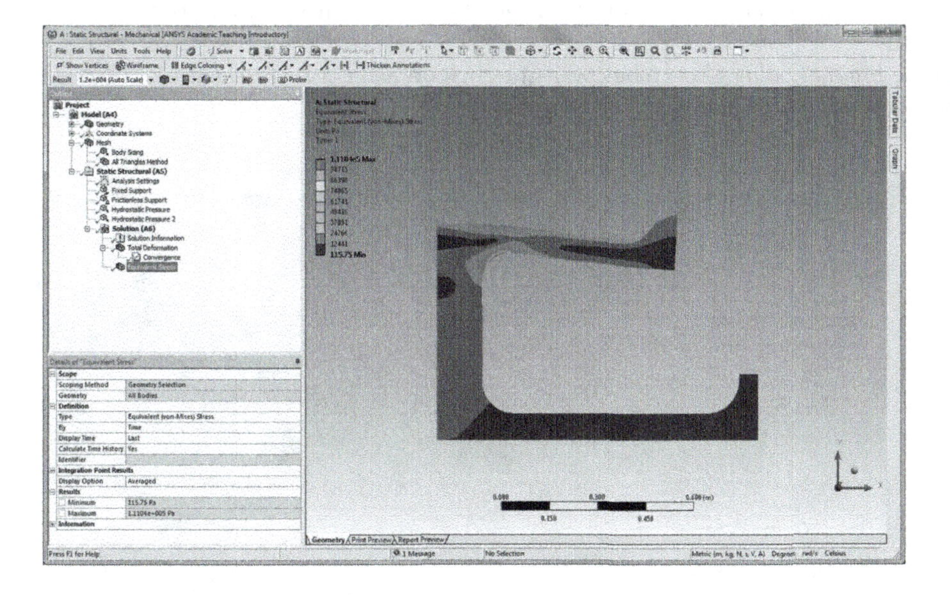

Modeling tips: A model may be subjected to body forces such as gravitational or radial centrifugal/inertia forces, in addition to the hydrostatic pressure load. To consider such forces, the density of the structure's material needs to be given as an input, and the forces are typically calculated as follows:

$$f_r = \rho r \omega^2, \quad \text{Equivalent radial centrifugal / inertial force}$$
$$f_z = -\rho g, \quad \text{Gravitational force}$$

where ρ is the mass density and g the gravitational acceleration (=9.8 m/s²).

Take the following steps to add body forces to the fountain model. First, double-click on *Density* in *Physical Properties Toolbox*. Enter *2.38e3* for *Density* in the *Properties of Outline Row 3: Concrete*. Click on the *Return to Project* button.

Next, refresh the *Model* cell of *Project Schematic* after the above change is made on the *Engineering Data* cell. Double-click on the *Model* cell to launch the *Static Structural* Program.

In the *Project Outline* shown below, right-click on *Static Structural (A5)*. Choose *Insert* and then *Standard Earth Gravity*.

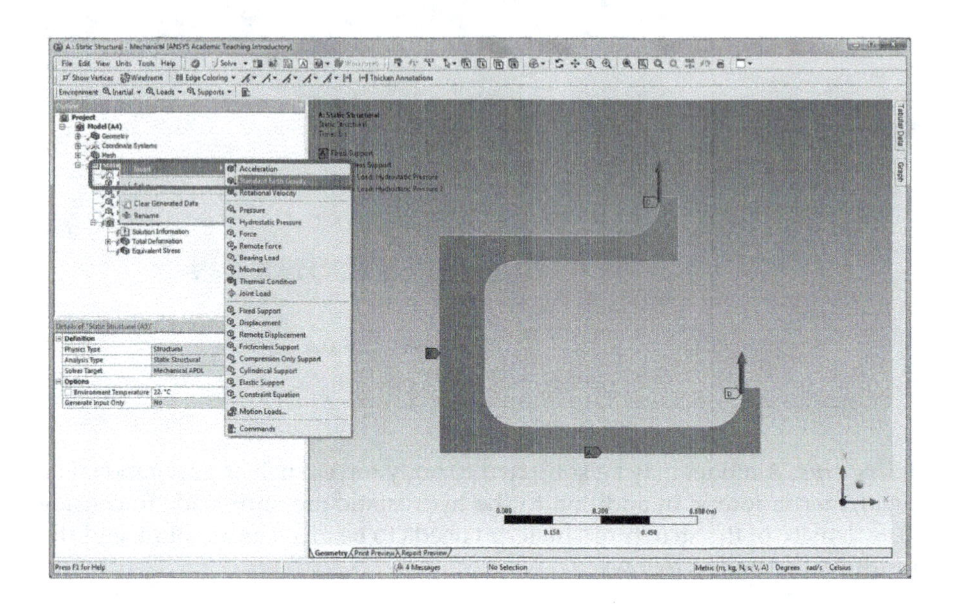

Right-click on *Static Structural (A5)* and choose *Insert* then *Rotational Velocity*.

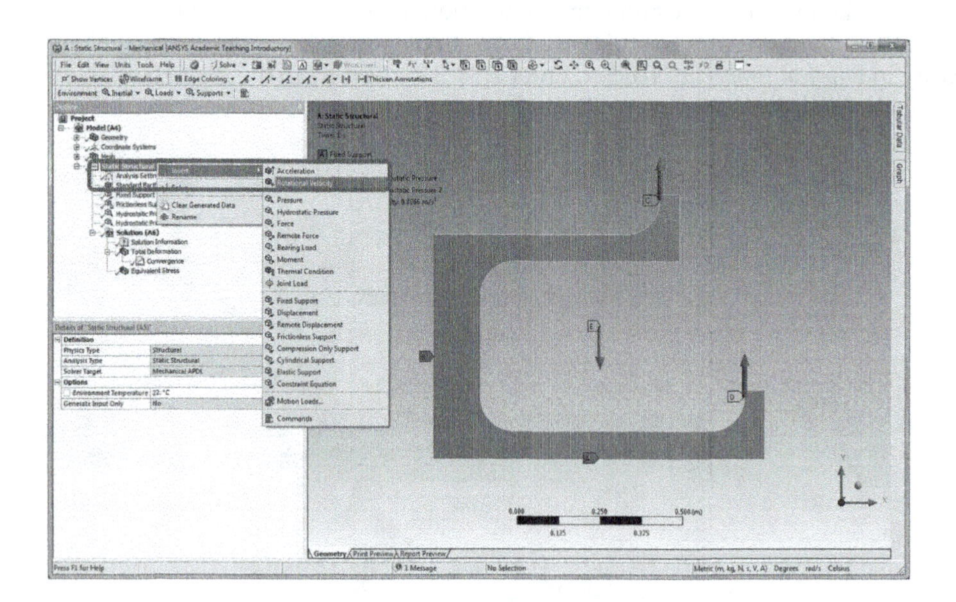

In the *Details of "Rotational Velocity,"* change *Define By* to *Components*. Enter 5 rad/s for *Y Component*.

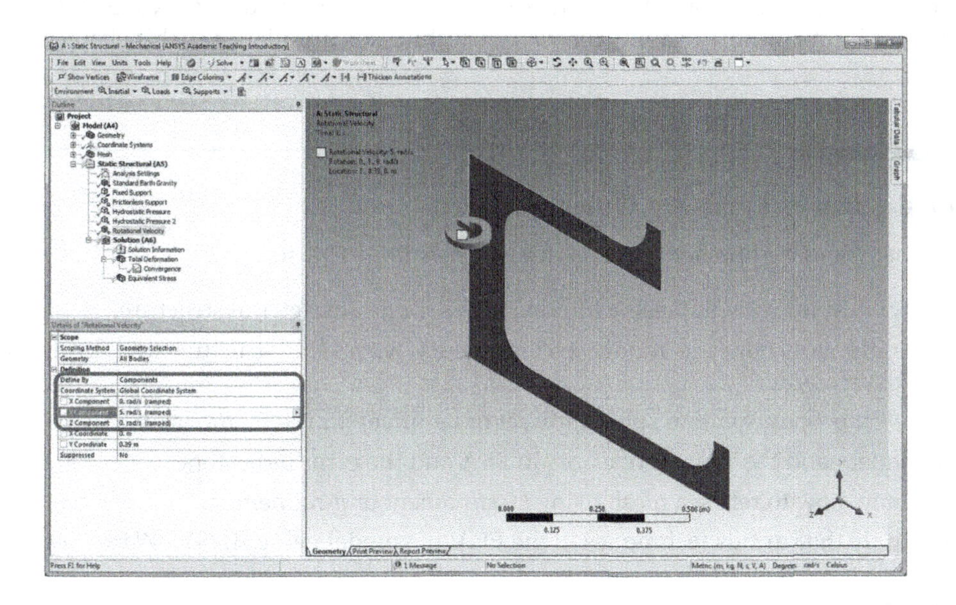

Right-click on *Solution (A6)* in the *Outline* tree, and select *Solve* to update the model results. The new deformation and stress results are shown below. Both the maximum deformation and the maximum von Mises stress values are shown to be slightly increased, as compared to the results considering only the hydrostatic pressure load.

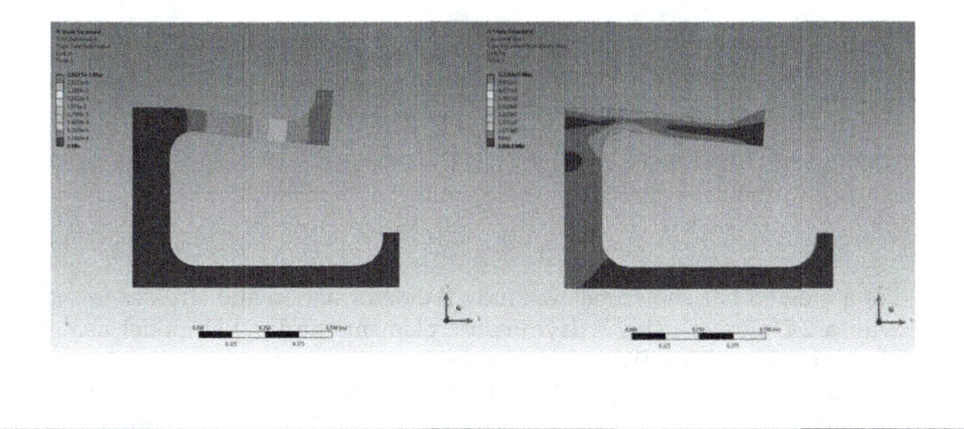

5.9 Summary

In this chapter, we discussed a few modeling techniques and options related to FEA solutions. For static analysis of symmetrical structures, the symmetry features should be explored in both the modeling (meshing) stage and the solution stage (if the BCs are symmetrical as well). Substructuring or using SEs is a useful technique for solving large-scale problems with constrained computing resources. Convergence of the FEA solutions is the important goal in FEA and should be monitored by using the error estimates and employing the adaptive FEA capabilities in the software. The showcase using *ANSYS Workbench* is introduced to show how to apply the various techniques introduced in this chapter.

5.10 Review of Learning Objectives

Now that you have finished this chapter you should be able to

1. Apply symmetry in meshing and analysis for a symmetrical structure.
2. Apply axisymmetric model and elements when the structure and loads are axisymmetric.
3. Select proper equation solvers (direct or iterative) for the FE model at hand.
4. Understand the sources of errors in FEA and the error indicators.
5. Know how to refine a mesh using h-refinement or p-refinement.
6. Apply the various techniques in the FEA of a model using *ANSYS Workbench.*

PROBLEMS

5.1 Consider the two plate structures shown below. The structures are symmetrically loaded on the x–y plane. Use symmetry to construct appropriate finite element models for the structures, indicating load and support conditions.

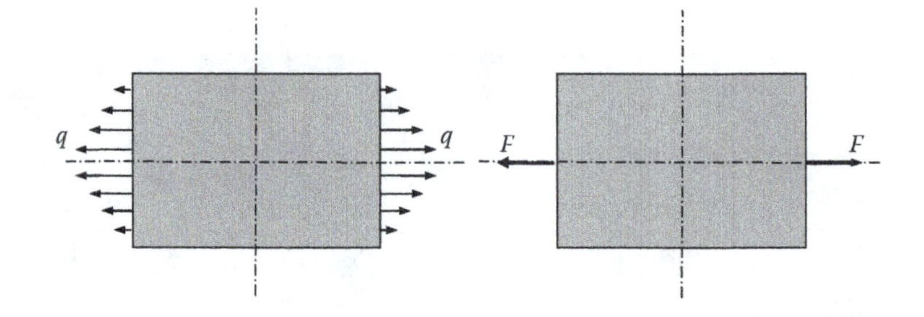

5.2 For a tapered bar shown below, study the deformation and stresses in the bar with a 2-D model using axisymmetric elements and a 1-D model using bar elements in *ANSYS Workbench.* Assume $R_1 = 1$ m, $R_2 = 0.5$ m, $L = 5$ m, force $F = 3000$ N, Young's modulus $E = 200$ GPa, and Poisson's ratio $\nu = 0.3$. The bar is fixed at the left end. Compare the results from the 2-D and 1-D models.

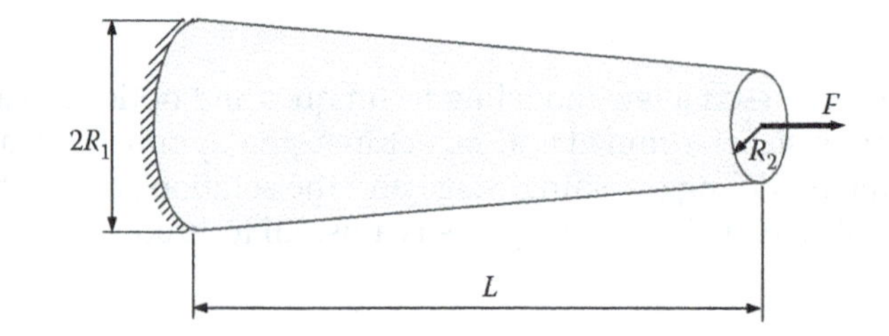

5.3 An open cylinder shown below has an inner radius $a = 1$ m, outer radius $b = 1.1$ m, length $L = 6$ m, and is applied with a pressure load $p = 10$ GPa on the inner surface. The outer surface of the cylinder is prevented from radial deformation, and the cylinder is free to deform along the length direction. Compute the stresses in the cylinder using a 2-D axisymmetric model in *ANSYS Workbench*. Assume Young's modulus $E = 200$ GPa and Poisson's ratio $\nu = 0.3$.

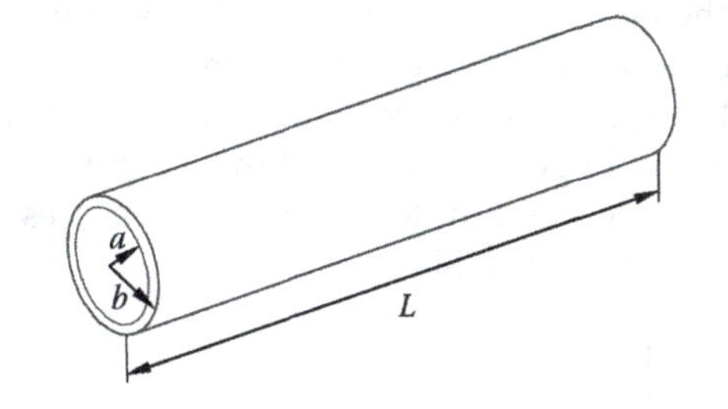

5.4 The rotating part sketched below is made of steel with Young's modulus $E = 200$ GPa, Poisson's ratio $\nu = 0.3$, and mass density $\rho = 7850$ kg/m^3. Assume that the part is rotating at a speed of 100 RPM about the z axis. Ignore the gravitational force. Compute the stresses in the part using an axisymmetric model in *ANSYS Workbench*.

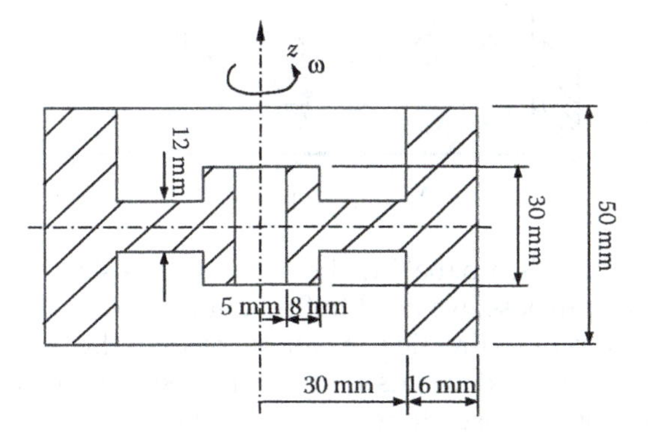

5.5 Determine the deformation and stress distributions for a solid cylinder under the influence of the gravity force and radial centrifugal force using *ANSYS Workbench*. The cylinder is made of aluminum with Young's modulus $E = 73$ GPa, Poisson's ratio $\nu = 0.3$, and mass density $\rho = 2800$ kg/m^3. Assume that the part is rotating at a speed of 50 rad/s about its central axis. Radius = 10 mm, height = 20 mm, angular velocity around cylinder central axis = 50 rad/s.

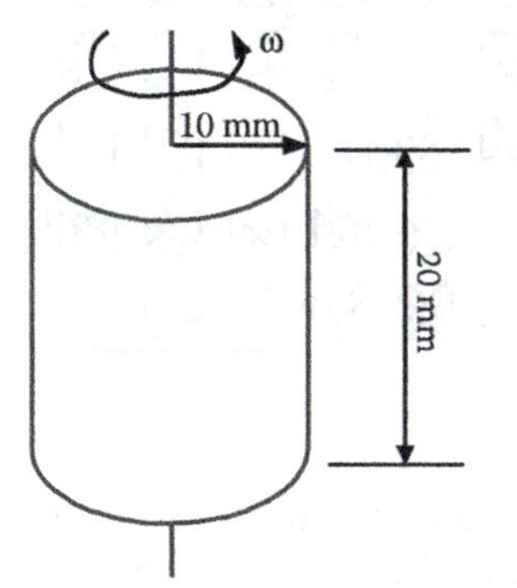

5.6 Suppose that we need to find out the in-plane effective modulus of a composite reinforced with long fibers aligned in the z-direction and distributed uniformly. A 2-D elasticity model shown below can be used for this study with *ANSYS Workbench*. The effective modulus can be estimated by using the formula $E_{eff} = \sigma_{x(ave)}/\varepsilon_{x(ave)}$, where the averaged stress and strain are evaluated along the vertical edge on the right side of the model. Assume for the matrix $E = 10$ GPa, $\nu = 0.35$, and for the fibers $E = 100$ GPa, $\nu = 0.3$. The unit cell has a dimension of $1 \times 1\ \mu m^2$ and the radius of the fibers is 0.2 μm.

Start with 1 × 1 cell, 2 × 2 cells, 3 × 3 cells, ... and keep increasing the number of the cells as you can. Report the value of the effective Young's modulus of the composite in the x direction. Employ symmetry features of the model in generating the meshes for your analysis.

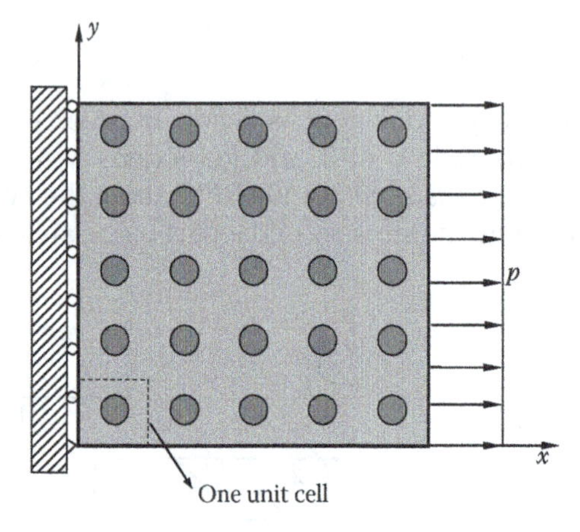

5.7 Suppose that a "meshed panel" will be used in a design in order to reduce the weight. For this purpose, we need to find out the in-plane effective modulus of this panel in the x- or y-direction. A sample piece of the panel similar to the one shown below can be used for this study. Employ symmetry and study the effects of the numbers of cells used in the model on the effective modulus of the panel in the x-direction.

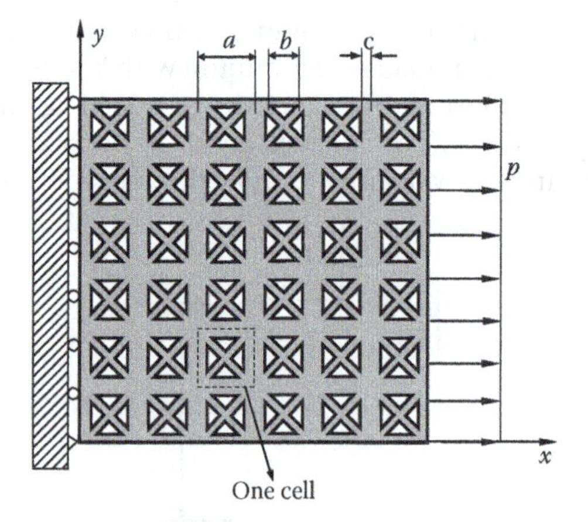

Assume the panel is made of aluminum with $E = 70$ GPa, $\nu = 0.35$, $a = 10$ mm, $b = 5$ mm, $c = 1.5$ mm, and thickness $t = 1$ mm.

6

Plate and Shell Analyses

6.1 Introduction

Many structure members can be categorized as plates and shells, which are extensions of the 1-D straight beams and curved beams to 2-D cases, respectively. Examples of structures that can be modeled as plates are shear walls, floor panels, and shelves, while those that can be modeled as shells include shell structures in nature (such as sea shells and egg shells), various containers, pipes, tanks, roofs of buildings (such as the superdome), and bodies of cars, boats, and aircrafts. Figure 6.1 shows an airplane (Boeing 787) that is constructed mainly using plate and shell structure members.

The advantages of using plate and shell structures are their light weight, superior load-carrying capabilities, and sometimes, simply their artistic appeals.

6.2 Review of Plate Theory

Plates are flat surfaces applied with lateral loading, with bending behaviors dominating the structural response. Shells are structures which span over curved surfaces; they carry both membrane and bending forces under lateral loading. The thickness t of a plate or a shell is much smaller than the other dimensions of the structure. Theories related to plates and shells are briefly reviewed below.

6.2.1 Force and Stress Relations in Plates

Consider an infinitesimally small element in a plate under lateral loading. The internal shear forces and bending moments acting on the element, and the induced stresses are shown below in Figure 6.2 and Figure 6.3, respectively.

The induced stresses are related to the internal moments and forces in the following manner:

Bending moments (per unit length):

$$M_x = \int_{-t/2}^{t/2} \sigma_x z dz, \quad (N \cdot m/m) \tag{6.1}$$

FIGURE 6.1
An airplane made of numerous plate and shell structures.

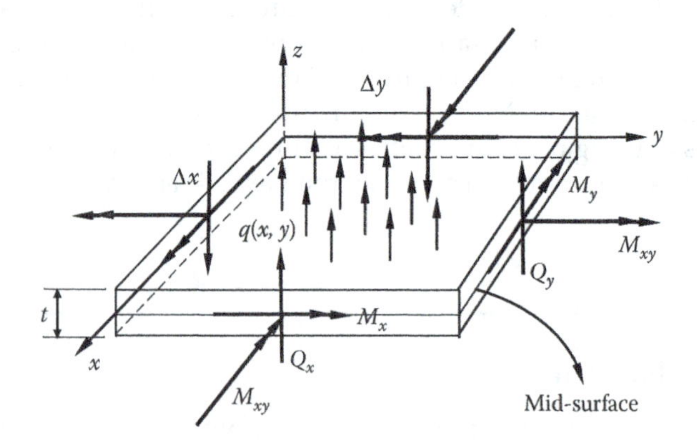

FIGURE 6.2
Forces and moments acting on an infinitesimally small element in a plate.

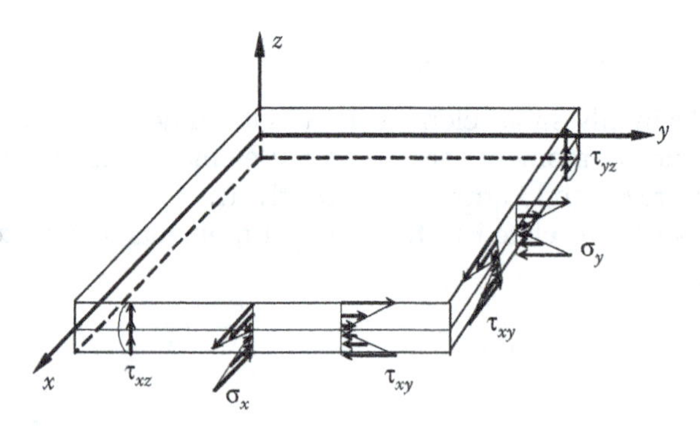

FIGURE 6.3
Stresses acting on the infinitesimally small element in the plate.

$$M_y = \int_{-t/2}^{t/2} \sigma_y z\, dz, \quad (N \cdot m/m) \tag{6.2}$$

Twisting moment (per unit length):

$$M_{xy} = \int_{-t/2}^{t/2} \tau_{xy} z\, dz, \quad (N \cdot m/m) \tag{6.3}$$

Shear Forces (per unit length):

$$Q_x = \int_{-t/2}^{t/2} \tau_{xz}\, dz, \quad (N/m) \tag{6.4}$$

$$Q_y = \int_{-t/2}^{t/2} \tau_{yz}\, dz, \quad (N/m) \tag{6.5}$$

Maximum bending stresses:

$$(\sigma_x)_{\max} = \pm \frac{6M_x}{t^2}, \quad (\sigma_y)_{\max} = \pm \frac{6M_y}{t^2} \tag{6.6}$$

Similar to the beam model, there is no bending stress at the mid-surface and the maximum/minimum stresses are always at $z = \pm t/2$.

6.2.2 Thin Plate Theory (Kirchhoff Plate Theory)

The thin plate theory is based on assumptions that a straight line normal to the mid-surface remains straight and normal to the deflected mid-surface after loading (Figure 6.4); that is

$$\gamma_{xz} = \gamma_{yz} = 0 \quad \text{(Negligible transverse shear deformations)}$$

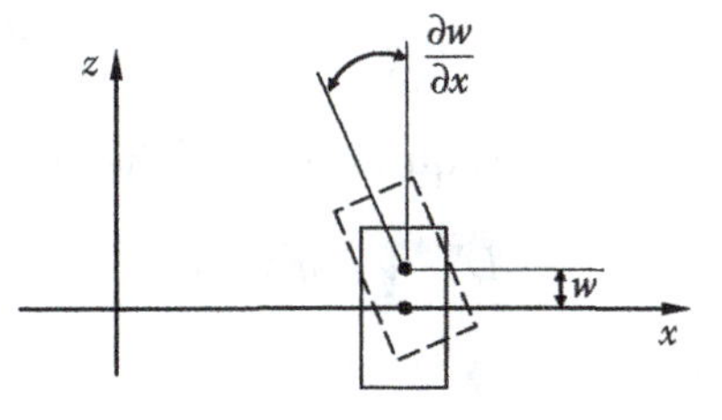

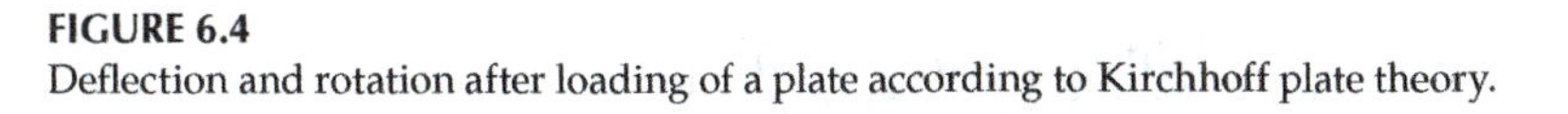
FIGURE 6.4
Deflection and rotation after loading of a plate according to Kirchhoff plate theory.

The displacement, strains, and stresses can be written in terms of the main variable—deflection $w = w(x, y)$, as follows:

Displacement:

$$\begin{aligned} w &= w(x,y), \quad (\textit{deflection}) \\ u &= -z\frac{\partial w}{\partial x} \\ v &= -z\frac{\partial w}{\partial y} \end{aligned} \tag{6.7}$$

Strains:

$$\begin{aligned} \varepsilon_x &= -z\frac{\partial^2 w}{\partial x^2} \\ \varepsilon_y &= -z\frac{\partial^2 w}{\partial y^2} \\ \gamma_{xy} &= -2z\frac{\partial^2 w}{\partial x \partial y} \end{aligned} \tag{6.8}$$

Note that there is no stretch of the mid-surface caused by the deflection of the plate.

Stresses (plane stress state):

$$\begin{Bmatrix} \sigma_x \\ \sigma_y \\ \tau_{xy} \end{Bmatrix} = \frac{E}{1-\nu^2}\begin{bmatrix} 1 & \nu & 0 \\ \nu & 1 & 0 \\ 0 & 0 & (1-\nu)/2 \end{bmatrix}\begin{Bmatrix} \varepsilon_x \\ \varepsilon_y \\ \gamma_{xy} \end{Bmatrix} \tag{6.9}$$

or,

$$\begin{Bmatrix} \sigma_x \\ \sigma_y \\ \tau_{xy} \end{Bmatrix} = -z\frac{E}{1-\nu^2}\begin{bmatrix} 1 & \nu & 0 \\ \nu & 1 & 0 \\ 0 & 0 & (1-\nu) \end{bmatrix}\begin{Bmatrix} \dfrac{\partial^2 w}{\partial x^2} \\ \dfrac{\partial^2 w}{\partial y^2} \\ \dfrac{\partial^2 w}{\partial x \partial y} \end{Bmatrix} \tag{6.10}$$

The following equation governs the equilibrium of the plate in the z-direction:

$$D\nabla^4 w = q(x,y) \tag{6.11}$$

where

$$\nabla^4 \equiv \left(\frac{\partial^4}{\partial x^4} + 2\frac{\partial^4}{\partial x^2 \partial y^2} + \frac{\partial^4}{\partial y^4}\right)$$

$$D = \frac{Et^3}{12(1-\nu^2)} \quad \text{(the bending rigidity of the plate)} \tag{6.12}$$

q = lateral distributed load (force per unit area).

Note that Equation 6.11 is of analogous form with the following 1-D equation for straight beam:

$$EI\frac{d^4w}{dx^4} = q(x) \tag{6.13}$$

The fourth-order partial differential equation, given in Equation 6.11 and in terms of the deflection $w(x,y)$, needs to be solved under certain given boundary conditions. Typical boundary conditions for plate bending include:

$$\text{Clamped:} \quad w = 0, \quad \frac{\partial w}{\partial n} = 0 \tag{6.14}$$

$$\text{Simply supported:} \quad w = 0, \quad M_n = 0 \tag{6.15}$$

$$\text{Free:} \quad Q_n = 0, \quad M_n = 0 \tag{6.16}$$

where n is the normal direction of the boundary (Figure 6.5). Note that the given values in the boundary conditions shown above can be nonzero values as well.

6.2.2.1 *Example: A Thin Plate*

A square plate (Figure 6.6) with four edges clamped or hinged, and under a uniform load q or a concentrated force P at the center C.

For this simple geometry, Equation 6.11 with the boundary condition of Equation 6.14 or 6.15 can be solved analytically. The maximum deflections are given in Table 6.1 for the four different cases. These values can be used to verify the FEA solutions.

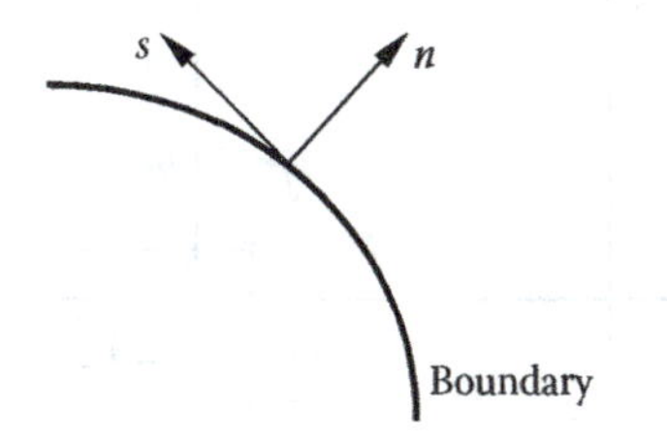

FIGURE 6.5
The boundary of a plate.

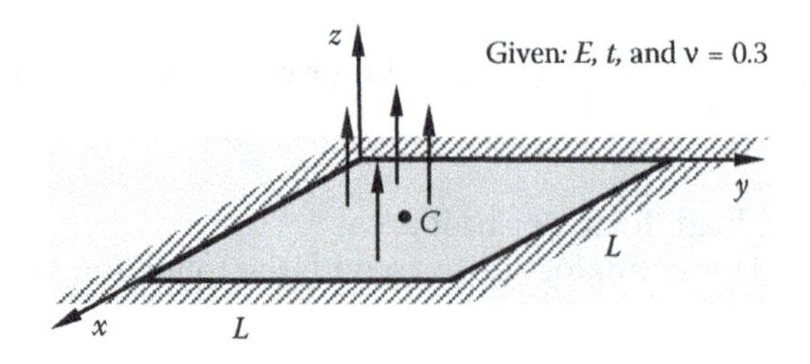

FIGURE 6.6
A square plate.

TABLE 6.1
Deflection at the Center (with $D = Et^3/(12(1 - v^2))$)

	Clamped	Simply Supported
Under uniform load q	$0.00126\, qL^4/D$	$0.00406\, qL^4/D$
Under concentrated force P	$0.00560\, PL^2/D$	$0.0116\, PL^2/D$

6.2.3 Thick Plate Theory (Mindlin Plate Theory)

If the thickness t of a plate is not small, for example, when $t/L \geq 1/10$ (L = a characteristic dimension of the plate main surface), then the thick plate theory by Mindlin should be applied. The theory accounts for the angle changes within a cross section, that is

$$\gamma_{xz} \neq 0, \qquad \gamma_{yz} \neq 0 \quad \text{(transverse shear deformations)}$$

This means that a line which is normal to the mid-surface before the deformation will not be so after the deformation (Figure 6.7).

The new independent variables θ_x and θ_y are rotation angles of a line, which is normal to the mid-surface before the deformation, about x- and y-axis, respectively.

The following new relations hold:

$$u = z\theta_y, \quad v = -z\theta_x \tag{6.17}$$

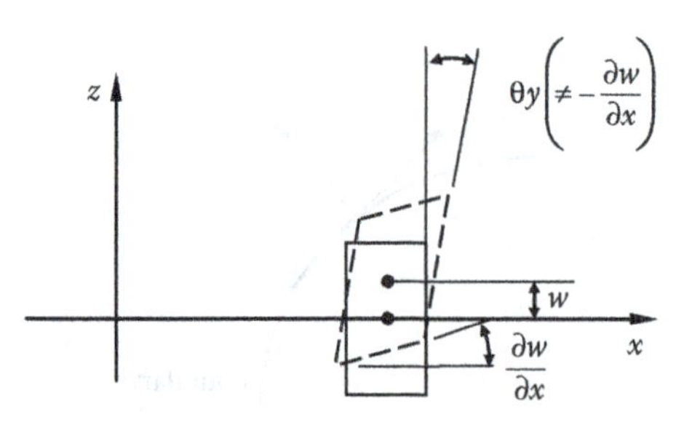

FIGURE 6.7
Displacement and rotation based on the Mindlin thick plate theory.

and

$$\varepsilon_x = z\frac{\partial \theta_y}{\partial x}, \quad \varepsilon_y = -z\frac{\partial \theta_x}{\partial y}, \quad \gamma_{xy} = z\left(\frac{\partial \theta_y}{\partial y} - \frac{\partial \theta_x}{\partial x}\right),$$
$$\gamma_{xz} = \frac{\partial w}{\partial x} + \theta_y, \quad \gamma_{yz} = \frac{\partial w}{\partial y} - \theta_x \tag{6.18}$$

Note that if we impose the conditions (or assumptions) that

$$\gamma_{xz} = \frac{\partial w}{\partial x} + \theta_y = 0, \quad \gamma_{yz} = \frac{\partial w}{\partial y} - \theta_x = 0 \tag{6.19}$$

then we can recover the relations applied in the thin plate theory.

The governing equations and boundary conditions can be established for thick plates based on the above assumptions, with the three main variables involved being $w(x, y)$, $\theta_x(x, y)$, and $\theta_y(x, y)$.

6.2.4 Shell Theory

Unlike the plate models, where only bending forces exist, there are two types of forces in shells, that is, membrane forces (in plane forces) and bending forces (out of plane forces) (Figures 6.8 and 6.9).

6.2.4.1 Shell Example: A Cylindrical Container

Similar to the plate theories, there are two types of theories for modeling shells, namely thin shell theory and thick shell theory. Shell theories are the most complicated ones to formulate and analyze in mechanics. Many of the contributions were made by Russian scientists in the 1940s and 1950s, due to the need to develop new aircrafts and other lightweight structures. Interested readers can refer to Reference [6] for in-depth studies on this subject. These theoretical works have laid the foundations for the development of various finite elements for analyzing shell structures.

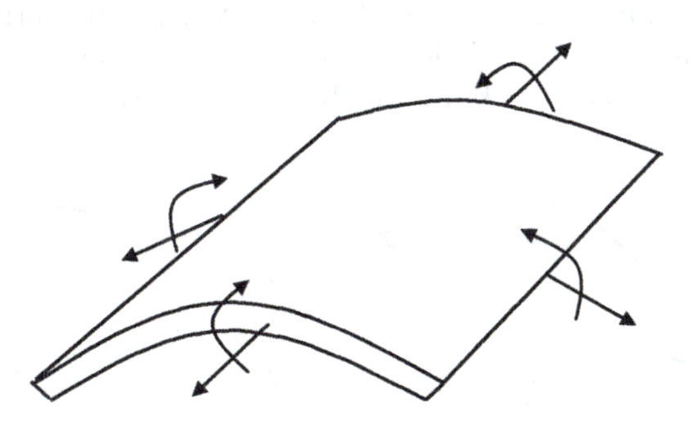

FIGURE 6.8
Forces and moments in a shell structure member.

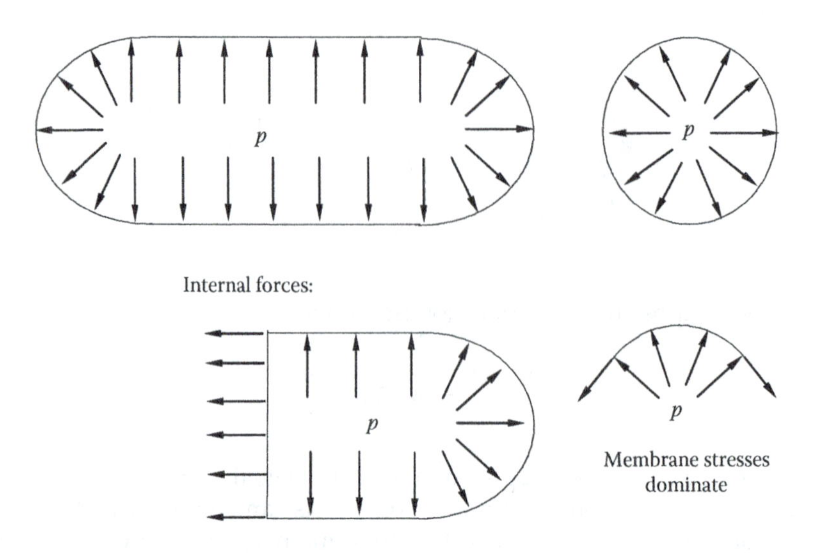

FIGURE 6.9
Internal forces acting in a cylinder under internal pressure load.

6.3 Modeling of Plates and Shells

Plates or shells can be modeled as flat or curved surfaces in space, with the thickness t assigned as a parameter (Figure 6.10). Discretization of the surfaces will involve the use of plate or shell elements, with the quality of the surface mesh improving with decreasing element size.

Figure 6.11 gives an example of a stamping part analyzed using shell elements. The bracket has a uniform thickness and is fixed at the four bolt hole positions. A load is applied through a pin passing through the two holes in the lower part of the bracket. Note that one layer of elements on the edge of each hole has been masked in the stress contour plot (Figure 6.11b), due to inaccurate stress results near the constraint locations. To reduce the true stress levels in the bracket, the thickness can be changed, the shape of the bracket can be modified, and the model is remeshed and reanalyzed, all of which are very easy to carry out with the shell elements.

In many cases, however, the plate and shell models may not be adequate for analyzing a structure member, even if it is considered thin. For example, the structure component has a nonuniform thickness (turbine blades, vessels with stiffeners, thin layered structures, etc.), see Figure 6.12, or has a crack for which detailed stress analysis is needed. In such

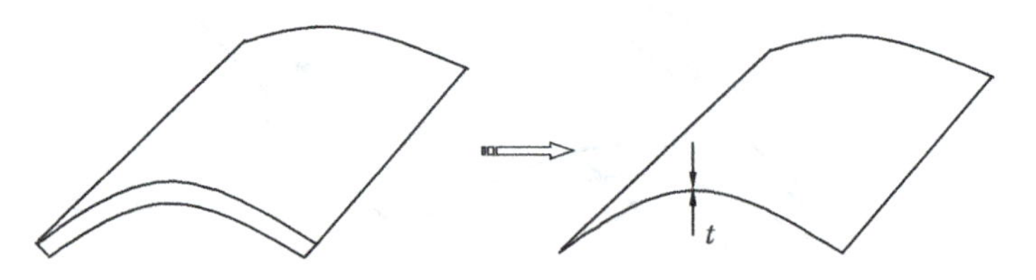

FIGURE 6.10
A shell structure member and its mathematical representation.

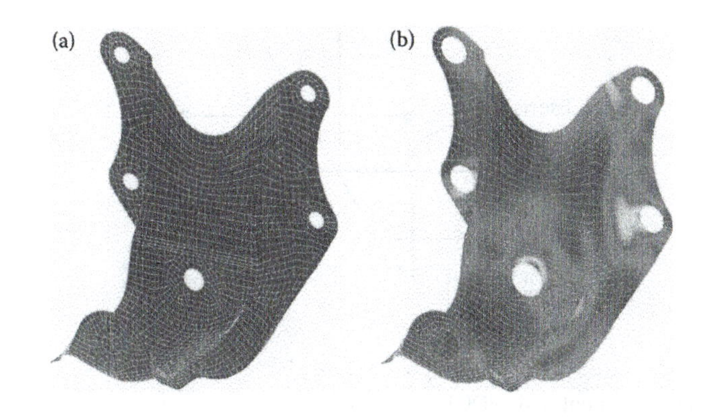

FIGURE 6.11
Stress analysis of a bracket using shell elements: (a) the FEA model; (b) stress contour plot.

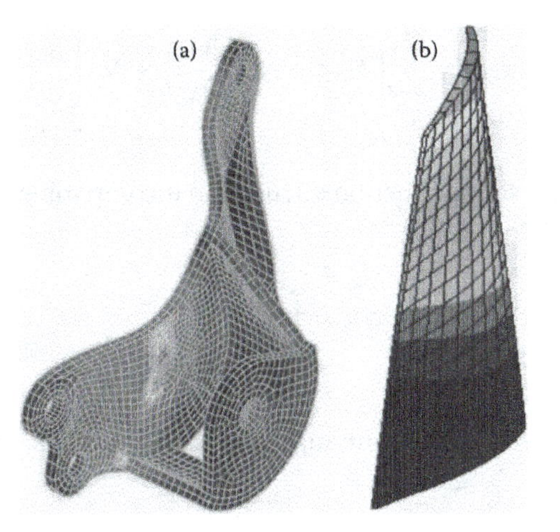

FIGURE 6.12
Cases in which shell elements are *not* adequate: (a) casting parts; (b) parts with nonuniform thickness. 3-D solid elements should be applied in such cases.

cases, one should turn to 3-D elasticity theory and apply solid elements which will be discussed in the next chapter.

6.4 Formulation of the Plate and Shell Elements

Several types of finite elements suitable for the analyses of plates and shells are introduced below.

6.4.1 Kirchhoff Plate Elements

The following four-node quadrilateral element has only bending capabilities. It has three DOFs at each node, that is, w, $\partial w/\partial y$, and $\partial w/\partial y$ (Figure 6.13).

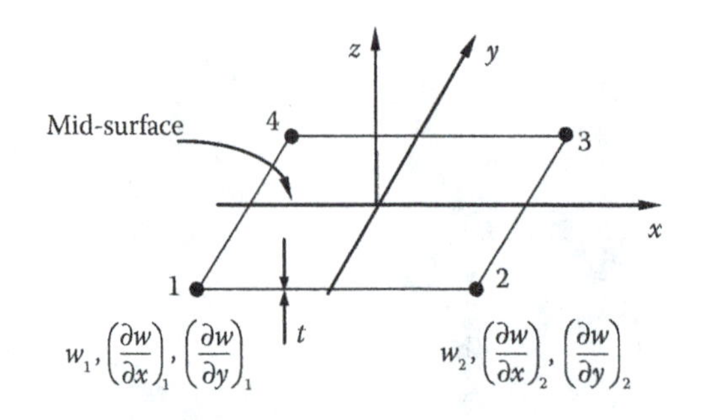

FIGURE 6.13
A four-node quadrilateral element with 3 DOFs at each node.

On each element, the deflection $w(x, y)$ is represented by

$$w(x,y) = \sum_{i=1}^{4} \left[N_i w_i + N_{xi} \left(\frac{\partial w}{\partial x} \right)_i + N_{yi} \left(\frac{\partial w}{\partial y} \right)_i \right] \tag{6.20}$$

where N_i, N_{xi}, and N_{yi} are shape functions. This is an incompatible element [7]. The stiffness matrix is still of the form

$$\mathbf{k} = \int_V \mathbf{B}^T \mathbf{E} \mathbf{B} \, dV \tag{6.21}$$

where **B** is the strain–displacement matrix and **E** Young's modulus (stress–strain) matrix.

6.4.2 Mindlin Plate Elements

The following two quadrilateral elements are Mindlin types with only bending capabilities (Figure 6.14). There are three DOFs at each node, that is, w, θ_x, and θ_y.

On each element, the displacement and rotations are represented by

$$\begin{aligned} w(x,y) &= \sum_{i=1}^{n} N_i w_i \\ \theta_x(x,y) &= \sum_{i=1}^{n} N_i \theta_{xi} \\ \theta_y(x,y) &= \sum_{i=1}^{n} N_i \theta_{yi} \end{aligned} \tag{6.22}$$

For these elements, there are three independent fields within each element. Deflection $w(x, y)$ is linear for Q4, and quadratic for Q8.

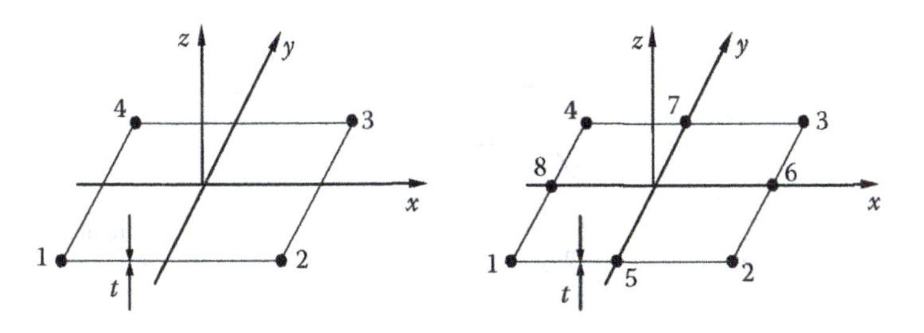

FIGURE 6.14
Four- and eight-node quadrilateral plate elements.

6.4.3 Discrete Kirchhoff Elements

This is a triangular element with only bending capabilities. First, start with a six-node triangular element (Figure 6.15). There are 5 DOFs at each corner node (w, $\partial w/\partial x$, $\partial w/\partial y$, θ_x, θ_y) 2 DOFs (θ_x, θ_y) at each mid-node, and a total of 21 DOFs for the six-node element.

Then, impose conditions $\gamma_{xz} = \gamma_{yz} = 0$, and so on, at selected nodes to reduce the DOFs (using relations in Equation 6.18), to obtain the following discrete Kirchhoff triangular (DKT) element (Figure 6.16).

For the three-node DKT element shown above, there are 3 DOFs at each node, that is, w, $\theta_x (= \partial w/\partial x)$, and $\theta_y (= \partial w/\partial y)$, and a total of 9 DOFs for the element. Note that $w(x, y)$ is incompatible for DKT elements [7]; however, its convergence is faster (w is cubic along each edge) and it is efficient.

6.4.4 Flat Shell Elements

A flat shell element can be developed by superimposing a plane stress element to a plate element (Figure 6.17).

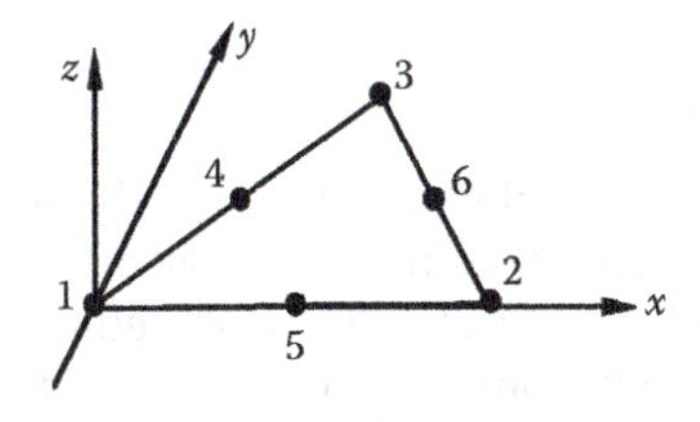

FIGURE 6.15
A six-node triangular element with 5 DOFs at each corner node and 2 DOFs at each mid-node.

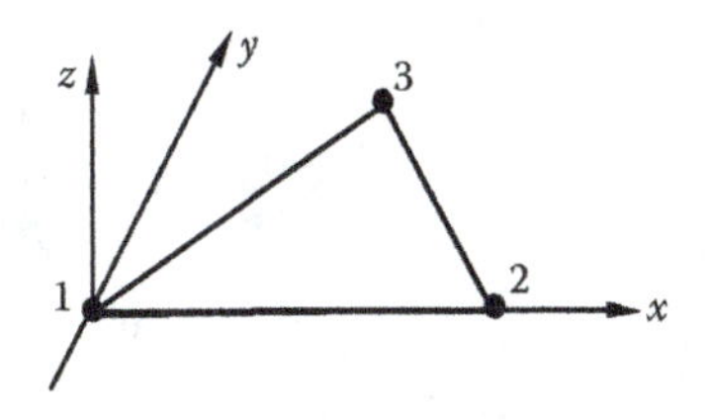

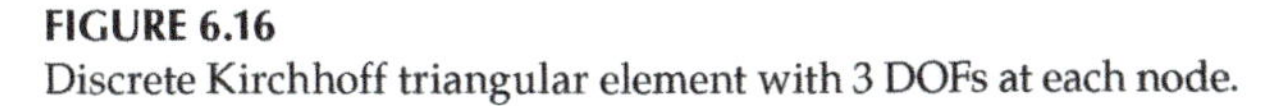
FIGURE 6.16
Discrete Kirchhoff triangular element with 3 DOFs at each node.

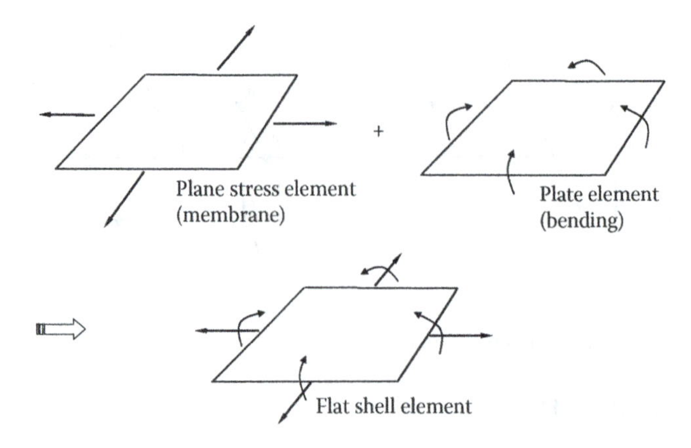

FIGURE 6.17
Combination of plane stress and plate bending elements yields a flat shell element.

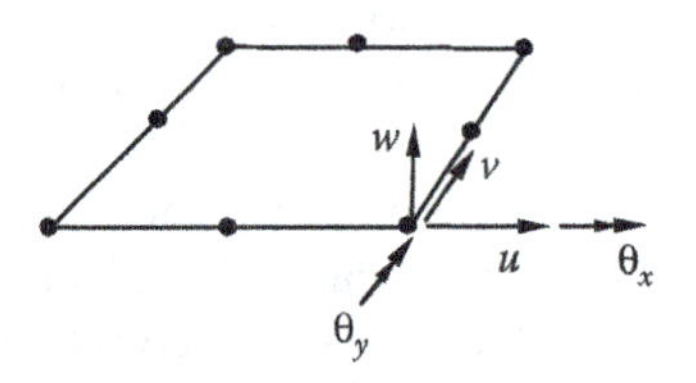

FIGURE 6.18
Q4 or Q8 shell elements.

This is analogous to the combination of a bar element and a simple beam element to yield a general beam element for modeling curved beams. A flat shell element, with the DOFs labeled at a typical node *i*, is shown in Figure 6.18.

6.4.5 Curved Shell Elements

Curved shell elements are based on the various shell theories. They are the most general shell elements (flat shell and plate elements are subsets). An eight-node curved shell element is illustrated in Figure 6.19, with the DOFs labeled at a typical node *i*. Formulations of the shell elements are relatively complicated. They are not discussed here and detailed derivations are available in References [7–9].

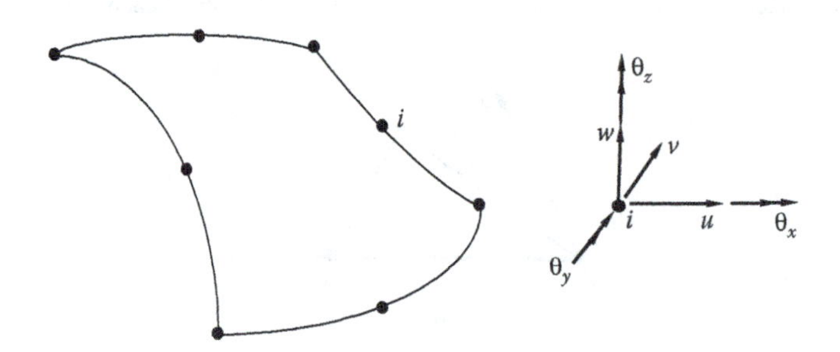

FIGURE 6.19
An eight-node curved shell element and the DOFs at a typical node *i*.

6.5 Case Studies with *ANSYS Workbench*

Problem Description: Vases are decorative pieces that can be of any artistic shapes. The figure below gives the dimensions of a flower vase made of glass. Assume that the vase has a uniform thickness of 4 mm. The water level reaches 100 mm below the opening of the vase. Determine the maximum deformation and von Mises stress in the vase under the hydrostatic pressure.

<u>*Material*</u>: Glass
$E = 70$ GPa
$\nu = 0.17$
<u>Boundary conditions:</u>
Bottom surface: fixed.

<u>Coordinates of construction points:</u>
A: (50, 0, and 0 mm)
B: (90, 40, and 0 mm)
C: (60, 120, and 0 mm)
D: (40, 180, and 0 mm)
E: (80, 280, and 0 mm)

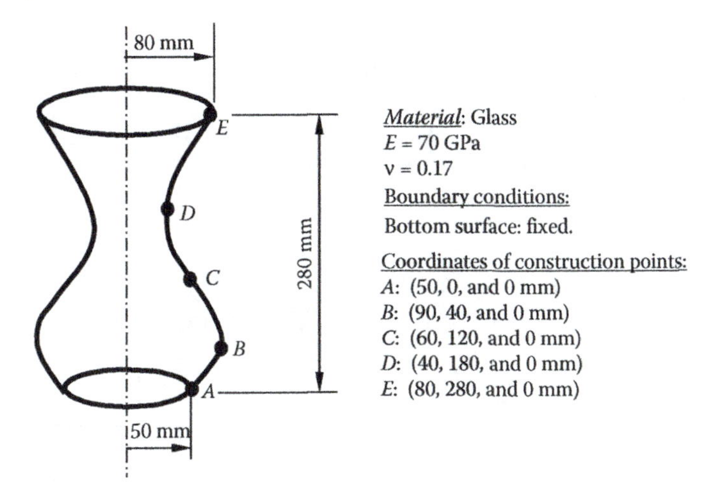

Solution

To solve the problem with *ANSYS® Workbench*, we employ the following steps:

Step 1: Start an *ANSYS Workbench Project*
Launch *ANSYS Workbench* and save the blank project as "*Vase.wbpj*."

Step 2: Create a *Static Structural (ANSYS)* Analysis System
Drag the *Static Structural (ANSYS)* icon from the *Analysis Systems Toolbox* window and drop it inside the highlighted green rectangle in the *Project Schematic* window to create a standalone static structural analysis system.

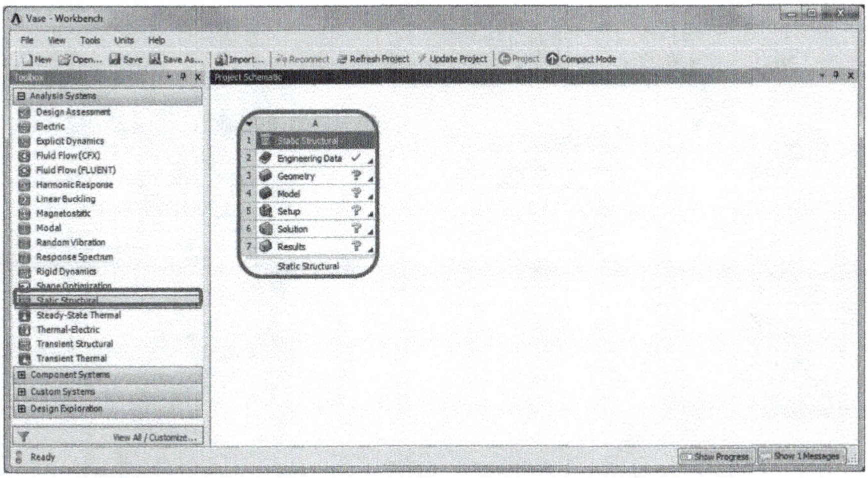

Step 3: Add a New Material

Double-click on the *Engineering Data* cell to add a new material. In the following *Engineering Data* interface which replaces the *Project Schematic*, type *"Glass"* as the name for the new material, and double-click *Isotropic Elasticity* under *Linear Elastic* in the leftmost *Toolbox* window. Enter *"70E9"* for *Young's Modulus* and *"0.17"* for *Poisson's Ratio* in the *Properties* window. Click the *Return to Project* button to go back to *Project Schematic*.

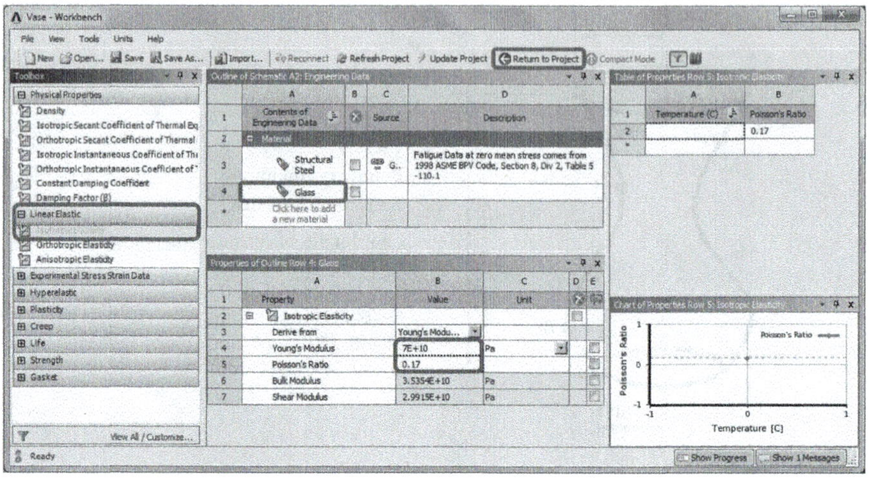

Step 4: Launch the *DesignModeler Program*

Ensure *Surface Bodies* is checked in the *Properties of Schematic A3: Geometry* window (select *Properties* from the *View* drop-down menu to enable display of this window). Choose *3D* as the *Analysis Type* in this *Properties* window. Double-click the *Geometry* cell to launch *DesignModeler*, and select *"Millimeter"* in the *Units* pop-up window.

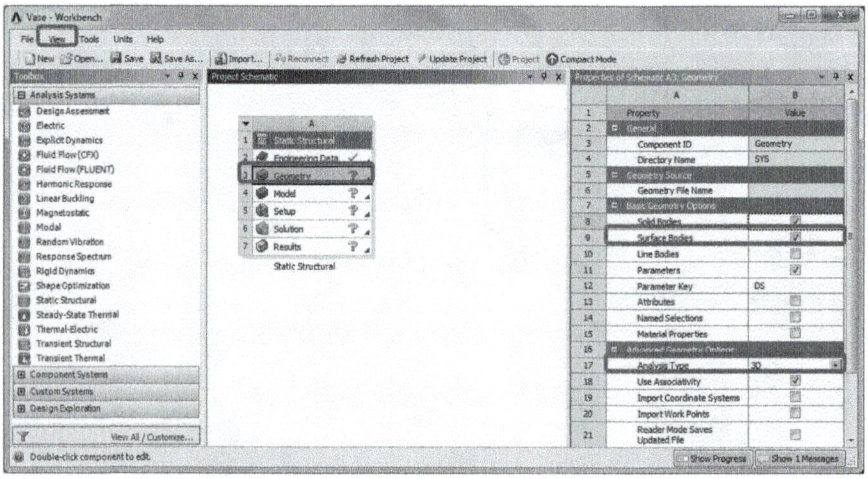

Step 5: Create a Profile Sketch

Click on the *Sketching* tab. Select the *Draw* toolbox and then *Construction Point*. Draw five construction points *A, B, C, D,* and *E* as shown below. Draw a horizontal line from the origin to point **A**. Next, draw a spline passing through points *A, B, C, D,* and *E*; right-click at the last construction point and choose *Open End* from the context menu to finish the spline creation.

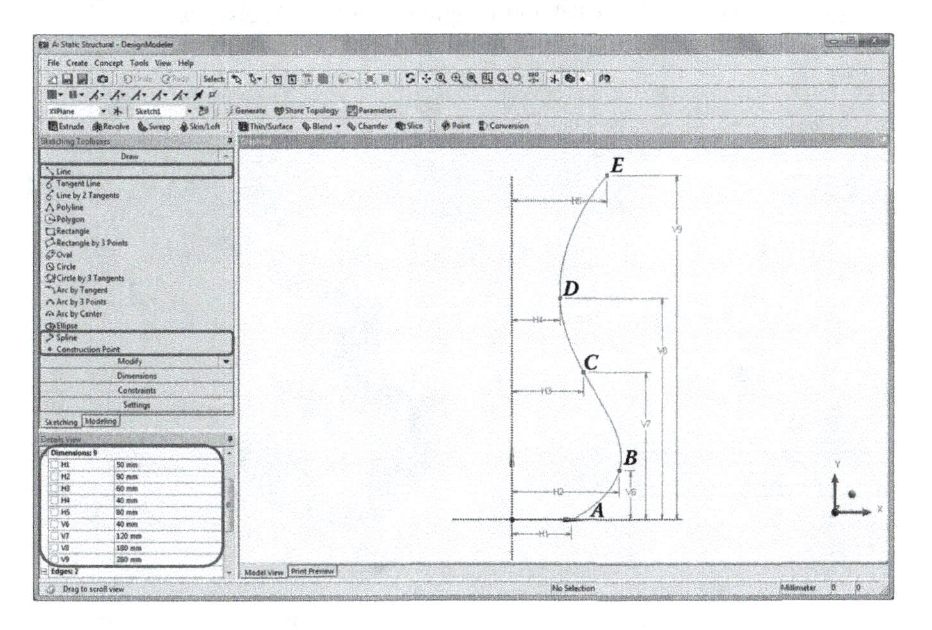

Step 6: Create a Surface Body

Switch to the *Modeling* tab and click on the *Revolve* feature.

The default *Base Object* is set as *Sketch1* in the *Details of Revolve1*. Click on the *y*-axis in the *Graphics* window and apply it to the *Axis* selection. Click *Generate*.

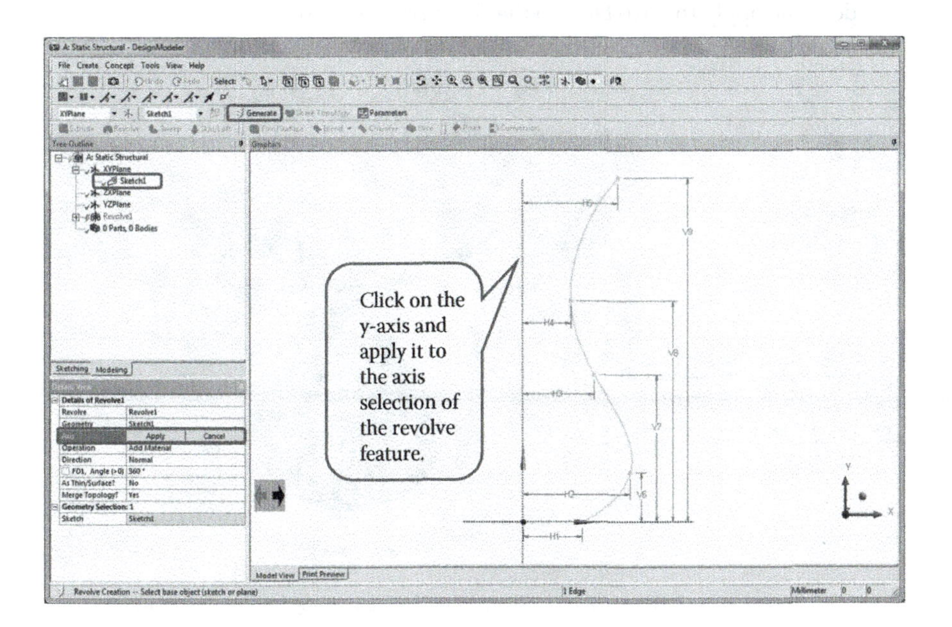

A *Surface Body* will be created from revolving the profile sketch. Exit the *DesignModeler* program.

Step 7: Launch the *Static Structural* Program

Double-click on the *Model* cell to launch the *Static Structural* program. Click on *Surface Body* under *Geometry* in the *Outline tree*. In the *Details of "Surface Body,"* set *Thickness* as *4e-3 m*. Click to the right of the *Material Assignment* field and select *Glass* from the drop-down menu.

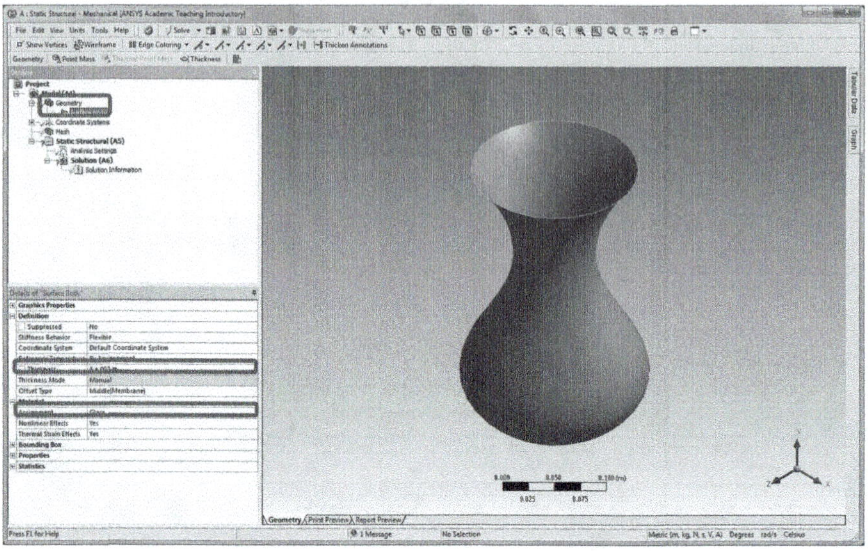

Step 8: Generate Mesh

Right click on *Mesh* in the *Project Outline*. Select *Insert* and then *Sizing* from the context menu. In the *Details of "Face Sizing,"* enter "2e-3 m" for the *Element Size*. Click on the side wall and the bottom surface of the vase in the *Graphics* window and apply the two faces to the *Geometry* selection.

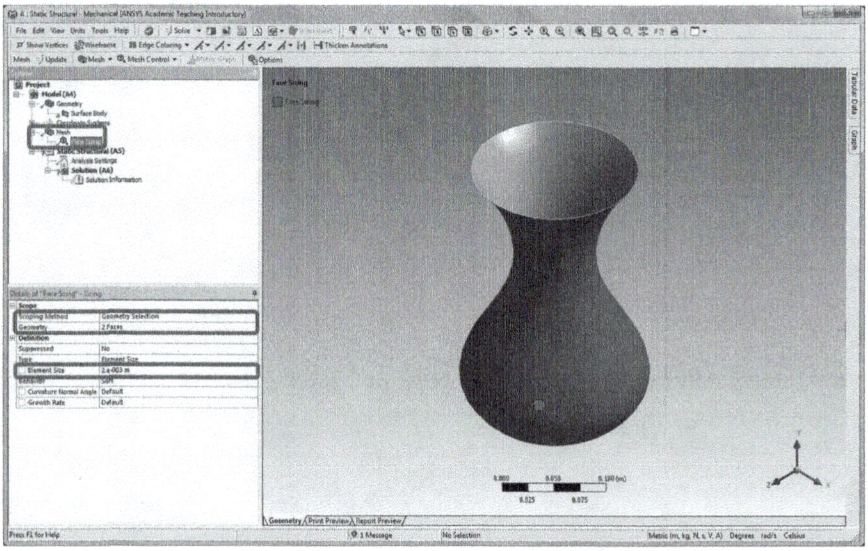

Right-click on *Mesh* and select *Generate Mesh.*

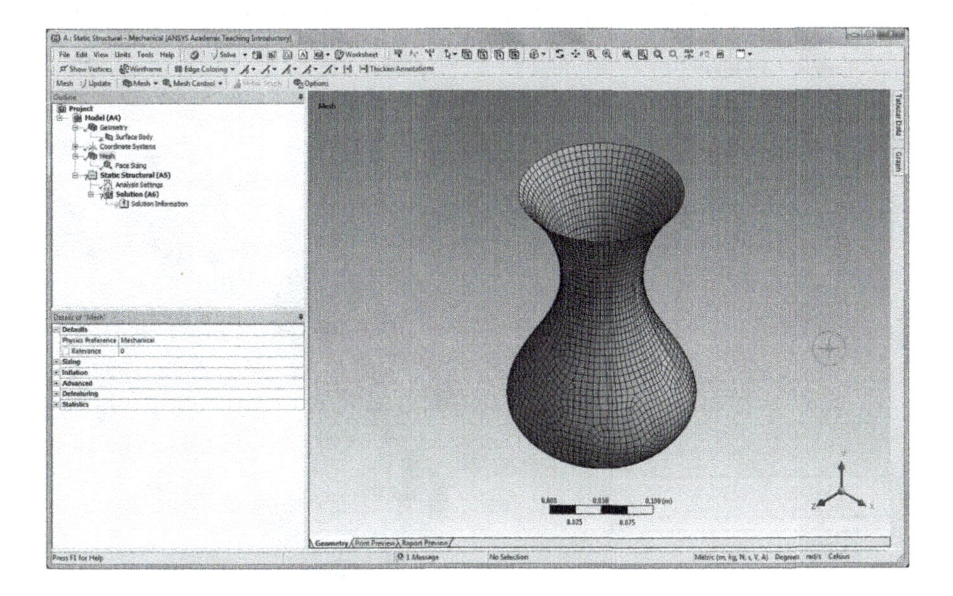

Step 9: Apply Boundary Conditions

Right-click on *Static Structural (A5).* Choose *Insert* and then *Fixed Support* from the context menu. Apply this support to the bottom surface of the vase.

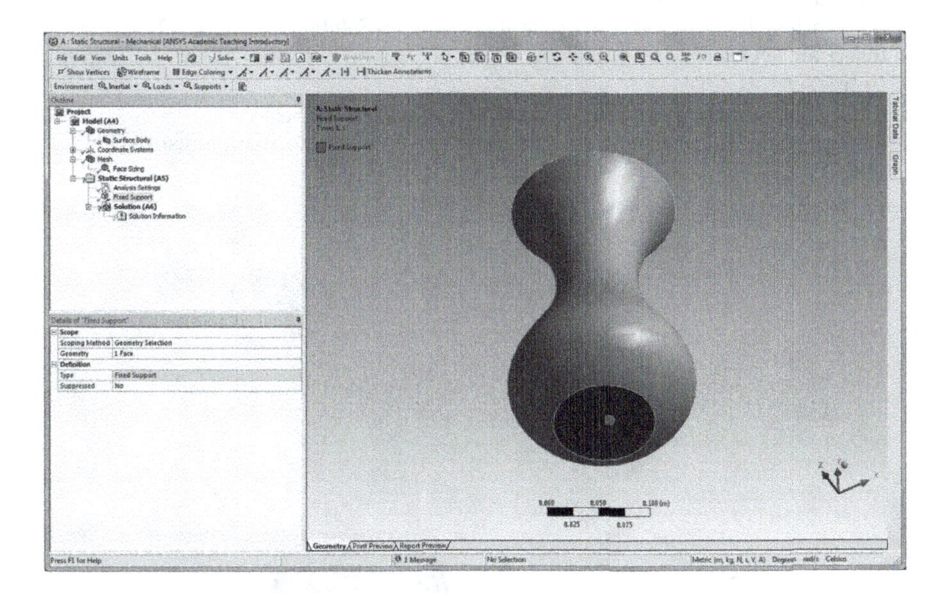

Step 10: Apply Loads

Right-click on *Static Structural (A5).* Choose *Insert* and then *Hydrostatic Pressure.* In the *Details of "Hydrostatic Pressure,"* Ctrl-click the side wall and the bottom surface of the vase, and apply the two faces to the *Geometry* selection.

Change the *Shell Face* to *Bottom,* because the hydrostatic pressure is applied to the inside of the shell surface. Enter *1000* kg/m³ for *Fluid Density.* Change the *Define By* selection to *Components,* and enter *9.8 m/s²* for the *Y Component* of *Hydrostatic Acceleration.* Enter *180e-3 m* in the field of *Y Coordinate* of the *Free Surface Location.*

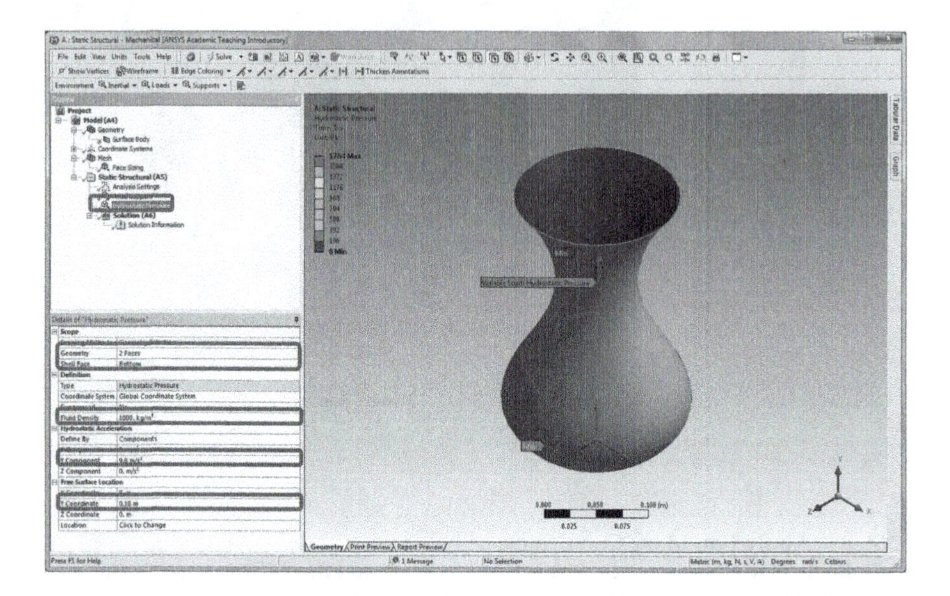

Step 11: Retrieve Solution

Insert a *Total Deformation* item by right-clicking on *Solution (A6)* in the *Outline* tree.

Insert an *Equivalent Stress* item by right-clicking on *Solution (A6)* in the *Outline* tree.

Right-click on *Solution (A6)* in the *Outline* tree and select *Solve.*

Click on *Total Deformation* in the *Outline* to review the deformation results.

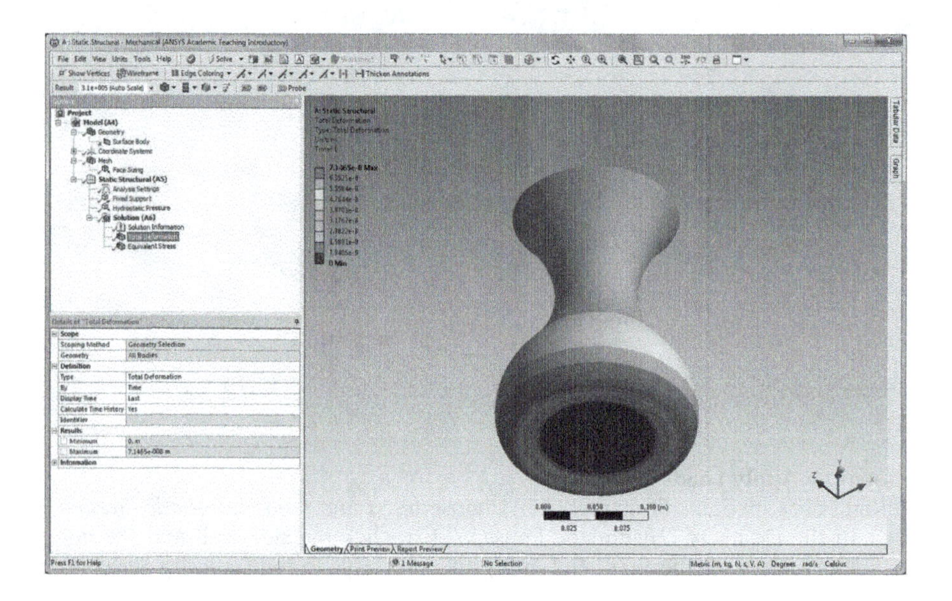

Click on *Equivalent Stress* in the *Outline* to review the stress results.

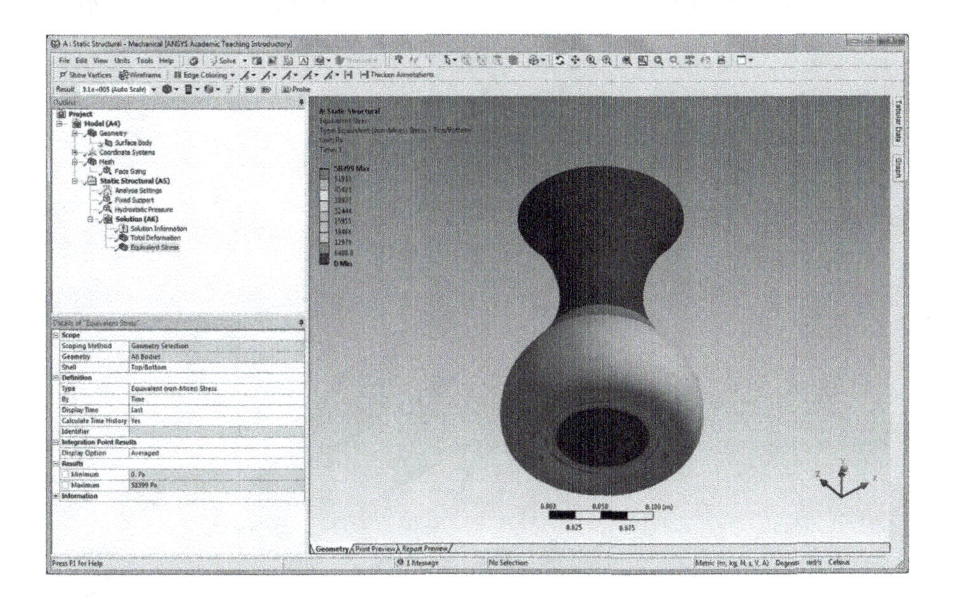

Modeling tips: Surface bodies can be created from planar 2D sketches, or by revolving, extruding, or sweeping lines or curves. In many cases, they can also be created from solid models, by using the *Mid-Surface* tool provided in *Workbench*.

For example, to create surface bodies from the solid model of a display shelf given below, select *Mid-Surface* from the drop-down menu of *Tools*.

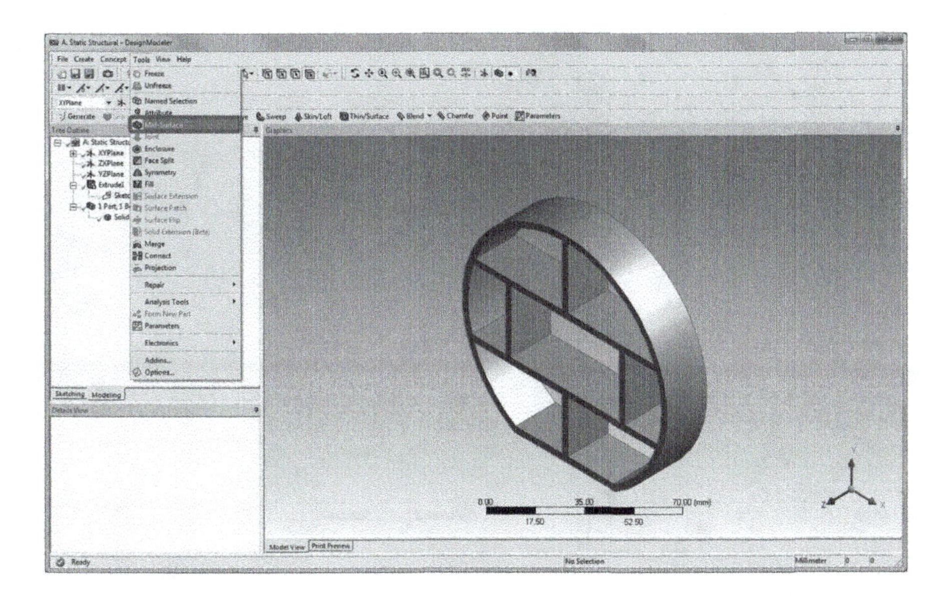

In the *"Details of MidSurf1,"* change the *Selection Method* to *Automatic*. The shelf panel has a thickness of 4 mm; set the *Minimum Threshhold* as *3 mm*, and the *Maximum Threshhold* as *5 mm*. In the field of *Find Face Pairs Now?*, choose *Yes* from the drop-down menu. Click *Generate*.

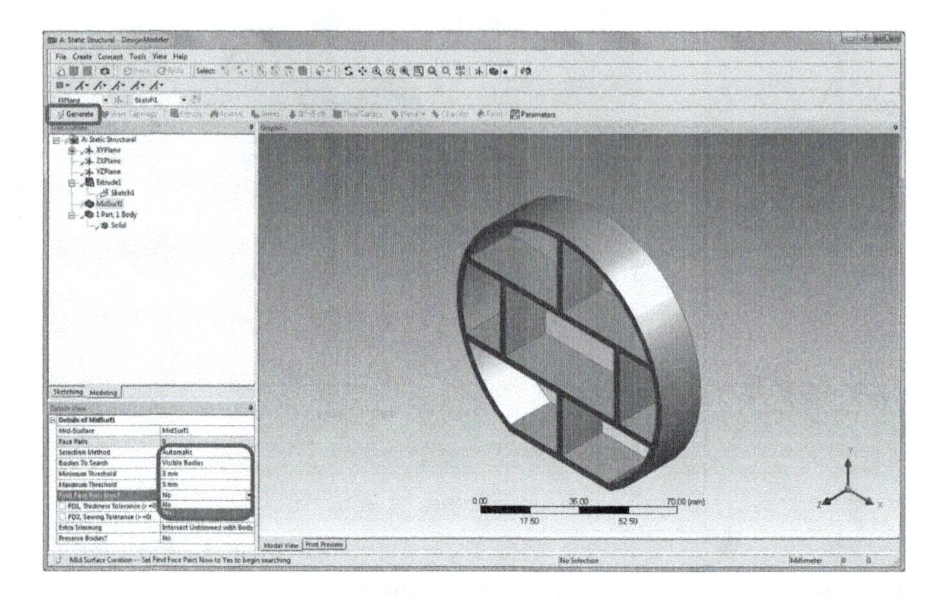

Eight surface bodies will be generated based on automatic face pair detection, as shown in the following figure. Exit *DesignModeler*, and the new model is now ready for subsequent simulation with plate or shell elements.

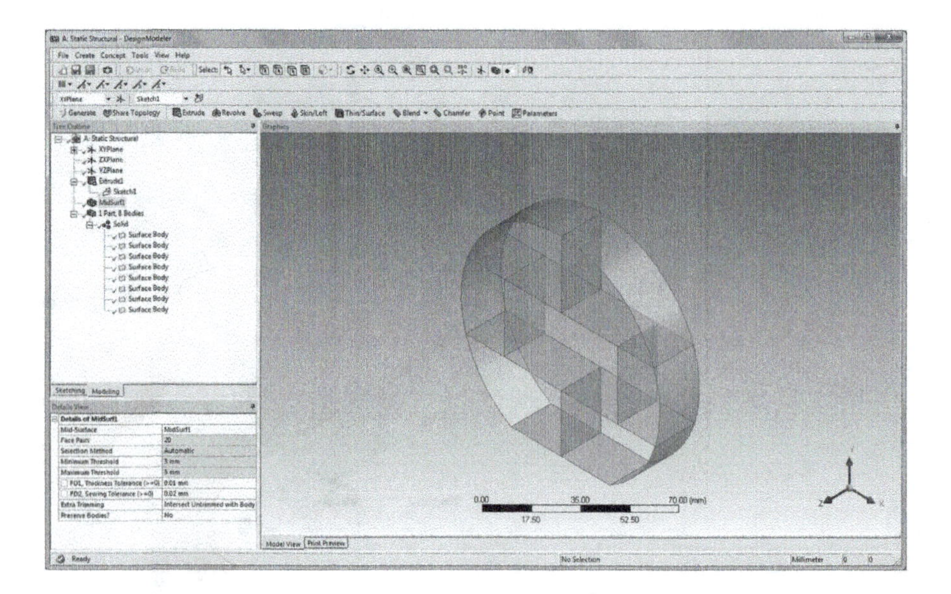

6.6 Summary

In this chapter, the main aspects of the plate and shell theories and the plate and shell elements used for analyzing plate and shell structures are discussed. Plates and shells can be regarded as the extensions of the beam elements from 1-D line elements to 2-D surface elements. Plates are usually applied in modeling flat thin structure members, while the shells are applied in modeling curved thin structure members. In applying the plate and shell elements, one should keep in mind the assumptions used in the development of these types of elements. In cases where these assumptions are no longer valid, one should turn to general 3-D theories and solid elements. A showcase using shell elements in modeling a vase is introduced using *ANSYS Workbench*.

6.7 Review of Learning Objectives

Now that you have finished this chapter you should be able to

1. Understand the assumptions used in the plate and shell theories.
2. Understand the behaviors of the plate and shell elements.
3. Know when *not* to use plate and shell elements in stress analysis (e.g., when plate or shell structures have nonuniform thickness or small features).
4. Create FEA models using plate and shell elements for deformation and stress analysis using *ANSYS Workbench*.

PROBLEMS

6.1 The square plate shown below has the following dimensions and material constants: $L = 1$ m, $t = 0.02$ m, $E = 30 \times 10^6$ N/m^2, and $\nu = 0.3$. Use plate elements to determine the maximum displacement and von Mises stress in the slab in the following cases: (a) The four edges are clamped and the slab is under a uniform load $q = 1$ N/m^2. (b) The four edges are clamped and the slab is under a concentrated force $P = 1$ N at the center C. (c) The four edges are hinged and the slab is under a uniform load $q = 1$ N/m^2. (d) The four edges are hinged and the slab is under a concentrated force $P = 1$ N at the center C. Verify the FEA solutions with the analytical solutions given in the Table 6.1.

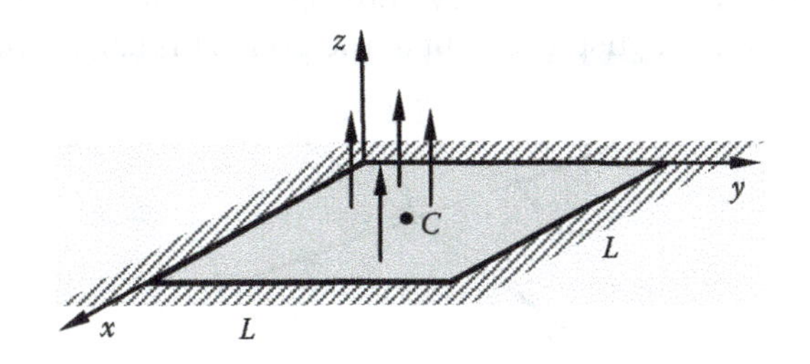

6.2 The roof structure shown below is loaded by its own weight with $q = 90\ \mathrm{lb_f/in^2}$. The dimensions and material constants are: $R = 25$ in., $L = 50$ in., $t = 0.25$ in., $E = 432 \times 10^6$ psi, and $\nu = 0.0$. The two straight edges are free, while the two curved edges have a "diaphragm" support (meaning that x and y DOFs are constrained, but z (along the length axis) and all rotational DOFs are unconstrained). Use shell elements to find the maximum displacement and von Mises stress in the structure. Verify your results (note that the value of the analytical solution for the displacement at the mid-point A of the straight edge is 0.3024 in.).

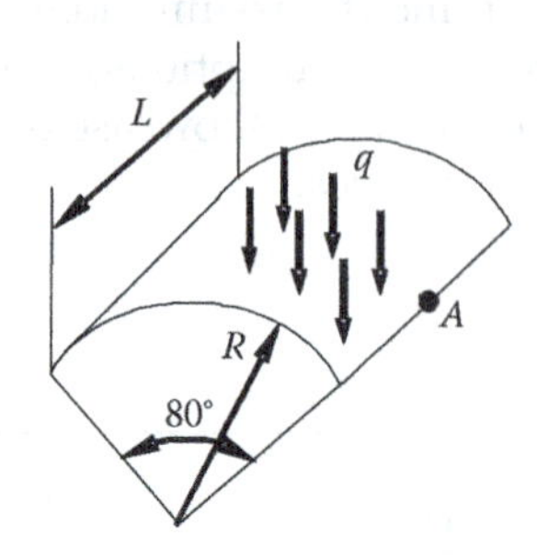

6.3 The pinched cylinder shown below is loaded by a force $F = 1$ N. The dimensions and material constants are: $R = 300$ mm, $L = 600$ mm, $t = 3$ mm, $E = 3 \times 10^6\ \mathrm{N/mm^2}$, and $\nu = 0.3$. Circular ends have diaphragm support (meaning that y and z DOFs are constrained, but x (along the length axis) and all rotational DOFs are unconstrained). Use shell elements to find the maximum displacement and von Mises stress in the structure. Verify your results (note that the value of the analytical solution for the displacement at the mid-point A is 0.1825×10^{-4} mm).

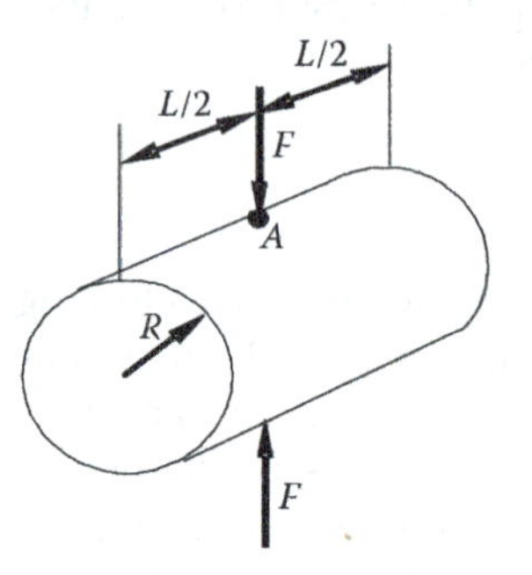

6.4 The pinched hemisphere shown below has radial loads $F = 2$ N at the equator. The dimensions and material constants are: $R = 10$ m, $t = 0.04$ m, $E = 68.25 \times 10^6\ \mathrm{N/m^2}$, and $\nu = 0.3$. The hemisphere has a free edge, and is restrained only against rigid body motion. Use shell elements to find the maximum displacement and von Mises stress in the structure. Verify your results (note that the value of the analytical solution for the displacement at the point A is 0.0924 m).

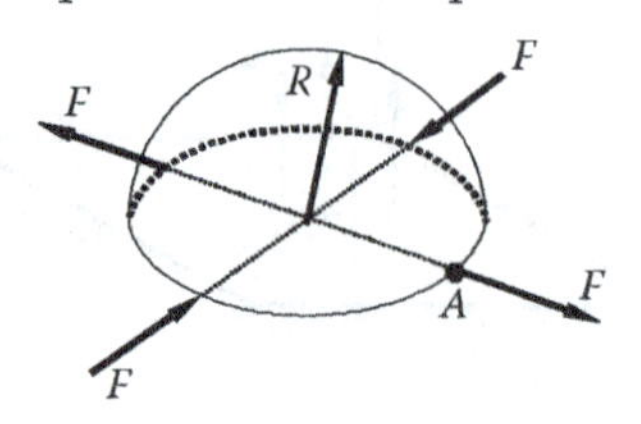

6.5 The twisted strip shown below has a load $F = 10^{-6}$ N. The dimensions and material constants are: $b = 1.1$ m, $L = 12$ m, $t = 0.0032$ m, $E = 29 \times 10^6$ N/m², and $\nu = 0.22$. The strip has a 90° twist over length L and is cantilevered. Use shell elements to find the maximum displacement and von Mises stress in the structure.

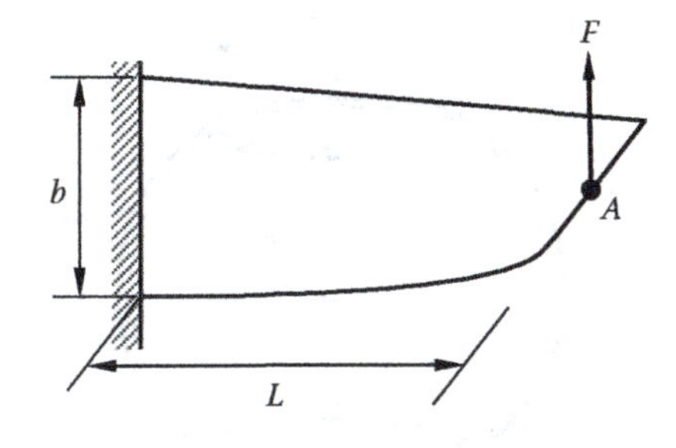

6.6 The twisted strip shown below has a load $F = 10^{-6}$ N. The dimensions and material constants are: $b = 1.1$ m, $L = 12$ m, $t = 0.0032$ m, $E = 29 \times 10^6$ N/m², and $\nu = 0.22$. The strip has a 90° twist over length L and is cantilevered. Use shell elements to find the maximum displacement and von Mises stress in the structure.

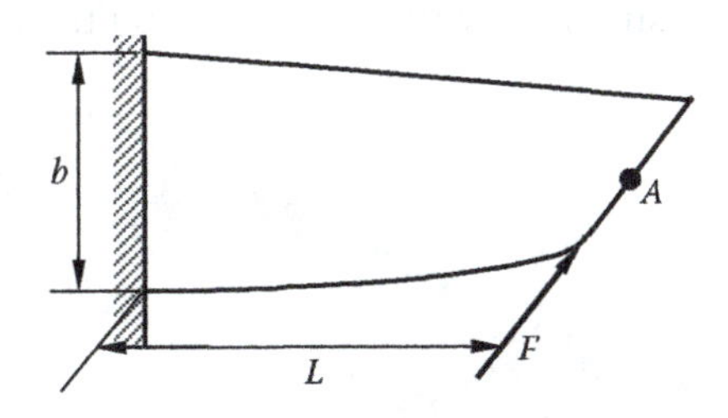

6.7 Consider a glass cup placed on a table, as shown in the figure. Using shell elements, find the maximum displacement and von Mises stress in the cup when the cup is applied with a pressure load of 10 N/mm² on the inner wall. Assume that the cup has a uniform thickness of 4 mm, $E = 70$ GPa, and $\nu = 0.17$.

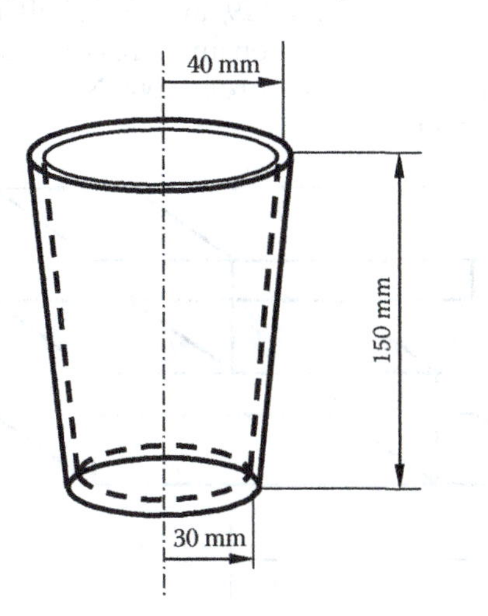

6.8 Consider a glass fish tank placed on a table, as shown in the figure. Using shell elements, find the maximum displacement and von Mises stress in the tank when the tank is full of water. Assume that the tank has a uniform thickness of 4 mm, $E = 70$ GPa, and $\nu = 0.17$.

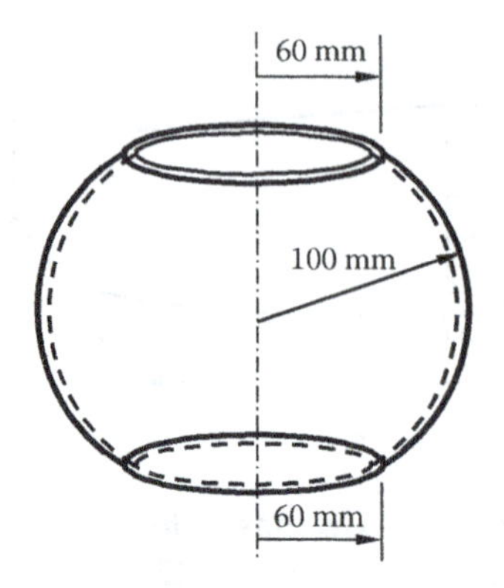

6.9 A fuel tank, with a total length = 5 m, diameter = 1 m, and thickness = 0.01 m, is shown below. Assume Young's modulus $E = 200$ GPa and Poisson's ratio $\nu = 0.3$. Using shell elements, find the deformation and stresses when the tank is applied with an internal pressure $p = 1$ MPa and placed on the ground.

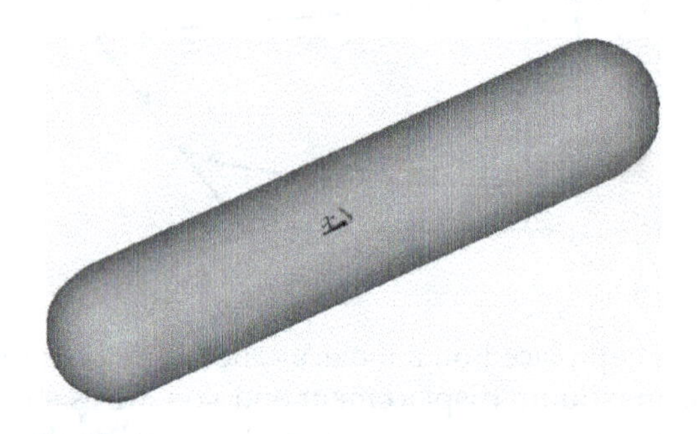

6.10 A bookshelf shown below is placed on the ground. Assume Young's modulus $E = 13.1$ GPa, Poisson's ratio $\nu = 0.29$, and the shelf thickness $t = 6$ mm. Using shell elements, find the deformation and stresses when the bookshelf is applied with a downward uniform pressure $p = 500\ \text{N/m}^2$ on the entire surface of the bookshelf middle layer.

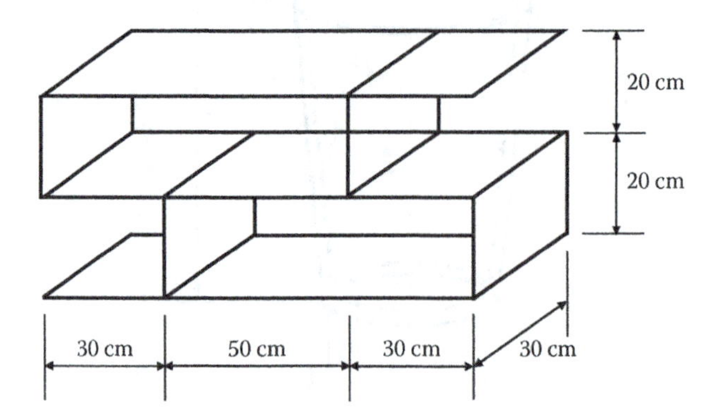

7

Three-Dimensional Elasticity

7.1 Introduction

Engineering designs involve 3-D structures that cannot be adequately represented using 1-D or 2-D models. Solid elements based on 3-D elasticity [10,11] are the most general elements for stress analysis when the simplified bar, beam, plane stress/strain, plate/shell elements are no longer valid or accurate. In general, 3-D structural analysis is one of the most important and powerful ways of providing insight into the behavior of an engineering design. In this chapter, we will review the elasticity equations for 3-D and then discuss a few types of finite elements commonly used for 3-D stress analysis. Several different types of supports, loads, and contact constraints will be introduced for 3-D structural modeling, followed by a case study on predicting the deformation and stresses in an assembly structure using *ANSYS® Workbench*.

7.2 Review of Theory of Elasticity

The state of stress at a point in a 3-D elastic body is shown in Figure 7.1.

In vector form, the six independent stress components determining the state of stress can be written as

$$\sigma = \{\sigma\} = \begin{Bmatrix} \sigma_x \\ \sigma_y \\ \sigma_z \\ \tau_{xy} \\ \tau_{yz} \\ \tau_{zx} \end{Bmatrix}, \quad \text{or} \quad [\sigma_{ij}] \tag{7.1}$$

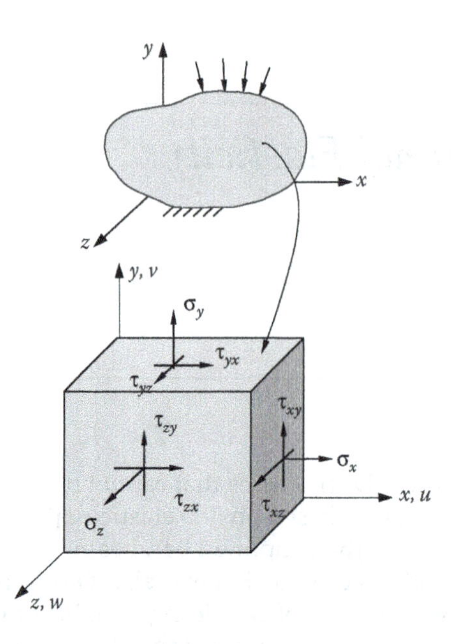

FIGURE 7.1
State of stress at a point in a 3-D elastic body.

Similarly, the six independent strain components in 3-D can be expressed as

$$\varepsilon = \{\varepsilon\} = \begin{Bmatrix} \varepsilon_x \\ \varepsilon_y \\ \varepsilon_z \\ \gamma_{xy} \\ \gamma_{yz} \\ \gamma_{zx} \end{Bmatrix}, \quad \text{or} \quad [\varepsilon_{ij}] \tag{7.2}$$

7.2.1 Stress–Strain Relation

The stress–strain relation in 3-D is given by

$$\begin{Bmatrix} \sigma_x \\ \sigma_y \\ \sigma_z \\ \tau_{xy} \\ \tau_{yz} \\ \tau_{zx} \end{Bmatrix} = \frac{E}{(1+v)(1-2v)} \begin{bmatrix} 1-v & v & v & 0 & 0 & 0 \\ v & 1-v & v & 0 & 0 & 0 \\ v & v & 1-v & 0 & 0 & 0 \\ 0 & 0 & 0 & \dfrac{1-2v}{2} & 0 & 0 \\ 0 & 0 & 0 & 0 & \dfrac{1-2v}{2} & 0 \\ 0 & 0 & 0 & 0 & 0 & \dfrac{1-2v}{2} \end{bmatrix} \begin{Bmatrix} \varepsilon_x \\ \varepsilon_y \\ \varepsilon_z \\ \gamma_{xy} \\ \gamma_{yz} \\ \gamma_{zx} \end{Bmatrix} \tag{7.3}$$

Or in a matrix form:

$$\sigma = \mathbf{E}\varepsilon$$

7.2.2 Displacement

The displacement field can be described as

$$\mathbf{u} = \begin{Bmatrix} u(x,y,z) \\ v(x,y,z) \\ w(x,y,z) \end{Bmatrix} = \begin{Bmatrix} u_1 \\ u_2 \\ u_3 \end{Bmatrix} \tag{7.4}$$

7.2.3 Strain–Displacement Relation

Strain field is related to the displacement field as given below

$$\begin{aligned} \varepsilon_x &= \frac{\partial u}{\partial x}, \quad \varepsilon_y = \frac{\partial v}{\partial y}, \quad \varepsilon_z = \frac{\partial w}{\partial z}, \\ \gamma_{xy} &= \frac{\partial v}{\partial x} + \frac{\partial u}{\partial y}, \gamma_{yz} = \frac{\partial w}{\partial y} + \frac{\partial v}{\partial z}, \gamma_{xz} = \frac{\partial u}{\partial z} + \frac{\partial w}{\partial x} \end{aligned} \tag{7.5}$$

These six equations can be written in the following index or tensor form:

$$\varepsilon_{ij} = \frac{1}{2}\left(\frac{\partial u_i}{\partial x_j} + \frac{\partial u_j}{\partial x_i}\right), \quad i, j = 1,2,3$$

Or simply,

$$\varepsilon_{ij} = \frac{1}{2}(u_{i,j} + u_{j,i}) \quad \text{(Tensor notation)}$$

7.2.4 Equilibrium Equations

The stresses and body force vector f at each point satisfy the following three equilibrium equations for elastostatic problems:

$$\begin{aligned} &\frac{\partial \sigma_x}{\partial x} + \frac{\partial \tau_{xy}}{\partial y} + \frac{\partial \tau_{xz}}{\partial z} + f_x = 0, \\ &\frac{\partial \tau_{yx}}{\partial x} + \frac{\partial \sigma_y}{\partial y} + \frac{\partial \tau_{yz}}{\partial z} + f_y = 0, \\ &\frac{\partial \tau_{zx}}{\partial x} + \frac{\partial \tau_{zy}}{\partial y} + \frac{\partial \sigma_z}{\partial z} + f_z = 0 \end{aligned} \tag{7.6}$$

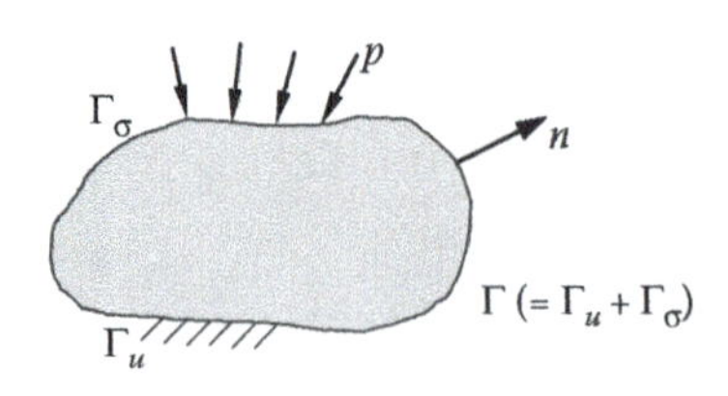

FIGURE 7.2
The boundary of a 3-D elastic domain.

Or in index or tensor notation:

$$\sigma_{ij,j} + f_i = 0$$

7.2.5 Boundary Conditions

At each point on the boundary Γ and in each direction, either displacement or traction (stress on the boundary) should be given, that is

$$\begin{aligned} u_i &= \bar{u}_i, \quad && \textit{on } \Gamma_u \textit{ (specified displacement);} \\ t_i &= \bar{t}_i, \quad && \textit{on } \Gamma_\sigma \textit{ (specified traction)} \end{aligned} \tag{7.7}$$

in which the barred quantities denote given values, and the traction (stress on a surface) is defined by $t_i = \sigma_{ij} n_j$ or in a matrix form:

$$\begin{Bmatrix} t_x \\ t_y \\ t_z \end{Bmatrix} = \begin{bmatrix} \sigma_x & \tau_{xy} & \tau_{xz} \\ \tau_{xy} & \sigma_y & \tau_{yz} \\ \tau_{xz} & \tau_{yz} & \sigma_z \end{bmatrix} \begin{Bmatrix} n_x \\ n_y \\ n_z \end{Bmatrix}$$

with n being the normal (Figure 7.2).

7.2.6 Stress Analysis

For 3-D stress analysis, one needs to solve equations in Equations 7.3, 7.5, and 7.6 under the BCs in Equation 7.7 in order to obtain the stress, strain, and displacement fields (15 equations for 15 unknowns for a 3-D problem). Analytical solutions are often difficult to find and thus numerical methods such as the FEA are often applied in 3-D stress analysis.

7.3 Modeling of 3-D Elastic Structures

3-D stress analysis using solid elements is one of the most challenging tasks in FEA. In the following, practical considerations in 3-D FEA modeling such as mesh and BCs are discussed. The types of contact used in assembly analyses are presented.

7.3.1 Mesh Discretization

Meshing structures with complicated geometries can be very tedious and time-consuming. Great care should be taken to ensure that the FEA mesh is in good quality (e.g., with no distorted elements). Computing cost is another factor. For structures with stress concentrations, large FEA models are often needed, which can run hours or days. A good CAE engineer should be able to decide where to apply a fine mesh and where not to, in order to strike a balance between the computational cost and accuracy for an FEA task.

Figure 7.3 shows an example of an FEA model using solid elements of a drag link in a car. Although the structure has a slender shape, it has a bended angle and holes. 3-D solid elements are needed for the stress analysis in this case. Great care is taken in meshing this part, where quadratic elements are used for better accuracy in the stress analysis. Buckling analysis may also be conducted for slender structures when they are under compressions. More information about buckling analysis using FEA can be found in Chapter 12.

Figure 7.4 shows a 3-D FEA of a gear coupling which is applied to transmit power through two aligned rotating shafts. Contact stresses and failure modes are to be determined based on detailed 3-D FE models. This analysis requires the use of nonlinear FEA options, which are readily available now in almost all FEA software packages.

7.3.2 Boundary Conditions: Supports

A number of support types are available for 3-D structural analysis. The following list includes three common support conditions.

- *Fixed support*: Prevent the geometry entity from moving or deforming.
- *Frictionless support*: Prevent the face geometry from moving or deforming in the normal direction relative to the face.
- *Cylindrical support*: Prevent the cylindrical face from moving or deforming in any combination of radial, axial, or tangential directions relative to the cylinder.

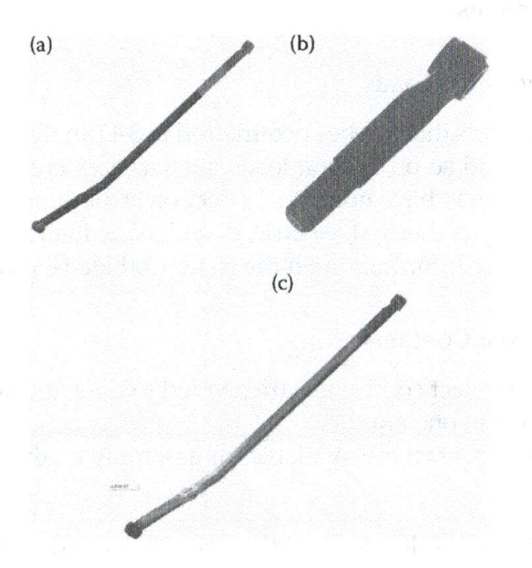

FIGURE 7.3
FEA for a drag link: (a) the model; (b) mesh for the right end; (c) stress distribution due to tension loads applied at the two ends.

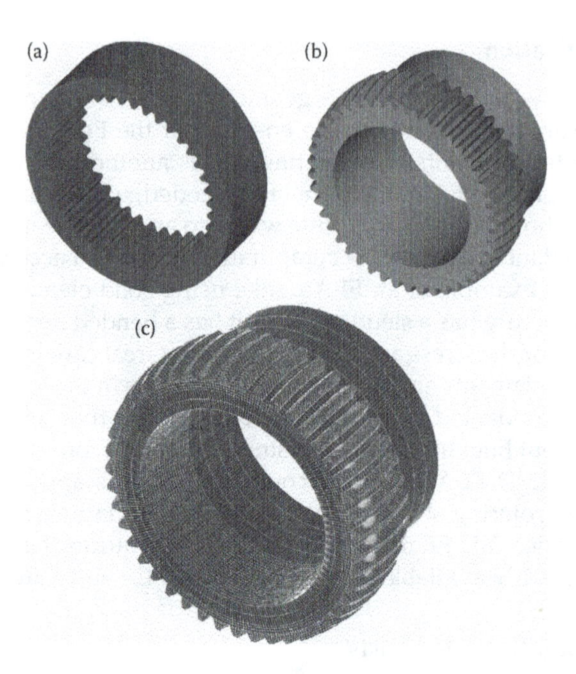

FIGURE 7.4
Analysis of a gear coupling: (a) ring gear; (b) hub gear; (c) high-contact stresses in the gear teeth obtained using nonlinear FEA.

Note that there are six rigid-body motions for 3-D bodies: three translations and three rotations. These rigid-body motions (causes of singularity of the system of equations) must be removed from the FEA model for stress analysis to ensure the accuracy of the analysis. On the other hand, over constraints can also cause inaccurate or unwarranted results. For more information on the support conditions, please refer to the *Mechanical User Guide* of the *ANSYS* help documents [12].

7.3.3 Boundary Conditions: Loads

The types of structural loads that can be encountered in 3-D analyses include force, moment, pressure, bearing load, and so on. Inertia loads such as acceleration, standard earth gravity, or rotational velocity may have nontrivial effect on structures' stress behaviors as well. Other loading types such as thermal, electric, or magnetic loads can also be involved, but are less common. For more information on the structural loads, please see Reference [12].

7.3.4 Assembly Analysis: Contacts

For assembly analysis, contact conditions are needed to describe how different contacting bodies can move relative to one another.

The following types of contact are available for assembly analysis, as listed in Reference [12]:

- *Bonded*: This is the default configuration. Bonded regions can be considered as glued together, allowing no sliding or separation between the contacting regions. For many applications, bonded contact is sufficient for stress calculations between bolted or welded parts in assemblies.

- *No separation*: This contact type allows frictionless sliding along the contact faces, but separation of faces in contact is not allowed.
- *Frictionless*: This contact model allows free sliding, assuming a zero coefficient of friction. Gap can form in between regions in contact.
- *Rough*: This model assumes an infinite friction coefficient between the bodies in contact. No sliding can occur.
- *Frictional*: This model allows bodies in contact to slide relative to each other, once an equivalent shear stress up to a certain magnitude is exceeded.

In most cases, contact regions can be automatically detected and generated in the FEA program. They can also be manually modified, if needed.

7.4 Formulation of Solid Elements

In this section, we will first summarize the FEA formulation for 3-D elasticity problems, which are straightforward extensions of the FEA formulations for 1-D bar and 2-D elasticity problems. We will then use an example of linear hexahedral (eight-node brick) element to examine the element formulation in detail.

7.4.1 General Formulation

As in the FEA formulations for 1-D and 2-D problems, we first interpolate the displacement fields within a 3-D element using shape functions N_i:

$$\begin{aligned} u &= \sum_{i=1}^{N} N_i u_i, \\ v &= \sum_{i=1}^{N} N_i v_i, \\ w &= \sum_{i=1}^{N} N_i w_i \end{aligned} \tag{7.8}$$

in which u_i, v_i, and w_i are nodal values of the displacement on the element, and N is the number of nodes on that element. In matrix form, we have:

$$\begin{Bmatrix} u \\ v \\ w \end{Bmatrix}_{(3\times1)} = \begin{bmatrix} N_1 & 0 & 0 & N_2 & 0 & 0 & \cdots \\ 0 & N_1 & 0 & 0 & N_2 & 0 & \cdots \\ 0 & 0 & N_1 & 0 & 0 & N_2 & \cdots \end{bmatrix}_{(3\times3N)} \begin{Bmatrix} u_1 \\ v_1 \\ w_1 \\ u_2 \\ v_2 \\ w_2 \\ \vdots \end{Bmatrix}_{(3N\times1)} \tag{7.9}$$

Or in a matrix form:

$$\mathbf{u} = \mathbf{N}\,\mathbf{d}$$

Using relations given in Equations 7.5 and 7.8, we can derive the strain vector to obtain:

$$\boldsymbol{\varepsilon} = \mathbf{B}\,\mathbf{d}$$

in which **B** is the matrix relating the nodal displacement vector **d** to the strain vector **ε**. Note that the dimensions of the **B** matrix are $6 \times 3N$.

Once the **B** matrix is found, one can apply the following familiar expression to determine the stiffness matrix for the element:

$$\mathbf{k} = \int_{v} \mathbf{B}^{T}\mathbf{E}\mathbf{B}\,dv \tag{7.10}$$

The dimensions of the stiffness matrix **k** are $3N \times 3N$. A numerical quadrature is often needed to evaluate the above integration, which can be expensive if the number of nodes is large, such as for higher-order elements.

7.4.2 Typical Solid Element Types

We can classify the type of elements for 3-D problems as follows (Figure 7.5) according to their shapes and the orders of the shape functions constructed on the elements:

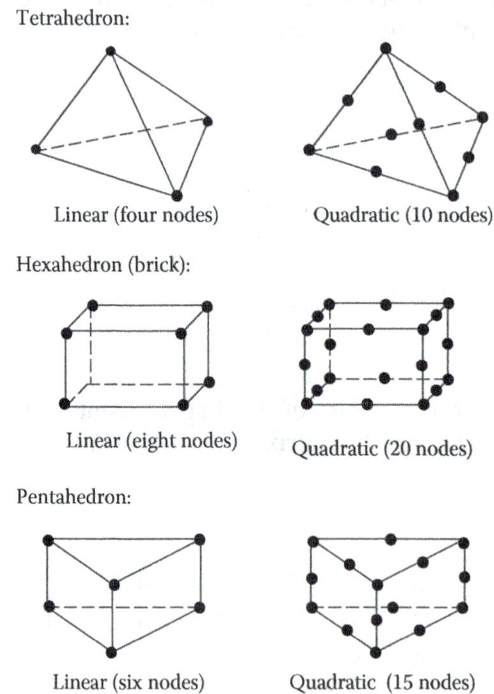

FIGURE 7.5
Different types of 3-D solid elements.

Whenever possible, one should try to apply higher-order (quadratic) elements, such as 10-node tetrahedron and 20-node brick elements for 3-D stress analysis. Avoid using the linear, especially the four-node tetrahedron elements in 3-D stress analysis, because they are inaccurate for such purposes. However, it is fine to use them for deformation analysis or in vibration analysis (see Chapter 8).

7.4.3 Formulation of a Linear Hexahedral Element Type

Displacement Field in the Element:

$$u=\sum_{i=1}^{8} N_i u_i, \quad v=\sum_{i=1}^{8} N_i v_i, \quad w=\sum_{i=1}^{8} N_i w_i \tag{7.11}$$

Shape Functions:

$$\begin{aligned}
N_1(\xi,\eta,\zeta) &= \frac{1}{8}(1-\xi)(1-\eta)(1-\zeta),\\
N_2(\xi,\eta,\zeta) &= \frac{1}{8}(1+\xi)(1-\eta)(1-\zeta),\\
N_3(\xi,\eta,\zeta) &= \frac{1}{8}(1+\xi)(1+\eta)(1-\zeta),\\
&\ \ \vdots \qquad\qquad \vdots\\
N_8(\xi,\eta,\zeta) &= \frac{1}{8}(1-\xi)(1+\eta)(1+\zeta)
\end{aligned} \tag{7.12}$$

Note that we have the following relations for the shape functions:

$$N_i(\xi_j,\eta_j,\zeta_j) = \delta_{ij}, \quad i,j = 1,2,\ldots,8.$$

$$\sum_{i=1}^{8} N_i(\xi,\eta,\zeta) = 1$$

Coordinate Transformation (Mapping):

$$x=\sum_{i=1}^{8} N_i x_i, \quad y=\sum_{i=1}^{8} N_i y_i, \quad z=\sum_{i=1}^{8} N_i z_i \tag{7.13}$$

That is, the same shape functions are used for the element geometry as for the displacement field. This kind of element is called an isoparametric element. The transformation between (ξ, η, ζ) and (x, y, z) described by Equation 7.13 is called isoparametric mapping (see Figure 7.6).

Jacobian Matrix:

$$\begin{Bmatrix} \dfrac{\partial u}{\partial \xi} \\ \dfrac{\partial u}{\partial \eta} \\ \dfrac{\partial u}{\partial \zeta} \end{Bmatrix} = \underbrace{\begin{bmatrix} \dfrac{\partial x}{\partial \xi} & \dfrac{\partial y}{\partial \xi} & \dfrac{\partial z}{\partial \xi} \\ \dfrac{\partial x}{\partial \eta} & \dfrac{\partial y}{\partial \eta} & \dfrac{\partial z}{\partial \eta} \\ \dfrac{\partial x}{\partial \zeta} & \dfrac{\partial y}{\partial \zeta} & \dfrac{\partial z}{\partial \zeta} \end{bmatrix}}_{\equiv\ \mathbf{J}\ \text{Jacobian matrix}} \begin{Bmatrix} \dfrac{\partial u}{\partial x} \\ \dfrac{\partial u}{\partial y} \\ \dfrac{\partial u}{\partial z} \end{Bmatrix} \tag{7.14}$$

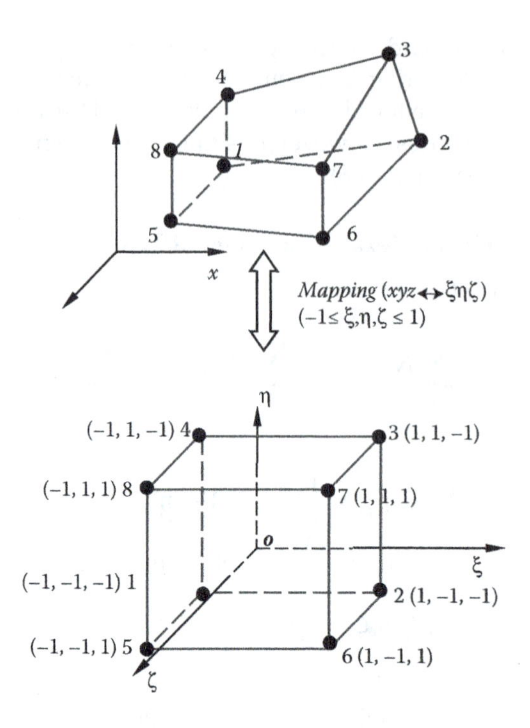

FIGURE 7.6
Mapping an element to the natural coordinate system.

Inverting this relation, we have:

$$\begin{Bmatrix} \dfrac{\partial u}{\partial x} \\ \dfrac{\partial u}{\partial y} \\ \dfrac{\partial u}{\partial z} \end{Bmatrix} = \mathbf{J}^{-1} \begin{Bmatrix} \dfrac{\partial u}{\partial \xi} \\ \dfrac{\partial u}{\partial \eta} \\ \dfrac{\partial u}{\partial \zeta} \end{Bmatrix}, \quad \text{with } \frac{\partial u}{\partial \xi} = \sum_{i=1}^{8} \frac{\partial N_i}{\partial \xi} u_i \text{ and so on} \tag{7.15}$$

and similarly for v and w. These relations lead to the following expression for the strain:

$$\varepsilon = \begin{Bmatrix} \varepsilon_x \\ \varepsilon_y \\ \varepsilon_z \\ \gamma_{xy} \\ \gamma_{yz} \\ \gamma_{zx} \end{Bmatrix} = \begin{Bmatrix} \dfrac{\partial u}{\partial x} \\ \dfrac{\partial v}{\partial y} \\ \dfrac{\partial w}{\partial z} \\ \dfrac{\partial v}{\partial x} + \dfrac{\partial u}{\partial y} \\ \dfrac{\partial w}{\partial y} + \dfrac{\partial v}{\partial z} \\ \dfrac{\partial u}{\partial z} + \dfrac{\partial w}{\partial x} \end{Bmatrix} = \cdots_{use\,(6.15)} = \mathbf{Bd}$$

where **d** is the nodal displacement vector, that is:

$$\varepsilon = \mathbf{B}\,\mathbf{d} \tag{7.16}$$

Strain energy is evaluated as

$$\begin{aligned} U &= \frac{1}{2}\int_V \sigma^T \varepsilon\, dV = \frac{1}{2}\int_V (\mathbf{E}\varepsilon)^T \varepsilon\, dV \\ &= \frac{1}{2}\int_V \varepsilon^T \mathbf{E}\,\varepsilon\, dV \\ &= \frac{1}{2}\mathbf{d}^T \left[\int_V \mathbf{B}^T \mathbf{E}\,\mathbf{B}\, dV\right]\mathbf{d} \end{aligned} \tag{7.17}$$

That is, the element stiffness matrix is

$$\mathbf{k} = \int_V \mathbf{B}^T \mathbf{E}\mathbf{B}\, dV \tag{7.18}$$

In $\xi\eta\xi$ coordinates:

$$dV = (\det \mathbf{J})\, d\xi\, d\eta\, d\zeta \tag{7.19}$$

Therefore,

$$\mathbf{k} = \int_{-1}^{1}\int_{-1}^{1}\int_{-1}^{1} \mathbf{B}^T \mathbf{E}\mathbf{B}(\det \mathbf{J})\, d\xi\, d\eta\, d\zeta \tag{7.20}$$

It is easy to verify that the dimensions of this stiffness matrix is 24×24.

Stresses:

To compute the stresses within an element, one uses the following relation once the nodal displacement vector is known for that element:

$$\sigma = \mathbf{E}\varepsilon = \mathbf{E}\mathbf{B}\mathbf{d}$$

Stresses are evaluated at selected points (Gaussian points or nodes) on each element. Stress values at the nodes are often discontinuous and less accurate. Averaging of the stresses from surrounding elements around a node is often employed to smooth the stress field results.

The von Mises stress for 3-D problems is given by

$$\sigma_e = \sigma_{VM} = \frac{1}{\sqrt{2}}\sqrt{(\sigma_1 - \sigma_2)^2 + (\sigma_2 - \sigma_3)^2 + (\sigma_3 - \sigma_1)^2} \tag{7.21}$$

where σ_1, σ_2, and σ_3 are the three principal stresses.

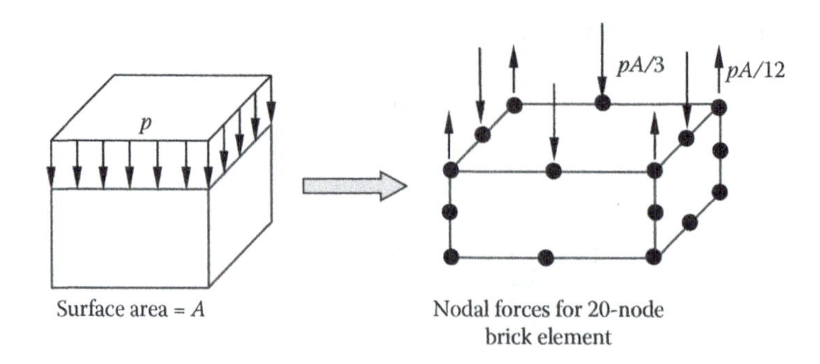

FIGURE 7.7
Equivalent nodal forces on a 20-node brick element for a pressure load p.

7.4.4 Treatment of Distributed Loads

Distributed loads need to be converted into nodal forces using the equivalent energy concept as discussed in earlier chapters. Figure 7.7 shows the result of a pressure load converted into nodal forces for a 20-node hexahedron element. Note the direction of the forces at the four corner nodes, which is not intuitive at all.

7.5 Case Studies with *ANSYS Workbench*

Problem Description: A base stand assembly includes a base, a holder, and a pin, as shown in the following figure. The stand assembly is made of structural steel. Assume a no-separation condition for all contact regions. Determine the deformation and von Mises stress distributions of the assembly under the given load and boundary conditions.

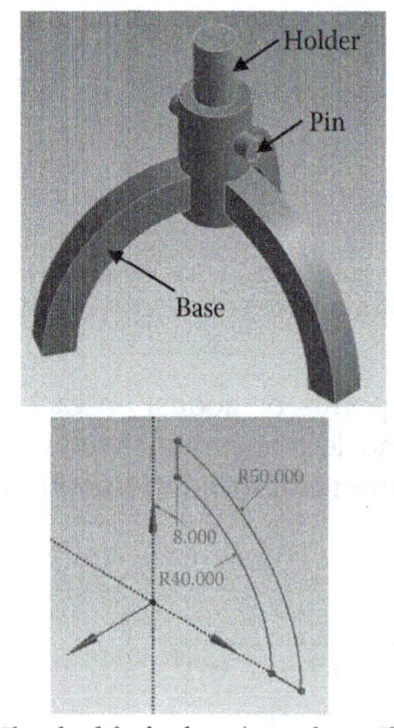

Material: Structural steel (E= 200 GPa, ν = 0.3)

Boundary conditions:

The bottom faces of the leg base are fixed.

A downward force of 1 kN is applied to the holder's top face.

Geometry construction:

The bottom of the hub base is 35 mm above the ground level.

The holder is 36 mm tall, 18 mm of which is in contact with the hub base.

Sketch of the leg base (extrude 5 mm on both sides)

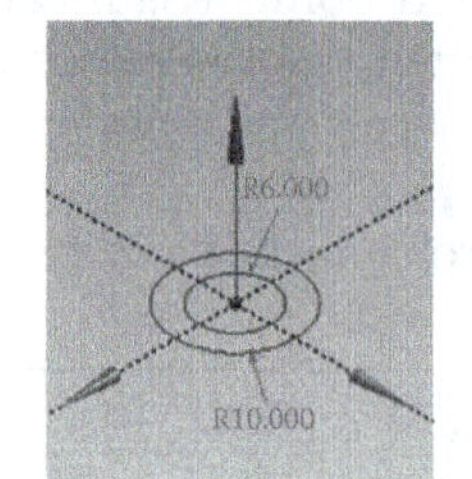

Sketch of the hub base (extrude 35 mm on one side)

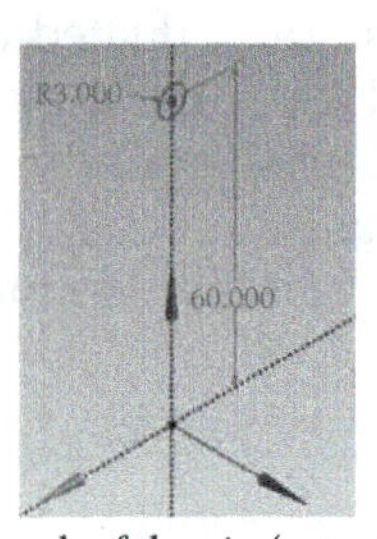

Sketch of the pin (extrude 15 mm on both sides)

Solution

To solve the problem with *ANSYS Workbench*, we employ the following steps:

Step 1: Start an *ANSYS Workbench* Project

Launch *ANSYS Workbench* and save the blank project as *"Assembly.wbpj."*

Step 2: Create a *Static Structural (ANSYS)* Analysis System

Drag the *Static Structural (ANSYS)* icon from the *Analysis Systems Toolbox* window and drop it inside the highlighted green rectangle in the *Project Schematic* window to create a standalone static structural analysis system.

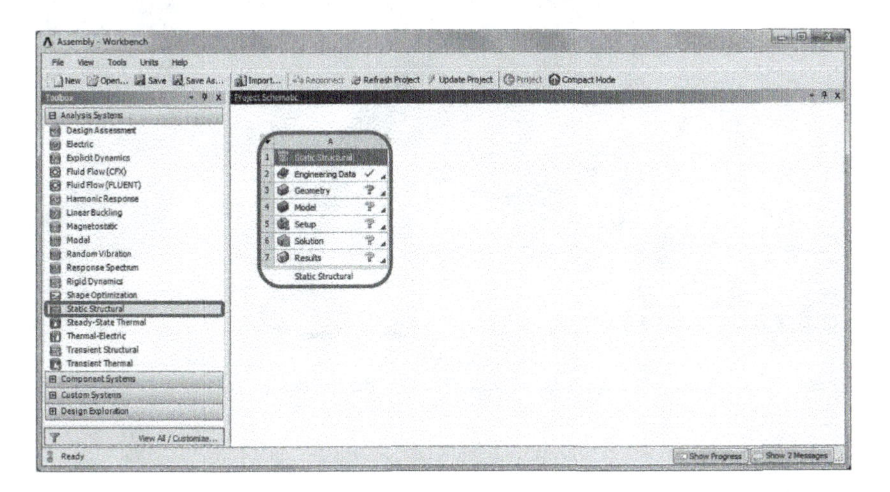

Step 3: Launch the *DesignModeler* Program

Double-click the *Geometry* cell to launch *DesignModeler*, and select *"Millimeter"* in the *Units* pop-up window.

Step 4: Create the Base Geometry

Click on the *Sketching* tab. Select the *Draw* toolbox and then *Line* and *Arc by Center*. Draw a sketch with two lines and two arcs on the *XY Plane*, as shown below. An entity named *Sketch1* will be shown underneath *XY Plane* of the model's *Tree Outline*.

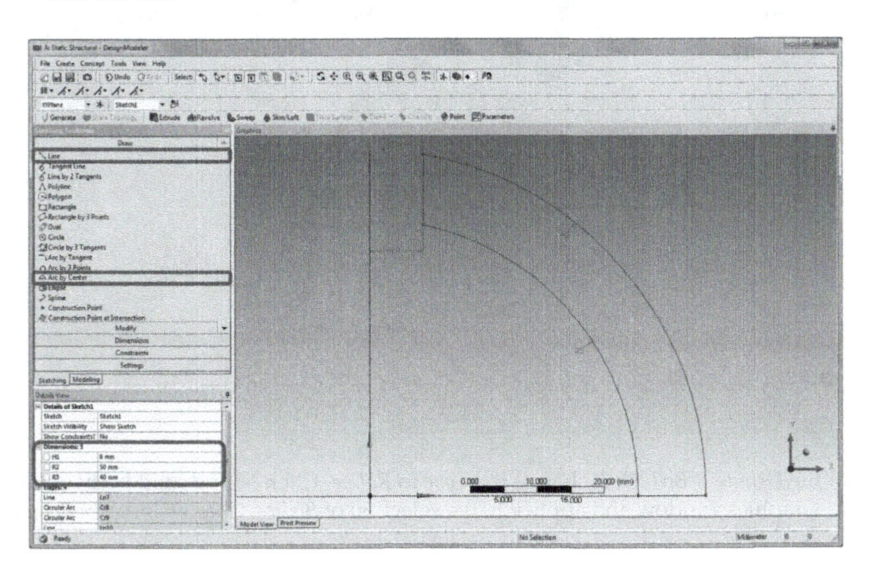

Extrude *Sketch1* to create a solid body. In *Details of Extrude1,* change *Direction* to *Both-Symmetric,* and enter *5 mm* for *FD1, Depth.* Click on *Generate.*

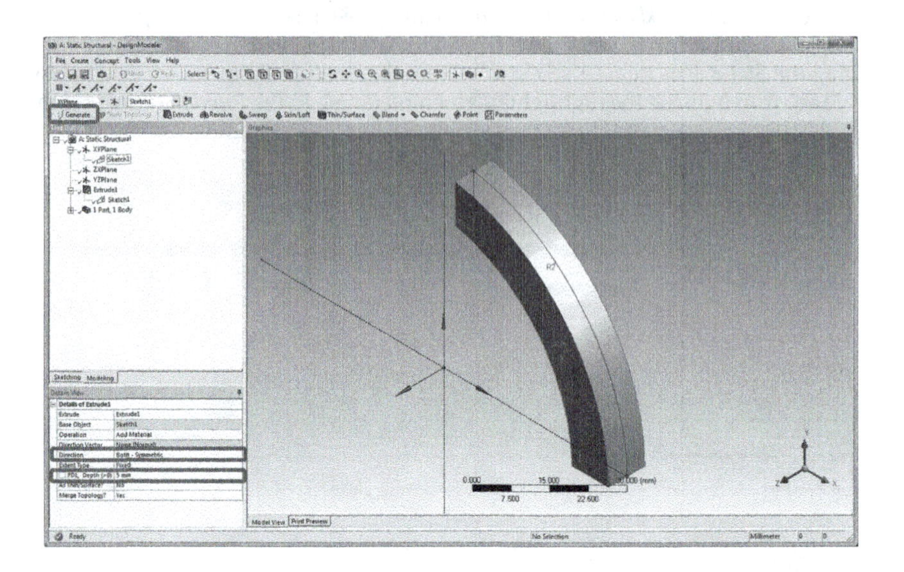

From the drop-down menu of *Create,* click on *Body Operation.*

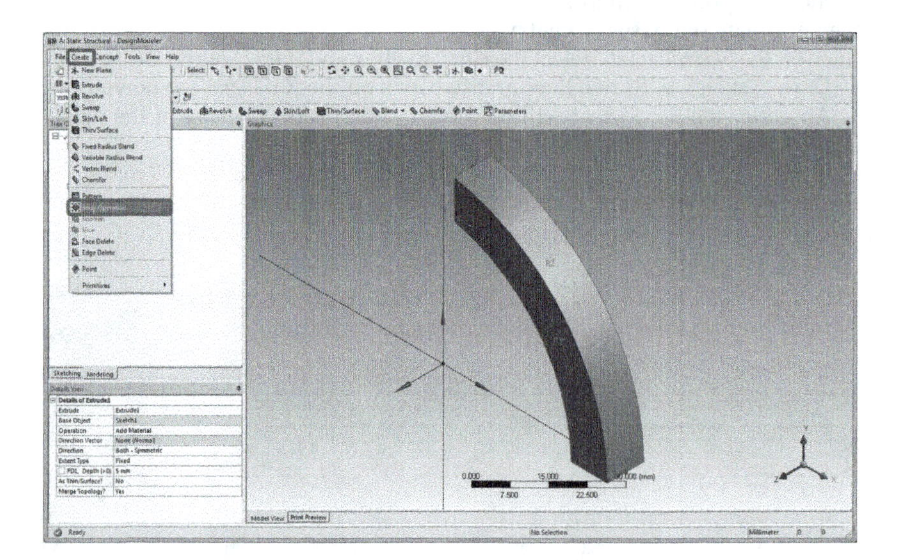

In the *Details of BodyOp1,* change the *Type* to *Rotate.* Click on the solid body in the Graphics window and apply it to the selection of *Bodies.* Change *Preserve Bodies*

to *Yes*. Click on the *Y-axis* in the Graphics window and apply it to the *Axis Selection*. Set the *FD9, Angle* to *120°*. Click on *Generate*.

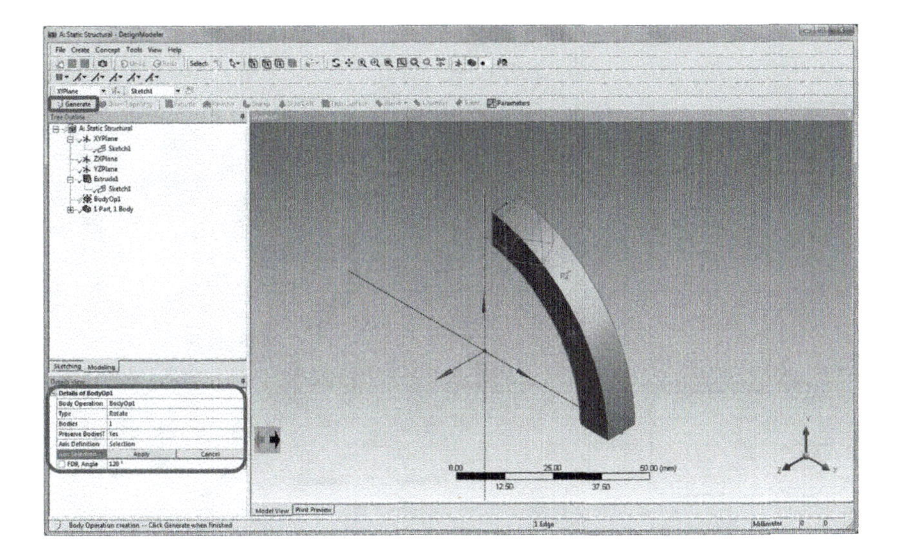

From the drop-down menu of *Create*, click on *Body Operation*. In the *Details of BodyOp2*, change the *Type* to *Rotate*. Click on the original solid body in the *Graphics* window and apply it to the selection of *Bodies*. Change *Preserve Bodies* to *Yes*. Click on the *Y-axis* in the Graphics window and apply it to the *Axis Selection*. Set the *FD9, Angle* to *–120°*. Click on *Generate*.

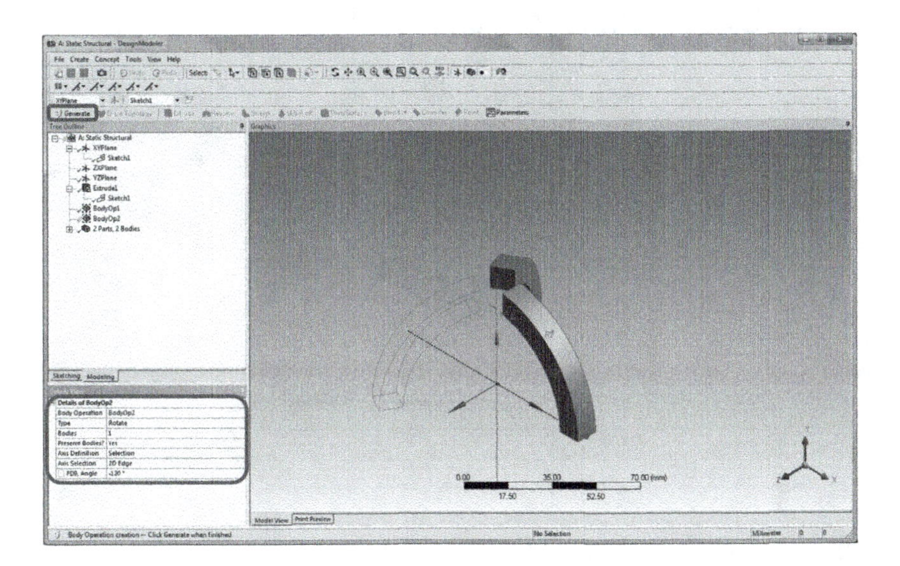

Click on *New Plane* to create the *Plane4*. In the *Details of Plane4*, Change the *Base Plane* to *ZXPlane*, and *Transform1 (RMB)* to *Offset Z*. Enter *35 mm* for *FD1, Value1*. Click on *Generate*.

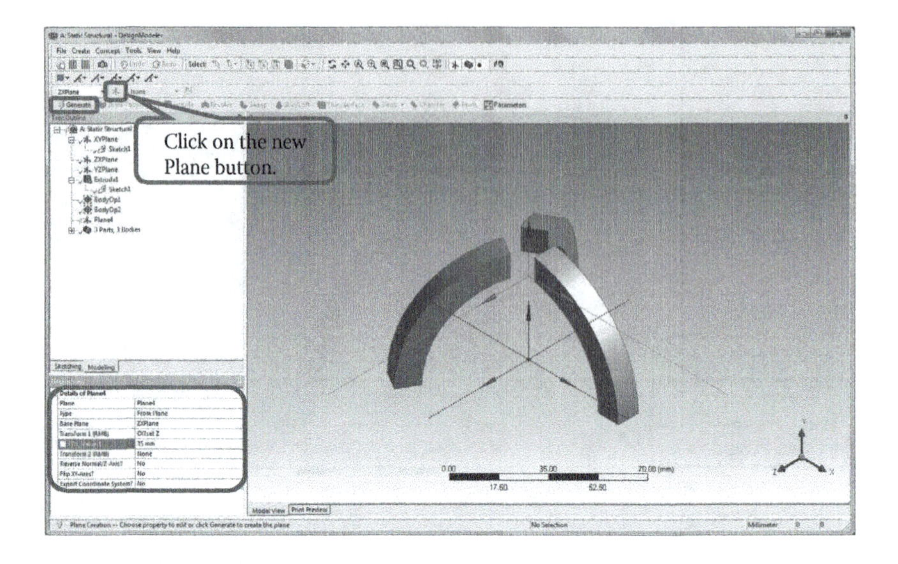

Click on *Look At Face/Plane/Sketch*. Draw two concentric circles as shown below, with a radius given as *6 mm* and *10 mm*, respectively. Click on *Generate* to finish creation of *Sketch2*.

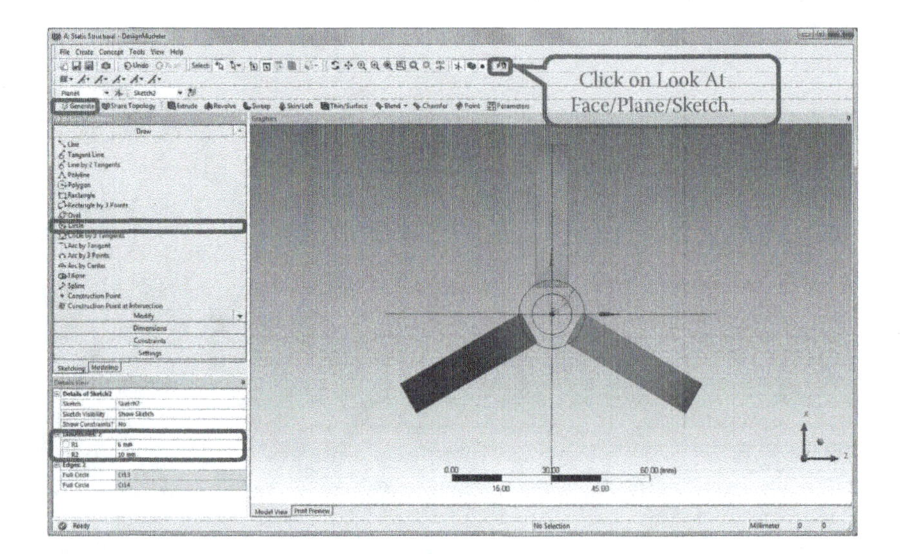

Extrude *Sketch2* along the normal direction for an *FD1, Depth* of *35 mm,* as shown below. Click on *Generate.* The new *Extrude2* will be combined with the other three parts into a single solid body.

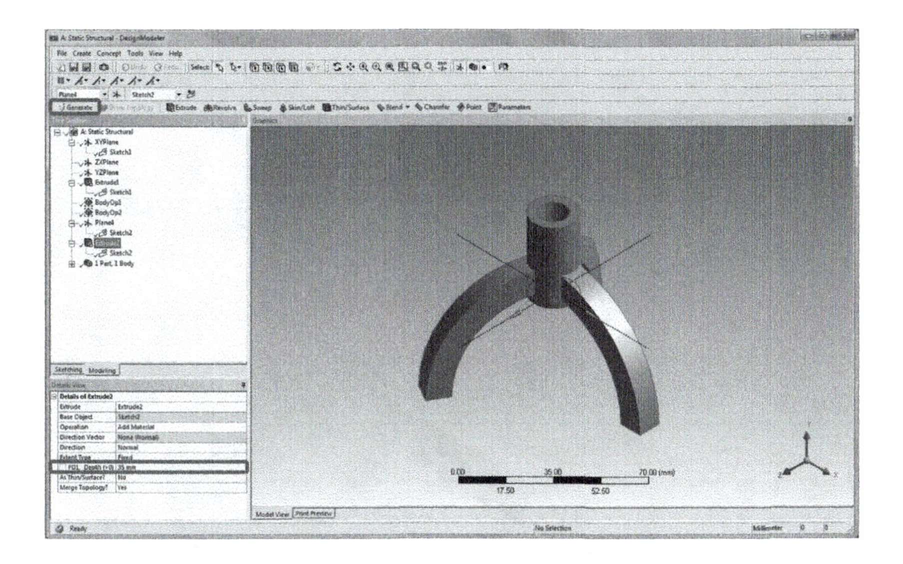

Step 5: Create the Holder Geometry

Click on *New Plane* to create the *Plane5.* In the *Details of Plane5,* change the *Type* to *From Face.* Click on the annulus face from the *Graphics* window and apply it to the selection of *Base Face.* Click on *Generate.*

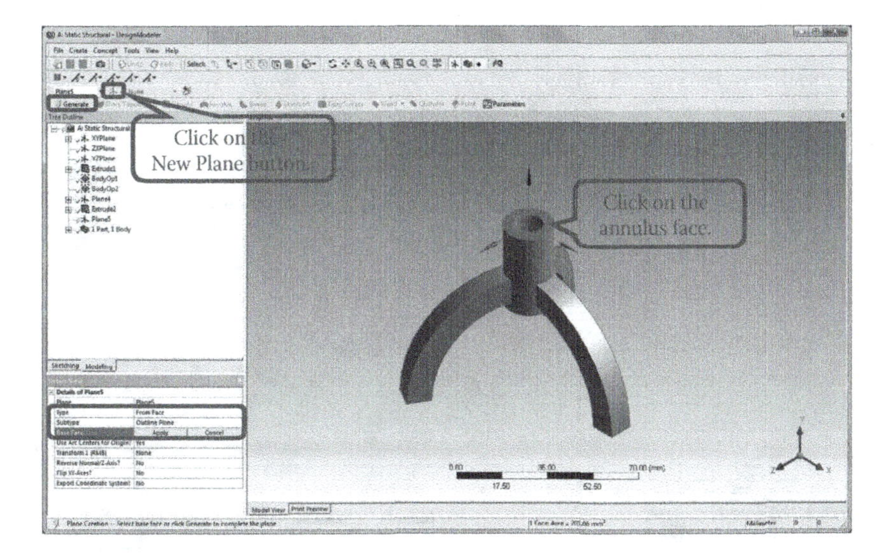

Create a *New Sketch* on *Plane5*. Draw a circle, the same size as the smaller of the two concentric circles. Click on *Generate*. A *Sketch3* will be created underneath *Plane5* in the *Tree Outline*.

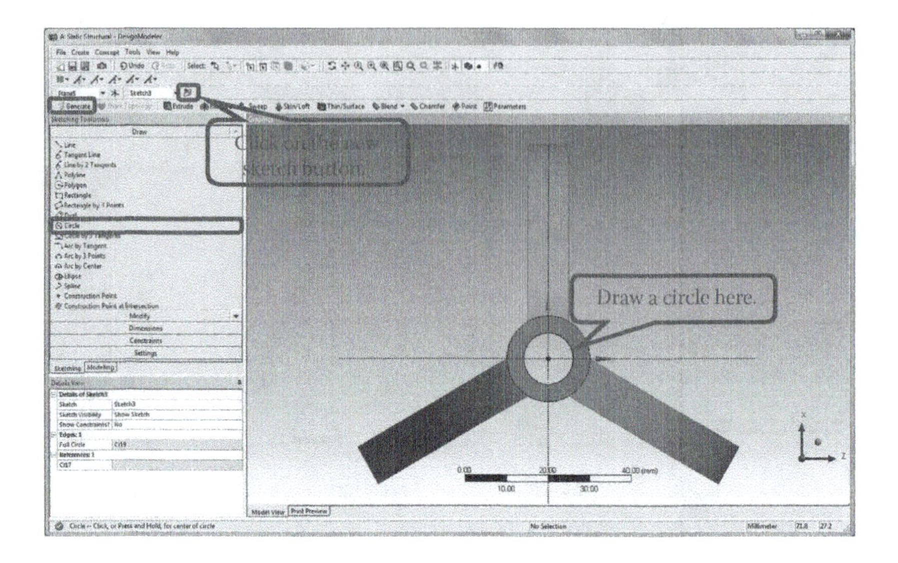

Click on *Extrude*. In the *Details of Extrude3*, change the *Direction* to *Both- Symmetric*, and set the *FD1, Depth* as *18 mm*. Click on *Generate*.

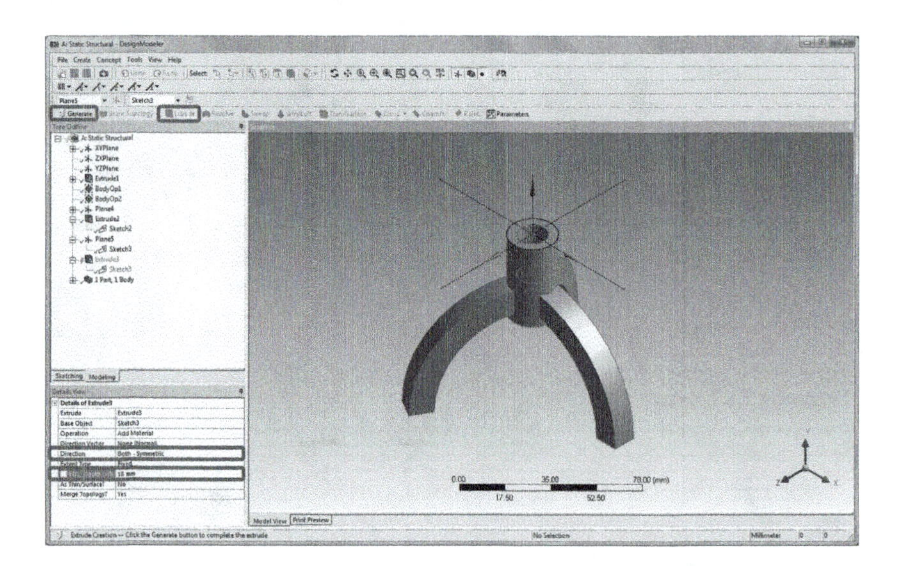

Create a *New Sketch* on the *YZPlane*. Draw a *circle* with a radius of *3 mm*, centered *60 mm* to the right of the origin, as shown below. Click on *Generate* to finish the creation of *Sketch4*.

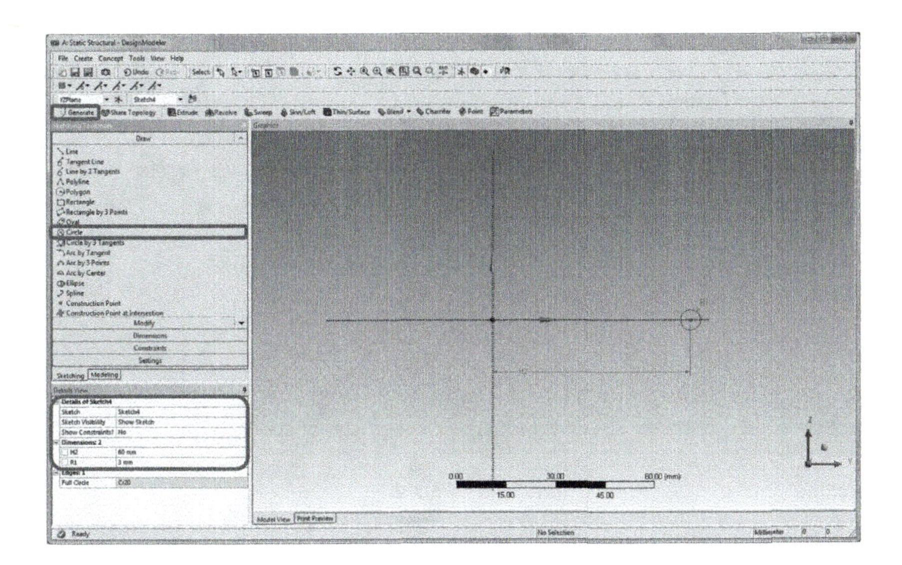

Extrude *Sketch4*. In the *Details of Extrude4*, change *Operation* to *Cut Material*, and the *Direction* to *Both- Symmetric*. Use the default *30 mm* for *FD1, Depth*. Click on *Generate*.

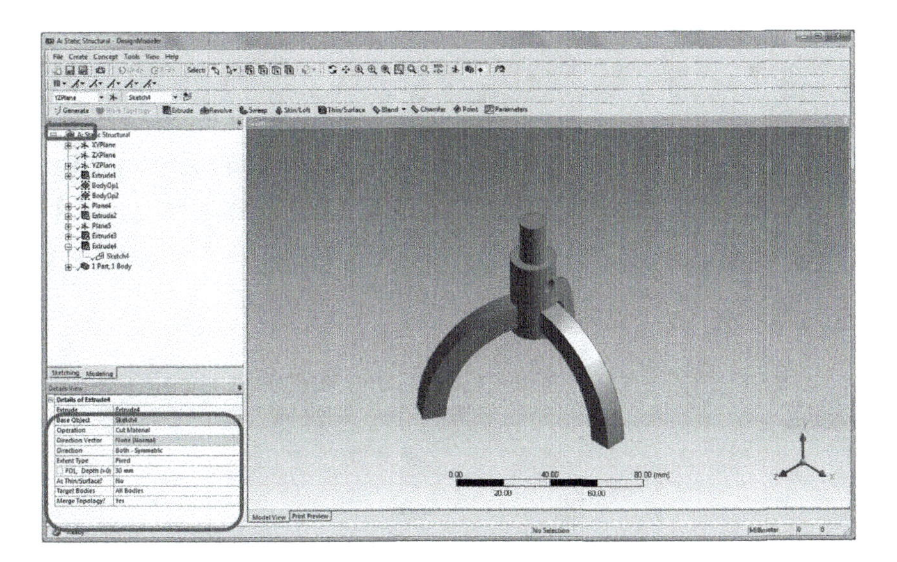

Click on *Freeze* from the drop-down menu of *Tools*. Freezing a body and then using a *Create Slice* can help physically separate merged bodies to create a multipart assembly.

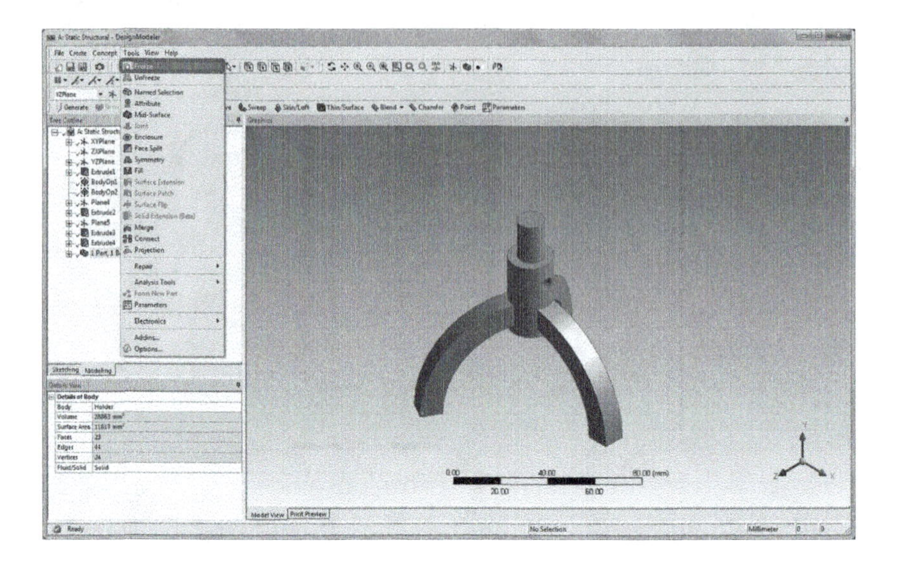

Then click on *Slice* from the drop-down menu of *Create*.

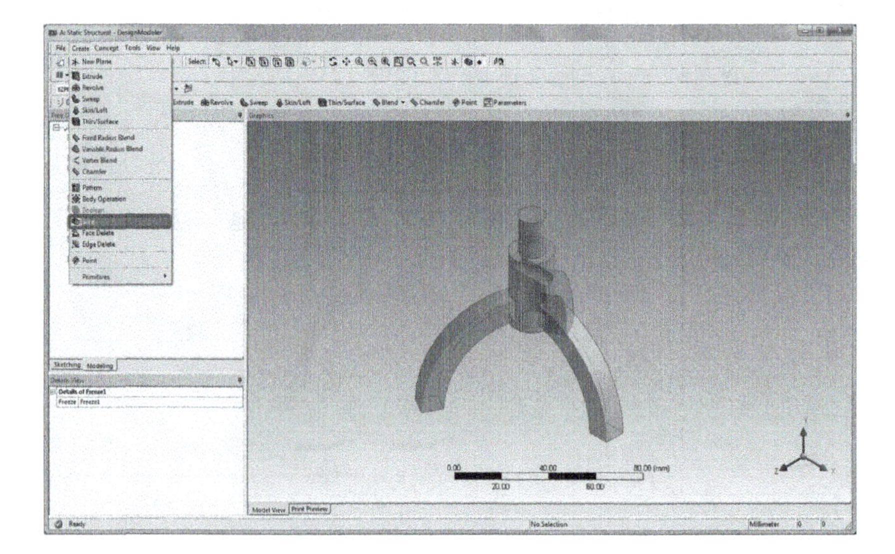

In the *Detailed of Slice1*, Change the *Slice Type* to *Slice by Surface*. Click on the cylindrical surface in the Graphics window and apply it to the selection of *Target Face*. Click on *Generate*.

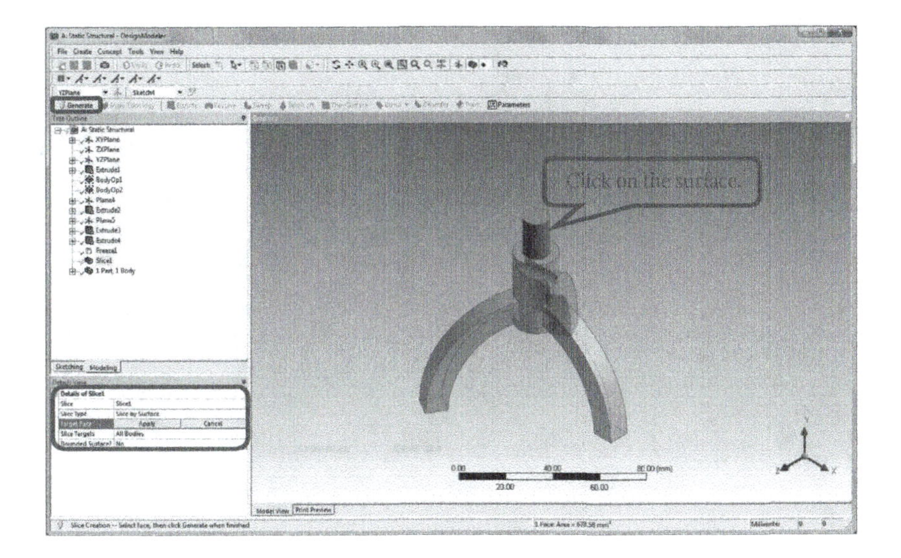

A two-part assembly will be created, as shown below. Right-click on the solid bodies from the *Tree Outline* to rename them as *Holder* and *Base*, respectively.

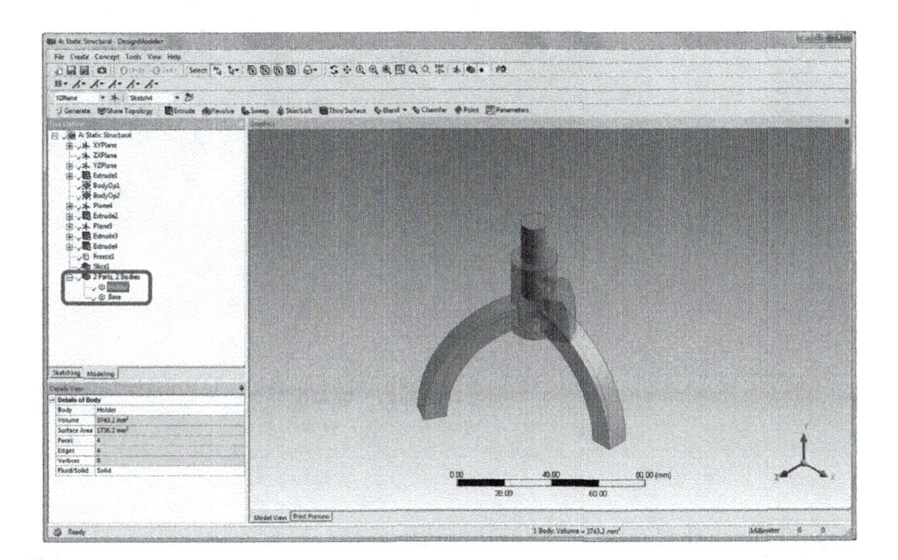

Step 6: Create the Pin Geometry

Extrude *Sketch4*. In the *Details of Extrude5*, change *Operation* to *Add Material*, and the *Direction* to *Both- Symmetric*. Set the *FD1, Depth* as *15 mm*.

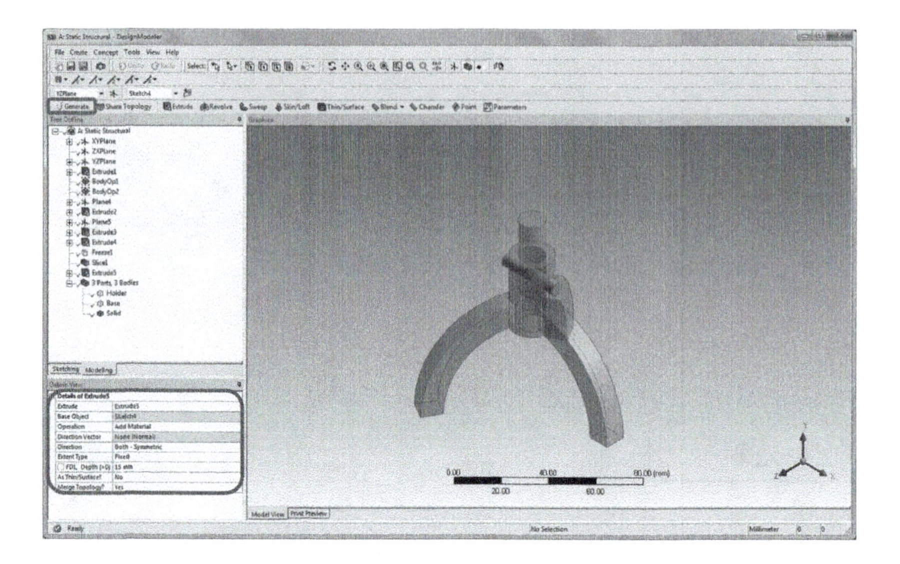

Right Click on *Solid* and rename it as *Pin* in the *Tree Outline*. A three-part assembly is now created.

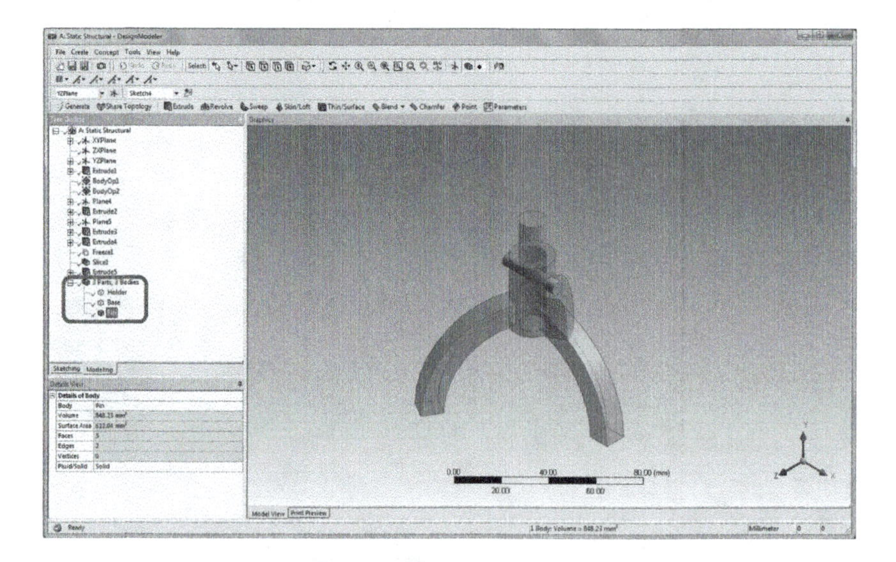

Step 7: Launch the *Static Structural* Program

Double-click on the *Model* cell to launch the *Static Structural* program. *ANSYS Workbench* automatically detects contacts between parts of an assembly, whether it is created in *DesignModeler* or imported from neutral geometry files. Click on *Contacts* under *Connections* in the *Outline tree*. In the *Details of "Contact Regions,"* change the *Type* to *No Separation* for all three contact regions. No separation contact prevents gap forming between the contact regions; a small amount of frictionless sliding may occur along the contact faces.

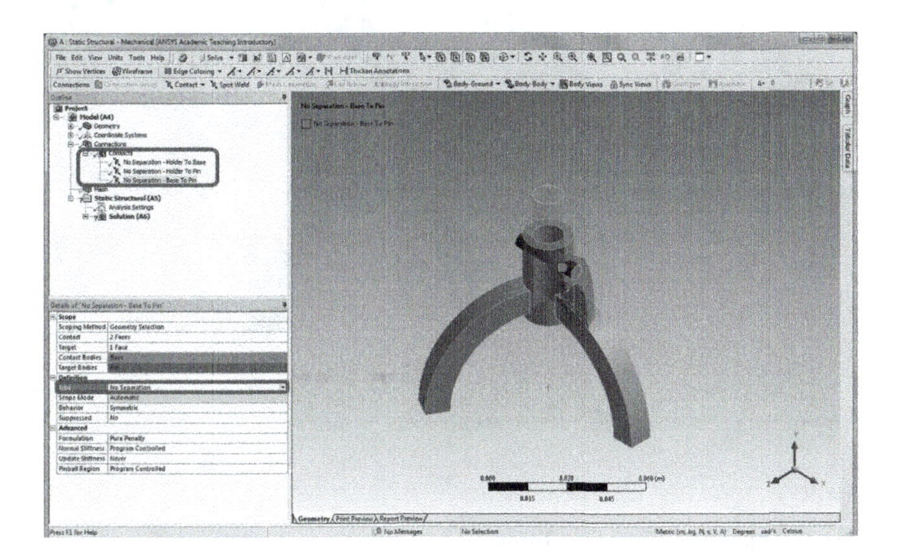

Step 8: Generate Mesh

Right click on *Mesh* in the *Project Outline*. Select *Insert* and then *Sizing* from the context menu. In the *Details of "Body Sizing,"* enter "*2.5e-3 m*" for the *Element Size*. Ctrl-click on the three parts in the *Graphics* window and apply the bodies to the *Geometry* selection.

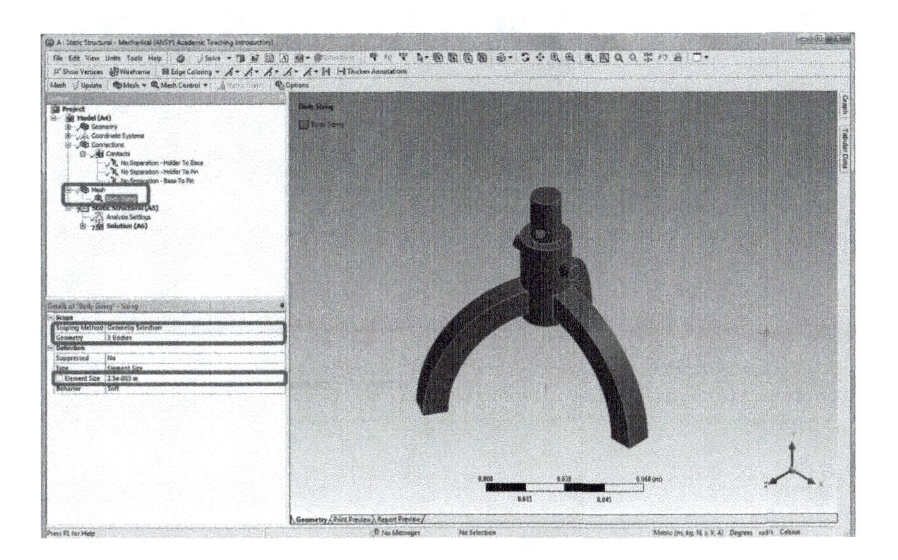

Right-click on *Mesh* and select *Generate Mesh*.

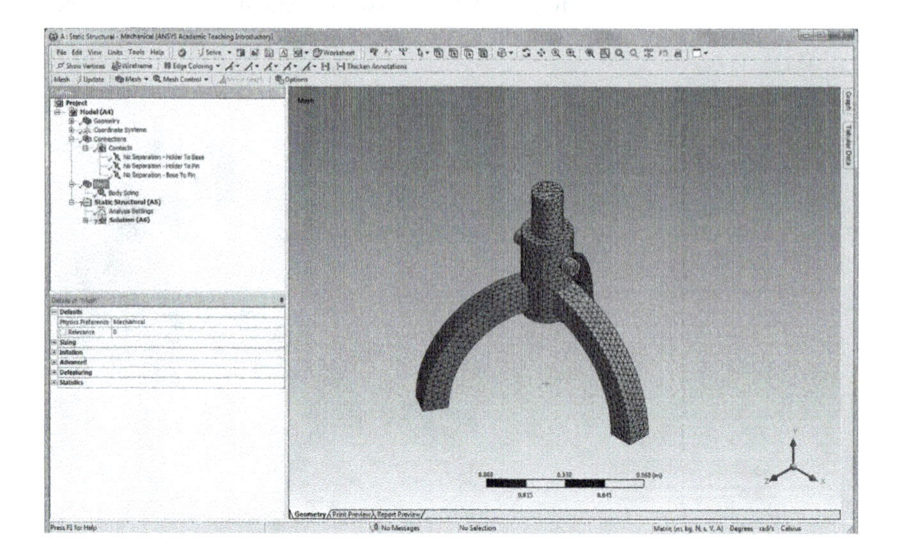

Step 9: Apply Boundary Conditions

Right-click on *Static Structural (A5)*. Choose *Insert* and then *Fixed Support* from the context menu. Ctrl-click on the three faces highlighted below and apply them to the *Geometry* selection in the *Details of "Fixed Support."*

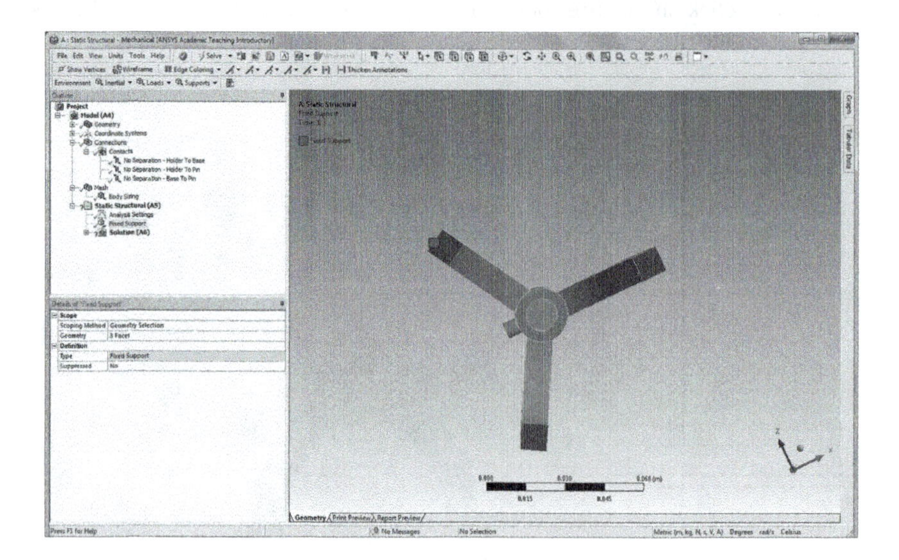

Step 10: Apply Loads

Right-click on *Static Structural (A5)*. Choose *Insert* and then *Force*. In the *Details of "Force,"* apply the circular face shown below to the *Geometry* selection. Change the *Define By* selection to *Components*, and enter *– 1000 N* for the *Y Component*.

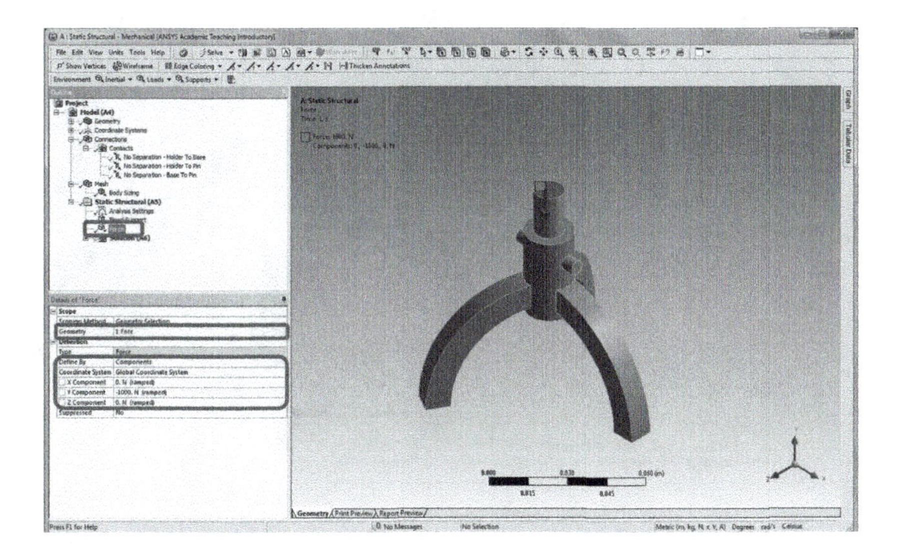

Step 11: Retrieve Solution

Insert a *Total Deformation* item by right-clicking on *Solution (A6)* in the *Outline* tree.
Insert an *Equivalent Stress* item by right-clicking on *Solution (A6)* in the *Outline*.
Right-click on *Solution (A6)* in the *Outline* and select *Solve*.
Click on *Total Deformation* in the *Outline* to review the deformation results.

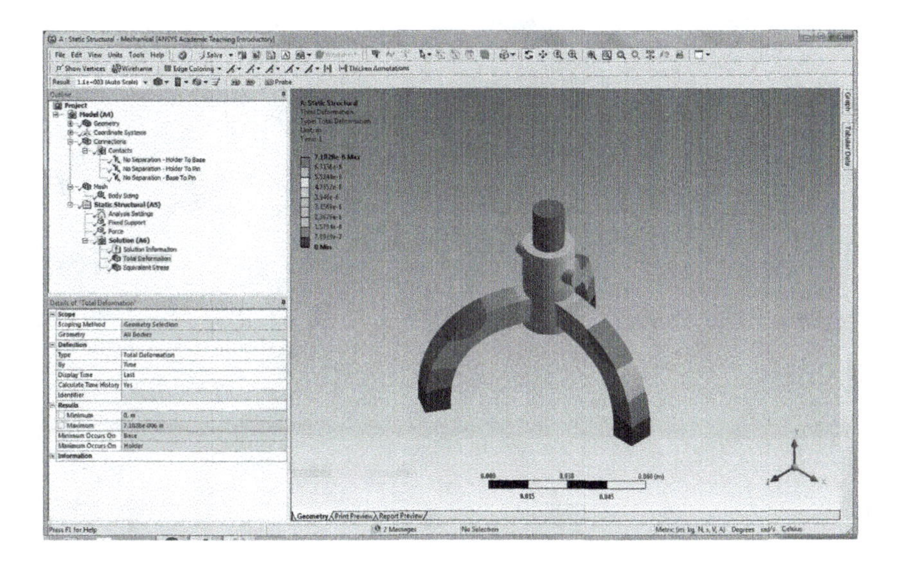

Click on *Equivalent Stress* in the *Outline* to review the stress results.

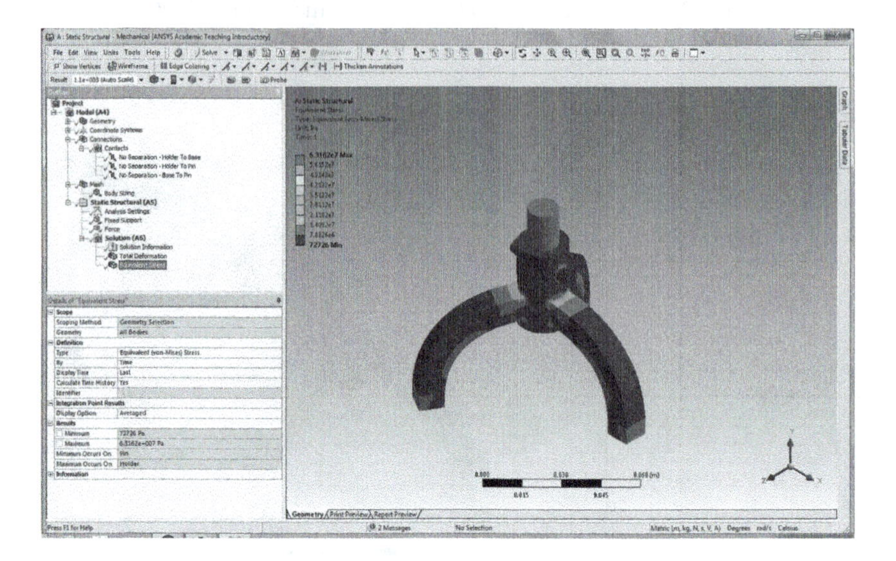

To view deformation of the pin, insert a *Total Deformation* item by right-clicking on *Solution (A6)* in the *Outline* tree. Click on the pin body in the *Graphics* window and apply it to the selection of *Geometry* in the *Details of "Total Deformation 2."*

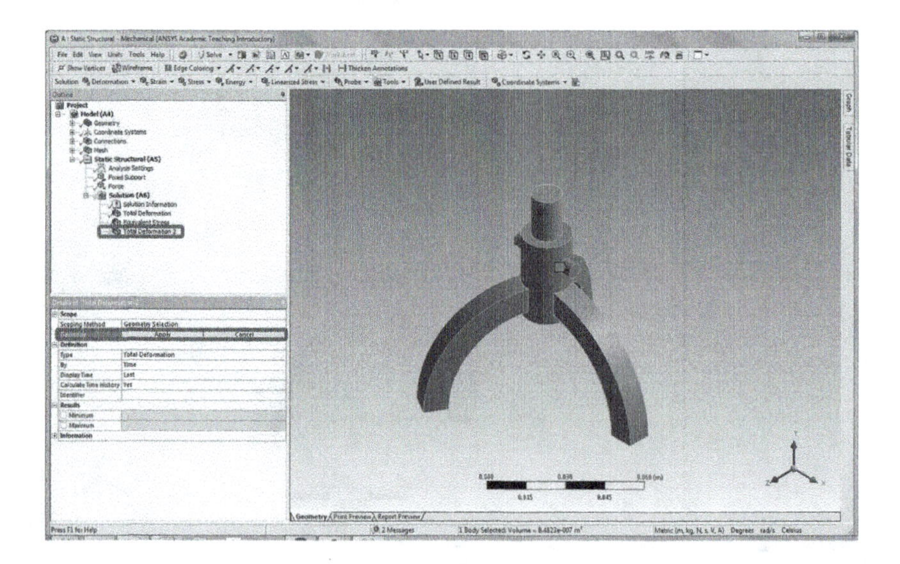

To view stress in the pin, insert an *Equivalent Stress* item by right-clicking on *Solution (A6)* in the *Outline* tree. Click on the pin body in the *Graphics* window and apply it to the selection of *Geometry* in the *Details of "Equivalent Stress 2."*

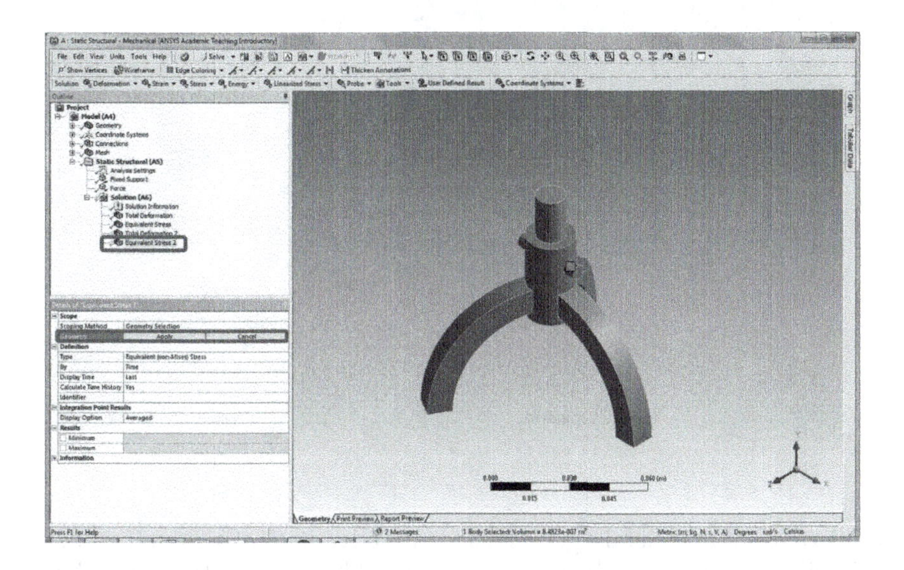

Repeat the last two steps to add a *Total Deformation 3* and an *Equivalent Stress 3* to the *Outline*. Select the holder body in the *Graphics* window for the *Geometry* in both *Details of "Total Deformation 3"* and *Details of "Equivalent Stress 3."*

Right-click on *Solution (A6)* in the *Outline* and select *Evaluate All Results*. Click on *Total Deformation 2* and *Total Deformation 3* in the *Outline* to review the deformation of the pin and the holder, respectively.

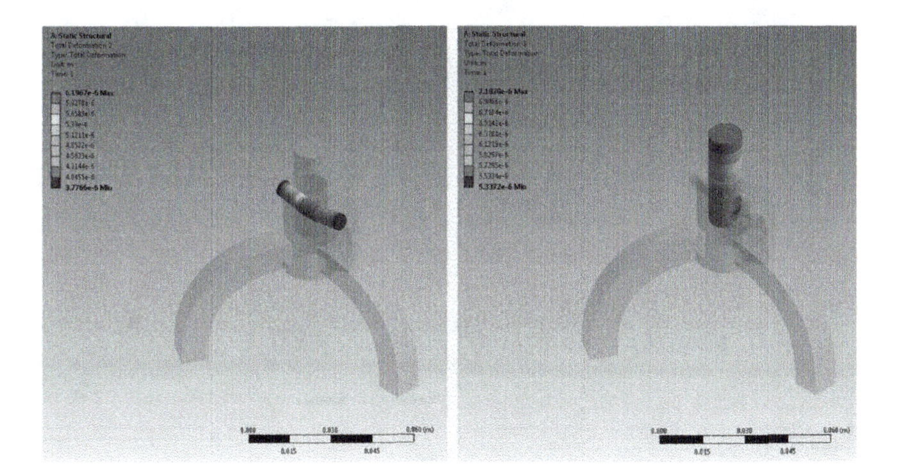

Click on *Equivalent Stress 2* and *Equivalent Stress 3* in the *Outline* to review the von Mises stress of the pin and the holder, respectively.

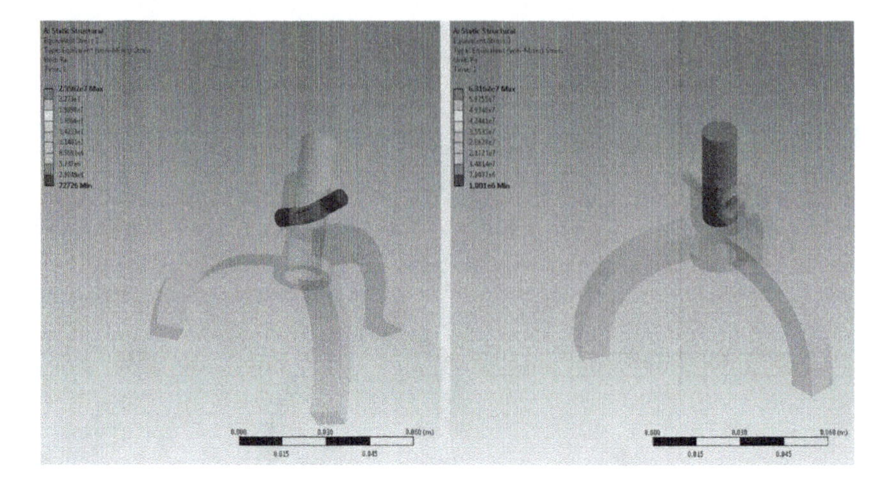

Modeling tips: It is in general a good idea to take advantage of symmetry whenever possible to simplify calculations. The base stand assembly is symmetric about the *XY*-plane. To create a half-symmetry model, open *"Assembly.wbpj"* and double-click on the *Geometry* cell in the *Project Schematic*. In the drop-down menu of *Tools* of the launched *DesignModeler*, select *Symmetry*, as shown below. An item named *Symmetry1* will be added to the *Tree Outline*.

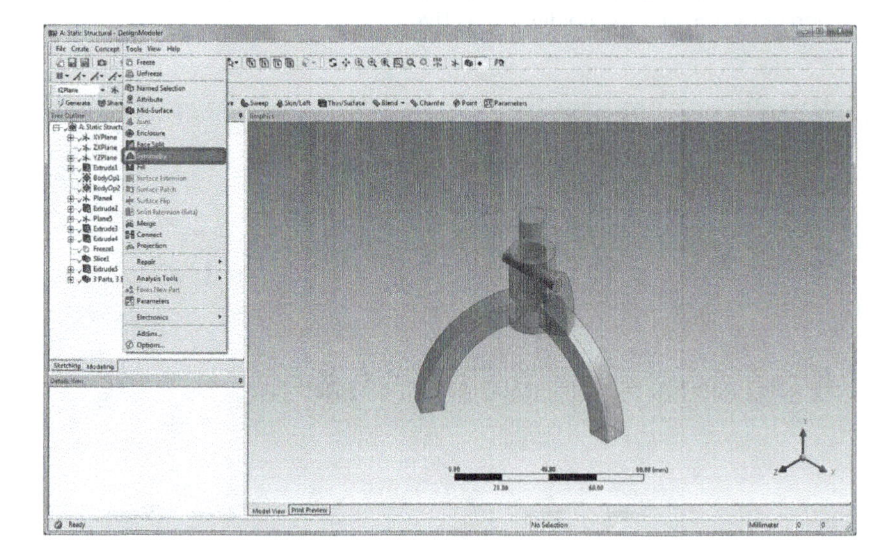

In the *Details of Symmetry1,* click on *XYPlane* in the *Tree Outline* and apply it to the selection of *Symmetry Plane1.* Click on *Generate.* A half-model will be created as shown below. Exit *DesignModeler.*

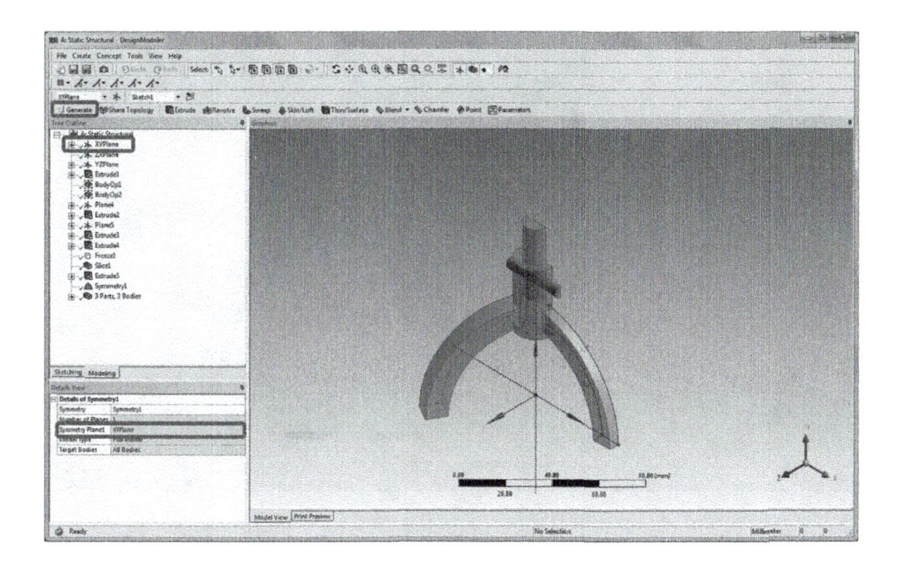

Double-click on *Model* cell in the *Project Schematic* to launch the *Static Structural–Mechanical* program. Click *Yes* on the popup menu to allow reading the modified upstream data. Note that a *Symmetry Region* is now added to the *Project Outline* as shown below.

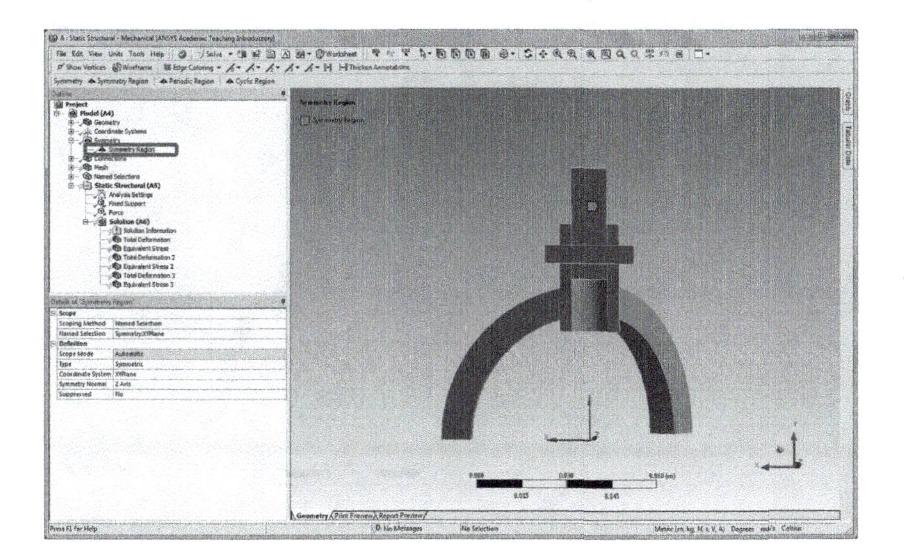

Right-click on *Mesh* in the *Outline* and select *Generate Mesh.*

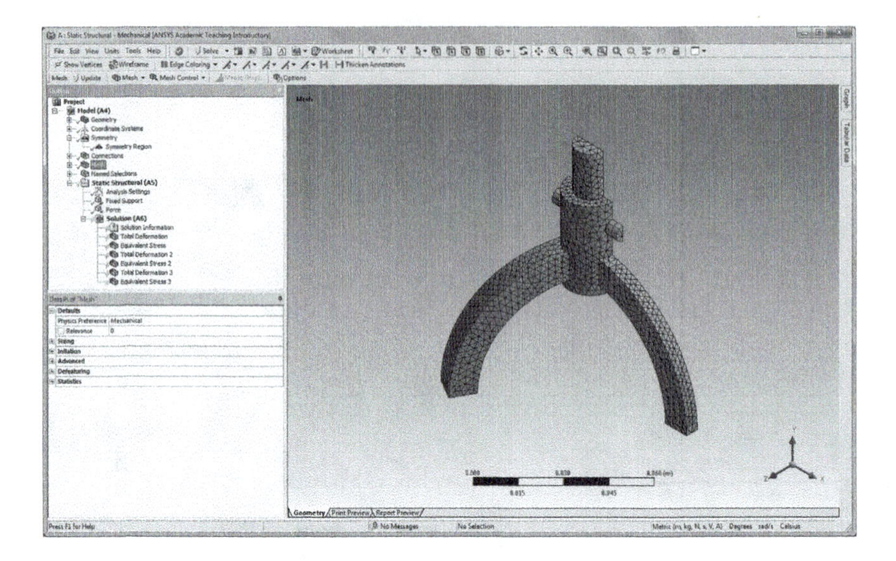

In the half-model, the magnitude of the force applied to a face should be divided by two. Change the *Y Component* of the force from *–1000N* to *–500N* in the *Details of "Force."*

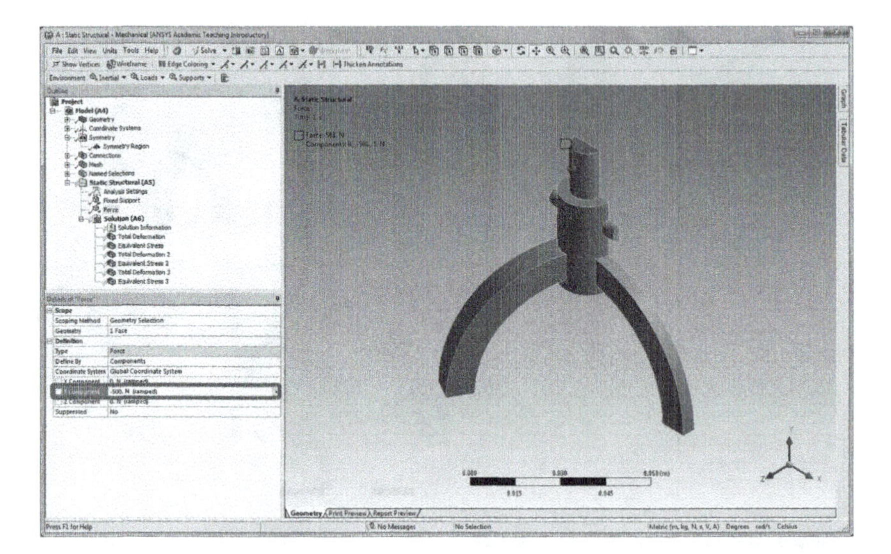

The frictionless support is equivalent to a symmetry condition. Right-click on *Static Structural (A5)*. Choose *Insert* and then *Frictionless Support* from the context menu. In the *Details of "Frictionless Support,"* Ctrl-click the seven faces in the *Graphics* window as shown below, and apply them to the *Geometry* selection field. Note that this step is especially important for imported geometry with cut faces at the symmetry plane, because the program cannot automatically detect symmetry regions as in the case of geometry created internally.

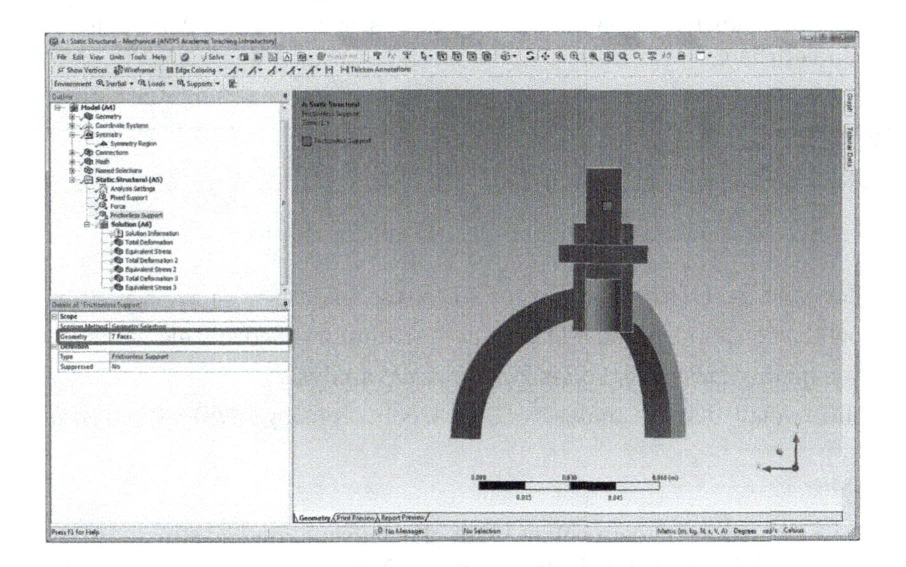

Right-click on *Static Structural (A5)* and select *Solve*. Click on the deformation and stress results under *Solution (A6)* and compare them with results from the full model.

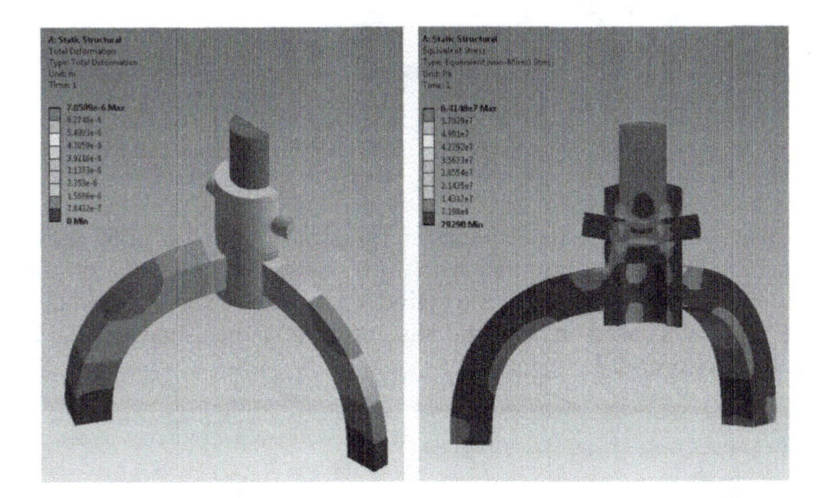

7.6 Summary

In this chapter, we discussed the 3-D solid elements for elasticity problems, that is, general 3-D deformation and stress analyses. Solid elements are the most accurate elements and should be applied when the bar, beam, plane stress/strain, and plate/shell elements are no longer valid or accurate. For stress concentration problems, higher-order solid elements, such as 10-node tetrahedron or 20-node hexahedron (brick) elements, should be employed in the FEA. For solids having symmetrical features, symmetric FEA models can be more effective and efficient.

7.7 Review of Learning Objectives

Now that you have finished this chapter you should be able to

1. Understand the FE formulations for 3-D stress analysis
2. Know the behaviors of the 3-D (tetrahedron and hexahedron) elements
3. Create quality mesh over solids for 3-D stress analysis
4. Perform detailed stress analysis of 3-D structures using *ANSYS Workbench*

PROBLEMS

7.1 For a tapered bar shown below, study the deformation and stresses in the bar with a 3-D model using solid elements and a 1-D model using 1-D bar elements. Assume $R_1 = 1$ m, $R_2 = 0.5$ m, $L = 5$ m, force $F = 3000$ N, Young's modulus $E = 200$ GPa, and Poisson's ratio $\nu = 0.3$. The bar is fixed at the left end. Compare the FEA results between the 3-D and 1-D models.

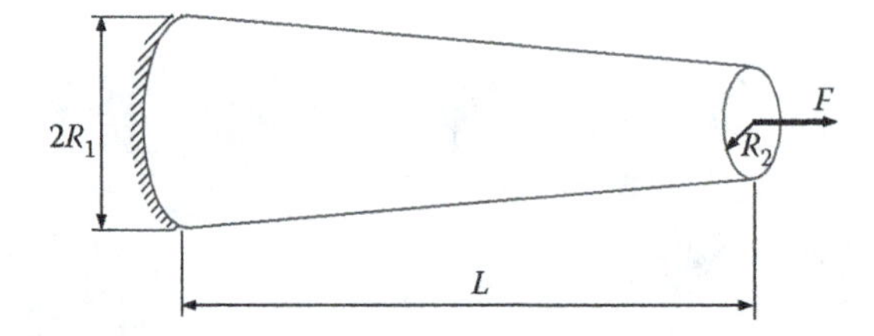

7.2 An open cylinder shown below has an inner radius $a = 1$ m, outer radius $b = 1.1$ m, length $L = 10$ m, and is applied with a pressure load $p = 10$ GPa on the inner surface. Compute the stresses in the cylinder using shell and solid models. Compare the FEA results based on these models. Assume Young's modulus $E = 200$ GPa and Poisson's ratio $\nu = 0.3$.

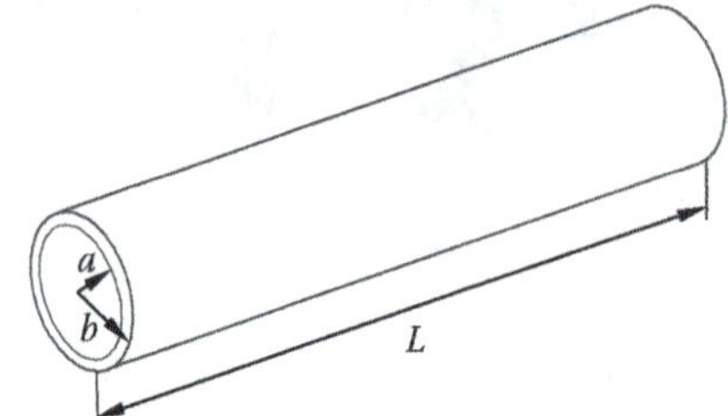

7.3 For the rotating part sketched below, assume that it is made of steel with Young's modulus $E = 200$ GPa, Poisson's ratio $\nu = 0.3$, and mass density $\rho = 7850$ kg/m^3. Assume that the part is rotating at a speed of 100 RPM about the z axis. Ignore the gravitational force. Compute the deformation in the part using the FEA with a full 3-D model. Compare the 3-D results with the axisymmetric modeling results obtained in Problem 5.4.

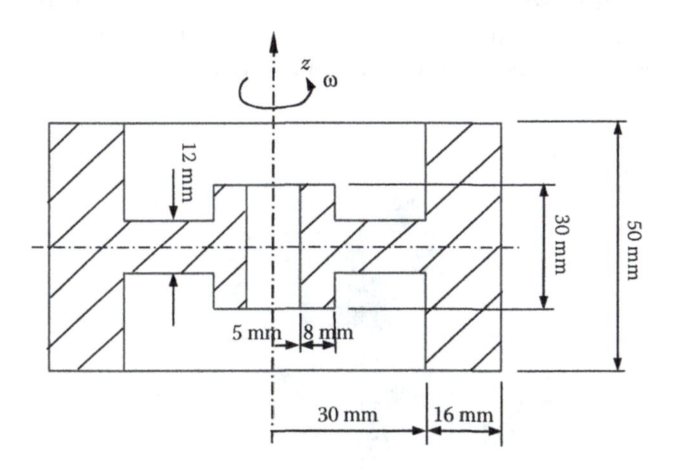

7.4 The following figure shows a simple bracket made of structural steel with Young's modulus $E = 200$ GPa and Poisson's ratio $\nu = 0.3$. The bracket has a cross-section profile shown below and an extrusion depth of 12 cm. A through-hole of diameter 1.5 cm is centered on the bracket's flat ends, one on each side, as shown in the figure. Suppose that the bracket is fixed on the bottom, and lifted by a pressure load of 1000 N/cm^2 from the inside surface of the top flat portion. Compute the deformation and stresses in the bracket using the FEA.

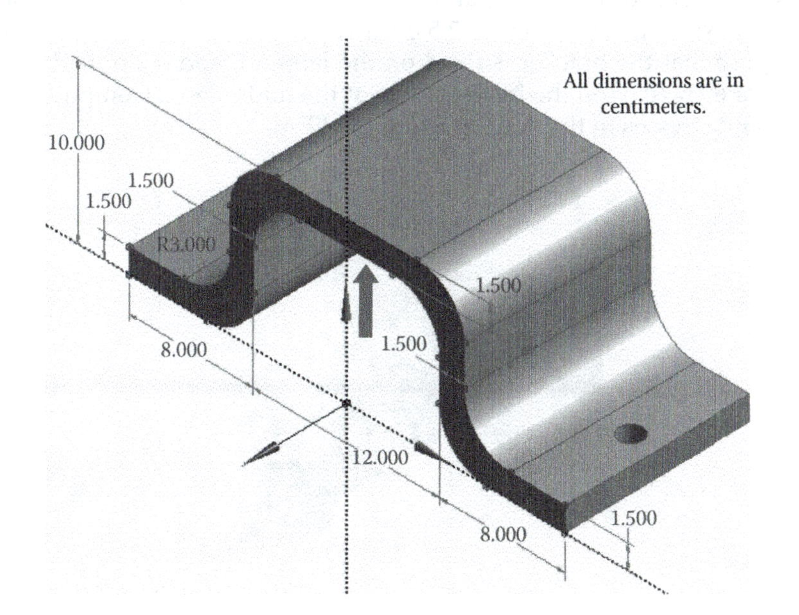

7.5 The following figure shows a lifting bracket made of structural steel with Young's modulus $E = 200$ GPa and Poisson's ratio $\nu = 0.3$. The base portion of the bracket is 20 cm long, 12 cm wide, and 3 cm deep. The fillet radius at the base corners is 1.75 cm. Four through-holes each of radius 1 cm are located symmetrically both length wise and width wise in the base, with the hole centers being 12 cm apart by 6 cm apart. The upper portion of the base has a profile sketch shown below and is 3 cm thick. Suppose that the bracket is fixed on the bottom, and lifted by a bearing load of 20 kN through the bolt hole located in the upper portion. Compute the deformation and stresses in the bracket using the FEA.

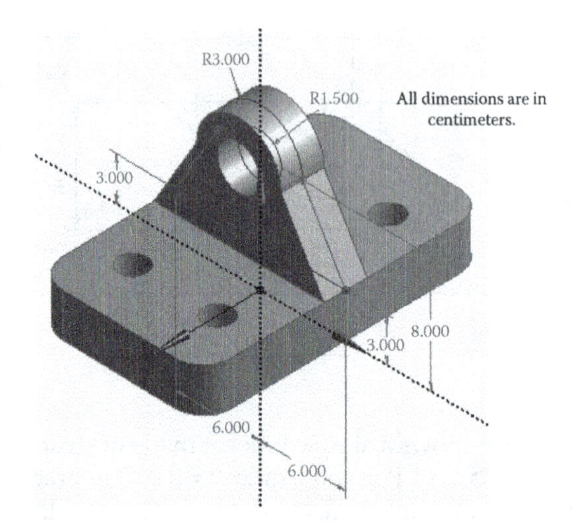

7.6 The following figure shows a bracket made of structural steel with Young's modulus $E = 200$ GPa and Poisson's ratio $\nu = 0.3$. The rectangular base of the bracket is 12 cm long, 6 cm wide, and 3 cm deep. The midsection has an overall extrusion depth of 1.6 cm and a profile sketch shown below. The hollow cylinder has an inner radius of 1 cm, an outer radius of 2 cm, and a height of 4 cm. The central axis of the cylinder is 12.5 cm away from the leftmost end of the bracket. Suppose that the bracket is fixed on the left end, and is applied by a pressure load of 60 N/cm^2 on the inclined face of the midsection. Compute the deformation and stresses in the bracket using the FEA.

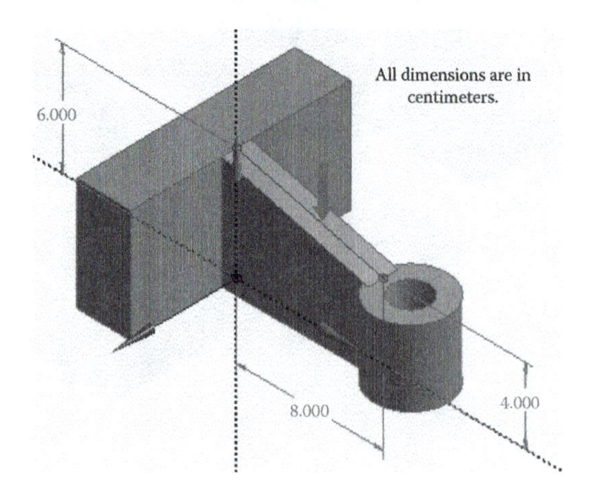

7.7 The following figure shows an S-shaped wooden block with Young's modulus $E = 70$ K N/mm^2 and Poisson's ratio $\nu = 0.3$. The block has a uniform thickness of 6 mm and a depth of 8 mm. It is fixed on the bottom and subjected to a surface pressure of 1 N/mm^2 on the top. Compute the deformation and stresses in the block using the FEA.

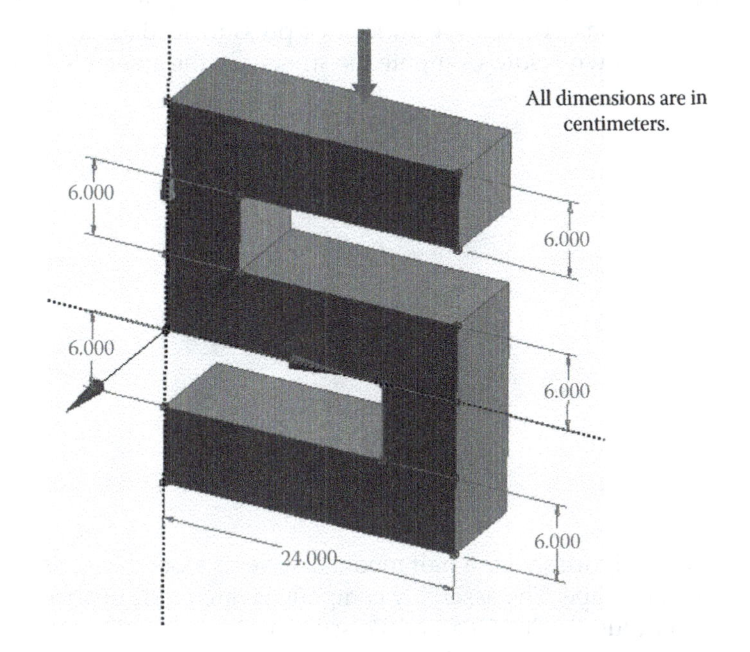

7.8 The following figure shows an L-shaped bracket made of structural steel with Young's modulus $E = 200$ GPa and Poisson's ratio $\nu = 0.3$. The 4 cm thick L-shape with a cross-section profile shown in the figure is rigidly connected to a rectangular base of size 3 cm × 8 cm × 1 cm. Suppose that the bracket is fixed on the right end at the base. A uniform pressure of 100 N/cm^2 is applied on the protruding surface of the L-shape as shown below. Compute the deformation and stresses in the bracket using the FEA.

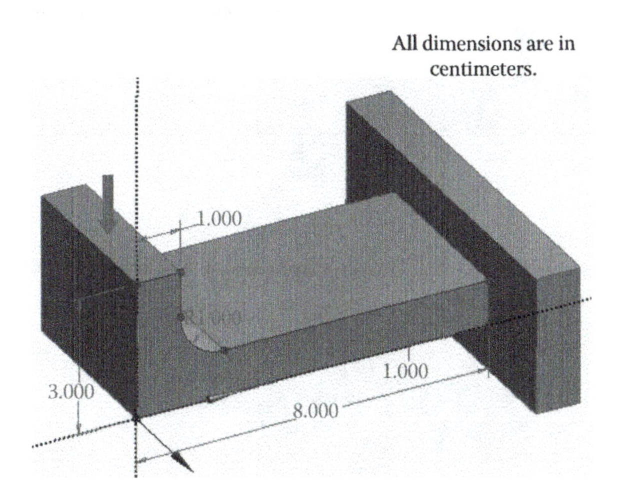

7.9 The following two-bar assembly consists of four components made of structural steel with Young's modulus $E = 200$ GPa, Poisson's ratio $\nu = 0.3$, and density $\rho = 7850$ kg/m^3. As shown in the figure below, the two bars are identical in size. Each bar has a diameter of 2 cm and a length of 16 cm. The two identical end blocks each has a thickness of 2 cm and a fillet radius of 1 cm at the round corners. The distance of the gap in between the two blocks is 8 cm. Suppose that the two blocks are fixed on the bottom. The bars are pulled by a pressure load of 100 N/cm^2 in opposite directions as shown below. Compute the stresses in the assembly using the FEA.

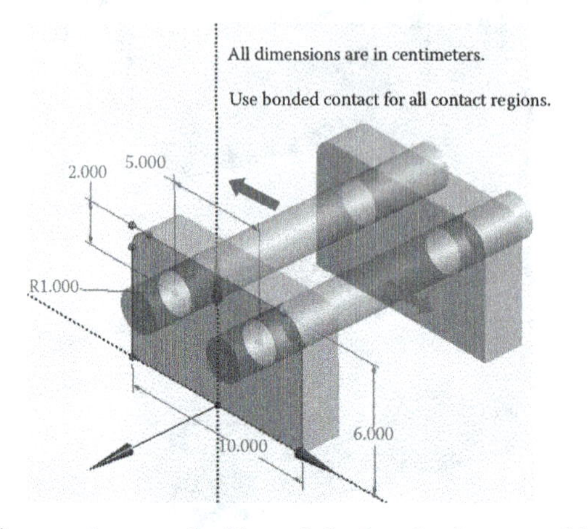

7.10 The following figure shows a half model of a clevis assembly consisting of a yoke, a pin, and a u-shape. The assembly components are made of structural steel with Young's modulus $E = 200$ GPa and Poisson's ratio $\nu = 0.3$. The profile sketches of the yoke and the u-shape are shown in the figure below. Both components have an extrusion depth of 5 cm. In the half model, the pin has a diameter of 2 cm and a length of 8 cm, and is centered 2.5 cm away from both the front and the side faces of the u-shape. Suppose that the yoke is fixed on the left end, and the u-shape is pulled by a pressure load of 100 N/cm^2 on the right end. Compute the stresses in the assembly using the FEA.

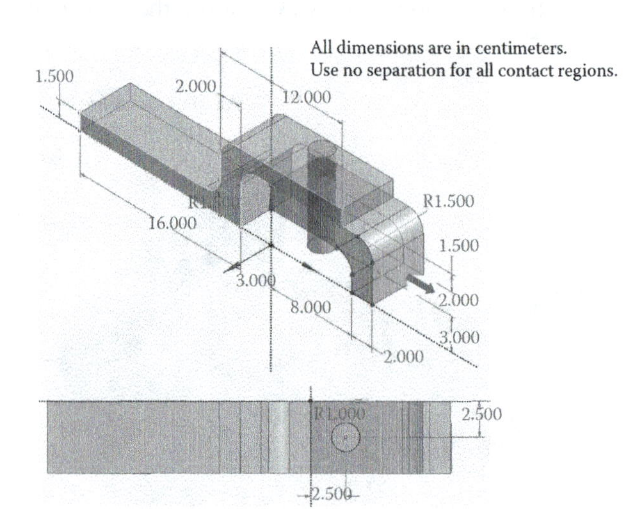

8

Structural Vibration and Dynamics

8.1 Introduction

Vibration and dynamics are fundamental subjects in engineering. Serious problems may arise from vibration when a structure is not carefully designed for its dynamic integrity. Vibration can cause malfunction or break down of machines that exhibit unbalance or misalignment. It can also lead to massive engineering failures such as the collapse of a bridge. Simulation plays an important role in our ability to understand a structure's dynamic behavior. Through modeling, the dynamic characteristics of a structure can be captured and improved before being put into actual use.

In this chapter, we first review the basic equations and their solutions for structural vibration and dynamic analysis. Then, we discuss the FEA formulations for solving vibration and dynamic responses. Guidelines in modeling and solving such problems are provided, along with a case study in *ANSYS® Workbench*.

A structure vibrates about an equilibrium position when excited by periodic or arbitrary inputs. There are three main types of problems for structural vibration and dynamic analyses:

- Modal analysis ($f(t) = 0$)
- Frequency response analysis ($f(t) = F \sin \omega t$)
- Transient response analysis ($f(t)$ is arbitrary)

where $f(t)$ is the dynamic force applied on the structure, t the time, and ω the circular frequency (Figure 8.1).

8.2 Review of Basic Equations

We begin with a review of vibration of a single-DOF system, consisting of a mass, a spring, and a damper as shown in Figure 8.2. Then we will review basic equations for a multi-DOF system, such as the discrete system resulting from the finite element discretization.

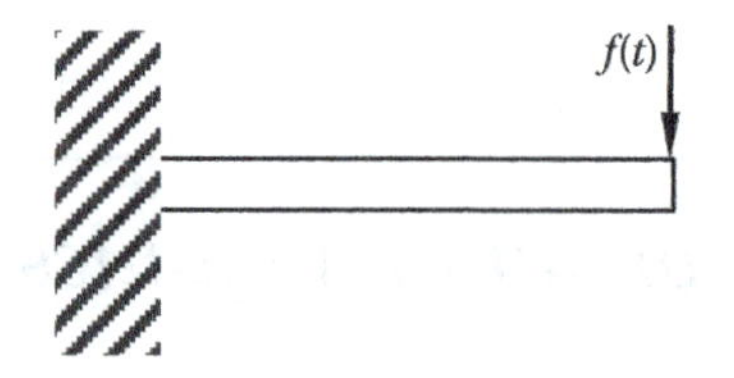

FIGURE 8.1
A dynamic force applied to the structure.

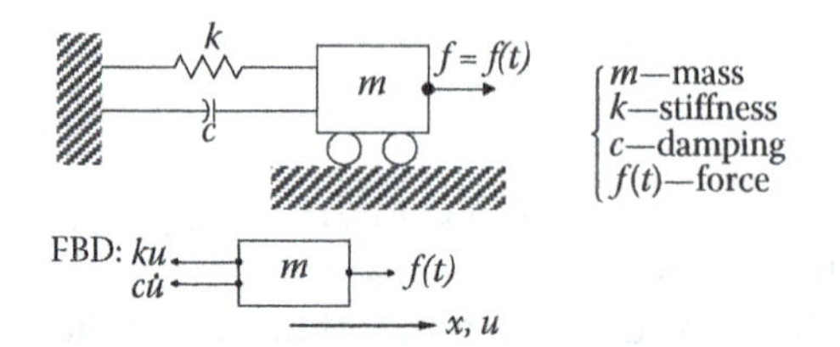

FIGURE 8.2
A single DOF system with damping.

8.2.1 A Single DOF System

From the free-body diagram (FBD) and Newton's law of motion ($ma = f$), we have:

$$m\ddot{u} = f(t) - ku - c\dot{u}$$

that is

$$m\ddot{u} + c\dot{u} + ku = f(t) \tag{8.1}$$

where u is the displacement, $\dot{u} = du/dt$ the velocity, and $\ddot{u} = d^2u/dt^2$ the acceleration.

Free Vibration (no applied force to the mass or f(t) = 0):

Free vibration occurs when a mass is moved away from its rest position due to initial conditions.

Assuming zero damping ($c = 0$) in a free vibration, Equation 8.1 becomes:

$$m\ddot{u} + ku = 0 \tag{8.2}$$

The physical meaning of this equation is: inertia force + elastic/stiffness force = 0.

Although there is no applied force, the mass can have nonzero displacement or experience vibrations under the initial conditions. To solve for such nontrivial solutions, we assume:

$$u(t) = \mathrm{U} \sin \omega t$$

where ω is the circular frequency of oscillation and U the amplitude. Substituting this into Equation 8.2 yields:

$$-U\omega^2.m \sin \omega t + kU\sin \omega t = 0$$

that is

$$[-\omega^2 m + k]U = 0$$

For nontrivial solutions for U, we must have:

$$[-\omega^2 m + k] = 0$$

which yields

$$\omega = \sqrt{\frac{k}{m}} \tag{8.3}$$

This is the circular *natural frequency* of the single DOF system (rad/s). The cyclic frequency ($1/s = Hz$) is $\omega/2\pi$.

Equation 8.3 is a very important result in free vibration analysis, which says that the natural frequency of a structure is proportional to the square-root of the stiffness of the structure and inversely proportional to the square-root of the total mass of the structure.

The typical response of the system in undamped free vibration is sketched in Figure 8.3.

For nonzero damping c, where

$$0 < c < c_c = 2m\omega = 2\sqrt{km} \quad (c_c = \text{critical damping}) \tag{8.4}$$

we have the *damped natural frequency*:

$$\omega_d = \omega\sqrt{1-\xi^2} \tag{8.5}$$

where

$$\xi = c/c_c \tag{8.6}$$

is called the damping ratio.

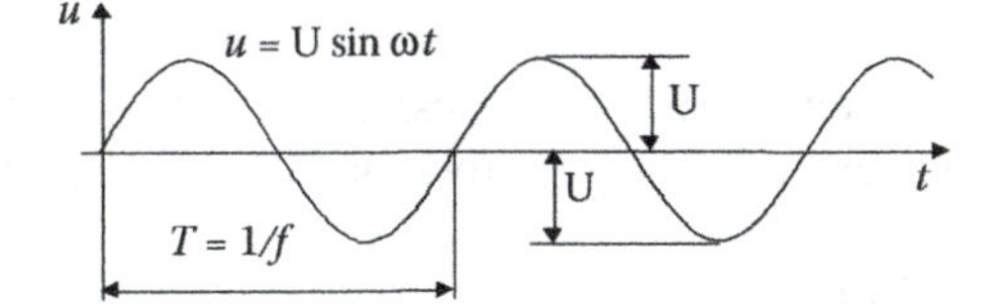

FIGURE 8.3
Typical response in an undamped free vibration.

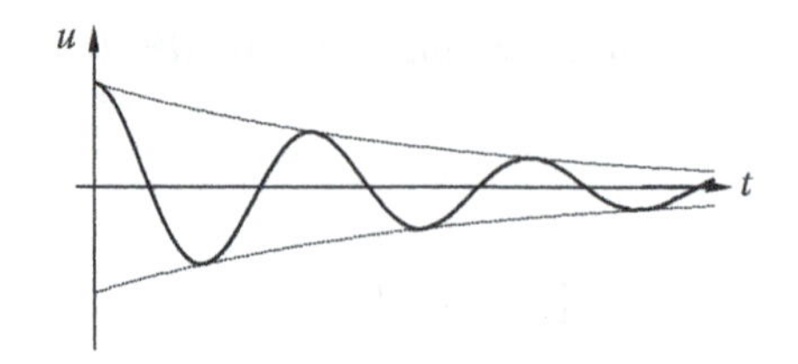

FIGURE 8.4
Typical response of a free vibration with a nonzero damping $c < c_c$.

For structural damping: $0 \le \xi < 0.15$ (usually 1 ~ 5%)

$$\omega_d \approx \omega \tag{8.7}$$

That is, we can ignore damping in normal mode analysis.

The typical response of the system in damped free vibration is sketched in Figure 8.4. We can see that damping has the effect of reducing the vibration of the system.

8.2.2 A Multi-DOF System

For a multi-DOF system, the equation of motion can be written as

$$\mathbf{M}\ddot{\mathbf{u}} + \mathbf{C}\dot{\mathbf{u}} + \mathbf{K}\mathbf{u} = \mathbf{f}(t) \tag{8.8}$$

in which:

u—nodal displacement vector;
M—mass matrix;
C—damping matrix;
K—stiffness matrix;
f—forcing vector.

The physical meaning of Equation 8.8 is

Inertia forces + Damping forces + Elastic forces = Applied forces

We already know how to determine the stiffness matrix **K** for a structure, as discussed in previous chapters. In vibration analysis, we also need to determine the mass matrix and damping matrix for the structure.

8.2.2.1 Mass Matrices

There are two types of mass matrices: *lumped mass matrices* and *consistent mass matrices*. The former is empirical and easier to determine, and the latter is analytical and more involved in their computing.

We use a bar element to illustrate the *lumped mass matrix* (Figure 8.5).

For this bar element, the *lumped mass matrix* for the element is found to be

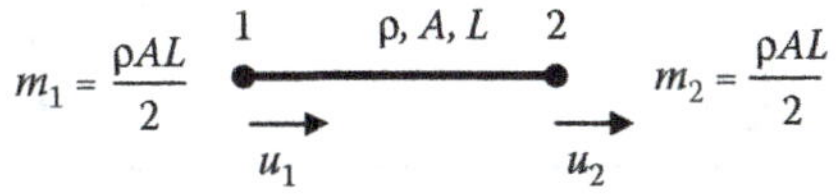

$$m_1 = \frac{\rho AL}{2} \quad \overset{1}{\bullet} \overset{\rho, A, L}{\text{———}} \overset{2}{\bullet} \quad m_2 = \frac{\rho AL}{2}$$

$u_1 \rightarrow \quad u_2 \rightarrow$

FIGURE 8.5
The lumped mass for a 1-D bar element.

$$\mathbf{m} = \begin{bmatrix} \frac{\rho AL}{2} & 0 \\ 0 & \frac{\rho AL}{2} \end{bmatrix}$$

which is a diagonal matrix and thus is easier to compute.

In general, we apply the following element *consistent mass matrix*:

$$\mathbf{m} = \int_V \rho \mathbf{N}^T \mathbf{N} \, dV \tag{8.9}$$

where $\mathbf{N}$ is the same shape function matrix as used for the displacement field and V is the volume of the element.

Equation 8.9 is obtained by considering the kinetic energy within an element:

$$\begin{aligned} K &= \frac{1}{2} \dot{\mathbf{u}}^T \mathbf{m} \dot{\mathbf{u}} \qquad\qquad (\text{cf. } \frac{1}{2} m v^2) \\ &= \frac{1}{2} \int_V \rho \dot{u}^2 \, dV = \frac{1}{2} \int_V \rho (\dot{u})^T \dot{u} \, dV \\ &= \frac{1}{2} \int_V \rho (\mathbf{N} \dot{\mathbf{u}})^T (\mathbf{N} \dot{\mathbf{u}}) \, dV \\ &= \frac{1}{2} \dot{\mathbf{u}}^T \underbrace{\int_V \rho \mathbf{N}^T \mathbf{N} \, dV}_{\mathbf{m}} \dot{\mathbf{u}} \end{aligned} \tag{8.10}$$

For the bar element (linear shape function), the *consistent mass matrix* is

$$\begin{aligned} \mathbf{m} &= \int_V \rho \mathbf{N}^T \mathbf{N} \, dV = \int_V \rho \begin{bmatrix} 1-\xi \\ \xi \end{bmatrix} \begin{bmatrix} 1-\xi & \xi \end{bmatrix} AL \, d\xi \\ &= \rho AL \begin{bmatrix} 1/3 & 1/6 \\ 1/6 & 1/3 \end{bmatrix} \begin{matrix} \ddot{u}_1 \\ \ddot{u}_2 \end{matrix} \end{aligned} \tag{8.11}$$

which is a nondiagonal matrix.

Similar to the formation of the global stiffness matrix $\mathbf{K}$, element mass matrices are established in local coordinates first, then transformed to global coordinates, and finally assembled together to form the global structure mass matrix $\mathbf{M}$.

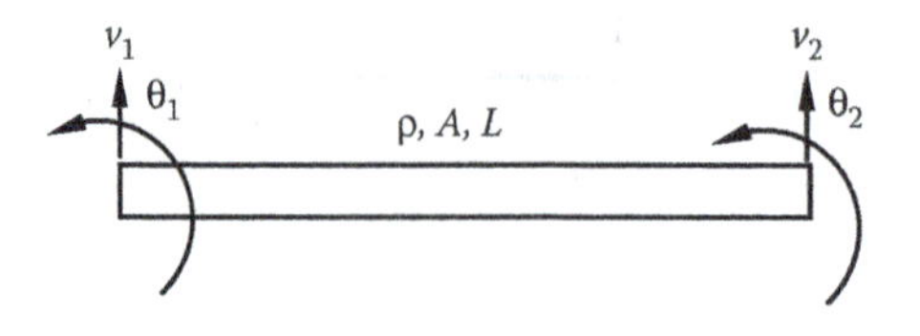

FIGURE 8.6
The consistent mass for a 1-D simple beam element.

For a simple beam element (Figure 8.6), the *consistent mass matrix* can be found readily by applying the four shape functions listed in Equation 3.6. We have:

$$\mathbf{m} = \int_V \rho \mathbf{N}^T \mathbf{N}\, dV$$

$$= \frac{\rho AL}{420}\begin{bmatrix} 156 & 22L & 54 & -13L \\ 22L & 4L^2 & 13L & -3L^2 \\ 54 & 13L & 156 & -22L \\ -13L & -3L^2 & -22L & 4L^2 \end{bmatrix} \begin{matrix} \ddot{v}_1 \\ \ddot{\theta}_1 \\ \ddot{v}_2 \\ \ddot{\theta}_2 \end{matrix} \tag{8.12}$$

8.2.2.2 *Damping*

There are two commonly used models for viscous damping: *proportional damping* (also called Rayleigh damping) and *modal damping*.

In the *proportional damping* model, the damping matrix **C** is assumed to be proportional to the stiffness and mass matrices in the following fashion:

$$\mathbf{C} = \alpha \mathbf{K} + \beta \mathbf{M} \tag{8.13}$$

where the constants α and β are found from the following two equations:

$$\xi_1 = \frac{\alpha\omega_1}{2} + \frac{\beta}{2\omega_1}, \quad \xi_2 = \frac{\alpha\omega_2}{2} + \frac{\beta}{2\omega_2} \tag{8.14}$$

with ω_1, ω_2, ξ_1 and ξ_2 (damping ratios) being specified by the user. The plots of the above two equations are shown in Figure 8.7.

In the *modal damping* model, the viscous damping is incorporated in the modal equations. The *modal damping* can be introduced as

$$\mathbf{C}_\phi = \begin{bmatrix} 2\xi_1\omega_1 & 0 & \cdots & 0 \\ 0 & 2\xi_2\omega_2 & & \\ \vdots & & \ddots & \vdots \\ 0 & & \cdots & 2\xi_n\omega_n \end{bmatrix} \tag{8.15}$$

where ξ_i is the damping ratio at mode i of a n-DOF system.

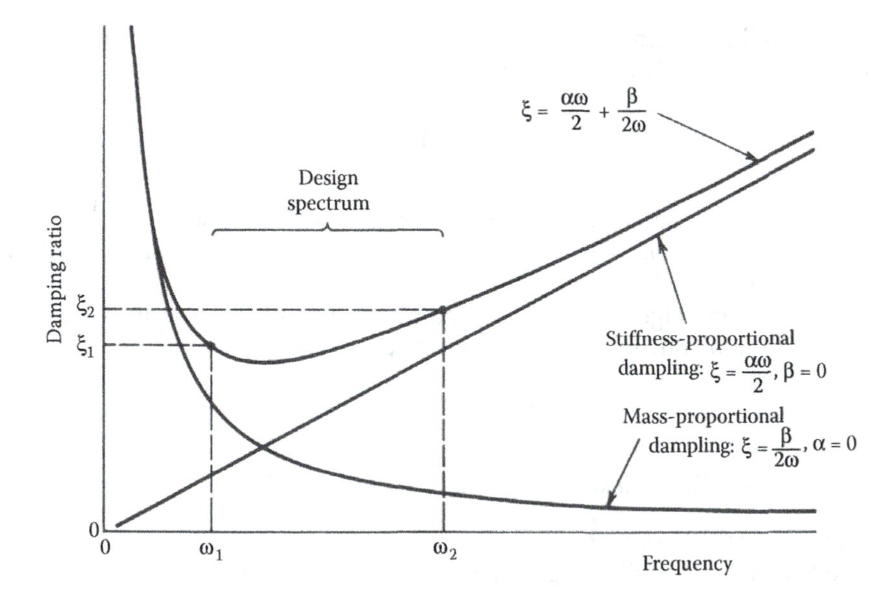

FIGURE 8.7
Two equations for determining the proportional damping coefficients. (R. D. Cook, *Finite Element Modeling for Stress Analysis*, 1995, Hoboken, NJ, Copyright Wiley-VCH Verlag GmbH & Co. KGaA. Reproduced with permission.)

8.3 Formulation for Modal Analysis

Modal analysis sets out to study the inherent vibration characteristics of a structure, including:

- Natural frequencies
- Normal modes (shapes)

Let $\mathbf{f}(t) = \mathbf{0}$ and $\mathbf{C} = \mathbf{0}$ (ignore damping) in the dynamic Equation 8.8 and obtain:

$$\mathbf{M}\ddot{\mathbf{u}} + \mathbf{K}\mathbf{u} = \mathbf{0} \tag{8.16}$$

Assume that displacements vary harmonically with time, that is:

$$\begin{aligned}\mathbf{u}(t) &= \bar{\mathbf{u}}\sin(\omega t),\\ \dot{\mathbf{u}}(t) &= \omega\bar{\mathbf{u}}\cos(\omega t),\\ \ddot{\mathbf{u}}(t) &= -\omega^2\bar{\mathbf{u}}\sin(\omega t)\end{aligned}$$

where $\bar{\mathbf{u}}$ is the vector of the amplitudes of the nodal displacements.

Substituting these into Equation 8.16 yields:

$$[\mathbf{K} - \omega^2\mathbf{M}]\bar{\mathbf{u}} = \mathbf{0} \tag{8.17}$$

This is a generalized eigenvalue problem (EVP). The trivial solution is $\bar{\mathbf{u}} = \mathbf{0}$ for any values of ω (not interesting). Nontrivial solutions ($\bar{\mathbf{u}} \neq \mathbf{0}$) exist only if:

$$\left|\mathbf{K} - \omega^2\mathbf{M}\right| = 0 \tag{8.18}$$

This is an n-th order polynomial of ω^2, from which we can find n solutions (roots) or eigenvalues ω_i ($i = 1, 2, \ldots, n$). These are the natural frequencies (or characteristic frequencies) of the structure.

The smallest nonzero eigenvalue ω_1 is called the *fundamental frequency*.

For each ω_i, Equation 8.17 gives one solution or eigen vector:

$$[\mathbf{K} - \omega_i^2\mathbf{M}]\bar{\mathbf{u}}_i = \mathbf{0}$$

$\bar{\mathbf{u}}_i$ ($i = 1, 2, \ldots, n$) are the *normal modes* (or *natural modes, mode shapes,* and so on).

Properties of the Normal Modes:
Normal modes satisfy the following properties:

$$\bar{\mathbf{u}}_i^T\mathbf{K}\bar{\mathbf{u}}_j = 0, \quad \bar{\mathbf{u}}_i^T\mathbf{M}\bar{\mathbf{u}}_j = 0, \quad \text{for} \quad i \neq j \tag{8.19}$$

if $\omega_i \neq \omega_j$. That is, modes are *orthogonal* (thus *independent*) to each other with respect to **K** and **M** matrices.

Normal modes are usually normalized:

$$\bar{\mathbf{u}}_i^T\mathbf{M}\bar{\mathbf{u}}_i = 1, \quad \bar{\mathbf{u}}_i^T\mathbf{K}\bar{\mathbf{u}}_i = \omega_i^2 \tag{8.20}$$

Notes:

- Magnitudes of displacements (modes) or stresses in normal mode analysis have no physical meaning.
- For normal mode analysis, no support of the structure is necessary.
- $\omega_i = 0$ means there are rigid-body motions of the whole or a part of the structure. This can be applied to check the FEA model (check to see if there are mechanisms or free elements in the FEA models).
- Lower modes are more accurate than higher modes in the FEA calculations (due to less spatial variations in the lower modes leading to that fewer elements/wavelengths are needed).

EXAMPLE 8.1

Consider the free vibration of a cantilever beam with one element as shown below.

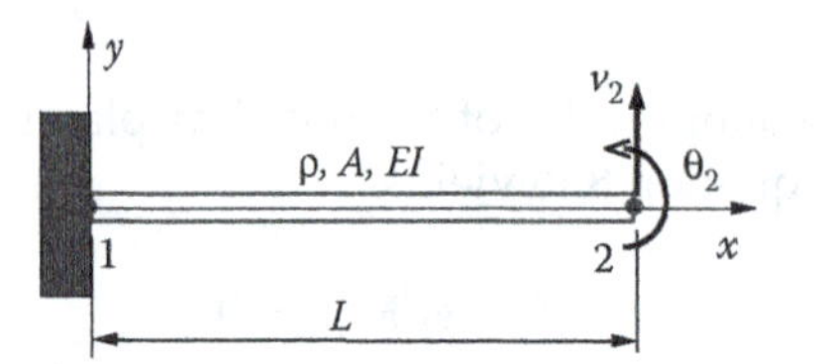

We have the following equation for the free vibration (EVP):

$$[\mathbf{K} - \omega^2\mathbf{M}]\begin{Bmatrix} \bar{v}_2 \\ \bar{\theta}_2 \end{Bmatrix} = \begin{Bmatrix} 0 \\ 0 \end{Bmatrix}$$

where

$$\mathbf{K} = \frac{EI}{L^3}\begin{bmatrix} 12 & -6L \\ -6L & 4L^2 \end{bmatrix}, \quad \mathbf{M} = \frac{\rho AL}{420}\begin{bmatrix} 156 & -22L \\ -22L & 4L^2 \end{bmatrix}$$

The equation for determining the natural frequencies is

$$\begin{vmatrix} 12 - 156\lambda & -6L + 22L\lambda \\ -6L + 22L\lambda & 4L^2 - 4L^2\lambda \end{vmatrix} = 0$$

in which $\lambda = \omega^2 \rho AL^4 / 420\, EI$.

Solving the EVP, we obtain:

$$\omega_1 = 3.533\left(\frac{EI}{\rho AL^4}\right)^{1/2}, \quad \begin{Bmatrix} \bar{v}_2 \\ \bar{\theta}_2 \end{Bmatrix}_1 = \begin{Bmatrix} 1 \\ 1.38/L \end{Bmatrix},$$

$$\omega_2 = 34.81\left(\frac{EI}{\rho AL^4}\right)^{1/2}, \quad \begin{Bmatrix} \bar{v}_2 \\ \bar{\theta}_2 \end{Bmatrix}_2 = \begin{Bmatrix} 1 \\ 7.62/L \end{Bmatrix}$$

The exact solutions of the first two natural frequencies for this problem are

$$\omega_1 = 3.516\left(\frac{EI}{\rho AL^4}\right)^{1/2}, \quad \omega_2 = 22.03\left(\frac{EI}{\rho AL^4}\right)^{1/2}$$

We can see that for the FEA solution with one beam element, mode 1 is calculated much more accurately than mode 2. More elements are needed in order to compute mode 2 more accurately. The first three mode shapes of the cantilever beam is shown in the insert above.

8.3.1 Modal Equations

Use the normal modes (modal matrices), we can transform the coupled system of dynamic equations to uncoupled system of equations or modal equations.

We have:

$$[\mathbf{K} - \omega_i^2\mathbf{M}]\bar{\mathbf{u}}_i = \mathbf{0}, \quad i = 1, 2, \ldots, n \tag{8.21}$$

where the normal modes $\bar{\mathbf{u}}_i$ satisfy:

$$\begin{cases} \bar{\mathbf{u}}_i^T \mathbf{K} \bar{\mathbf{u}}_j = 0, \\ \bar{\mathbf{u}}_i^T \mathbf{M} \bar{\mathbf{u}}_j = 0, \end{cases} \quad \text{for} \quad i \neq j$$

and

$$\begin{cases} \bar{\mathbf{u}}_i^T \mathbf{M} \bar{\mathbf{u}}_i = 1, \\ \bar{\mathbf{u}}_i^T \mathbf{K} \bar{\mathbf{u}}_i = \omega_i^2, \end{cases} \quad \text{for} \quad i = 1, 2, \ldots, n$$

Form the *modal matrix*:

$$\Phi_{(n \times n)} = \left[\bar{\mathbf{u}}_1 \; \bar{\mathbf{u}}_2 \cdots \; \bar{\mathbf{u}}_n\right] \tag{8.22}$$

We can verify that:

$$\begin{aligned} \Phi^T \mathbf{K} \Phi = \Omega \;&= \begin{bmatrix} \omega_1^2 & 0 & \cdots & 0 \\ 0 & \omega_2^2 & & \vdots \\ \vdots & & \ddots & 0 \\ 0 & \cdots & 0 & \omega_n^2 \end{bmatrix} \quad \text{(Spectral matrix)} \\ \Phi^T \mathbf{M} \Phi = \mathbf{I}. & \end{aligned} \tag{8.23}$$

Transformation for the displacement vector:

$$\mathbf{u} = z_1 \bar{\mathbf{u}}_1 + z_2 \bar{\mathbf{u}}_2 + \cdots + z_n \bar{\mathbf{u}}_n = \Phi \mathbf{z} \tag{8.24}$$

where

$$\mathbf{z} = \begin{Bmatrix} z_1(t) \\ z_2(t) \\ \vdots \\ z_n(t) \end{Bmatrix}$$

are called the principal coordinates.

Substitute Equation 8.24 into the dynamic Equation 8.8 and obtain:

$$\mathbf{M}\Phi\ddot{\mathbf{z}} + \mathbf{C}\Phi\dot{\mathbf{z}} + \mathbf{K}\Phi\mathbf{z} = \mathbf{f}(t)$$

Premultiply this result by Φ^T, and apply Equation 8.23:

$$\ddot{\mathbf{z}} + \mathbf{C}_\varphi \dot{\mathbf{z}} + \Omega \mathbf{z} = \mathbf{p}(t) \tag{8.25}$$

where $\mathbf{C}_\phi = \alpha \mathbf{I} + \beta \Omega$ if proportional damping is applied, and $\mathbf{p} = \Phi^T \mathbf{f}(t)$.

If we employ *modal damping*:

$$\mathbf{C}_\phi = \begin{bmatrix} 2\xi_1\omega_1 & 0 & \cdots & 0 \\ 0 & 2\xi_2\omega_2 & & \\ \vdots & & \ddots & \vdots \\ 0 & & \cdots & 2\xi_n\omega_n \end{bmatrix}$$

where ξ_i is the damping ratio at mode i, Equation 8.25 becomes:

$$\ddot{z}_i + 2\xi_i\omega_i\dot{z}_i + \omega_i^2 z_i = p_i(t), \quad i = 1, 2, \ldots, n \tag{8.26}$$

Equations in Equation 8.25 with modal damping, or in Equation 8.26, are called *modal equations*. These equations are uncoupled, second-order differential equations, that are much easier to solve than the original dynamic equation which is a coupled system.

To recover **u** from **z**, apply the transformation in Equation 8.24 again, once **z** is obtained from Equation 8.26.

Notes:

- Only the first few modes may be needed in constructing the modal matrix $\mathbf{\Phi}$ (i.e., $\mathbf{\Phi}$ could be an $n \times m$ rectangular matrix with $m < n$). Thus, significant reduction in the size of the system can be achieved.
- Modal equations are best suited for structural vibration problems in which higher modes are not important (i.e., for structural vibrations, but not for structures under impact or shock loadings).

8.4 Formulation for Frequency Response Analysis

For frequency response analysis (also referred to as harmonic response analysis), the applied dynamic load is a sine or cosine function. In this case, the equation of motion is

$$\mathbf{M}\ddot{\mathbf{u}} + \mathbf{C}\dot{\mathbf{u}} + \mathbf{K}\mathbf{u} = \underbrace{\mathbf{F}\sin\omega t}_{\text{Harmonic loading}} \tag{8.27}$$

8.4.1 Modal Method

In this approach, we apply the modal equations, that is

$$\ddot{z}_i + 2\xi_i\omega_i\dot{z}_i + \omega_i^2 z_i = p_i \sin\omega t \quad i = 1, 2, \ldots, m \tag{8.28}$$

These are uncoupled equations. The solutions for **z** are in the form:

$$z_i(t) = \frac{p_i/\omega_i^2}{\sqrt{(1-\eta_i^2)^2 + (2\xi_i\eta_i)^2}} \sin(\omega t - \theta_i) \tag{8.29}$$

where

$$\begin{cases} \theta_i = \arctan\dfrac{2\xi_i\eta_i}{1-\eta_i^2}, & \text{phase angle;} \\ \eta_i = \omega/\omega_i; & \\ \xi_i = \dfrac{c_i}{c_c} = \dfrac{c_i}{2m\omega_i}, & \text{damping ratio} \end{cases}$$

The response of each mode Z_i is similar to that of a single DOF system. Once the natural coordinate vector **z** is known, we can recover the real displacement vector **u** from **z** using Equation 8.24.

8.4.2 Direct Method

In this approach, we solve Equation 8.27 directly, that is, compute the inverse of the coefficient matrix, which is in general much more expensive than the modal method.

Using complex notation to represent the harmonic response, we have $\mathbf{u} = \bar{\mathbf{u}} e^{i\omega t}$ and Equation 8.27 becomes:

$$[\mathbf{K} + i\omega\mathbf{C} - \omega^2\mathbf{M}]\bar{\mathbf{u}} = \bar{\mathbf{F}} \tag{8.30}$$

Inverting the matrix $[\mathbf{K} + i\omega\mathbf{C} - \omega^2\mathbf{M}]$, we can obtain the displacement amplitude vector $\bar{\mathbf{u}}$. However, this equation is expensive to solve for large systems and the matrix $[\mathbf{K} + i\omega\mathbf{C} - \omega^2\mathbf{M}]$ can become ill-conditioned if ω is close to any natural frequency ω_i of the structure. Therefore, the direct method is only applied when the system of equations is small and the frequency is away from any natural frequency of the structure.

8.5 Formulation for Transient Response Analysis

In transient response analysis (also referred to as dynamic response/time-history analysis), we are interested in computing the responses of the structures under *arbitrary* time-dependent loading (Figure 8.8).

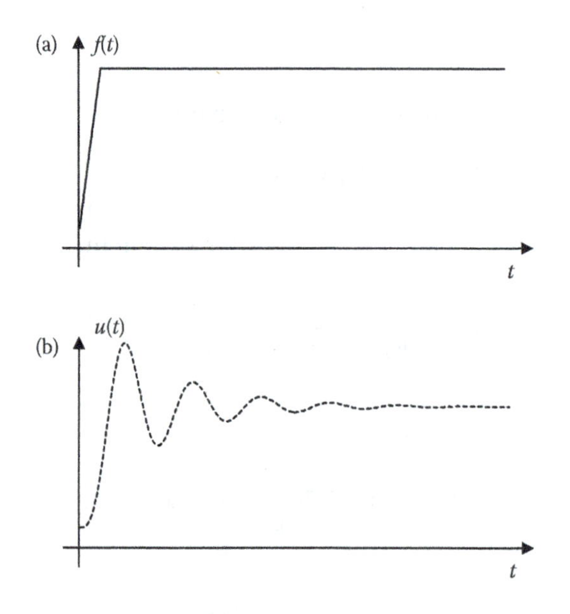

FIGURE 8.8
(a) A step type of loading; (b) structural response to the step loading.

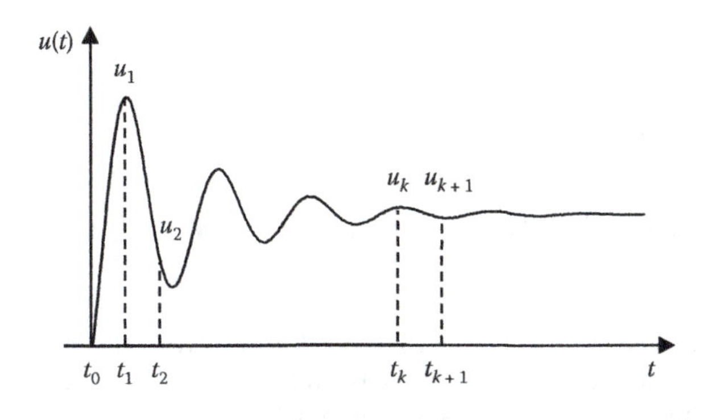

FIGURE 8.9
Computing the responses by integration through time.

To compute the transient responses, integration through time is employed (Figure 8.9). We write the equation of motion at instance t_k, $k = 0, 1, 2, 3, \ldots$, as

$$\mathbf{M}\ddot{\mathbf{u}}_k + \mathbf{C}\dot{\mathbf{u}}_k + \mathbf{K}\mathbf{u}_k = \mathbf{f}_k \tag{8.31}$$

Then, we introduce time increments: $\Delta t = t_{k+1} - t_k$, $k = 0, 1, 2, 3, \ldots$, and integrate through the time.

There are two categories of methods for transient analysis as described in the following sections.

8.5.1 Direct Methods (Direct Integration Methods)

Central Difference Method:
Approximate the velocity and acceleration vectors by using finite difference:

$$\begin{aligned} \dot{\mathbf{u}}_k &= \frac{1}{2\Delta t}(\mathbf{u}_{k+1} - \mathbf{u}_{k-1}), \\ \ddot{\mathbf{u}}_k &= \frac{1}{(\Delta t)^2}(\mathbf{u}_{k+1} - 2\mathbf{u}_k + \mathbf{u}_{k-1}) \end{aligned} \tag{8.32}$$

Dynamic equation becomes,

$$\mathbf{M}\left[\frac{1}{(\Delta t)^2}(\mathbf{u}_{k+1} - 2\mathbf{u}_k + \mathbf{u}_{k-1})\right] + \mathbf{C}\left[\frac{1}{2\Delta t}(\mathbf{u}_{k+1} - \mathbf{u}_{k-1})\right] + \mathbf{K}\mathbf{u}_k = \mathbf{f}_k$$

which yields

$$\mathbf{A}\mathbf{u}_{k+1} = \mathbf{F}(t) \tag{8.33}$$

where

$$\begin{cases} \mathbf{A} = \dfrac{1}{(\Delta t)^2}\mathbf{M} + \dfrac{1}{2\Delta t}\mathbf{C}, \\ \mathbf{F}(t) = \mathbf{f}_k - \left[\mathbf{K} - \dfrac{2}{(\Delta t)^2}\mathbf{M}\right]\mathbf{u}_k - \left[\dfrac{1}{(\Delta t)^2}\mathbf{M} - \dfrac{1}{2\Delta t}\mathbf{C}\right]\mathbf{u}_{k-1} \end{cases}$$

We compute $\mathbf{u}_{k+1}$ from $\mathbf{u}_k$ and $\mathbf{u}_{k-1}$, which are known from the previous time step. The solution procedure is repeated or marching from $t_0, t_1, \ldots t_k, t_{k+1}, \ldots$, until the specified maximum time is reached. This method is unstable if Δt is too large.

Newmark Method:
We use the following approximations:

$$\begin{aligned} \mathbf{u}_{k+1} &\approx \mathbf{u}_k + \Delta t\dot{\mathbf{u}}_k + \frac{(\Delta t)^2}{2}[(1-2\beta)\ddot{\mathbf{u}}_k + 2\beta\ddot{\mathbf{u}}_{k+1}], \rightarrow (\ddot{\mathbf{u}}_{k+1} = \cdots) \\ \dot{\mathbf{u}}_{k+1} &\approx \dot{\mathbf{u}}_k + \Delta t[(1-\gamma)\ddot{\mathbf{u}}_k + \gamma\ddot{\mathbf{u}}_{k+1}] \end{aligned} \tag{8.34}$$

where β and γ are chosen constants. These lead to the following equation:

$$\mathbf{A}\mathbf{u}_{k+1} = \mathbf{F}(t) \tag{8.35}$$

where

$$\begin{aligned} &\mathbf{A} = \mathbf{K} + \frac{\gamma}{\beta\Delta t}\mathbf{C} + \frac{1}{\beta(\Delta t)^2}\mathbf{M}, \\ &\mathbf{F}(t) = f(\mathbf{f}_{k+1}, \gamma, \beta, \Delta t, \mathbf{C}, \mathbf{M}, \mathbf{u}_k, \dot{\mathbf{u}}_k, \ddot{\mathbf{u}}_k) \end{aligned}$$

This method is unconditionally stable if

$$2\beta \geq \gamma \geq \frac{1}{2}$$

For example, we can use $\gamma = 1/2$, $\beta = 1/4$, which gives the constant average acceleration method.

Direct methods can be expensive, because of the need to compute A^{-1}, repeatedly for each time step if nonuniform time steps are used.

8.5.2 Modal Method

In this method, we first do the transformation of the dynamic equations using the modal matrix before the time marching:

$$\begin{aligned} &\mathbf{u} = \sum_{i=1}^{m} \bar{\mathbf{u}}_i z_i(t) = \mathbf{\Phi z}, \\ &\ddot{z}_i + 2\xi_i\omega_i\dot{z}_i + \omega_i z_i = p_i(t), \quad i = 1, 2, \ldots, m. \end{aligned} \tag{8.36}$$

TABLE 8.1
Comparisons of the Methods

Direct Methods	Modal Method
Small models	Large models
More accurate (with small Δt)	Higher modes ignored
Single loading	Multiple loading
Shock loading	Periodic loading
...	...

Then, solve the uncoupled equations using an integration method. We can use, for example, 10% of the total modes ($m = n/10$). The advantages of the modal method are as follows:

- Uncoupled system
- Fewer equations
- No inverse of matrices
- More efficient for large problems

However, the modal method is less accurate if higher modes are important, which is the case for structures under impact or shock loading. Table 8.1 summarizes the advantages and disadvantages of the direct and modal methods for transient response analysis.

8.6 Modeling Examples

8.6.1 Modal Analysis

Take a bumper system as an example. Figure 8.10 shows a front bumper and the supporting brackets in a car. The model is applied to study the dynamic responses of the bumper. Shell elements were used for this study to obtain the natural frequencies and vibration modes. Figure 8.11 shows the first mode of the bumper when it is constrained at the bracket locations.

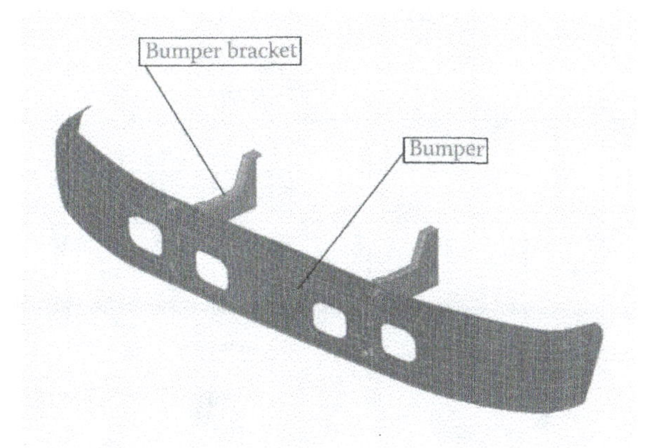

FIGURE 8.10
FEA model of a front bumper and supporting brackets.

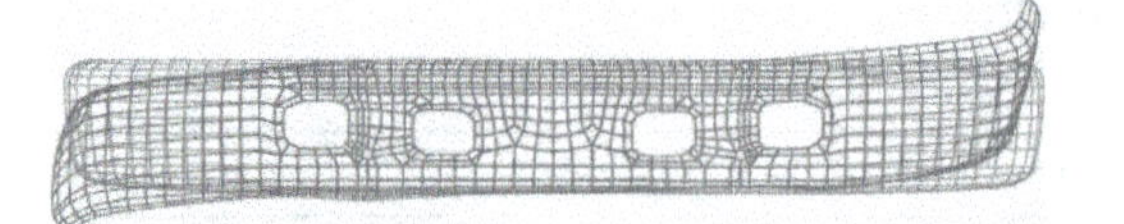

FIGURE 8.11
The first vibration mode of the bumper.

8.6.2 Frequency Response Analysis

Figure 8.12 shows the frequency response of the bumper system by the leftmost curve and with the acceleration of the two brackets as the input. Several modifications of the bumper design are also studied with the goal to increase the base natural frequency (e.g., from below 30 Hz to above 35 Hz) and to reduce the magnitudes of the frequency responses. The improved responses are shown by the other three curves in Figure 8.12.

8.6.3 Transient Response Analysis

One of the most interesting applications of the transient analysis with FEA is to conduct crash analysis and virtual drop tests of various products. Figure 8.13 is an example of crash analysis of a car using the FEA, showing that the bumper system absorbs the shock energy to help reduce the damage to the car.

8.6.4 Cautions in Dynamic Analysis

- Symmetry model should not be used in the dynamic analysis (normal modes, etc.) because symmetric structures can have nonsymmetric modes. However,

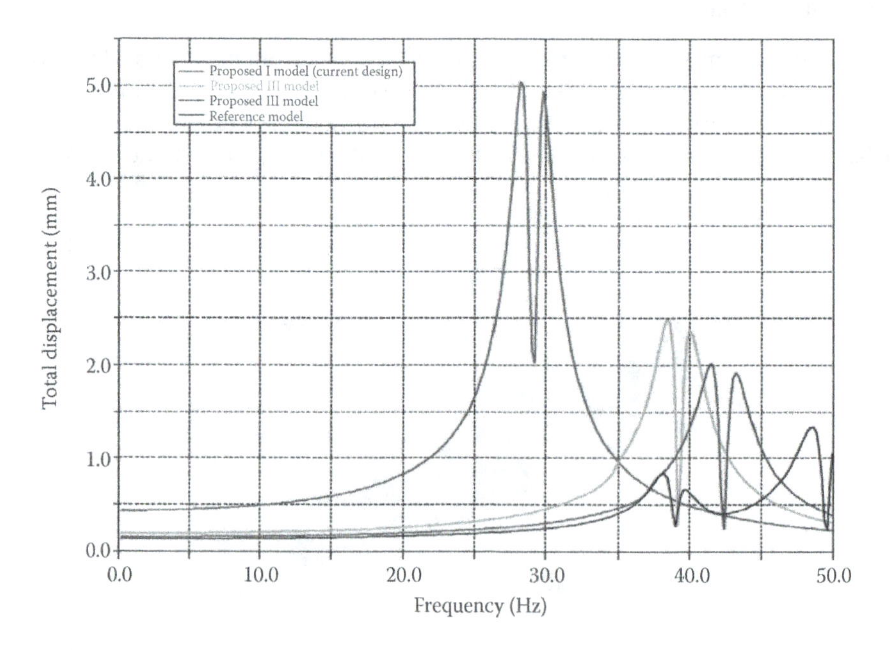

FIGURE 8.12
Frequency response of the bumper from 0 to 50 Hz.

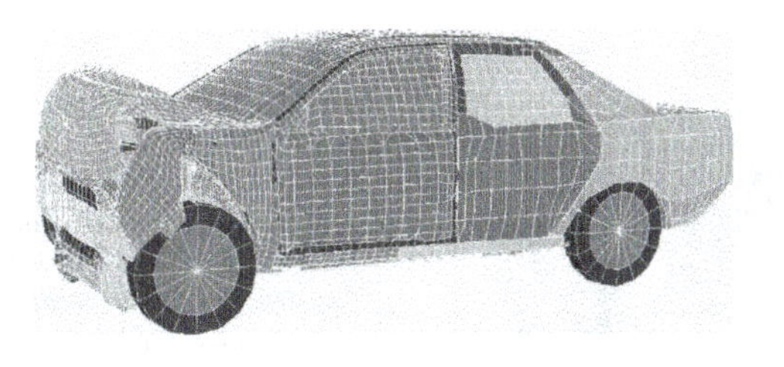

FIGURE 8.13
Car crash analysis using the FEA. (From LS-Dyna website, http://www.dynaexamples.com/)

TABLE 8.2
Units in Dynamic Analysis

	Choice I	Choice II
t (time)	s	s
L (length)	m	mm
m (mass)	kg	Mg
a (accel.)	m/s^2	mm/s^2
f (force)	N	N
ρ (density)	kg/m^3	Mg/mm^3

symmetry can still be applied in creating the FEA model of a symmetric structure.

- Mechanism or rigid body motion means $\omega = 0$. Can use this to check FEA models to see if they are properly connected and/or supported.
- Input for FEA: Loading $F(t)$ or $F(\omega)$ can be very complex and data can be enormous in real engineering applications (e.g., the load data for a car) and thus they often need to be filtered first before being used as input for FEA.
- Selecting a proper unit system is very important in vibration or dynamic analysis. Two choices of the units are listed in Table 8.2. Make sure they are consistent in the FEA models.

8.7 Case Studies with *ANSYS Workbench*

Problem Description: Musical instruments such as acoustic guitars create sound by means of vibration and resonance. The body of an acoustic guitar acts as a resonating chamber when the strings are set into oscillation at their natural frequencies. The following figure gives the dimensions of a simplified acoustic guitar model. The guitar has a wall thickness of 3 mm, and is made of Douglas fir wood ($E = 13.1$ GPa, Poison's ratio $\nu = 0.3$, density = 470 kg/m^3). Assuming the back surface of the guitar is fixed, find the first 10 natural frequencies and plot the first five vibration modes of the guitar. Suppose a harmonic pressure loading of magnitude 1 MPa is applied to a side wall of the guitar. Plot the frequency response of the z displacement (along the surface normal direction) of the front surface.

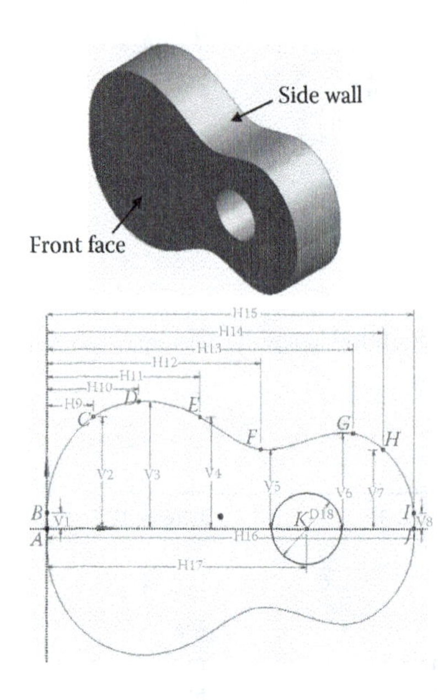

Material:

Douglas fir (E = 13.1 GPa, ν = 0.3, density = 470 kg/m^3)

Boundary conditions:

A fixed back face; harmonic pressure of 1 MPa applied to a side wall.

Construction point coordinates:

Point	*x* (mm)	*y* (mm)
A	0	0
B	0	10
C	30	70
D	60	80
E	100	70
F	140	50
G	200	60
H	220	50
I	240	10
J	240	0
K	170	0

The guitar profile is a spline that goes through points A through J.
A circular hole centered at K has a diameter of 45 mm.

Step 1: Start an *ANSYS Workbench* Project

Launch *ANSYS Workbench* and save the blank project as *"Guitar.wbpj."*

Step 2: Create a *Modal* Analysis System

Drag the *Modal* icon from the *Analysis Systems Toolbox* window and drop it inside the highlighted green rectangle in the *Project Schematic* window to create a standalone modal analysis system.

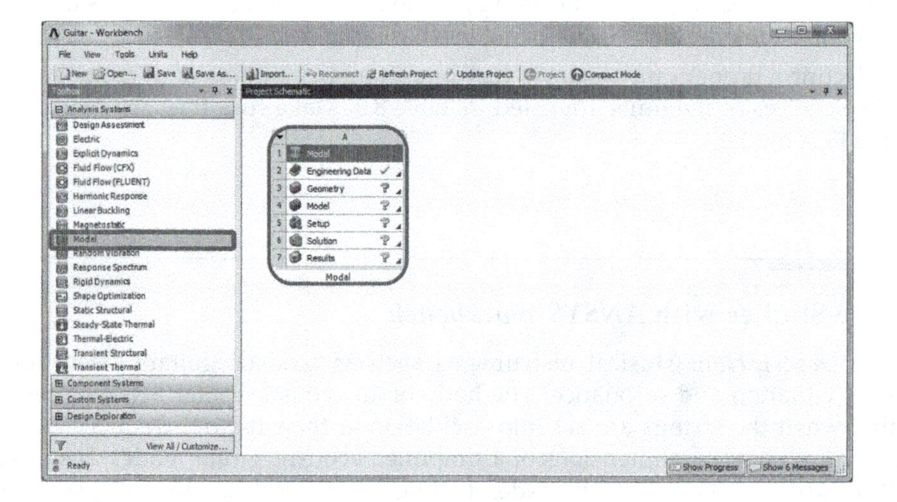

Step 3: Add a New Material

Double-click on the *Engineering Data* cell to add a new material. In the following *Engineering Data* interface which replaces the *Project Schematic*, type *"Wood"* as the name for the new material, and double-click *Isotropic Elasticity* under *Linear*

Elastic in the leftmost *Toolbox* window. Enter "*13.1E9*" for *Young's Modulus* and "*0.3*" for *Poisson's Ratio* in the *Properties* window. Double-click *Density* under *Physical Properties*. Enter "*470*" for *Density* in the *Properties* window. Click the *Return to Project* button to go back to *Project Schematic.*

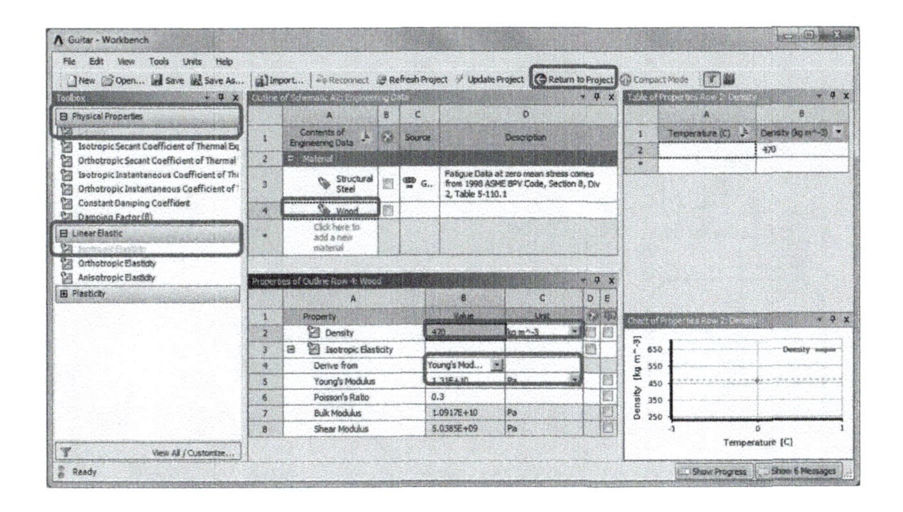

Step 4: Launch the *DesignModeler* Program

Ensure *Surface Bodies* is checked in the *Properties of Schematic A3: Geometry* window (select *Properties* from the *View* drop-down menu to enable display of this window). Choose *3D* as the *Analysis Type* in this *Properties* window. Double-click the *Geometry* cell to launch *DesignModeler,* and select "*Millimeter*" in the *Units* pop-up window.

Step 5: Create a Profile Sketch

Click on the *Sketching* tab. Select the *Draw* toolbox and then *Construction Point.* Draw 10 construction points *A through J,* as shown below. Draw a spline passing through points *A through J*; right-click at the last construction point and choose *Open End* from the context menu to finish the spline creation.

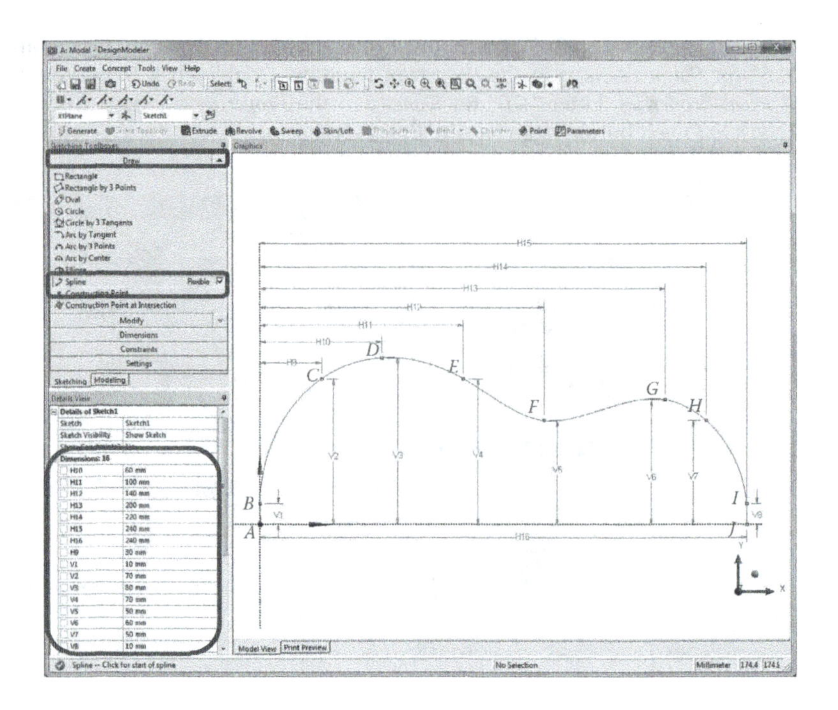

Step 6: Create a Replicate Curve

Select the *Modify* toolbox and then *Replicate*. Click on the spline from the *Graphics* window. Right-click anywhere in the *Graphics* to show the context menu. Select *End/Use Plane Origin as Handle* as shown below. A replicate spline will appear in the *Graphics* window.

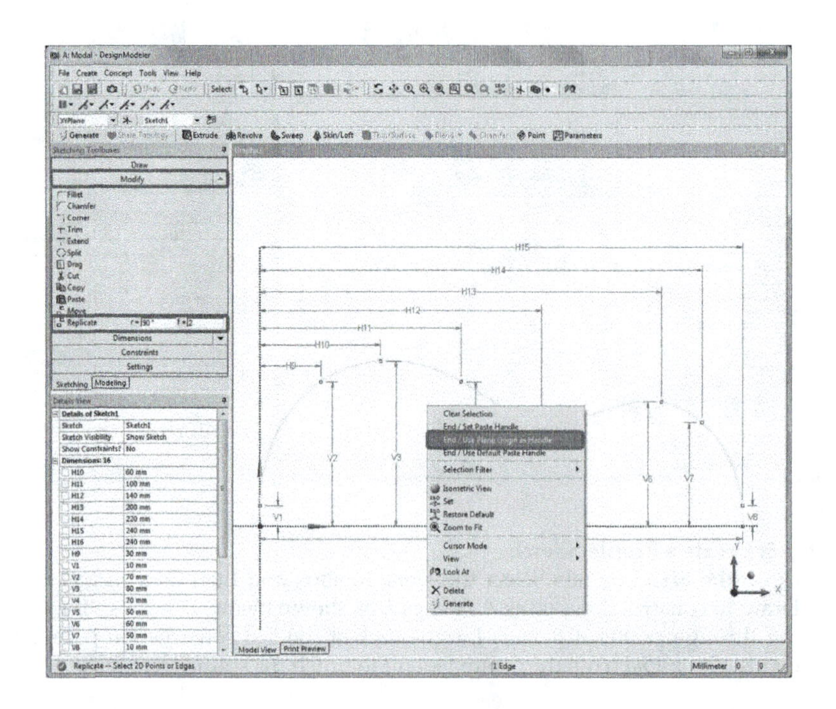

Next, right-click anywhere in the *Graphics*, and select *Flip Vertical* in the context menu. A vertically flipped spline will appear.

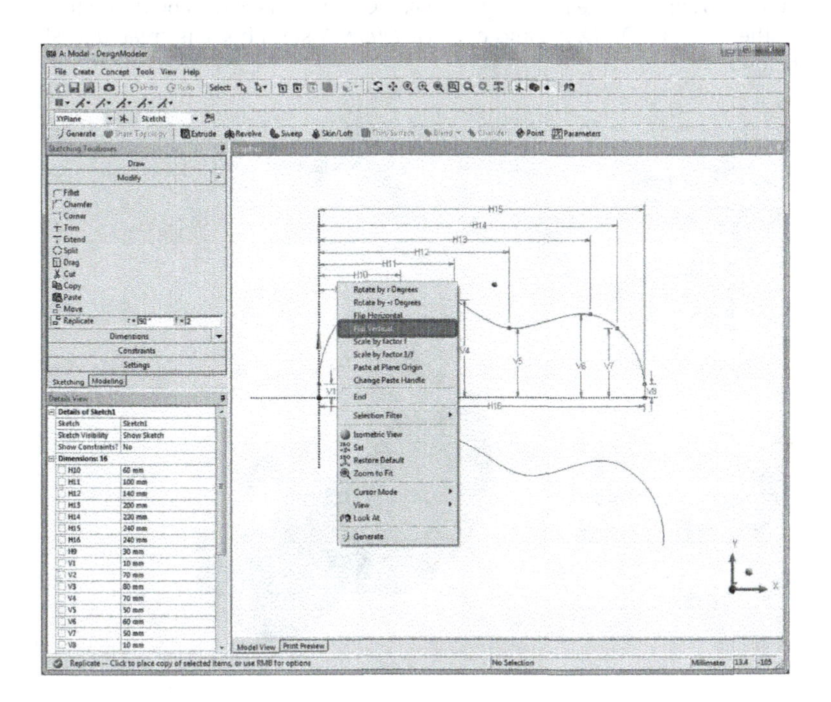

Click on the origin point in the *Graphics* to place the flipped spline, and press *Esc* to end the operation. A closed-loop curve is now formed as shown below.

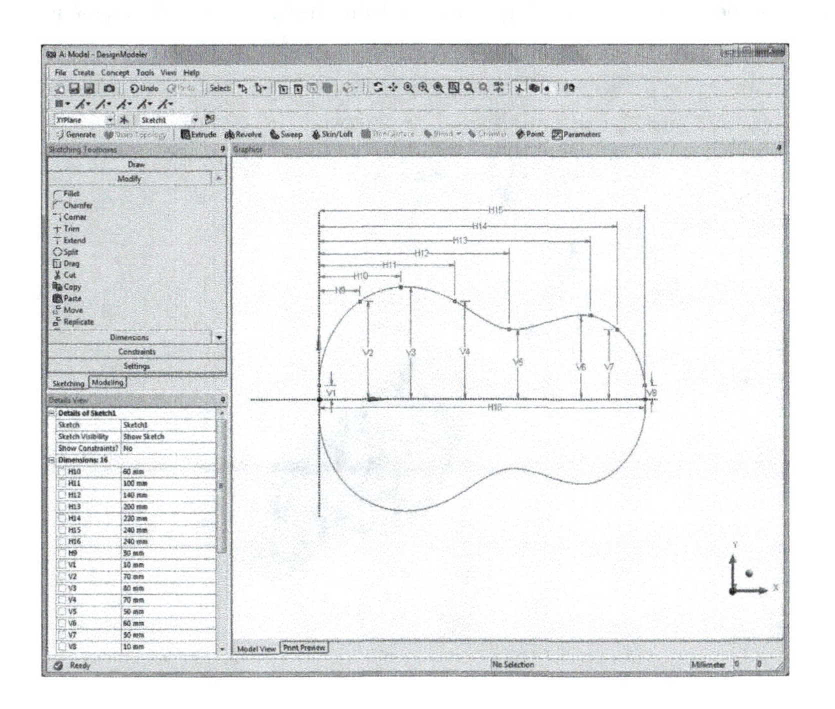

Step 7: Create an Extruded Body

Switch to the *Modeling* tab and click on the *Extrude* feature. The default *Base Object* is set as *Sketch1* in the *Details of Extrude1*. Change the extrusion depth to *50 mm* in the field of *FD1, Depth* and click *Generate*. A solid body is created as shown below.

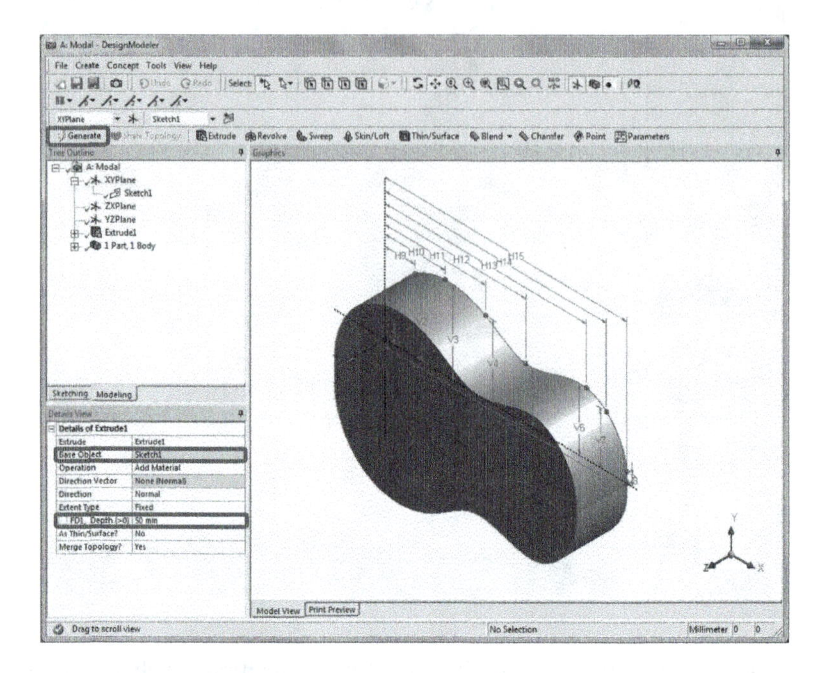

Step 8: Create an Extruded Cut on the Front Face

Create a new plane by selecting *New Plane* from the *Create* drop-down menu.

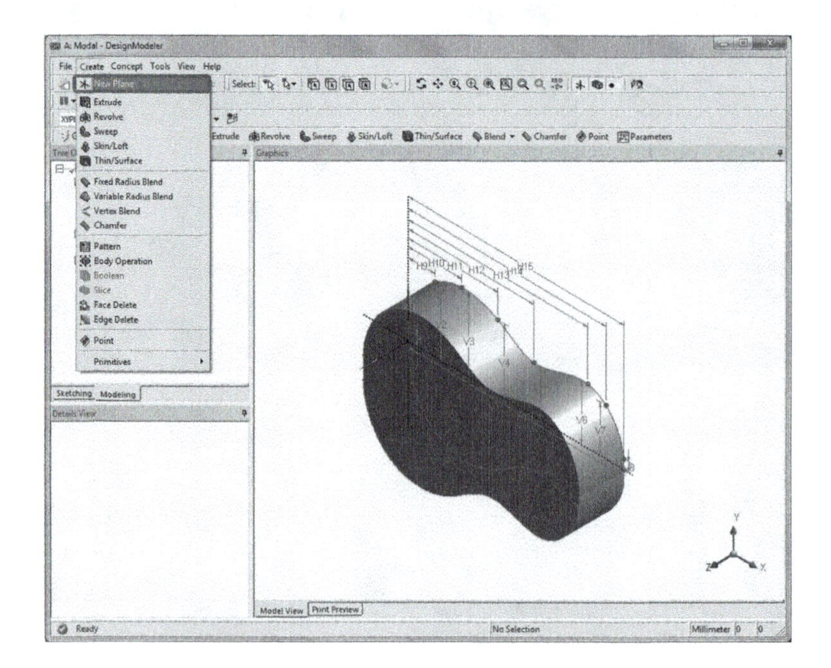

A new plane named *Plane4* is now added to the *Tree Outline*. In the *Details of Plane4*, set the *Type* to *From Face*. Click the front face of the guitar from the *Graphics* window, and apply it to the *Base Face* selection in the *Details of Plane4*. Click *Generate*.

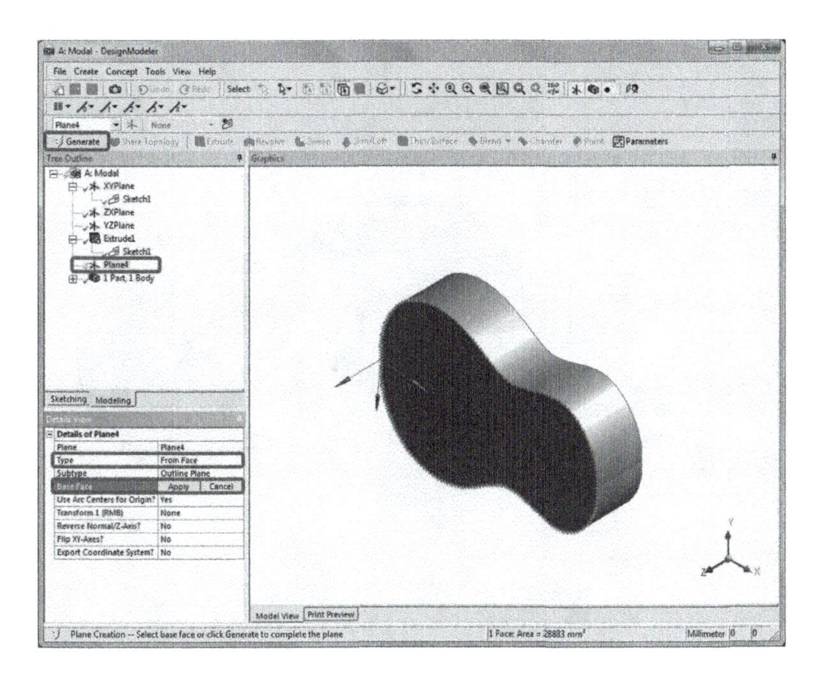

To create a new sketch under *Plane4* in the *Tree Outline*, click on the *New Sketch* icon.

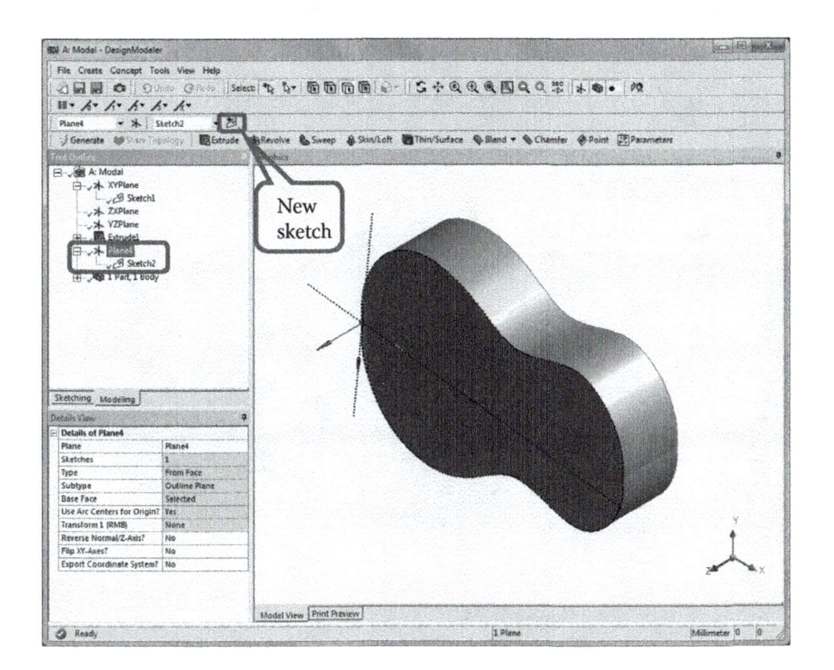

Switch to the *Sketching* tab for *Sketch2*. In the sketch, draw a horizontal line by connecting points *A* and *B* as shown below. Then draw a *circle* of diameter *45 mm* centered at point C, located *170 mm* to the left of point *A* along line *AB*.

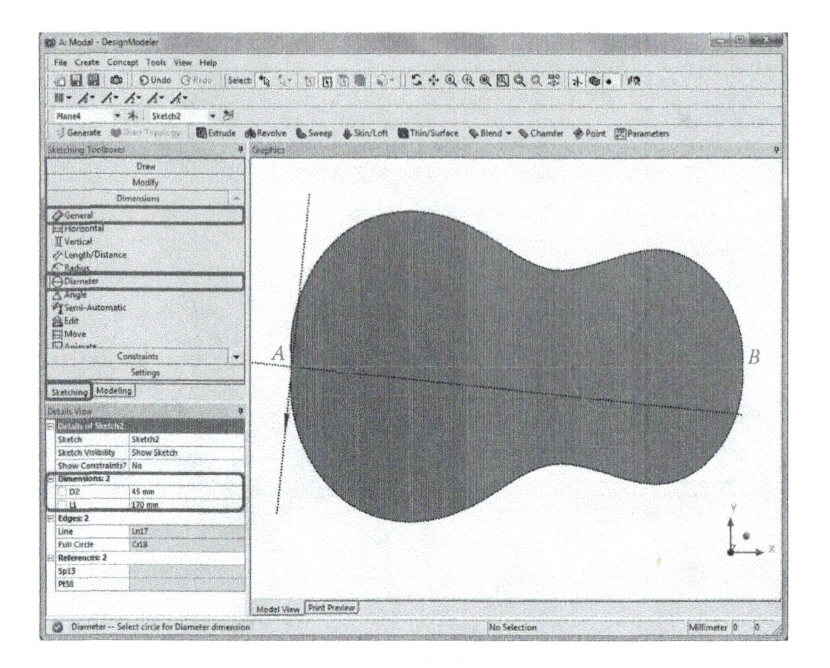

Next, choose *Trim* under the *Modify* tab, and click on line *AB* in the *Graphics* window. The sketch line *AB* will disappear. Click *Generate*.

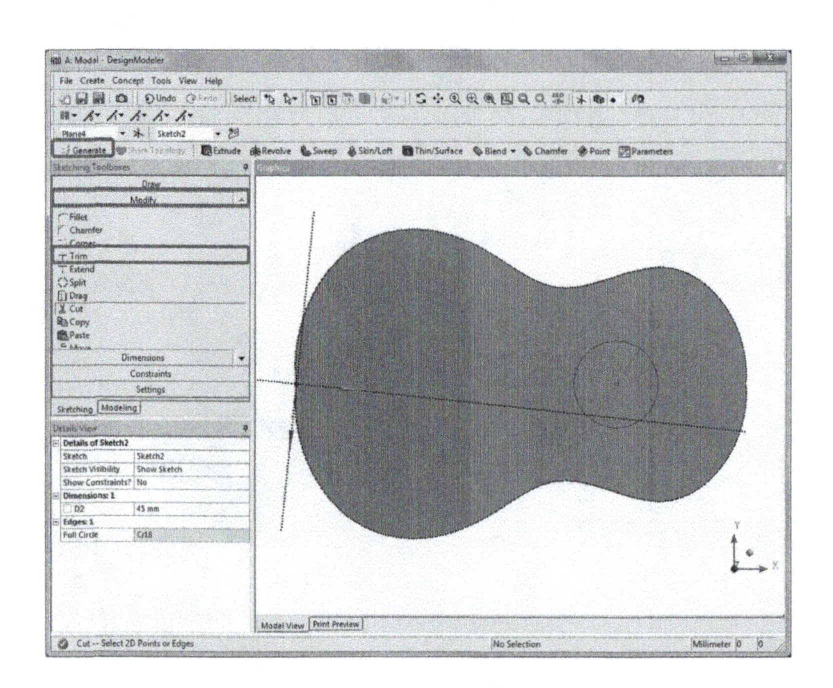

Switch to the *Modeling* tab, and click on the *Extrude* feature. The default *Base Object* is set as *Sketch2* in the *Details of Extrude2*. Set the *Operation* to *Cut Material*. Enter an extrusion depth of *10 mm* in the field of *FD1, Depth* and click *Generate*. An extruded cut feature is now added to the front face as shown below.

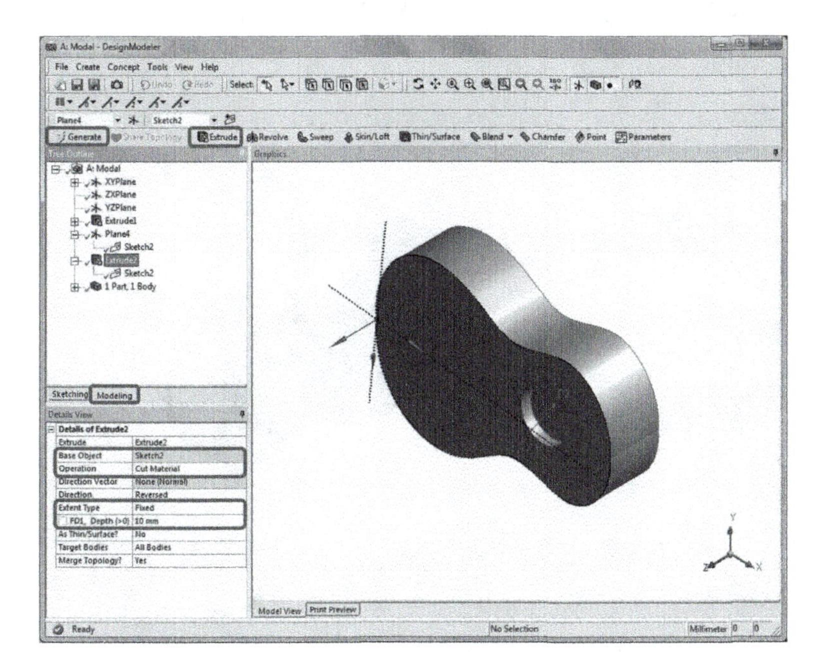

Step 9: Create a Surface Body

Select *Surface from Faces* from the *Concept* drop-down menu. In the *Graphics* window, Ctrl-click to select four faces, that is, the front, back, top, and bottom faces that enclose the solid body as shown below.

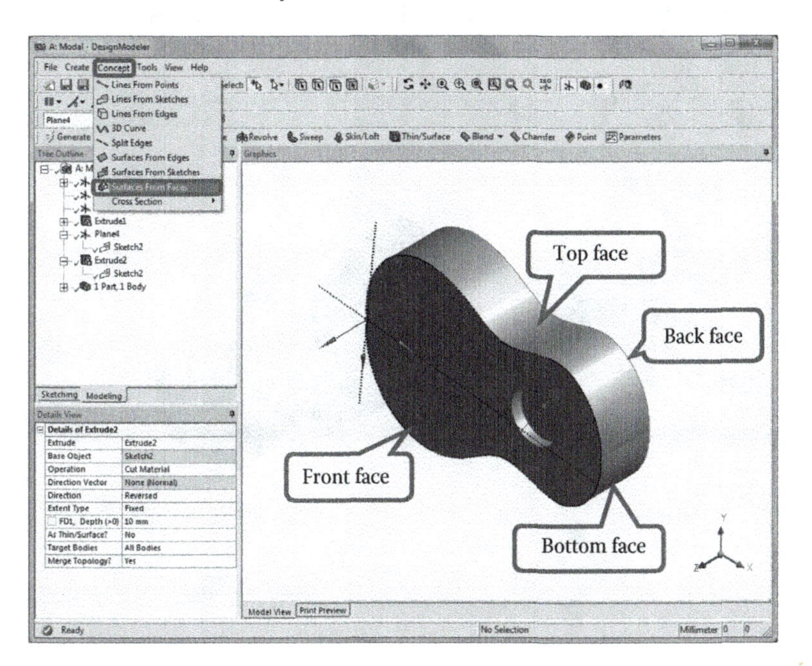

Click *Apply* next to *Faces* in the *Details of SurfFromFaces1*. Then click *Generate*. A surface body will be generated in the *Tree Outline* under *2 Parts, 2 Bodies*.

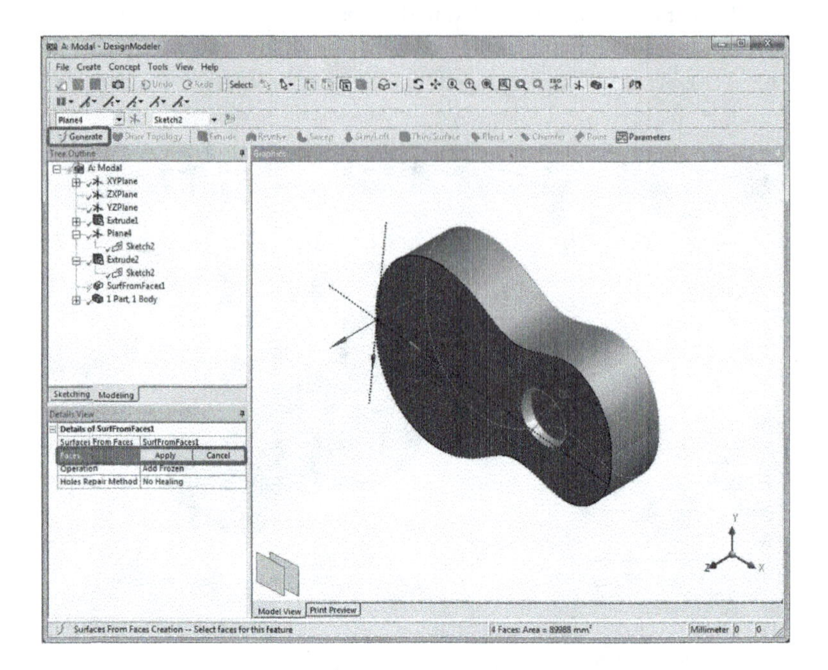

Right-click on *Solid* under *2 Parts, 2 Bodies* in the *Tree Outline*. In the context menu, select *Suppress Body*.

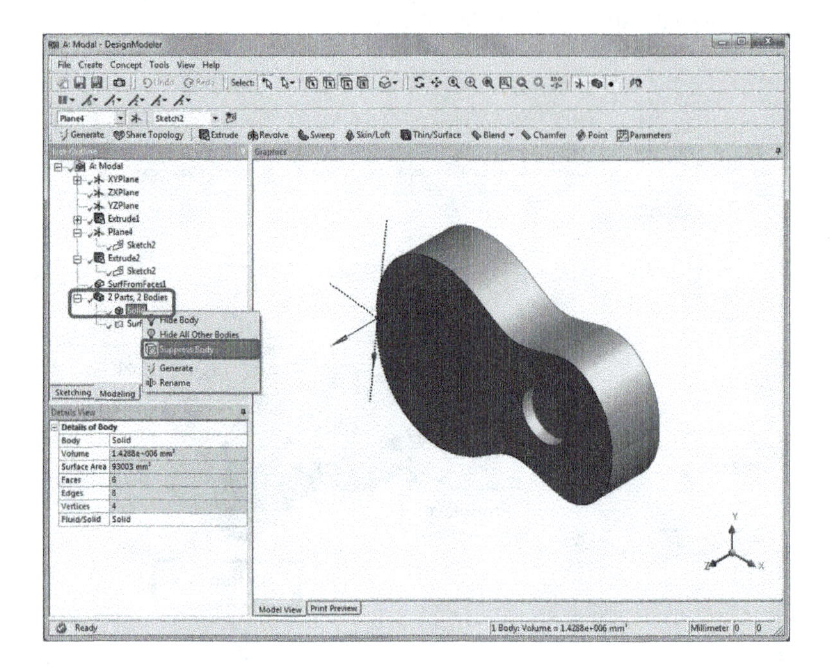

Click on *Surface Body* under *2 Parts, 2 Bodies* in the *Tree Outline*. Change the *Thickness* to *3 mm* in the *Details of Surface Body*. Save and exit the *DesignModeler*.

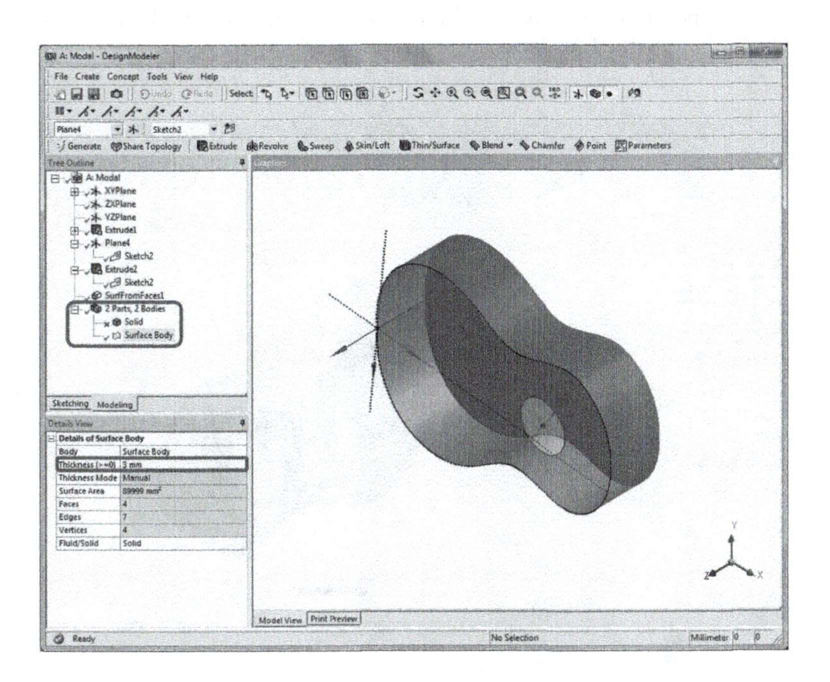

Step 10: Launch the *Modal–Mechanical* Program

Double-click on the *Model* cell to launch the *Modal–Mechanical* program. Click on the *Surface Body* under *Geometry* in the *Outline tree*. In the *Details of "Surface Body,"* click to the right of the *Material Assignment* field and select *Wood* from the drop-down menu.

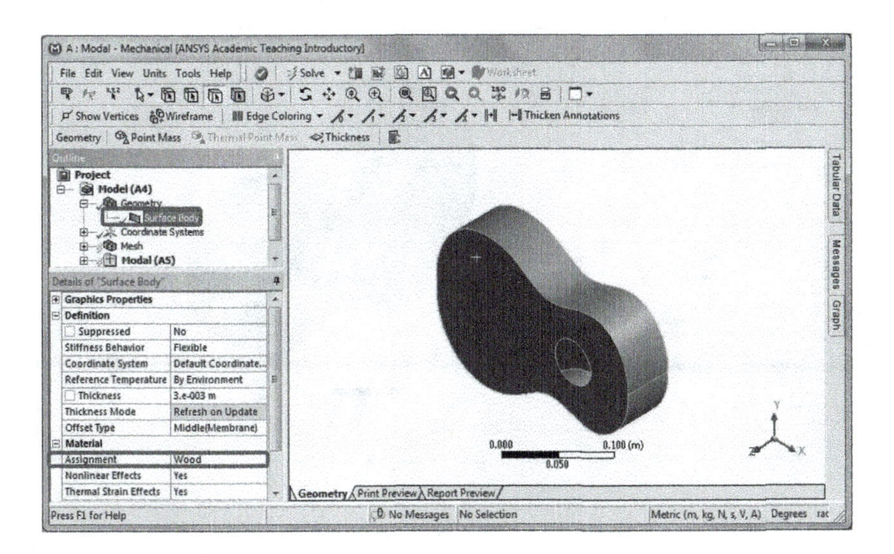

Step 11: ***Generate Mesh***

Right click on *Mesh* in the *Project Outline.* Select *Insert* and then *Sizing* from the context menu. In the *Details of "Face Sizing,"* enter *"5e-4 m"* for the *Element Size.* Click on the front, back, top, and bottom faces of the guitar in the *Graphics* window and apply the four faces to the *Geometry* selection.

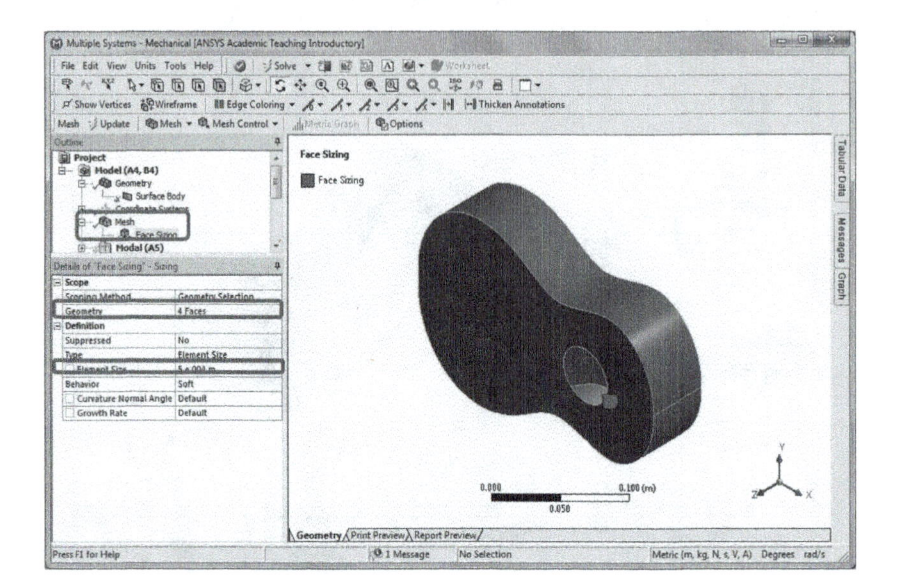

Right-click on *Mesh* in the *Outline,* and select *Generate Mesh* from the context menu.

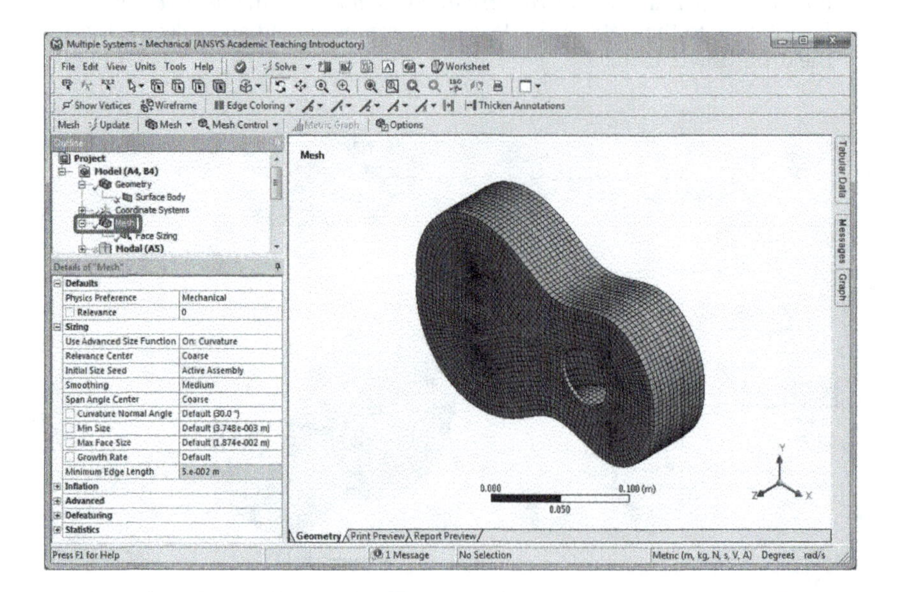

Step 12: Set Up Modal Analysis and Apply Boundary Conditions

Click on *Analysis Settings* under *Modal* in the *Outline* tree. Change the *Max Modes to Find* to *10* in the *Details of "Analysis Settings."*

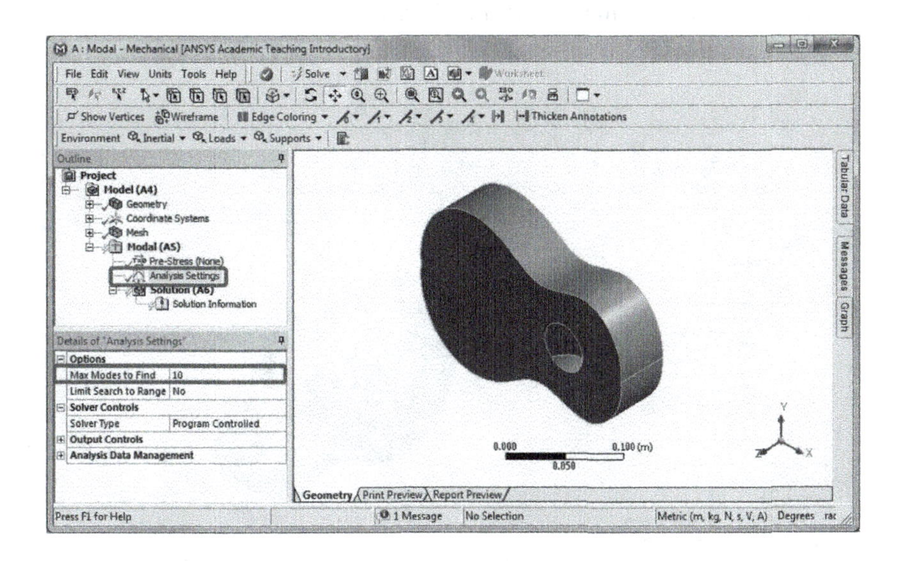

Right-click on *Modal(A5)*. Choose *Insert* and then *Fixed Support* from the context menu. Apply the back face to the *Geometry* selection.

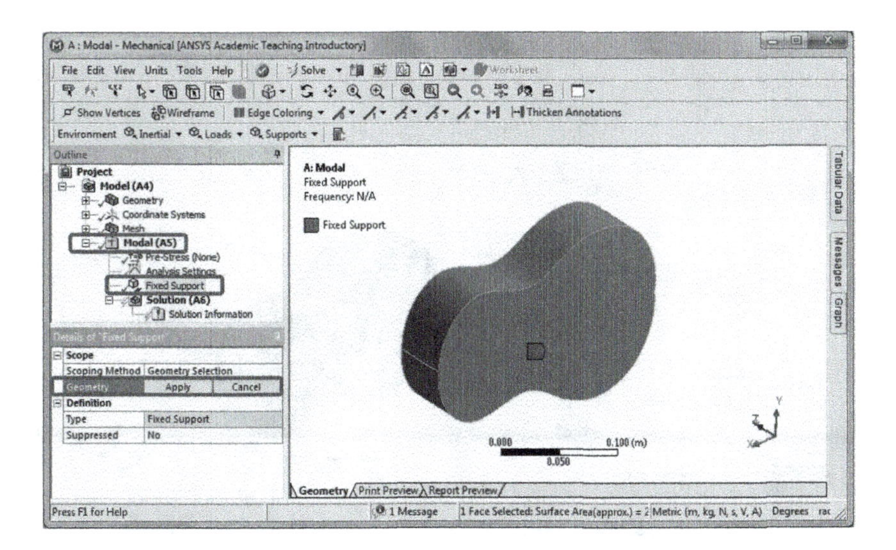

Step 13: Retrieve Results from Modal Analysis

Insert *Total Deformation* by right-clicking on *Solution (A6)* in the *Outline*. In the *Details of "Total Deformation,"* set *Mode* to *1*. Insert another *Total Deformation* item. In the *Details of "Total Deformation 2,"* set *Mode* to 2. Repeat this step three more times. Set *Mode* to *3, 4, and 5*, respectively, for each new insertion. Right-click on *Solution (A6)* in the *Outline* and *Solve*

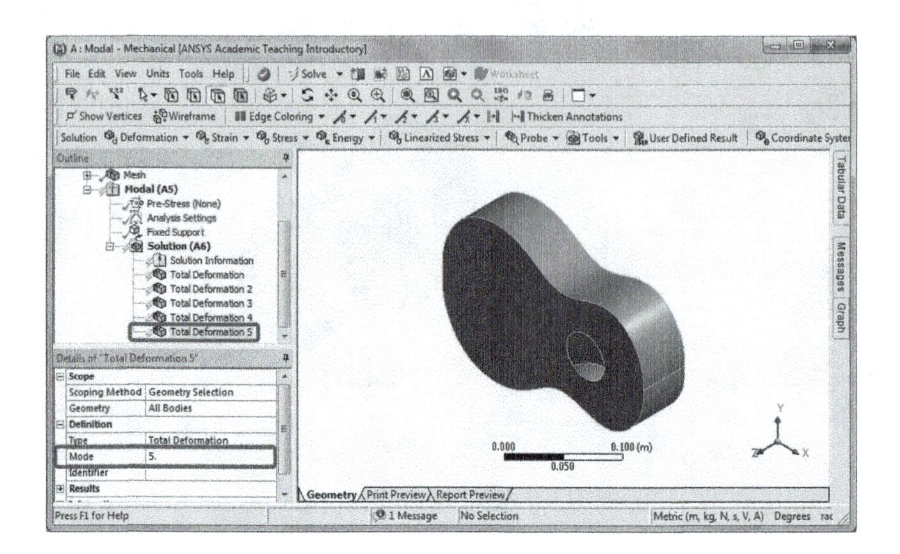

Click on *Total Deformation* in the *Outline* to review results. The results below show the first natural frequency of 1036.8 Hz and the corresponding mode shape.

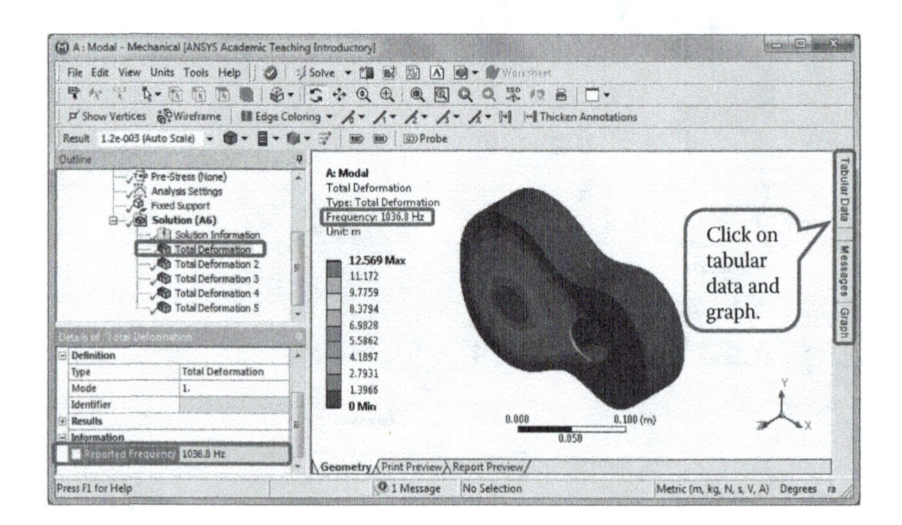

Click on *Tabular Data* and *Graph* on the right edge of the *Graphics* window, and then click on the push pin labeled *AutoHide* to display the *Tabular data* and the *Graph* as shown below.

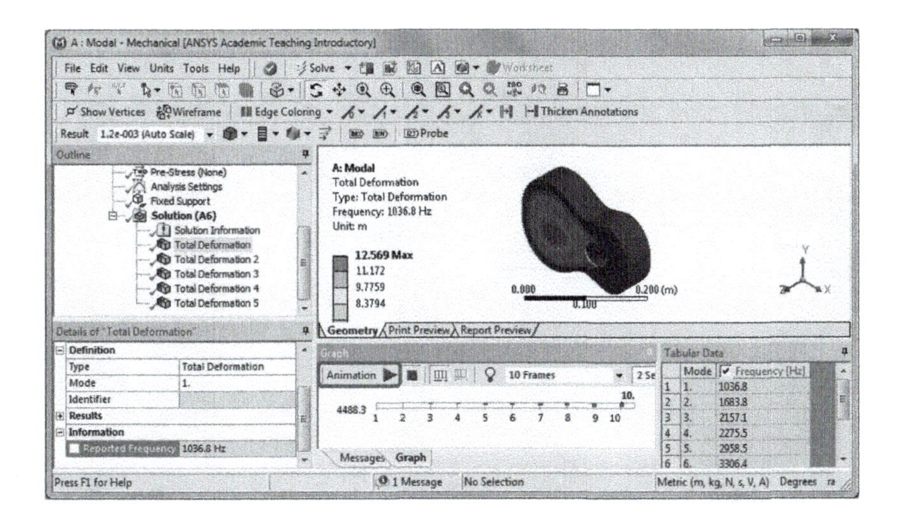

The *Tabular data* gives the first 10 natural frequencies of the guitar under the fixed bottom boundary condition. The *Play/Stop* control interface in the *Graph* window allows animation of mode shapes. Click on each different *Total Deformation* item in the *Outline* to review results, for example, the following figure shows the fourth mode shape, and then exit the *Modal–Mechanical* program.

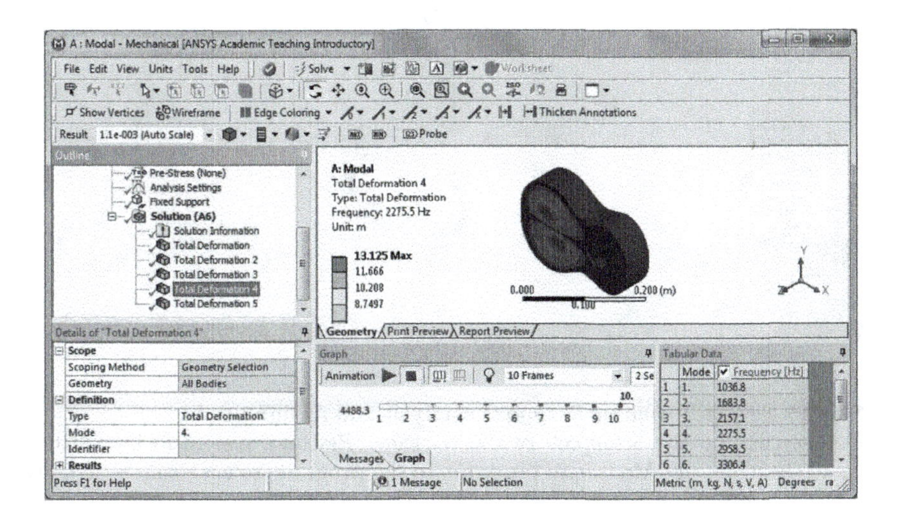

Step 14: Create a *Harmonic Response* Analysis System

Drag the *Harmonic Response* icon from the *Analysis Systems Toolbox* window and drop it onto the *Solution* cell of the highlighted *Modal* system in the *Project Schematic.*

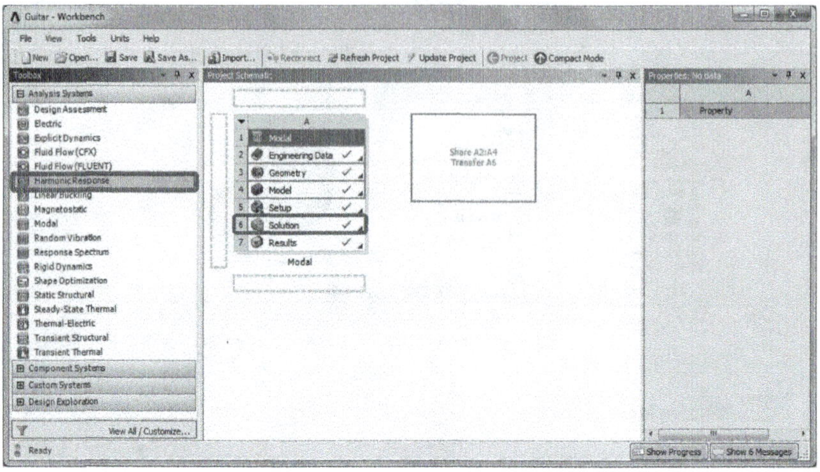

This creates a *Harmonic Response* system that shares data with *Modal* system as shown below.

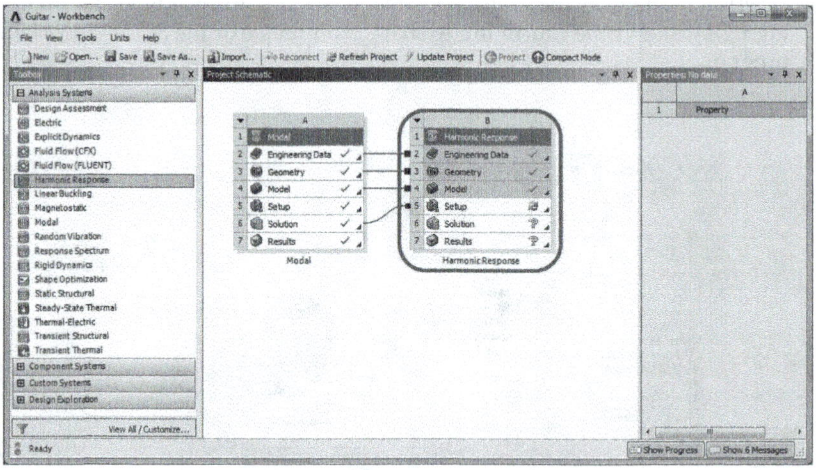

Step 15: Set Up *Harmonic Response* Analysis and Assign Loads

Double-click on the *Setup* cell of the *Harmonic Response* system to launch the *Multiple Systems–Mechanical* program. In the program, select *Analysis Settings* from the *Outline*. Set the *Range Minimum* to *1000 Hz*, *Range Maximum* to *4000 Hz*, and *Solution Intervals* to *300*.

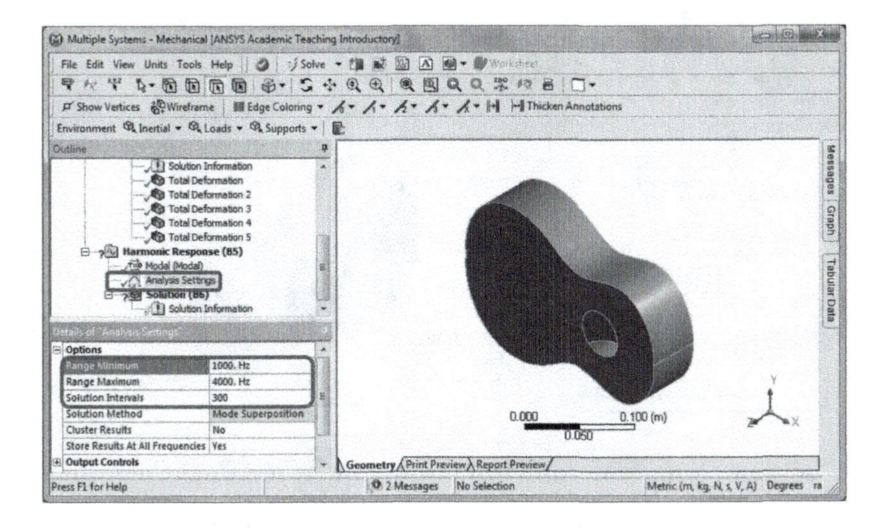

Right-click on *Harmonic Response (B5)*. Choose *Insert* and then *Pressure* from the context menu. In the *Details of "Pressure,"* set *magnitude* as *1 MPa*, and apply the top face to the *Geometry* selection.

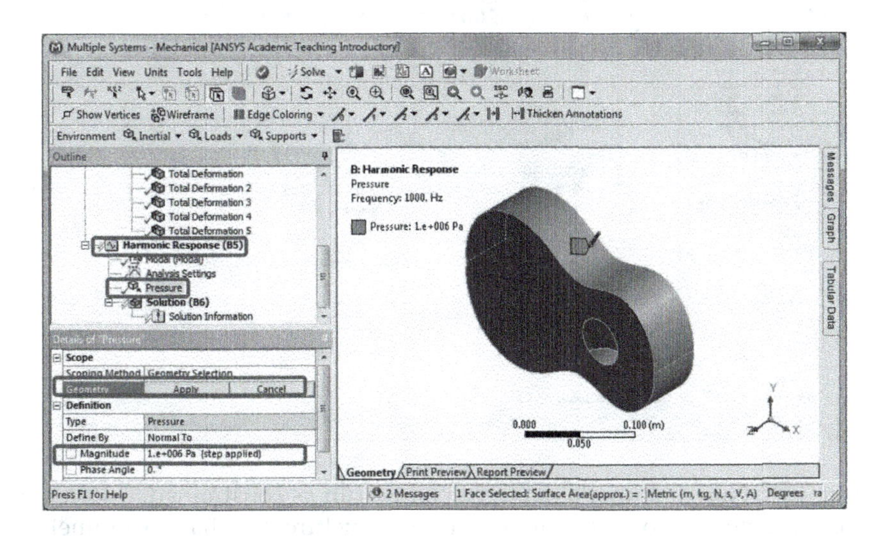

Step 16: Retrieve Results from *Harmonic* Response

Right-click on *Solution (B6)*. Choose *Insert* and *Frequency Response* and then *Deformation* from the context menu. In the *Details of "Frequency Response,"* set the *Orientation* of the directional deformation to *Z-Axis*. Click on the front face and apply it to the *Geometry* selection.

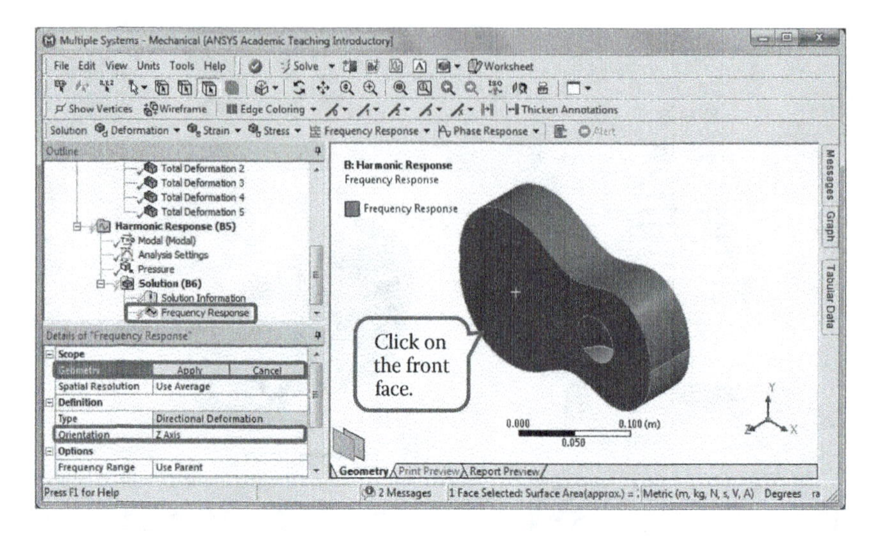

Right-click on *Solution (B6)* and select *Solve*. After solution is done, click on *Frequency Response* in the *Outline* to review the harmonic response of the guitar.

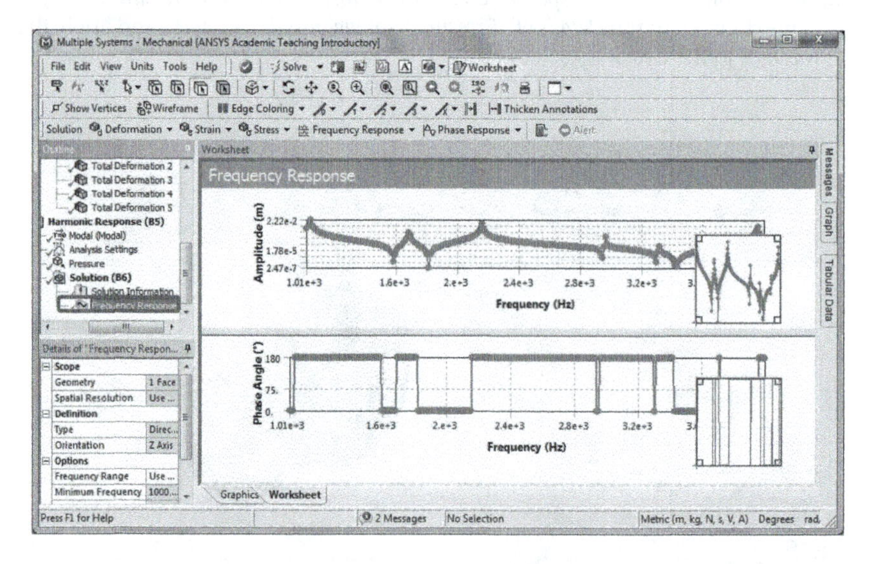

Modeling tips: Note that modal analysis can be run as constrained, unconstrained, or partially constrained. Also note that symmetric structure may have asymmetric modes, and thus, it is not recommended to take advantage of symmetry to simplify models for modal analysis. In the following, we are going to show a result comparison between an unconstrained model and the fixed model. For the unconstrained model, the first six mode shapes obtained from simulation are rigid body modes that allow the structure to move freely. They are not considered as structural modes. The comparison indicates that a free floating guitar has a different set of natural frequencies and mode shapes from that of a fixed guitar. In general, constraint conditions have an effect on the vibration characteristics of a structure and should be considered carefully when setting up a model.

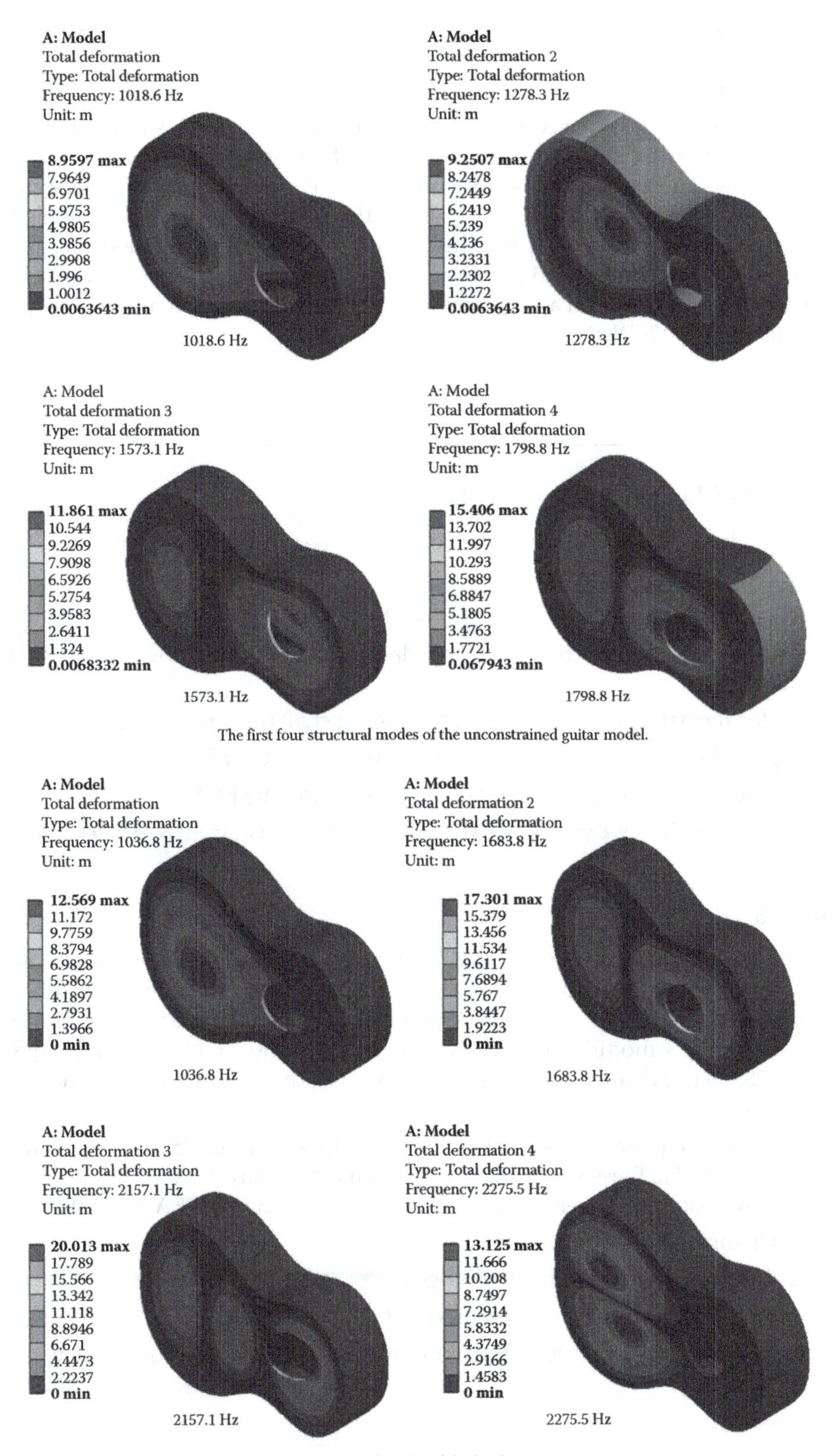
A: Model
Total deformation
Type: Total deformation
Frequency: 1018.6 Hz
Unit: m
8.9597 max
7.9649
6.9701
5.9753
4.9805
3.9856
2.9908
1.996
1.0012
0.0063643 min
1018.6 Hz
A: Model
Total deformation 2
Type: Total deformation
Frequency: 1278.3 Hz
Unit: m
9.2507 max
8.2478
7.2449
6.2419
5.239
4.236
3.2331
2.2302
1.2272
0.0063643 min
1278.3 Hz
A: Model
Total deformation 3
Type: Total deformation
Frequency: 1573.1 Hz
Unit: m
11.861 max
10.544
9.2269
7.9098
6.5926
5.2754
3.9583
2.6411
1.324
0.0068332 min
1573.1 Hz
A: Model
Total deformation 4
Type: Total deformation
Frequency: 1798.8 Hz
Unit: m
15.406 max
13.702
11.997
10.293
8.5889
6.8847
5.1805
3.4763
1.7721
0.067943 min
1798.8 Hz

The first four structural modes of the unconstrained guitar model.

A: Model
Total deformation
Type: Total deformation
Frequency: 1036.8 Hz
Unit: m
12.569 max
11.172
9.7759
8.3794
6.9828
5.5862
4.1897
2.7931
1.3966
0 min
1036.8 Hz
A: Model
Total deformation 2
Type: Total deformation
Frequency: 1683.8 Hz
Unit: m
17.301 max
15.379
13.456
11.534
9.6117
7.6894
5.767
3.8447
1.9223
0 min
1683.8 Hz
A: Model
Total deformation 3
Type: Total deformation
Frequency: 2157.1 Hz
Unit: m
20.013 max
17.789
15.566
13.342
11.118
8.8946
6.671
4.4473
2.2237
0 min
2157.1 Hz
A: Model
Total deformation 4
Type: Total deformation
Frequency: 2275.5 Hz
Unit: m
13.125 max
11.666
10.208
8.7497
7.2914
5.8332
4.3749
2.9166
1.4583
0 min
2275.5 Hz

The first four structural modes of the fixed guitar model.

8.8 Summary

In this chapter, we first reviewed the equations of motion for both single DOF and multiple DOF systems and discussed how to compute the mass and damping matrices in the FE formulations. Then, we discussed the methods for solving normal modes, harmonic responses, and transient responses for structural vibration and dynamic problems. The advantages and disadvantages of the direct method and modal method are discussed. Several examples of vibration and dynamic analyses are also discussed to show the applications of the FEA in vibration and dynamic analyses. The vibration modes of a guitar are analyzed using *ANSYS Workbench*.

8.9 Review of Learning Objectives

Now that you have finished this chapter you should be able to

1. Understand the FE formulations for vibration and dynamic analysis of structures
2. Know how to compute the mass and damping matrices in the FE equations of motion
3. Solve for the natural frequencies and modes of structures using the FEA
4. Compute the harmonic response of structures using the FEA
5. Compute the transient response of structures using the FEA
6. Perform vibration or dynamic analysis of structures using *ANSYS Workbench*

PROBLEMS

8.1 For the cantilever beam studied in Example 8.1, apply more elements and investigate the FEA solutions for the first 10 natural frequencies and normal modes of the beam structure. Assume the beam is made of structural steel with Young's modulus $E = 200$ GPa, Poisson's ratio $\nu = 0.3$, and mass density $\rho = 7850$ kg/m^3, and has a length of 100 cm with a 10 cm × 10 cm square cross section.

8.2 For a square plate with edge length = 1 m, thickness = 0.005 m, Young's modulus $E = 70$ GPa, Poisson's ratio $\nu = 0.3$, and mass density $\rho = 2800$ kg/m^3, find the first five natural frequencies and normal modes using the FEA when the plate is

a. Clamped at the four edges
b. Simply supported at the four edges
c. Free at the four edges (not supported)

Discuss the effects of the support on the natural frequencies of the plate.

8.3 A fuel tank, with a total length = 5 m, diameter = 1 m, and thickness = 0.01 m, is shown below. Using the FEA, find the first six natural frequencies and corresponding normal modes, when:

a. The tank is not supported at all.

b. The tank is constrained along the circumferences in the radial directions at the two locations 1 m away from the two ends.

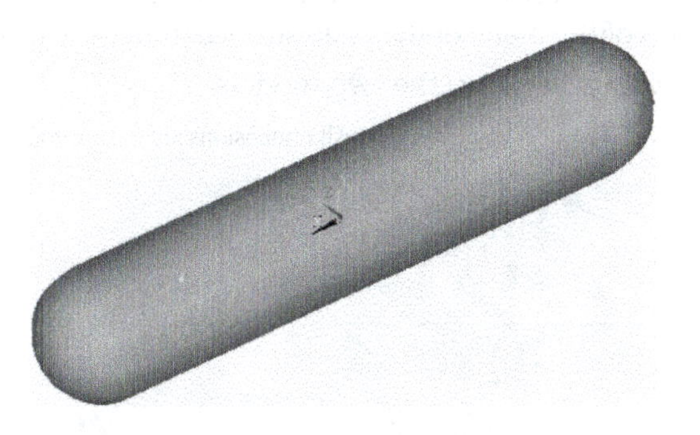

Carefully build your FE mesh (using shell elements) so that the symmetry of the tank is reserved and the boundary conditions in part (b) can be applied readily. Assume the tank is made of steel with Young's modulus $E = 200$ GPa, Poisson's ratio $\nu = 0.3$, and mass density $\rho = 7850$ kg/m^3.

8.4 For the rotating part sketched below, assume that it is made of steel with Young's modulus $E = 200$ GPa, Poisson's ratio $\nu = 0.3$, and mass density $\rho = 7850$ kg/m^3. Ignore the gravitational force. Compute the first five natural frequencies and normal modes in the part using the FEA when the part is

a. Fixed on the inner cylindrical surface of the hole

b. Fixed on the bottom surface

Compare the results between the two models.

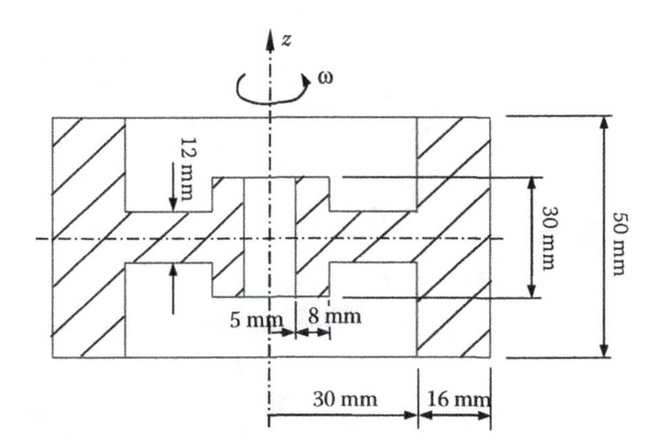

8.5 The following figure shows a simple plate made of structural steel with Young's modulus $E = 200$ GPa, Poisson's ratio $\nu = 0.3$, and density $\rho = 7850$ kg/m³. The plate has a thickness of 5 mm and is fixed on the left side. Suppose a harmonic force of magnitude 100 N is applied to the right edge along the plate's surface normal direction. Use a solid model and a shell model to compute:

a. The first five natural frequencies and the corresponding normal modes
b. The frequency response of the z directional deformation (along the surface normal direction of the plate) of the plate surface under the given harmonic load

Compare the results between the two models.

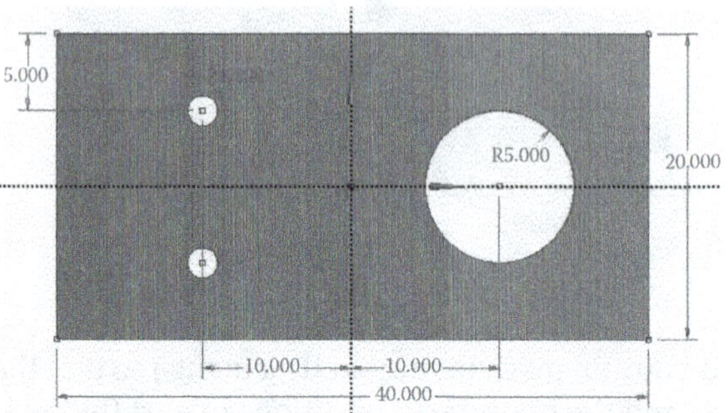

8.6 The following figure shows a simple bracket made of structural steel with Young's modulus $E = 200$ GPa, Poisson's ratio $\nu = 0.3$, and density $\rho = 7850$ kg/m³. The bracket has a front face profile shown below and an extrusion depth of 12 cm. A through-hole of diameter 1.5 cm is centered on the bracket's flat ends, one on each side. Suppose that the bracket is fixed on the bottom, and is subjected to a harmonic pressure load of magnitude 1000 N/cm² on the top face. Use a solid model to compute:

a. The first five natural frequencies and the corresponding normal modes
b. The frequency response of the z directional deformation (along the front face normal direction) of the front face of the bracket under the given harmonic load

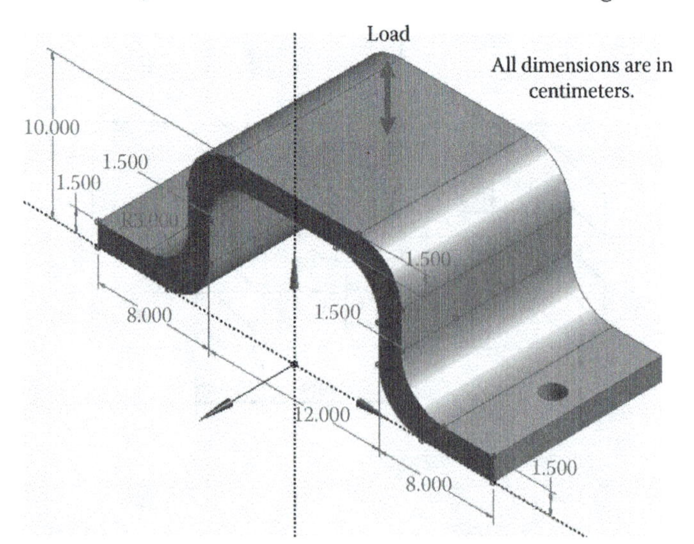

8.7 The following figure shows an L-shaped bracket made of structural steel with Young's modulus $E = 200$ GPa, Poisson's ratio $\nu = 0.3$, and density $\rho = 7850$ kg/m^3. The 4 cm thick L-shape with a cross-section profile shown in the figure is rigidly connected to a rectangular base of size 3 cm × 8 cm × 1 cm. Suppose that the bracket is fixed on the right end at the base. A harmonic pressure of magnitude 100 N/cm^2 is applied on the protruding surface of the L-shape as shown below. Using the FEA, compute:

a. The first five natural frequencies and the corresponding normal modes
b. The frequency response of the y directional deformation (along the load direction) of the bracket's protruding top face under the given harmonic load

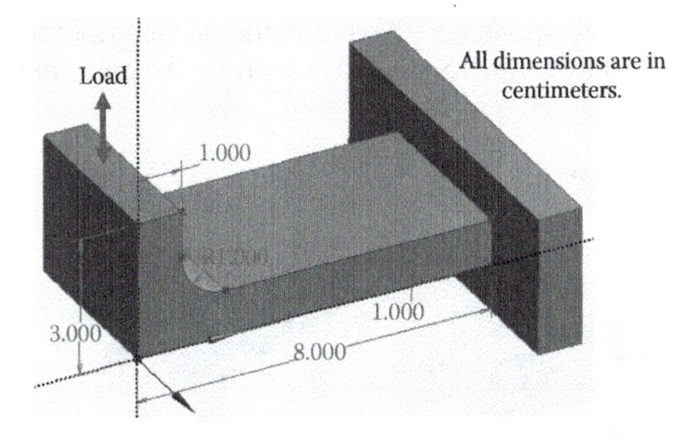

8.8 The following two-bar assembly consists of four components made of structural steel with Young's modulus $E = 200$ GPa, Poisson's ratio $\nu = 0.3$, and density $\rho = 7850$ kg/m^3. As shown in the figure below, the two bars are identical in size. Each bar has a diameter of 2 cm and a length of 16 cm. The two identical end blocks each has a thickness of 2 cm and a fillet radius of 1 cm at the round corners. The distance of the gap in between the two blocks is 8 cm. Suppose that the two blocks are fixed on the bottom. The bars are applied a harmonic pressure load of magnitude 100 N/cm^2 as shown below. Using the FEA, compute:

a. The first five natural frequencies and normal modes of the assembly
b. The frequency response of the x directional deformation (along the load direction) of the cylindrical bar surface under the given harmonic load

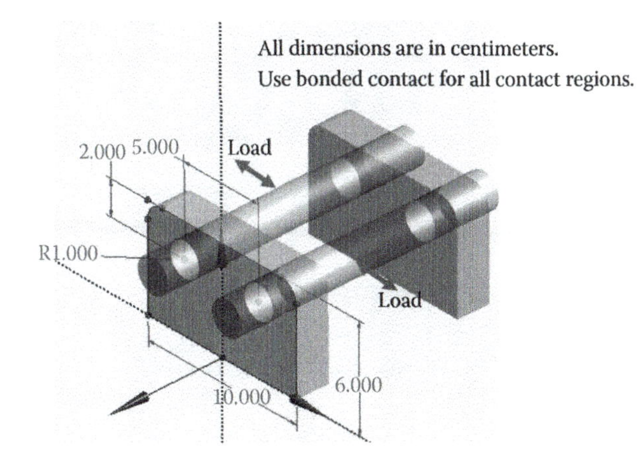

8.9 The following clevis assembly consists of a yoke, a pin, and a u-shape. The assembly components are made of structural steel with Young's modulus $E = 200$ GPa, Poisson's ratio $\nu = 0.3$, and density $\rho = 7850$ kg/m^3. The profile sketches of the yoke and the u-shape are shown below. Both components have an extrusion depth of 5 cm. The pin has a diameter of 2 cm and a length of 16 cm, and is centered 2.5 cm away from both the front and the side faces of the u-shape. Assume that the yoke is fixed on the left end, and the u-shape is applied a harmonic pressure load of magnitude 100 N/cm^2 on the right end. Using the FEA, compute:

a. The first five natural frequencies and normal modes of the assembly
b. The frequency response of the x directional deformation (along the load direction) of the cylindrical pin surface under the given harmonic load

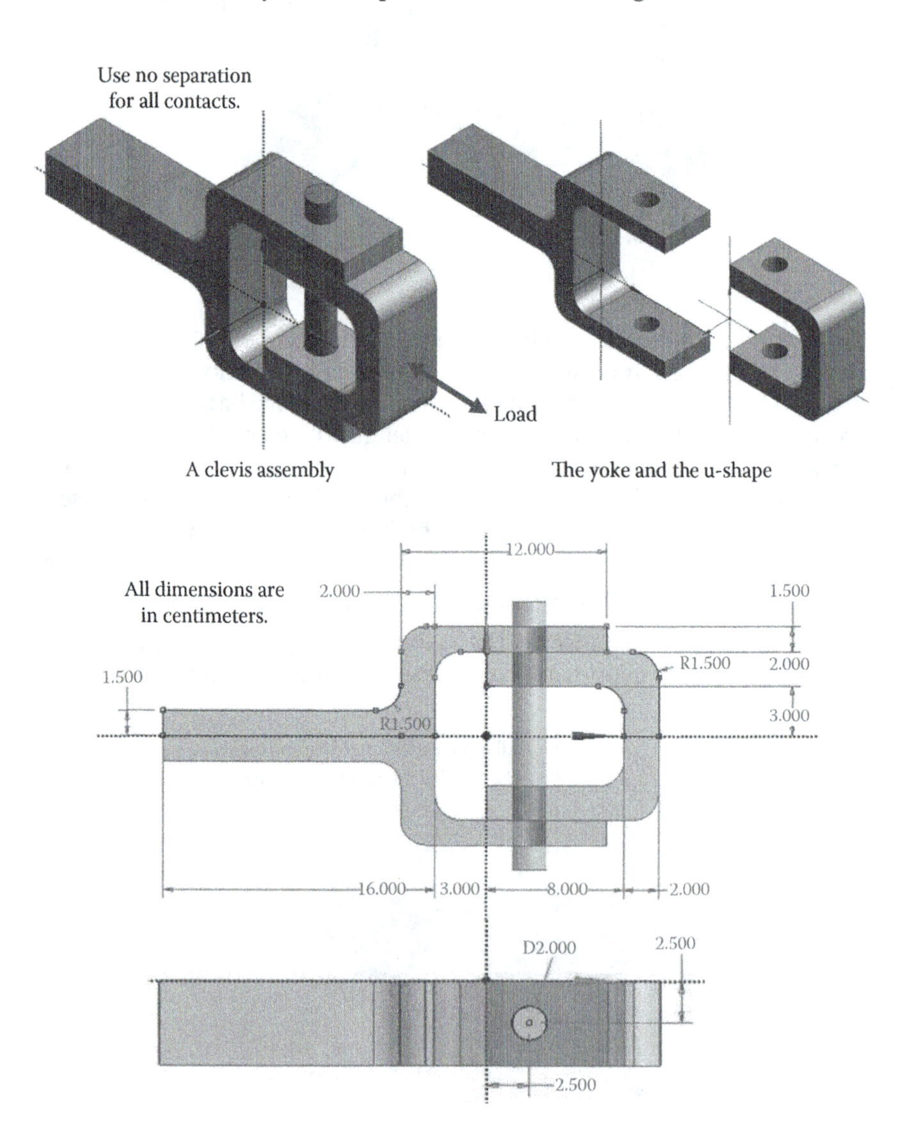

8.10 The following rotor assembly consists of a rotor, a shaft, and a base. The assembly components are made of structural steel with Young's modulus $E = 200$ GPa, Poisson's ratio $\nu = 0.3$, and density $\rho = 7850$ kg/m^3. The base of the assembly is fixed on the bottom. Suppose a harmonic pressure load of magnitude 100 N/cm^2 is applied to the side surface of the base as shown below. Using the FEA, compute:

a. The first five natural frequencies and normal modes of the assembly
b. The frequency response of the y directional deformation (in the vertical direction) of the cylindrical shaft surface under the given harmonic load

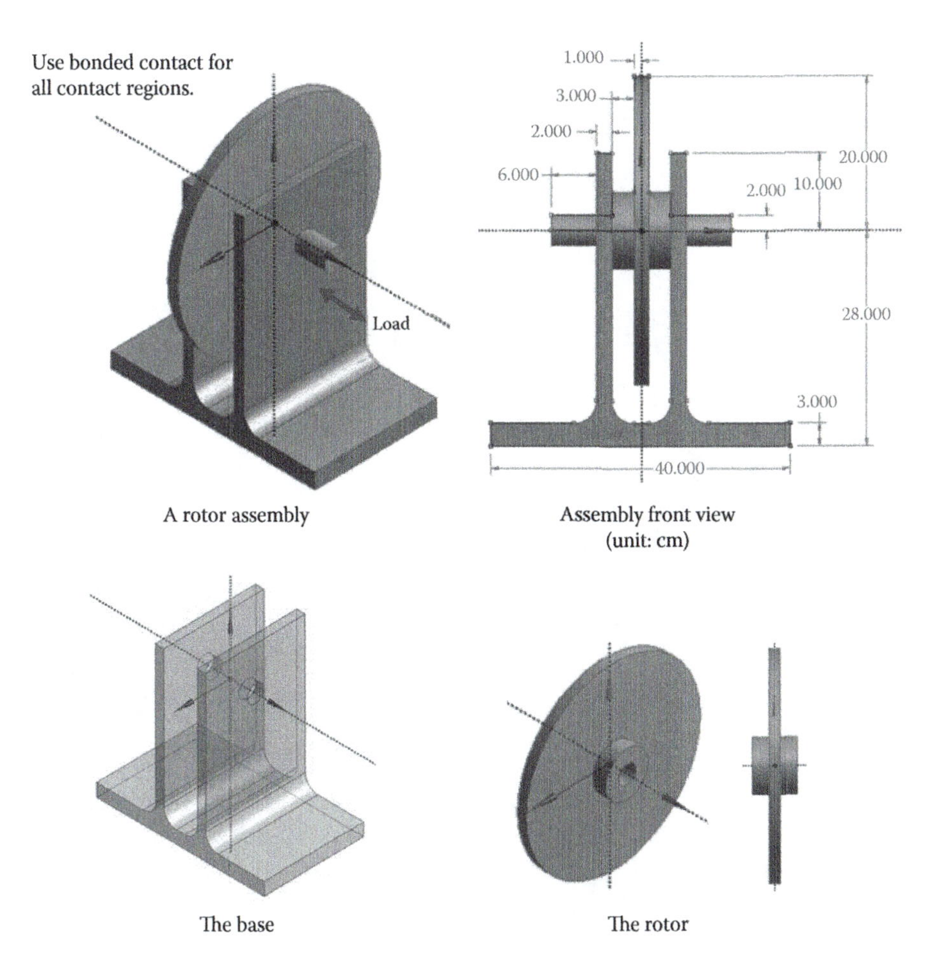

A rotor assembly

Assembly front view
(unit: cm)

The base

The rotor

9

Thermal Analysis

9.1 Introduction

Many engineering problems are thermal problems in nature. Devices such as appliances, advanced electronics, engines, and heating, ventilation, and air conditioning systems need to be evaluated for their thermal performance during the design process. In this chapter, we will discuss thermal analysis using the FEA. The objective of thermal analysis is to understand response and behavior of a structure with thermal loading. The resulting temperature distribution, heat flux distribution, and structural response under different thermal loading conditions constitute important knowledge in assuring design success of thermal engineering products. Both steady-state and transient thermal problems are introduced and a heat sink model is analyzed using *ANSYS® Workbench*. Thermal stress analysis, which is to find the structural response due to change of temperatures, is also included and discussed.

9.2 Review of Basic Equations

To study heat transfer within an object or between objects, one may conduct thermal analysis, from which thermal quantities such as the temperature, thermal gradient, and heat flux distributions can be determined. There are two types of thermal analyses: *steady-state thermal* and *transient thermal*. Steady-state thermal analysis aims to find the temperature or heat flux distribution in structures when a thermal equilibrium is reached, and transient thermal analysis sets out to determine the time history of how the temperature profile and other thermal quantities change with time. In addition, thermal expansion or contraction of engineering materials often leads to thermal stress in structures, which can be examined by conducting *thermal stress* analysis. The basic equations for thermal and thermal stress analyses are given as follows.

9.2.1 Thermal Analysis

For temperature field in a 1-D space, such as a bar (Figure 9.1), we have the following Fourier heat conduction equation:

$$f_x = -k\frac{\partial T}{\partial x} \tag{9.1}$$

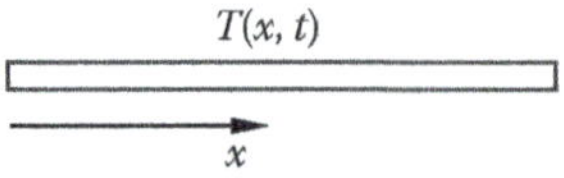

$T(x, t)$

x

FIGURE 9.1
The temperature field $T(x, t)$ in a 1-D bar model.

where,

f_x = heat flux per unit area
k = thermal conductivity
$T = T(x, t)$ = temperature field

For 3-D case, we have:

$$\begin{Bmatrix} f_x \\ f_y \\ f_z \end{Bmatrix} = -\mathbf{K} \begin{Bmatrix} \partial T/\partial x \\ \partial T/\partial y \\ \partial T/\partial z \end{Bmatrix} \tag{9.2}$$

where, f_x, f_y, f_z = heat flux in the x, y, and z direction, respectively. In the case of isotropic materials, the conductivity matrix is:

$$\mathbf{K} = \begin{bmatrix} k & 0 & 0 \\ 0 & k & 0 \\ 0 & 0 & k \end{bmatrix} \tag{9.3}$$

The equation of heat flow is given by

$$-\left[\frac{\partial f_x}{\partial x} + \frac{\partial f_y}{\partial y} + \frac{\partial f_z}{\partial z}\right] + q_v = c\rho\frac{\partial T}{\partial t} \tag{9.4}$$

in which,

q_v = rate of internal heat generation per unit volume
c = specific heat
ρ = mass density

For steady-state case ($\partial T/\partial t = 0$) and isotropic materials, we can obtain:

$$k\nabla^2 T = -q_v \tag{9.5}$$

This is a Poisson equation, which needs to be solved under given boundary conditions. Boundary conditions for steady-state heat conduction problems are (Figure 9.2):

$$T = \overline{T}, \quad \text{on } S_T \tag{9.6}$$

$$Q \equiv -k\frac{\partial T}{\partial n} = \overline{Q}, \quad \text{on } S_q \tag{9.7}$$

Note that at any point on the boundary $S = S_T \cup S_q$, only one type of BCs can be specified.

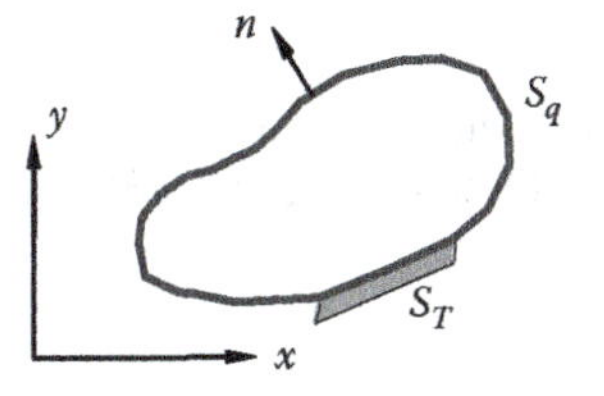

FIGURE 9.2
Boundary conditions for heat conduction problems.

9.2.1.1 Finite Element Formulation for Heat Conduction

For heat conduction problems, we can establish the following FE equation:

$$\mathbf{K}_T \mathbf{T} = \mathbf{q} \tag{9.8}$$

where,
$\mathbf{K}_T$ = conductivity matrix
$\mathbf{T}$ = vector of nodal temperature
$\mathbf{q}$ = vector of thermal loads

The element conductivity matrix is given by

$$\mathbf{k}_T = \int_V \mathbf{B}^T \mathbf{K} \mathbf{B} dV \tag{9.9}$$

This is obtained in a similar way as for the structural analysis, that is, by starting with the interpolation $T = \mathbf{N}\mathbf{T}_e$ for the temperature field (with $\mathbf{N}$ being the shape function matrix and T_e the nodal temperature). Note that there is only one DOF at each node for the thermal problems.

For transient (unsteady state) heat conduction problems, we have:

$$\frac{\partial T}{\partial t} \neq 0$$

In this case, we need to apply finite difference schemes (use time steps and integrate in time), as in the transient structural analysis, to obtain the transient temperature fields.

9.2.2 Thermal Stress Analysis

To determine the thermal stresses due to temperature changes in structures, we can proceed to

- Solve Equation 9.8 first to obtain the temperature (change) fields.
- Apply the temperature change ΔT as initial strains (or initial stresses) to the structure to compute the thermal stresses due to the temperature change.

9.2.2.1 1-D Case

To understand the stress–strain relations in cases where solids undergo temperature changes, we first examine the 1-D case (Figure 9.3). We have for the thermal strain (or initial strain):

$$\varepsilon_o = \alpha \Delta T \tag{9.10}$$

in which

α = the coefficient of thermal expansion
$\Delta T = T_2 - T_1$ = change of temperature

Total strain is given by

$$\varepsilon = \varepsilon_e + \varepsilon_o \tag{9.11}$$

with ε_e being the elastic strain due to mechanical load.

That is, the total strain can be written as

$$\varepsilon = E^{-1}\sigma + \alpha \Delta T \tag{9.12}$$

Or, inversely, the stress is given by

$$\sigma = E(\varepsilon - \varepsilon_o) \tag{9.13}$$

EXAMPLE 9.1

Consider the bar under thermal load ΔT as shown in Figure 9.3.

(a) If no constraint on the right-hand side, that is, the bar is free to expand to the right, then we have:

$$\varepsilon = \varepsilon_o, \quad \varepsilon_e = 0, \quad \sigma = 0$$

from Equation 9.13. That is, there is no thermal stress in this case.

(b) If there is a constraint on the right-hand side, that is, the bar cannot expand to the right, then we have:

$$\varepsilon = 0, \quad \varepsilon_e = -\varepsilon_o = -\alpha \Delta T, \quad \sigma = -E\alpha \Delta T$$

from Equations 9.11 and 9.13. Thus, thermal stress exists.

From this simple example, we see that the way in which the structure is constrained has a critical role in inducing the thermal stresses.

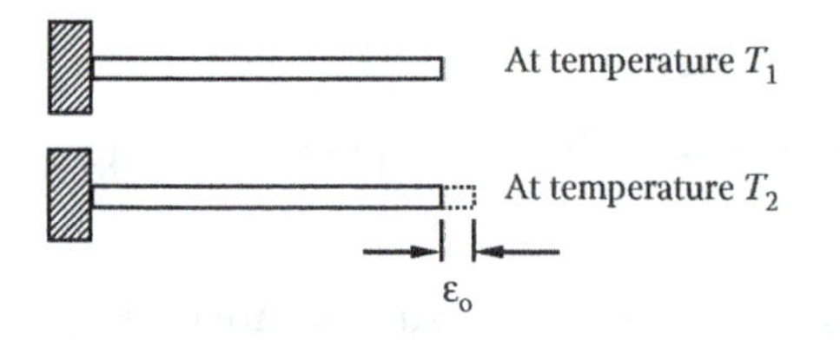

FIGURE 9.3
Expansion of a bar due to increase in temperature.

9.2.2.2 2-D Cases

For plane stress, we have:

$$\varepsilon_o = \begin{Bmatrix} \varepsilon_x \\ \varepsilon_y \\ \gamma_{xy} \end{Bmatrix}_o = \begin{Bmatrix} \alpha\Delta T \\ \alpha\Delta T \\ 0 \end{Bmatrix} \quad (9.14)$$

For plane strain, we have:

$$\varepsilon_o = \begin{Bmatrix} \varepsilon_x \\ \varepsilon_y \\ \gamma_{xy} \end{Bmatrix}_o = \begin{Bmatrix} (1+\nu)\alpha\Delta T \\ (1+\nu)\alpha\Delta T \\ 0 \end{Bmatrix} \quad (9.15)$$

in which, ν is Poisson's ratio.

9.2.2.3 3-D Case

$$\varepsilon_o = \begin{Bmatrix} \varepsilon_x \\ \varepsilon_y \\ \varepsilon_z \\ \gamma_{xy} \\ \gamma_{yz} \\ \gamma_{zx} \end{Bmatrix}_o = \begin{Bmatrix} \alpha\Delta T \\ \alpha\Delta T \\ \alpha\Delta T \\ 0 \\ 0 \\ 0 \end{Bmatrix} \quad (9.16)$$

Observation: Temperature changes do not yield shear strains.
In both 2-D and 3-D cases, the total strain can be given by the following vector equation:

$$\varepsilon = \varepsilon_e + \varepsilon_o \quad (9.17)$$

And the stress–strain relation is given by

$$\sigma = \mathbf{E}\varepsilon_e = \mathbf{E}(\varepsilon - \varepsilon_o) \quad (9.18)$$

9.2.2.4 Notes on FEA for Thermal Stress Analysis

Need to specify α for the structure and ΔT on the related elements (which experience the temperature change).

- Note that for linear thermoelasticity, same temperature change will yield same stresses, even if the structure is at two different temperature levels.
- Differences in the temperatures during the manufacturing and working environment are the main cause of thermal (residual) stresses.

A more comprehensive review of thermal problems, their governing equations and boundary conditions can be found in the references, such as References [13,14].

9.3 Modeling of Thermal Problems

Heat transfers in three ways through conduction, convection, and radiation. In FEM, conduction is modeled by solving the resulting heat balance equations for the nodal temperatures under specified thermal boundary conditions. Convection is modeled as a surface load with a user-specified heat transfer coefficient and a given bulk temperature of the surrounding fluid. Radiation effects, which are nonlinear, are typically modeled by using the radiation link elements or surface effect elements with the radiation option. Material properties such as density, thermal conductivity, and specific heat are needed as input parameters for transient thermal analysis, while steady-state thermal analysis needs only thermal conductivity as the material input. For thermal stress analysis, material input parameters include Young's modulus, Poisson's ratio, and thermal expansion coefficient. In the following, modeling of thermal problems is briefly illustrated with the aid of two examples.

9.3.1 Thermal Analysis

First, we use a heat sink model taken from Reference [14] for thermal analysis. A heat sink is a device commonly used to dissipate heat from a CPU in a computer. In this heat sink model, a given temperature field ($T = 120$) is specified on the bottom surface and a heat flux condition ($Q \equiv -k\partial T/\partial n = -0.2$) is specified on all the other surfaces. An FE mesh with a total node of 127,149 is created as shown in Figure 9.4. Using the steady-state thermal analysis system in *ANSYS*, the computed temperature distribution on the heat sink is calculated as shown in Figure 9.5. The cooling effect of the heat sink is most evident.

9.3.2 Thermal Stress Analysis

Next, we study the thermal stresses in structures due to temperature changes. For this purpose, we employ the same model of a plate with a center hole (Figure 9.6) as used in Chapters 4 and 5 to show the relation between the thermal stresses and constraints. We assume that the plate is made of steel with Young's modulus $E = 200$ GPa, Poisson's ratio $\nu = 0.3$, and thermal expansion coefficient $\alpha = 12 \times 10^{-6}/°C$. The plate is applied with a uniform temperature increase of 100°C.

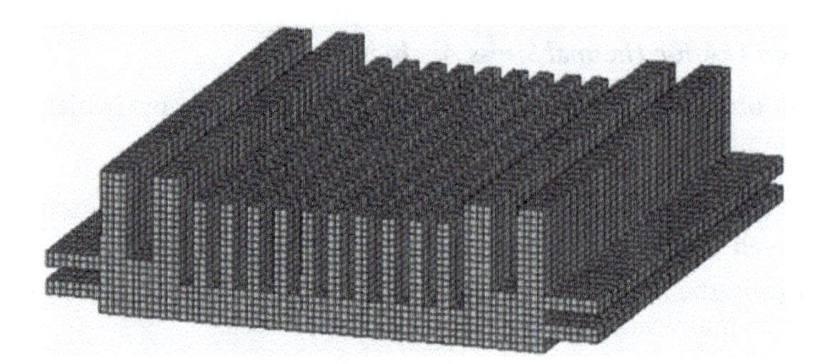

FIGURE 9.4
A heat-sink model used for heat conduction analysis.

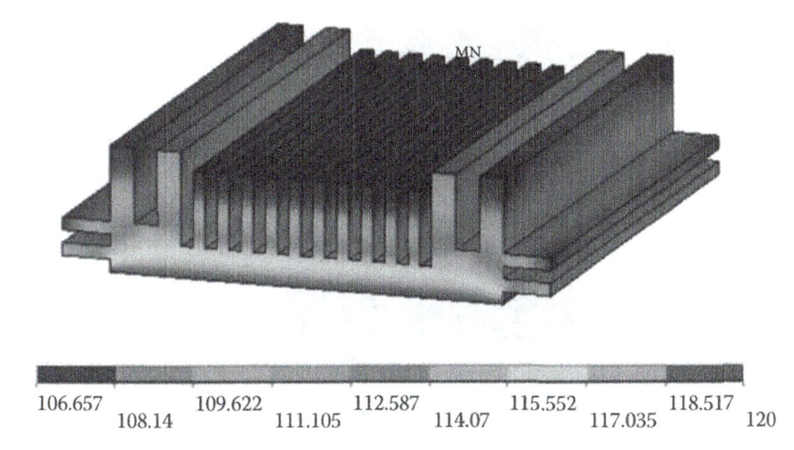

FIGURE 9.5
Computed temperature distribution in the heat-sink.

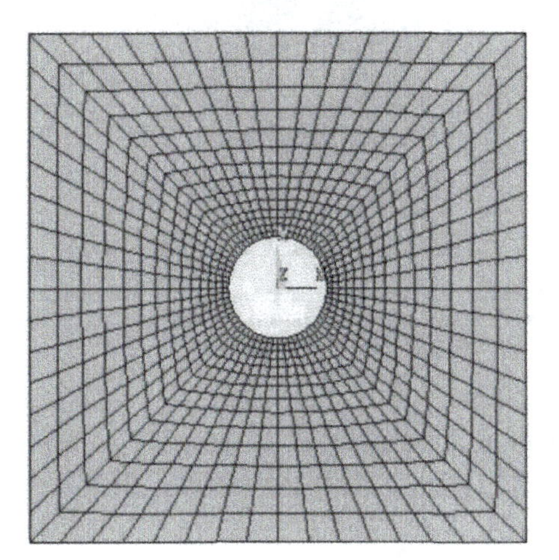

FIGURE 9.6
A square plate with a center hole and under a uniform temperature load.

Figure 9.7 shows the computed thermal stresses in the plate under two different types of constraints. When the plate is constrained (roller support) at the left side only, the plate expands uniformly in both the x and y directions, which causes no thermal stresses (Figure 9.7a, note that the numbers, ranging from 10^{-6} to 10^{-3}, are actually machine zeros). However, when the plate is constrained at both the left and right sides, the plate can expand only in the y direction and significant thermal stresses are induced (Figure 9.7b), especially near the edge of the hole.

In many cases, the changes of the material properties of a structure should be considered as well when the temperature changes are significant, especially when the structure is exposed to high temperatures such as in an aircraft engine. Cyclic temperature fields can also cause thermal fatigue of structures and lead to failures. All these phenomena can be modeled with the FEA and interested readers can consult with the documents of the FEA software at hand.

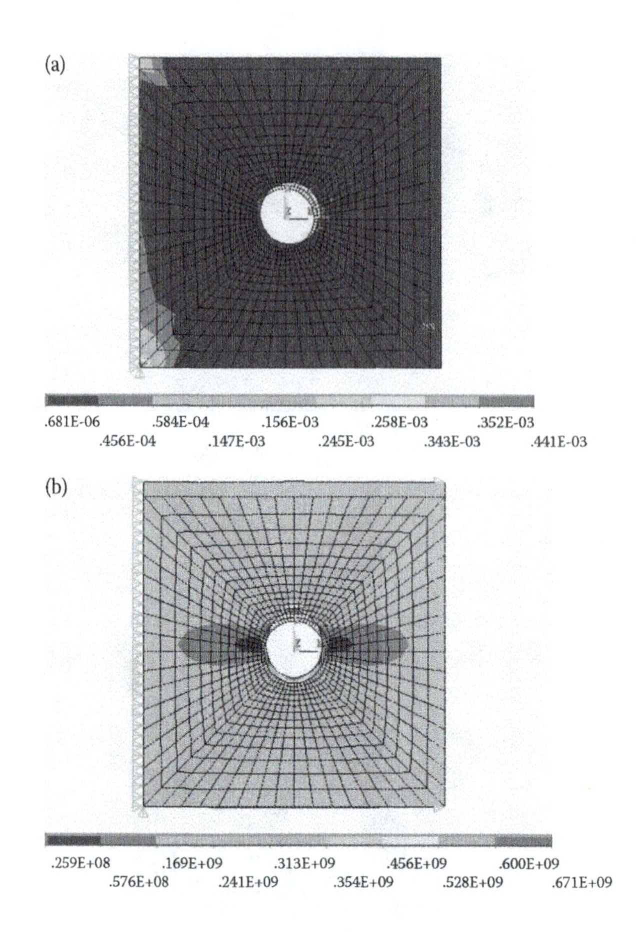

FIGURE 9.7
Thermal (von Mises) stresses in the plate: (a) When the plate is constrained at the left side only (thermal stresses = 0); (b) when the plate is constrained at both the left and right sides.

9.4 Case Studies with *ANSYS Workbench*

Problem Description: Heat sinks are commonly used to enhance heat dissipation from electronic devices. In the case study, we conduct thermal analysis of a heat sink made of aluminum with thermal conductivity $k = 170$ W/(m K), density $\rho = 2800$ kg/m^3, specific heat $c = 870$ J/(kg K), Young's modulus $E = 70$ GPa, Poisson's ratio $\nu = 0.3$, and thermal expansion coefficient $\alpha = 22 \times 10^{-6}$/°C. A fan forces air over all surfaces of the heat sink except for the base, where a heat flux q' is prescribed. The surrounding air is 28°C with a heat transfer coefficient of $h = 30$ W/(m^2°C). *(Part A):* Study the steady-state thermal response of the heat sink with an initial temperature of 28°C and a constant heat flux input of $q' = 1000$ W/m^2. *(Part B):* Suppose the heat flux is a square wave function with period of 90 s and magnitudes transitioning between 0 and 1000 W/m^2. Study the transient thermal response of the heat sink in 180 s by using the steady-state solution as the initial condition. *(Part C):* Suppose the base of the heat sink is fixed. Study the thermal stress response of the heat sink by using the steady-state solution as the temperature load.

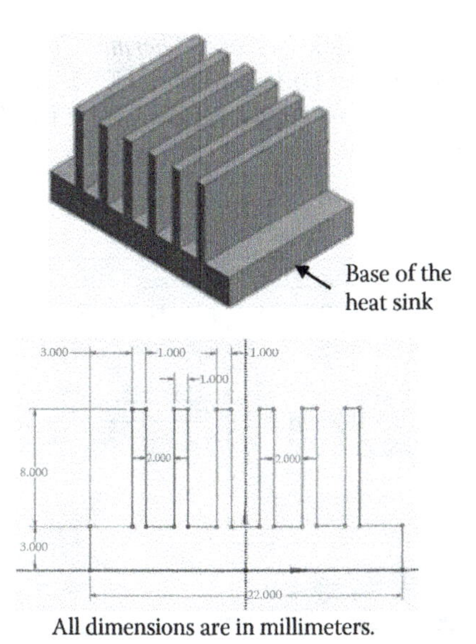

All dimensions are in millimeters.

Material: Aluminum

$k = 170\ \mathrm{W/(m \cdot K)}$

$\rho = 2800\ \mathrm{kg/m^3}$; $c = 870\ \mathrm{J/(kg \cdot K)}$

$E = 70$ GPa; $\nu = 0.3$

$\alpha = 22 \times 10^{-6}/°\mathrm{C}$

Boundary conditions:

Air temperature of 28°C; $h = 30\ \mathrm{W/(m^2 \cdot °C)}$.

Steady state: $q' = 1000\ \mathrm{W/m^2}$ on the base.

Transient: Square wave heat flux on the base.

Initial conditions:

Steady state: Uniform temperature of 28°C.
Transient: Steady-state temperature results.

Part A: Steady-State Thermal Analysis

Step 1: Start an *ANSYS Workbench* Project

Launch *ANSYS Workbench* and save the blank project as *HeatSink.wbpj*.

Step 2: Create a *Steady-State Thermal* Analysis System

Drag the *Steady-State Thermal* icon from the *Analysis Systems Toolbox* window and drop it inside the highlighted green rectangle in the *Project Schematic* window.

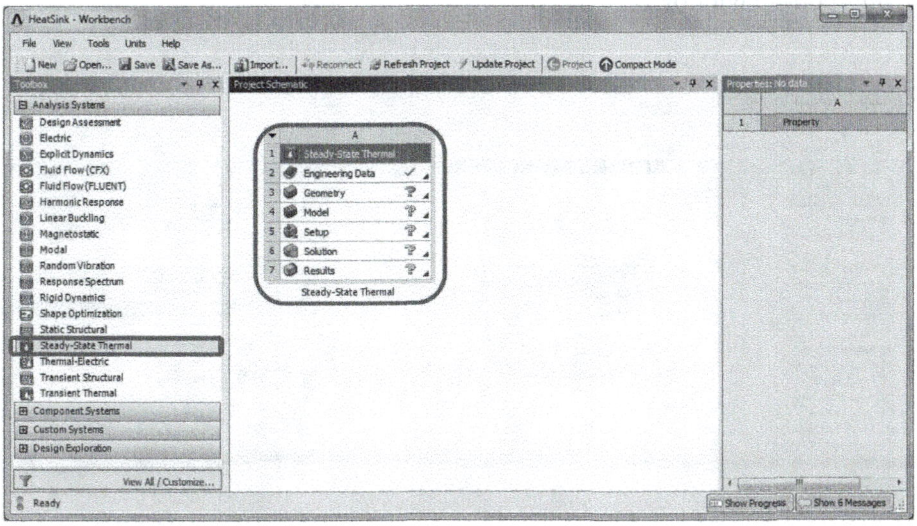

Step 3: Add a New Material

Double-click on the *Engineering Data* cell to add a new material. In the following *Engineering Data* interface which replaces the *Project Schematic*, type *"Aluminum"* as the name for the new material, and double-click *Isotropic Thermal Conductivity* under *Thermal* in the *Toolbox* window. Change the *Unit* to *"Wm^-1K^-1"* and enter *"170"* for *Isotropic Thermal Conductivity* in the *Properties* window. Click the *Return to Project* button to go back to *Project Schematic.*

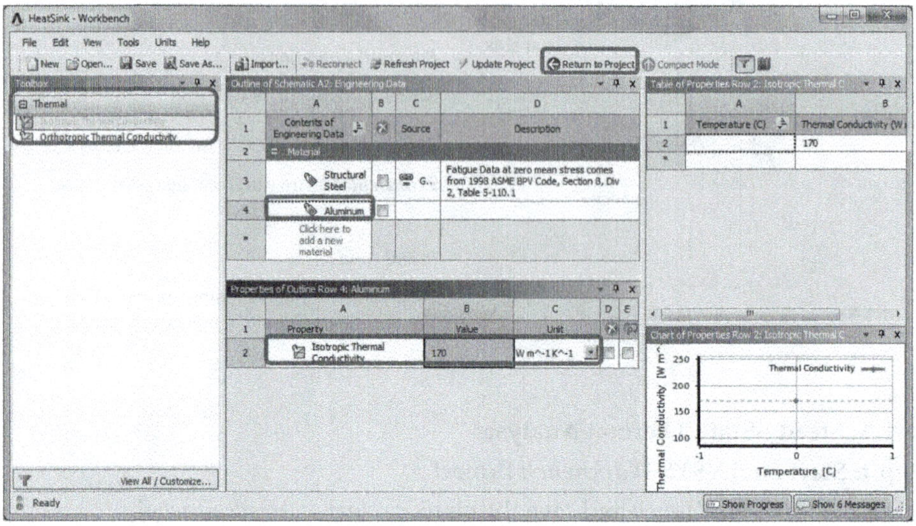

Step 4: Launch the *DesignModeler* Program

Double-click the *Geometry* cell to launch *DesignModeler,* and select *"Millimeter"* in the *Units* pop-up window.

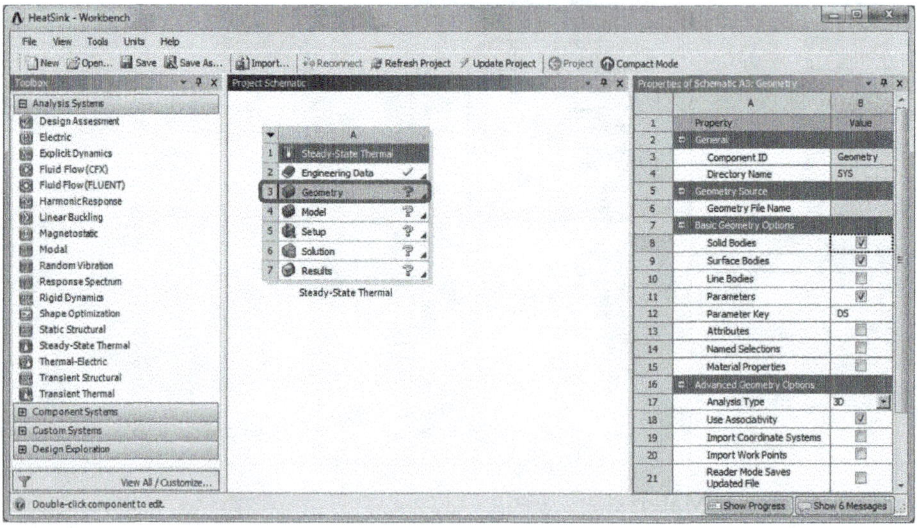

Step 5: Create a Profile Sketch

Click on the *Sketching* tab. Select the *Draw* toolbox and draw a 2D profile as shown below.

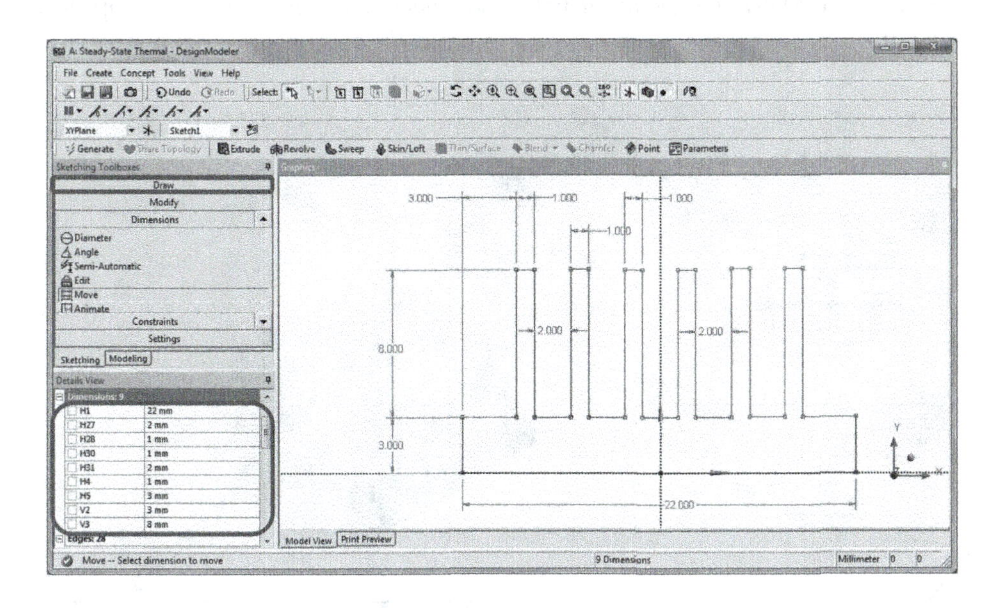

Step 6: Create an Extruded Body

Switch to the *Modeling* tab and click on the *Extrude* feature. The default *Base Object* is set as *Sketch1* in the *Details of Extrude1*. Change the extrusion depth to *15 mm* in the field of *FD1, Depth* and click *Generate*. A solid body is created as shown below.

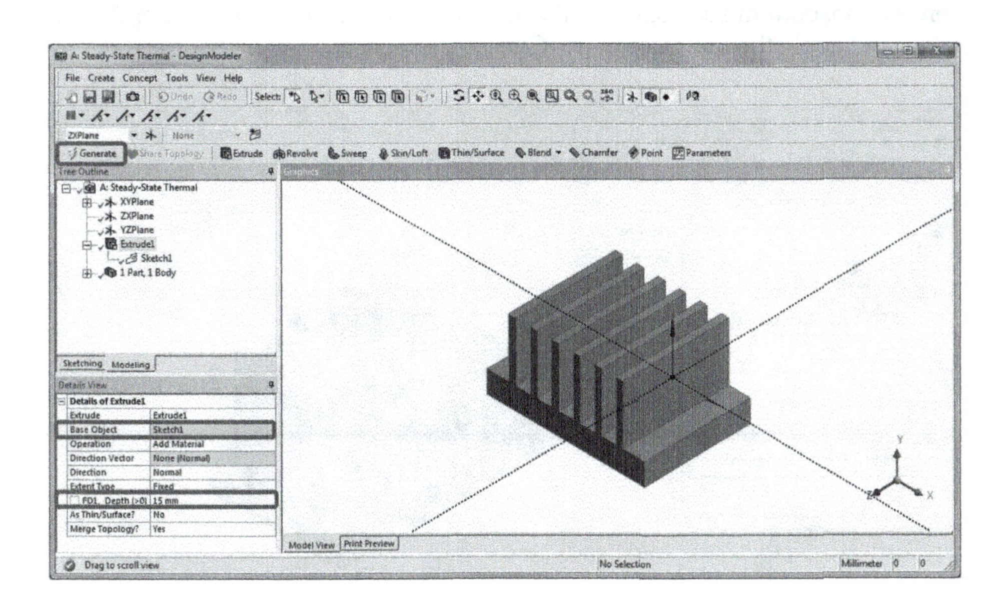

Step 7: Launch the *Steady-State Thermal* Program

Double-click on the *Model* cell to launch the *Steady-State Thermal* program. Click on the *Solid* under *Geometry* in the *Outline tree*. In the *Details of "Solid,"* click to the right of the *Material Assignment* field and select *Aluminum* from the drop-down menu.

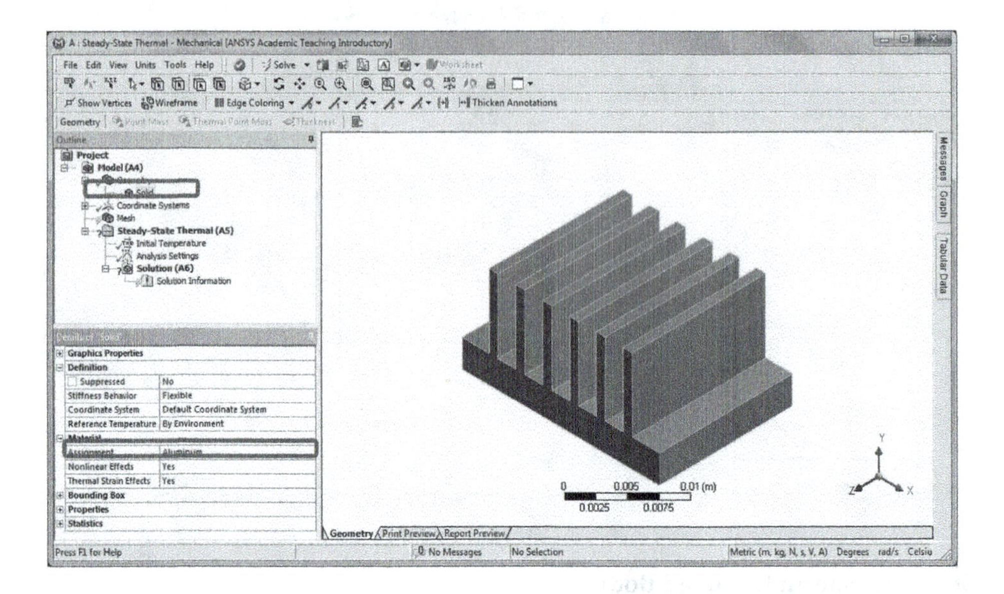

Step 8: Generate Mesh

Right click on *Mesh* in the *Project Outline*. Select *Insert* and then *Sizing* from the context menu. In the *Details of "Face Sizing,"* enter *"5e-4 m"* for the *Element Size*. Click on the top, bottom surfaces, and the side walls of the guitar in the *Graphics* window and apply the five faces to the *Geometry* selection.

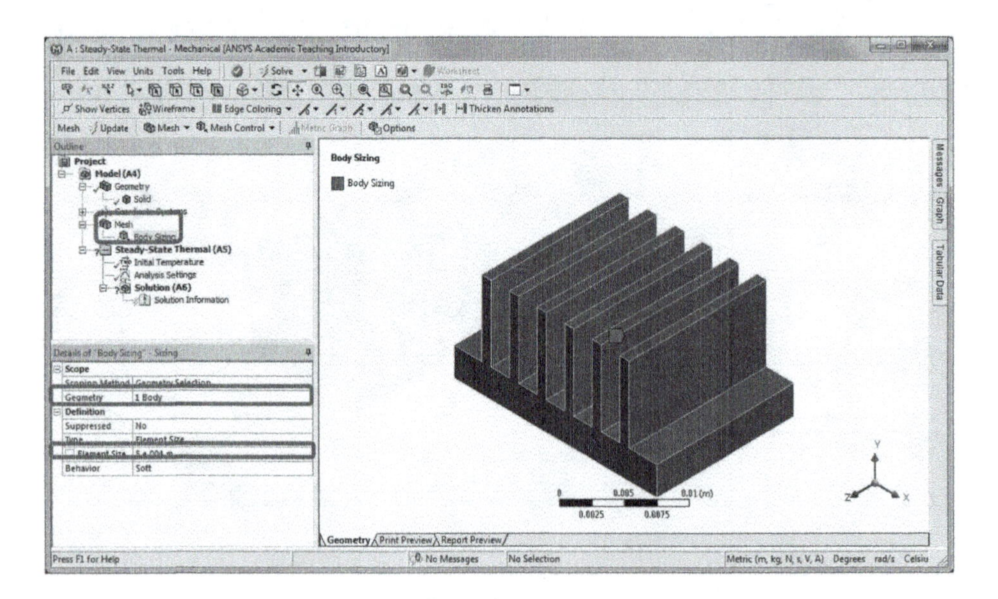

In the *Details of "Mesh,"* select *Dropped* for the *Element Midside Nodes* under the *Advanced* option. This helps reduce the total number of nodes to an acceptable level not exceeding the requested resources of educational licenses. Right-click on *Mesh* and select *Generate Mesh.*

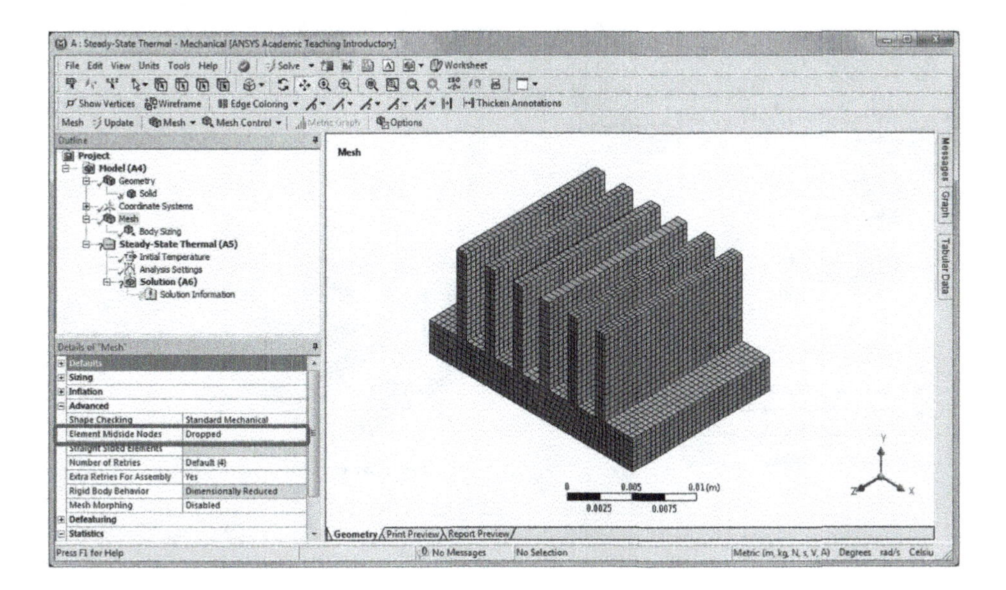

Step 9: Apply Boundary Conditions

Right-click on *Steady-State Thermal(A5).* Choose *Insert* and then *Heat Flux* from the context menu. Apply a heat flux of *1000 W/m*$^{-2}$ to the base of the heat sink.

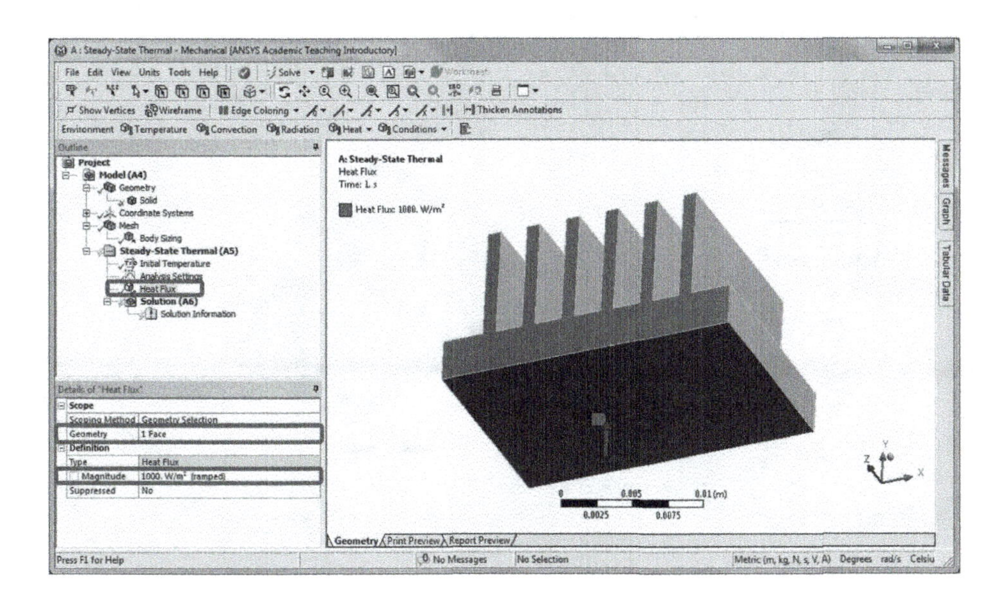

Right-click on *Steady-State Thermal(A5)*. Choose *Insert* and then *Convection* from the context menu. In the *Details of "Convection,"* enter *30 W/(m²°C)* for *Film Coefficient* and *28°C* for *Ambient Temperature* to all surfaces (a total of 29 faces) except for the base of the heat sink.

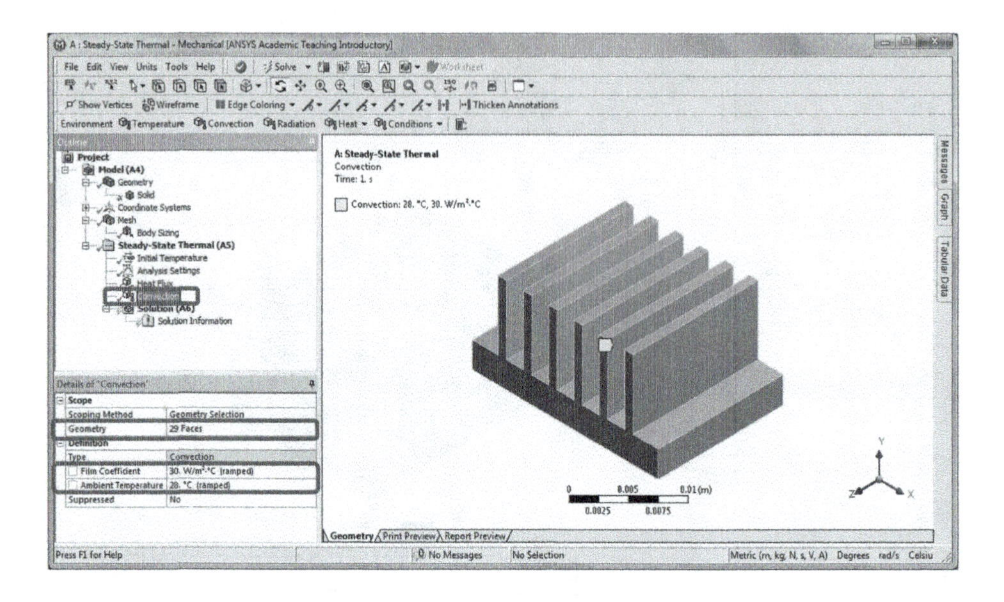

Step 10: Solve and Retrieve Results

Right-click on *Solution (A6)* in the *Outline*, and insert *Temperature, Total Heat Flux*, and *Directional Heat Flux* to the solution outline. In the *Details of "Directional Heat Flux,"* set the *Orientation* to *Y-axis*. Right-click on *Solution (A6)* and click *Solve*.

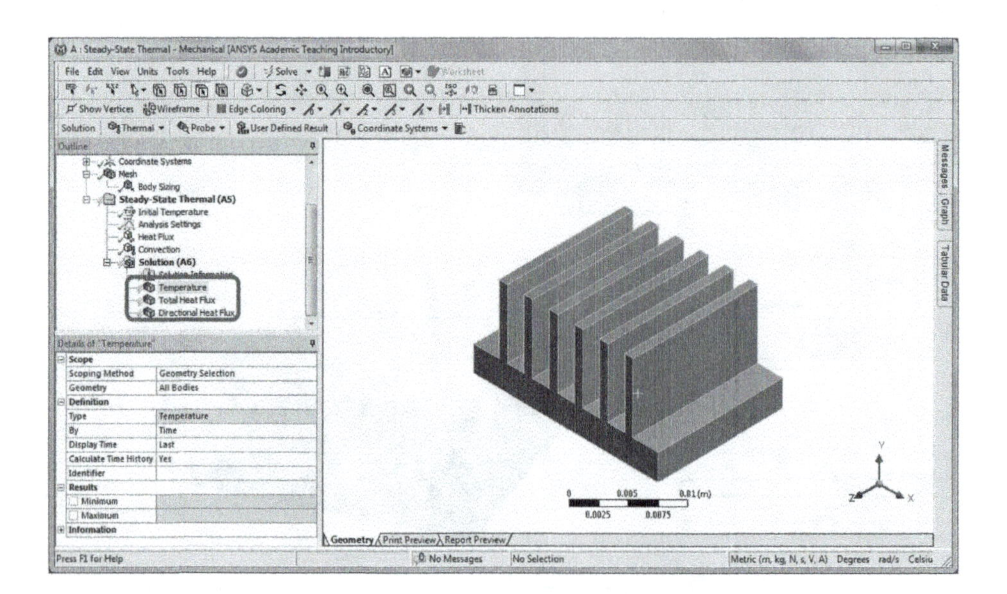

Click on *Temperature* in the *Outline* to review the temperature distribution.

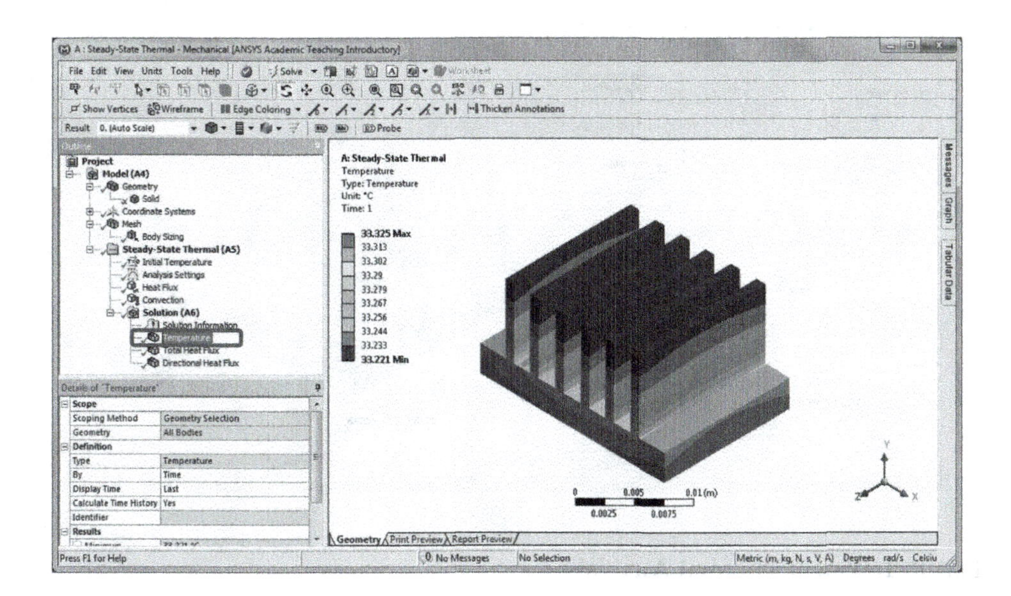

Click on *Total Heat Flux* to display the heat flux with directional arrows.

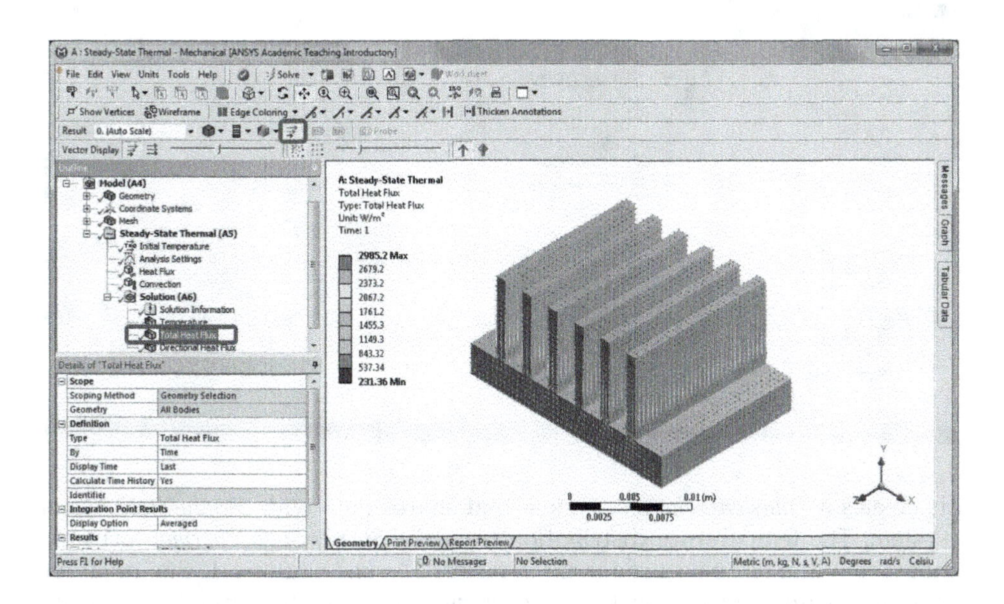

Click on *Directional Heat Flux* to review the heat flux isolines along *Y-axis.*

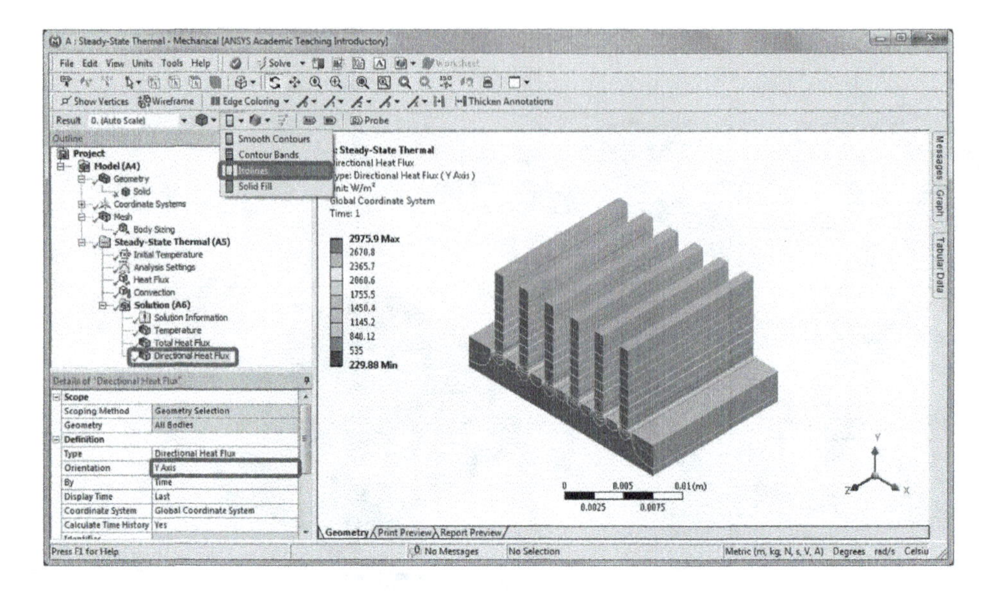

Part B: Transient Thermal Analysis

Step 1: Add a *Transient Thermal* Analysis System

Drag the *Transient Thermal* icon from the *Analysis Systems Toolbox* window and drop it onto the *Solution* cell of the highlighted *Steady-State* system in the *Project Schematic.*

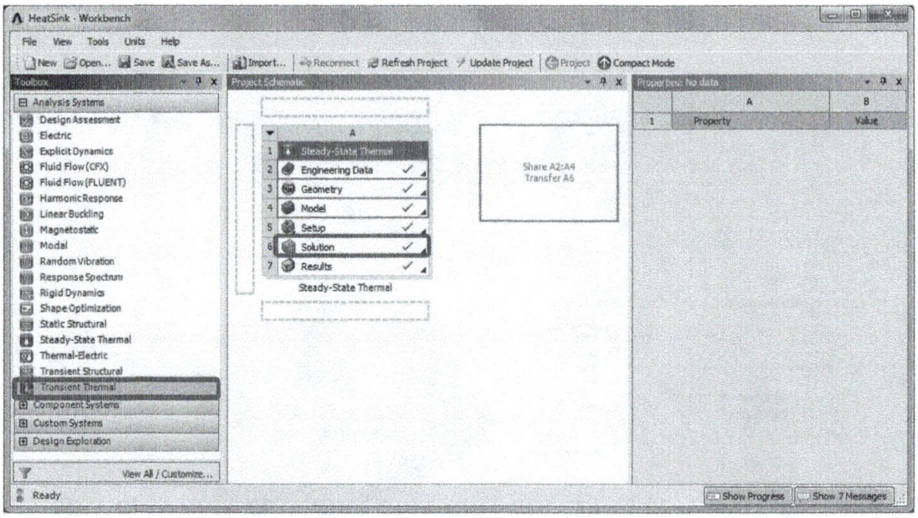

This creates a *Transient Thermal* system that shares data with *Steady-State Thermal* system. The temperature distribution from the *Steady-State Thermal* analysis is now the initial temperature for the *Transient Thermal* analysis. If the initial temperature is uniform for the *Transient Thermal* analysis, then this data sharing is not needed.

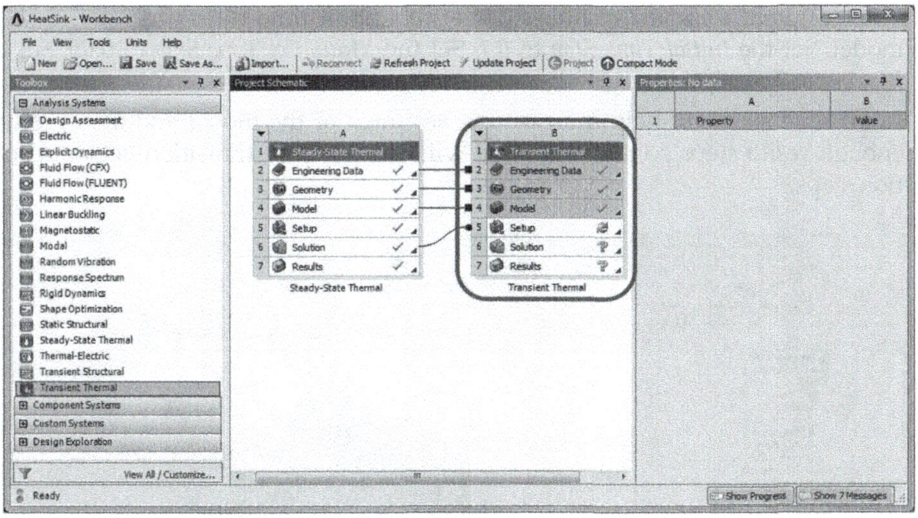

Step 2: Add Material Properties

Double-click on *Engineering Data* of *Steady-State Thermal*. Add a *Density* of *2800 kg/m³* and *Specific Heat* of *870 J/(kg K)* to the Properties of Aluminum. Click *Return to Project*.

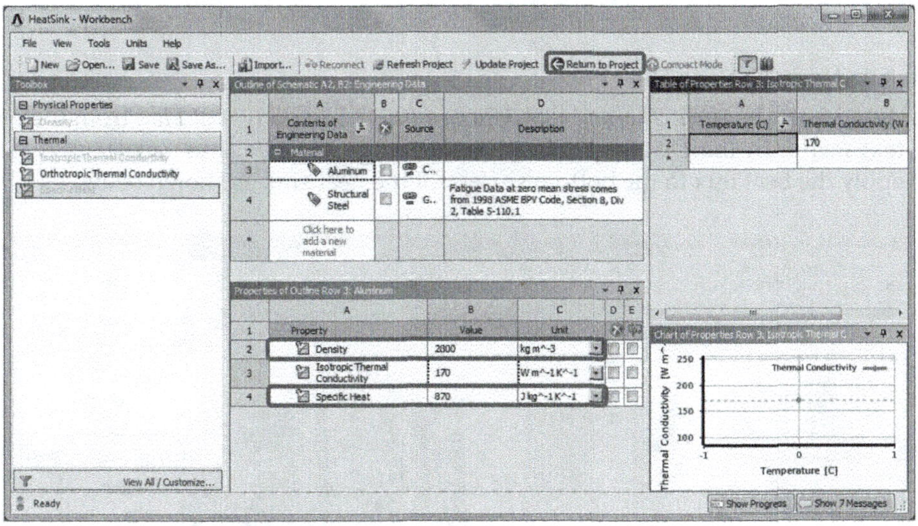

Step 3: Set Up *Transient Thermal* Analysis

Double-click on the *Setup* cell of the *Transient Thermal* system to launch the *Multiple Systems–Mechanical* program. Click *Yes* on the pop-up window to read the modified upstream data.

Select *Analysis Settings* from the *Outline* tree. In the *Details of "Analysis Settings"*, under *Step Controls*, set the *Step End Time* to *180s*. Change the *Auto Time Stepping* to *On* from *Program Controlled*. Change *Defined By* to *Time*.

The default values for the initial and the maximum time steps are small for this model. Set the *Initial Time Step* to *0.1*. Set the *Minimum Time Step* to *0.05*. Set the *Maximum Time Step* to *5*.

A small time step will help increase the accuracy of the model and also produce enough result steps so the animation will have smooth transition between solution steps.

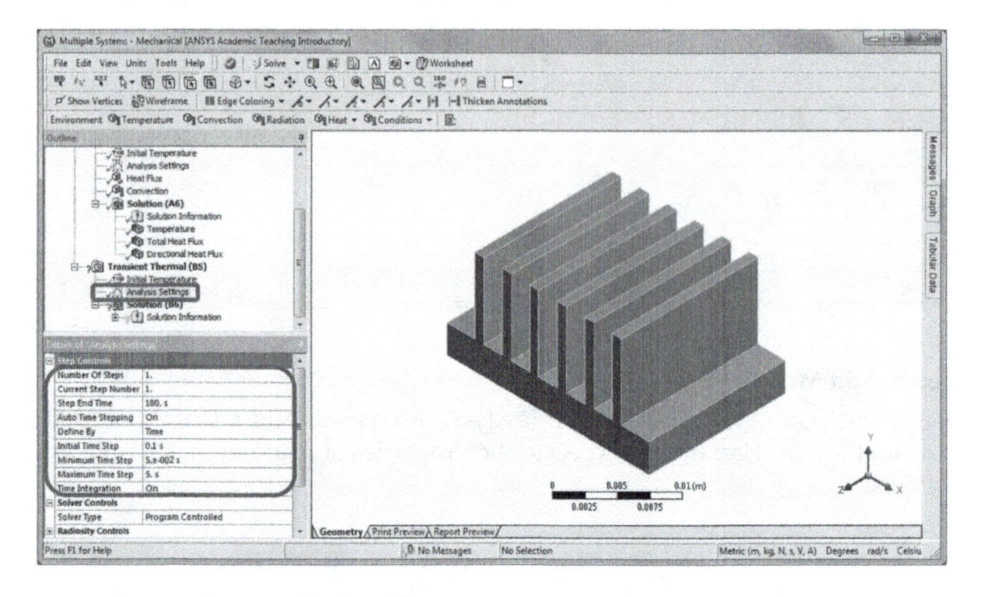

Step 4: Apply Cycling Heat Flux

Right-click on *Transient Thermal (B5)*. Choose *Insert* and then *Heat Flux* from the context menu. In the *Details of "Heat Flux,"* change *Magnitude* to *Tabular Data*, and apply the heat flux to the bottom surface of the heat sink as shown below.

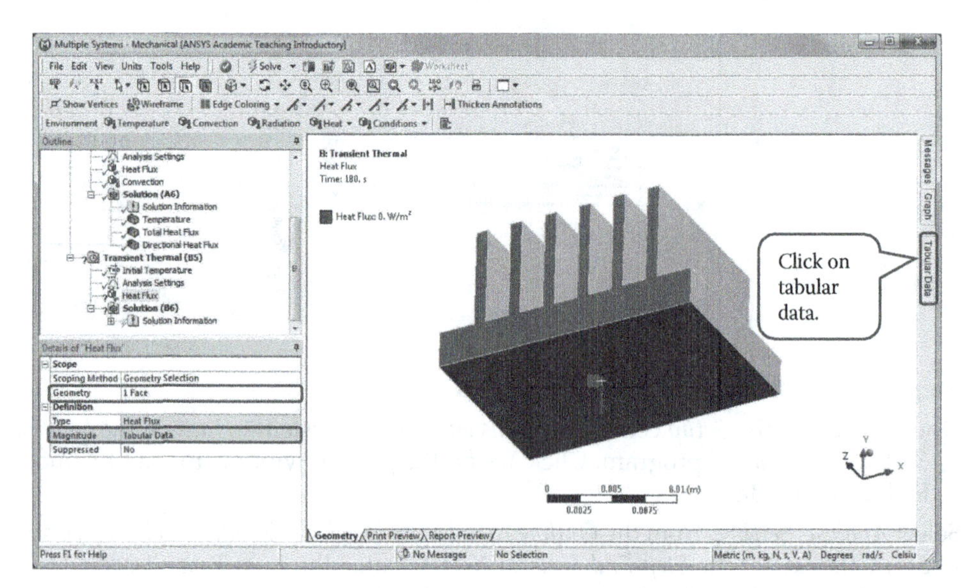

Click on *Tabular Data* on the right edge of the *Graphics* window, and then click on the push pin labeled *AutoHide* to display the *Tabular Data* window as shown below. Enter the following values in the *Tabular Data* table.

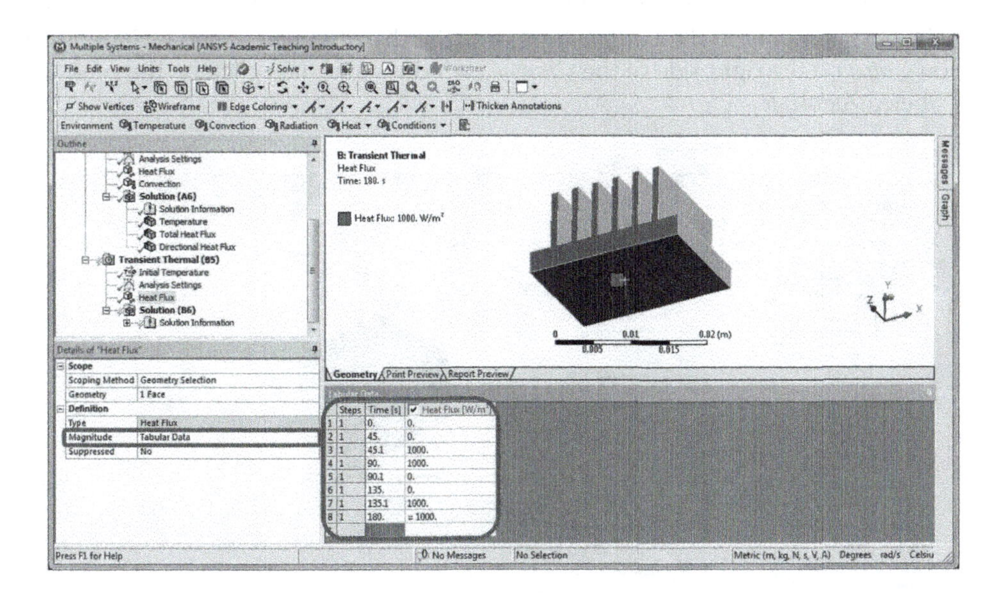

	Steps	Time [s]	Heat Flux [W/m²]
1	1	0.	0.
2	1	45.	0.
3	1	45.1	1000.
4	1	90.	1000.
5	1	90.1	0.
6	1	135.	0.
7	1	135.1	1000.
8	1	180.	= 1000.

Step 5: Apply Convective Boundary Condition

Right-click on *Transient Thermal (B5)*. Choose *Insert* and then *Convection* from the context menu. In the *Details of "Convection,"* enter *30 W/(m²°C)* for *Film Coefficient* and *28°C* for *Ambient Temperature* to all surfaces (a total of 29 faces) except for the base of the heat sink.

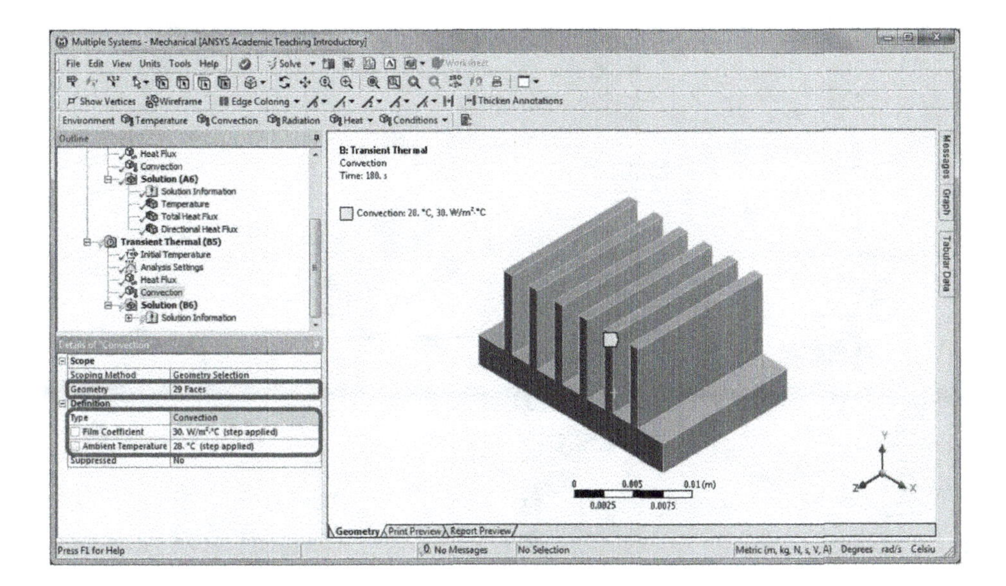

Step 6: Solve and Retrieve Results

Right-click on *Solution (B6)* in the *Outline*, and insert *Temperature*, *Total Heat Flux*, and *Directional Heat Flux*. In the *Details of "Directional Heat Flux,"* set the *Orientation* to *Y-axis*. Right-click on *Solution (B6)* and click *Solve*.

Click on *Temperature* in the *Outline* to review the distribution. To show *Graph*, click on *Graph* on the right side of the *Graphics* window, and then click on the push pin labeled *AutoHide*.

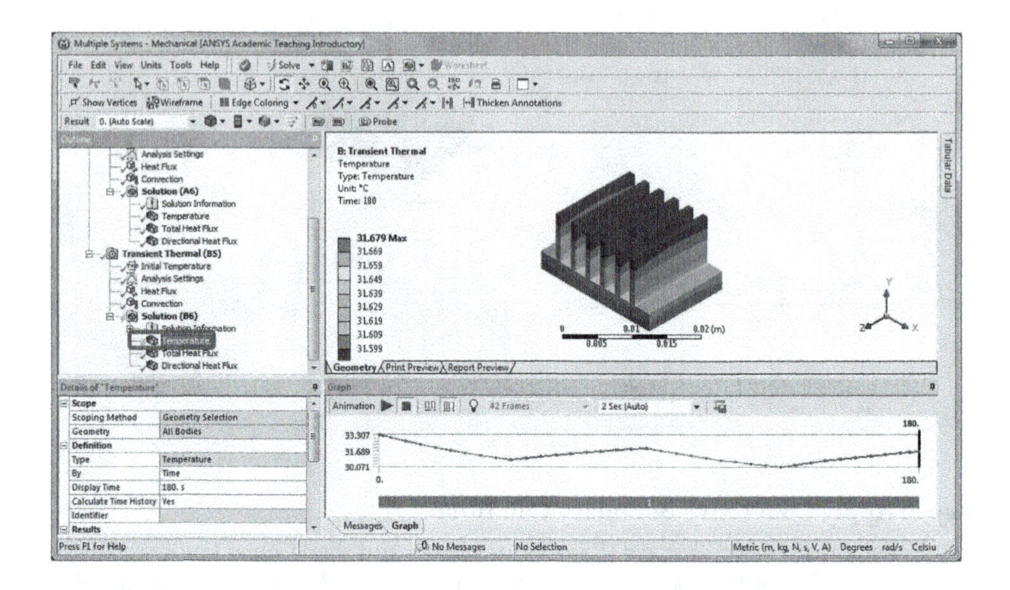

Click on *Total Heat Flux* to display the heat flux with directional arrows.

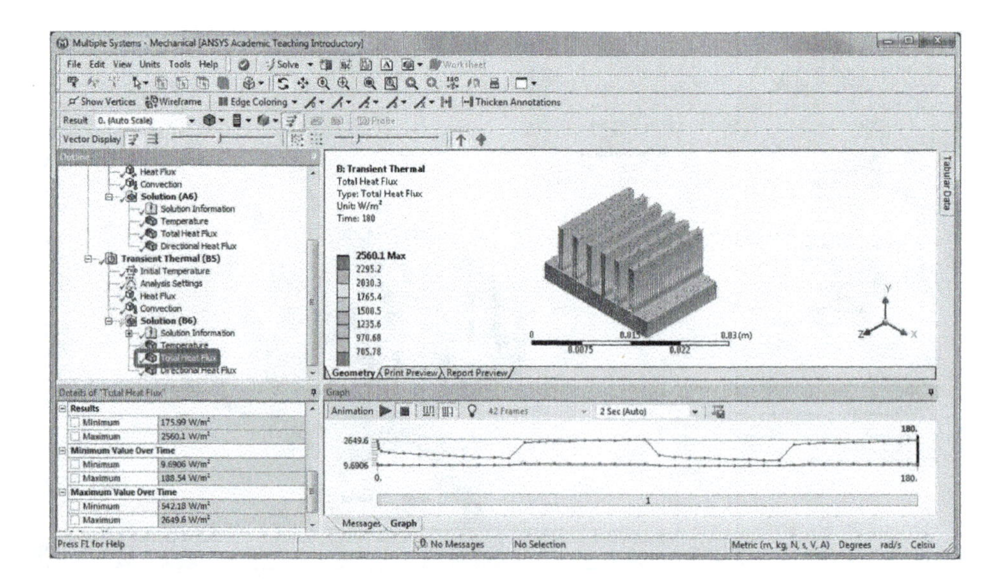

Click on *Directional Heat Flux* to review the heat flux isolines along *Y-axis*.

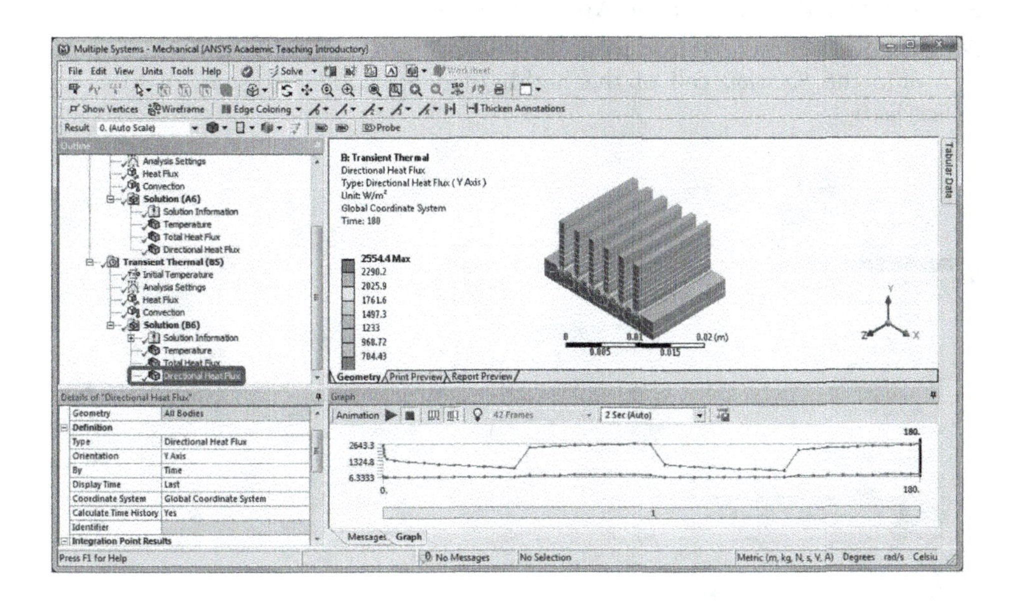

Modeling tips: In the case study, an end time of 180 s is specified in the transient setup. By default, the last set of results (solution at 180 s) from the transient analysis is used for graphics window display. To display results at a different time point, for example, temperature distribution at 40 s, you may change the *Display Time* from *Last* to *40s* in the *Details of "Temperature."* Then right click on *Solution (B6)* and select *Evaluate All Results*. A result at the specified time will be displayed at the *Graphics* window as shown below.

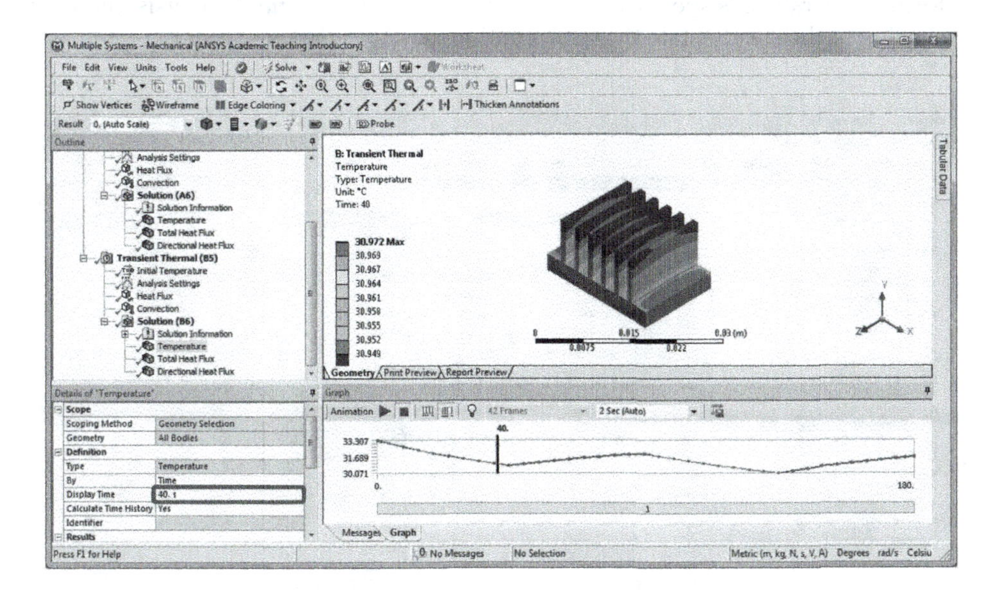

Part C: Thermal Stress Analysis

Step 1: Add a *Static Structural* Analysis System

Drag the *Static Structural* icon from the *Analysis Systems Toolbox* window and drop it onto the *Solution* cell of the highlighted *Steady-State* system in the *Project Schematic*.

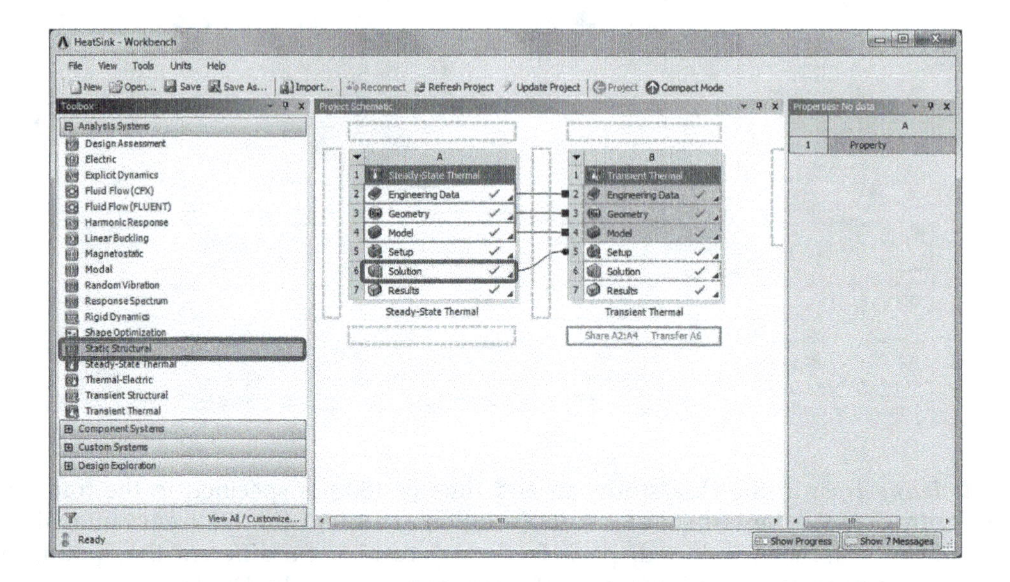

This creates a *Static Structural* system that shares data with the *Steady-State Thermal* system as shown below. The temperature distribution from the *Steady-State Thermal* analysis is now the load input for the *Static Structural* analysis. If a uniform temperature is specified as a load for the *Static Structural* analysis, then data sharing of the steady-state thermal solution is not needed.

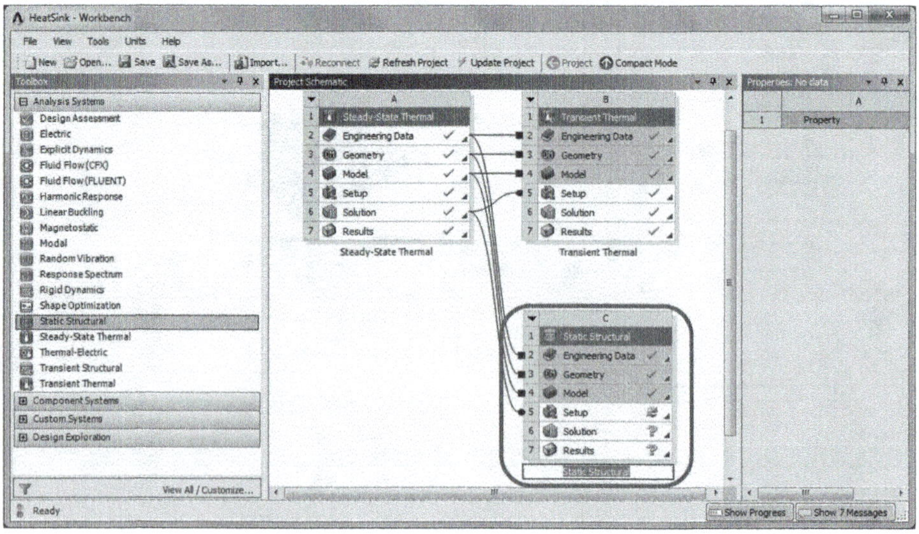

Step 2: Add Material Properties

Double-click on *Engineering Data* of *Steady-State Thermal*. Add *Young's Modulus* of *70 GPa*, *Poisson's ratio* of *0.3*, and *Isotropic Instantaneous Coefficient of Thermal Expansion* of *2.2E-5 1/°C* to the Properties of Aluminum as shown below. Click *Return to Project*.

Step 3: Set Up *Static Structural* Analysis

Double-click on the *Setup* cell of the *Static Structural* system to launch the *Multiple Systems–Mechanical* program. Click *Yes* on the pop-up window to read the modified upstream data.

Note an *Imported Load* item is automatically added to *Static Structural (C5)* in the *Outline* tree. Right-click on *Static Structural (C5)* and insert a *Fixed Support* to the *Outline*. Apply the fixed support to the bottom face of the heat sink.

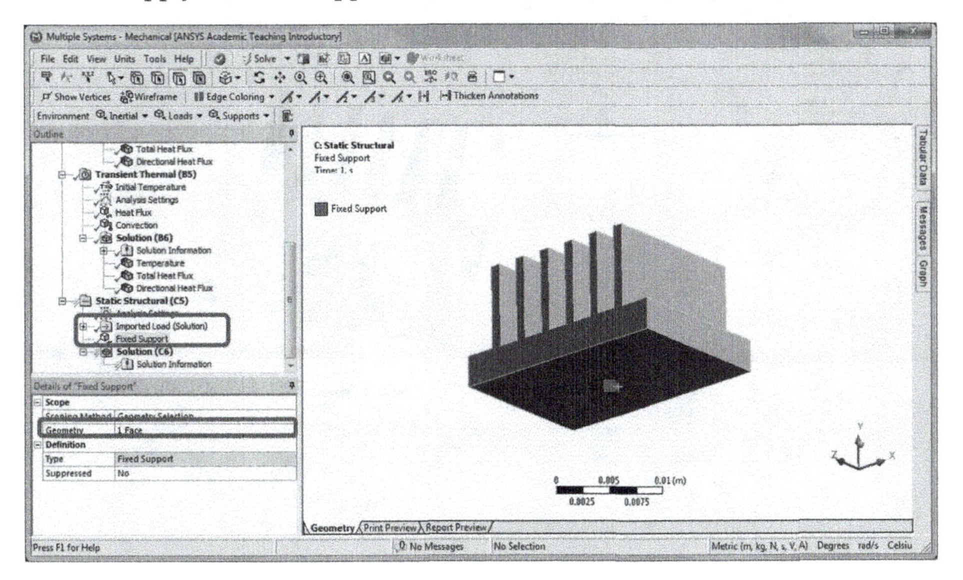

Step 4: Solve and Retrieve Results

Right-click on *Solution (C6)* in the *Outline,* and insert *Total Deformation* and *Equivalent Stress* to the outline. Then right-click on *Solution (C6)* and click *Solve.*

Click on *Total Deformation* in the *Outline* to review displacement results.

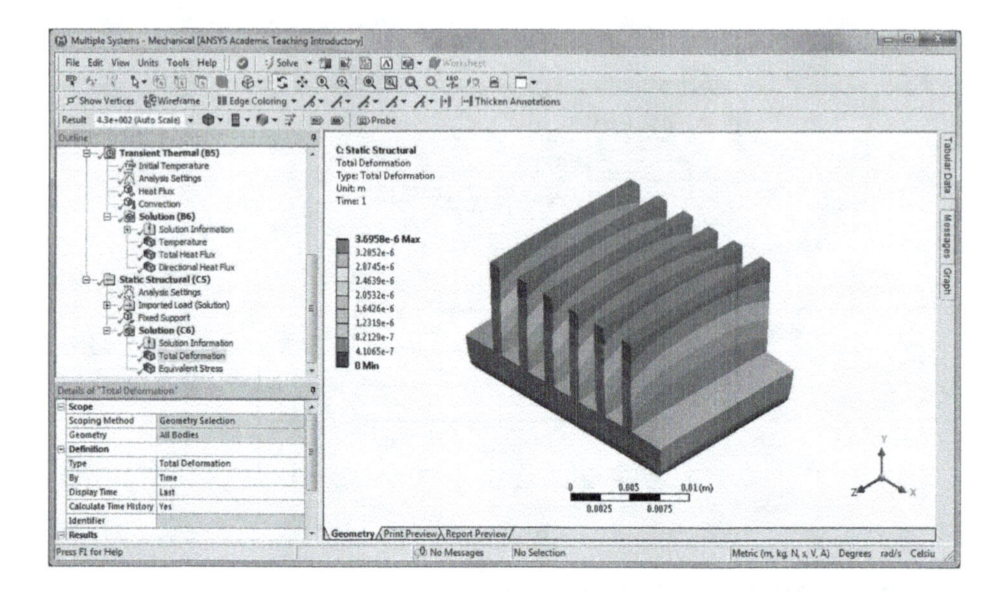

Click on *Equivalent Stress* in the *Outline* to review von Mises stress results.

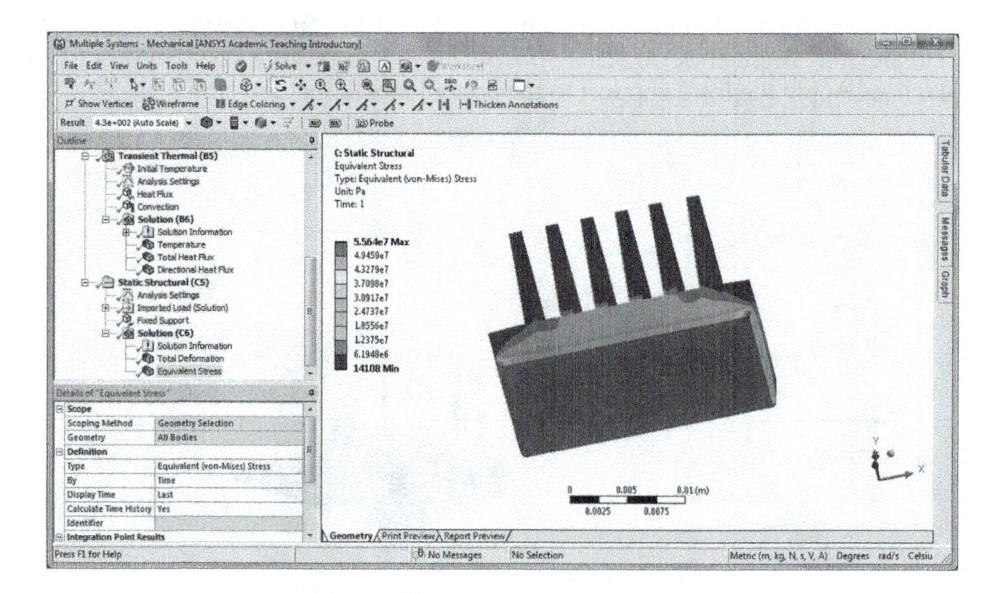

Modeling tips: To apply a uniform temperature load to the heat sink, you may add a *Static Structural* system that shares data with the *Steady-State Thermal* system at the model and above levels. And then insert a constant *Thermal Condition* load to the heat sink in the *Static Structural* analysis.

To do this, first drag the *Static Structural* icon from the *Analysis Systems Toolbox* window and drop it onto the *Model* cell of the highlighted *Steady-State* system in the *Project Schematic.*

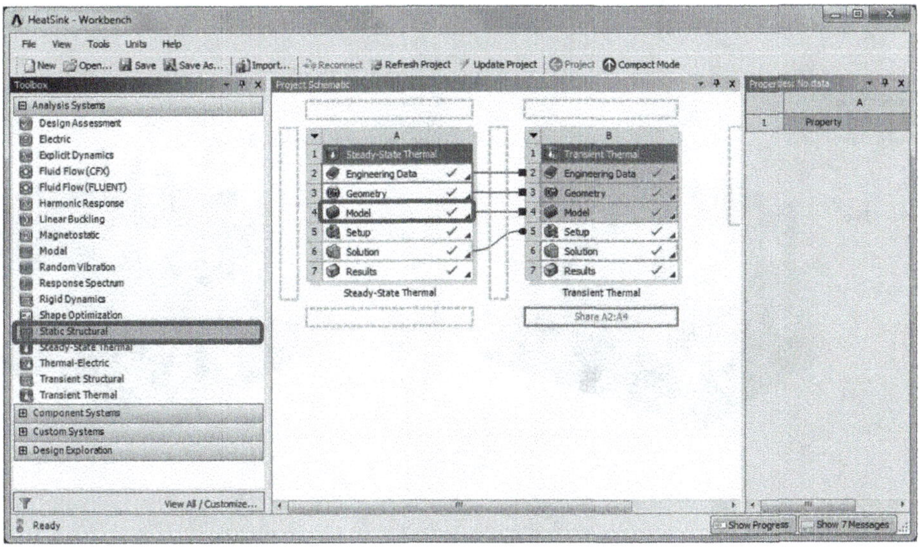

This creates a schematic where *Static Structural* is sharing data with *Steady-State Thermal* at the model and above levels as shown below.

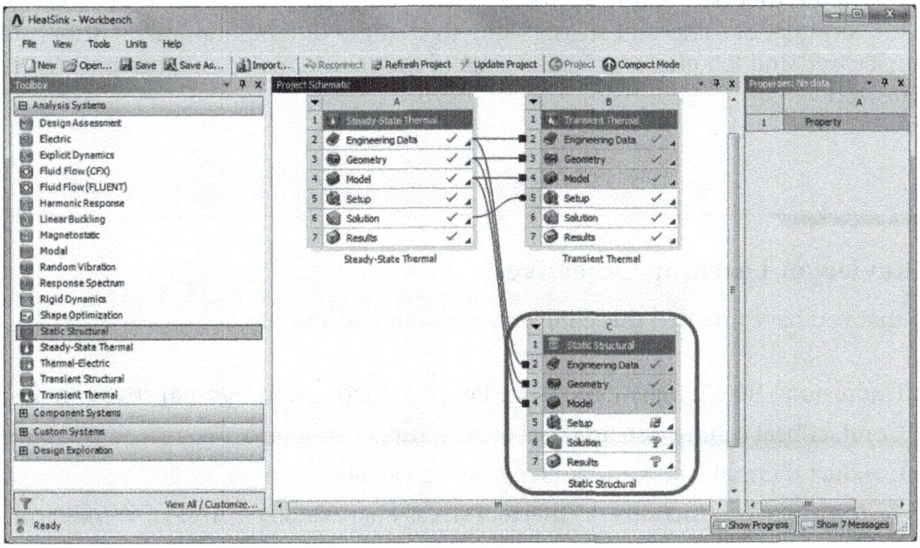

Next, add Young's Modulus, Poisson's ratio, and Coefficient of Thermal Expansion data to the Properties of Aluminum in the *Steady-State Thermal* analysis template following Step 2 of Part C. Set up the *Static Structural* analysis by following Step 3 of Part C. Insert a *Thermal Condition* load to the *Static Structural (C5)* in the *Outline* tree and apply a *Fixed Support* to the base of the heat sink. The total deformation and the equivalent stress distributions of the heat sink with a temperature increase of 10°C can be obtained as follows.

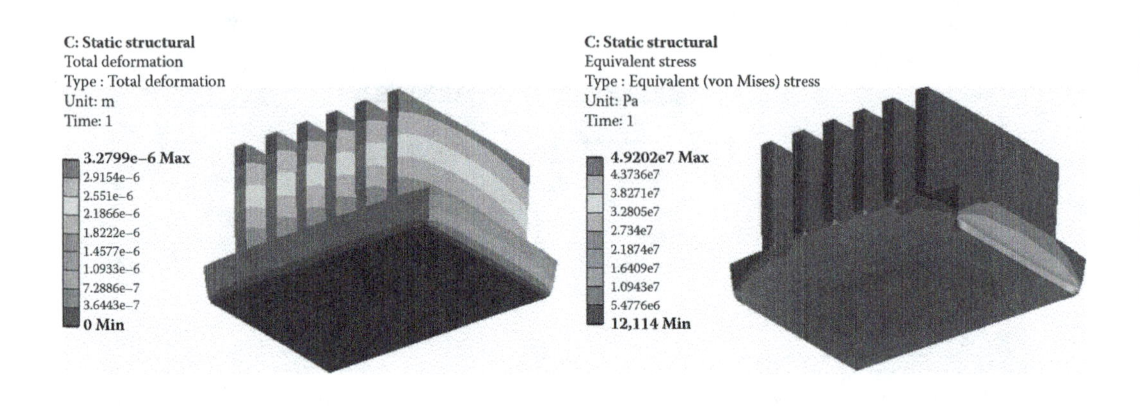

9.5 Summary

In this chapter, we discussed the governing equations for heat conduction problems and the FEA formulation. Thermal stresses due to changes of temperatures in structures are also discussed and the effects of constraints of the structures on the thermal stresses are emphasized.

9.6 Review of Learning Objectives

Now that you have finished this chapter you should be able to

1. Understand the FE formulations for heat conduction and thermal stress analysis
2. Conduct heat conduction analysis of structures using FEA
3. Conduct thermal stress analysis of structures using FEA
4. Perform heat conduction and thermal stress analysis using *ANSYS Workbench*

PROBLEMS

9.1 Study the heat conduction problem in a simple annular region shown below, using the FEA. Assume $a = 1$ m, $b = 2$ m, $T_a = 100$°C, and heat flux $Q_b = 200$ W/m^2. Using structural steel with thermal conductivity $k = 60.5$ W/(m°C) for the region, determine the temperature field and heat flux in this region and compare your FEA results with the analytical solution.

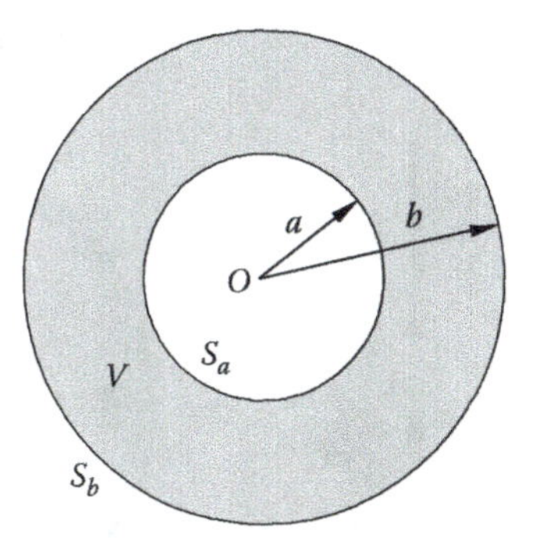

9.2 For the glass cup shown in the figure below, determine the thermal stresses when the inner surfaces of the cup experience a temperature change from 20°C to 60°C while all other surfaces are kept at 20°C. For glass, use Young's modulus $E = 70$ GPa, Poisson's ratio $\nu = 0.17$, thermal conductivity $k = 1.4$ W/(m°C), and coefficient of thermal expansion $\alpha = 8.0 \times 10^{-6}$/°C.

All dimensions are in millimeters.

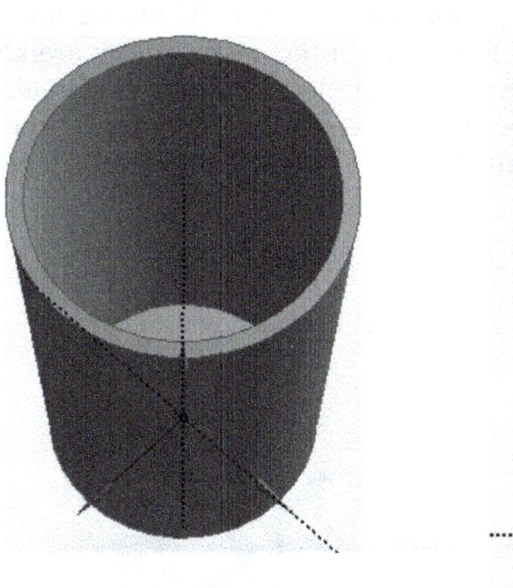

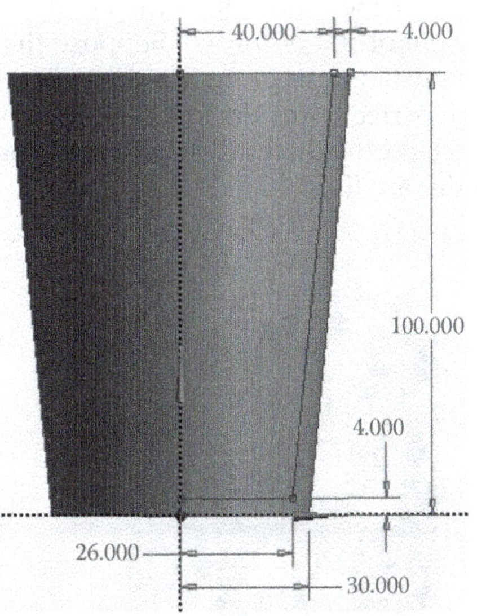

9.3 Determine the temperature and heat flux distributions inside the following heat transfer pipe with internal ribs. A bulk temperature of 80°C and 20°C with a film coefficient of 100 W/(m^2°C) and 30 W/(m^2°C) is specified for the interior and exterior convective heat transfer, respectively. Assume that the pipe is 30 cm long with a thermal conductivity of 230 W/(m K) and that the two ends of the pipe are perfectly insulated.

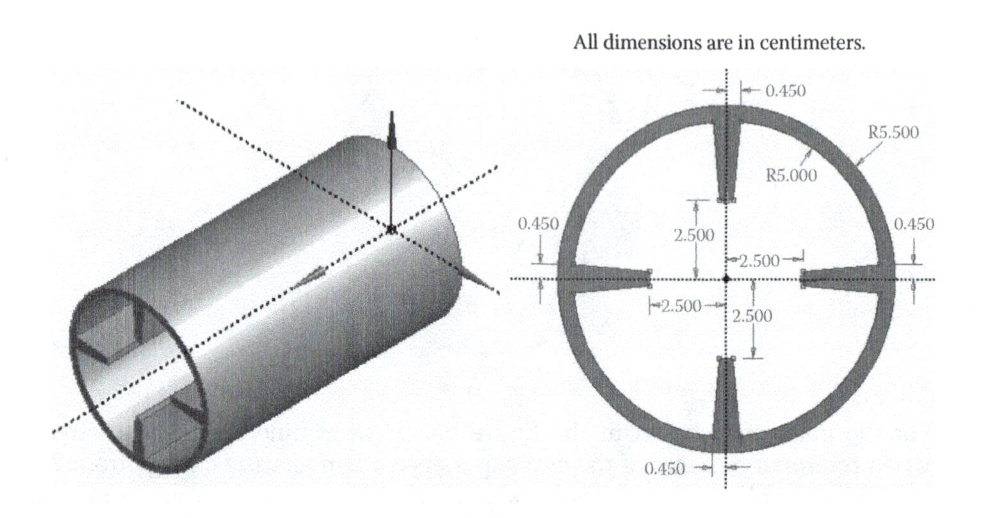

9.4 A 10 cm thick perforated concrete slab is shown below. The concrete has a thermal conductivity $k = 1.0$ W/(m K), Young's modulus $E = 29$ GPa, Poisson's ratio $\nu = 0.15$, and thermal expansion coefficient $\alpha = 11 \times 10^{-6}$/°C. The bulk temperature of the fluid inside the circular perforated holes is 60°C with a film coefficient of 100 W/(m^2°C). Suppose the four sides of the concrete slab are kept at a constant temperature of 20°C. The front and back faces of the perforated slab are perfectly insulated. (1) Determine the temperature distribution in the slab. (2) Find the thermally induced deformation and stresses in the slab if the four sides are fixed.

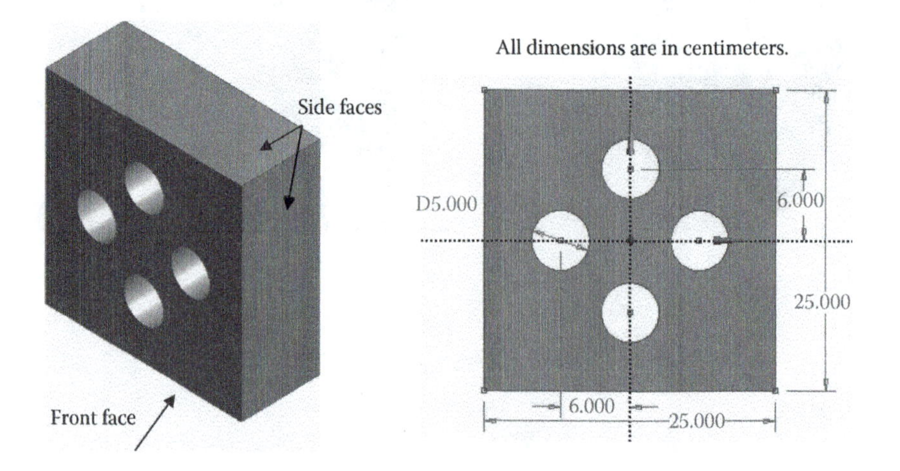

9.5 The following thermal bimorph is made of Silicon with a thermal conductivity $k = 130$ W/(m°C), Young's modulus $E = 200$ GPa, Poisson's ratio $\nu = 0.27$, and thermal expansion coefficient $\alpha = 2.6 \times 10^{-6}$/°C. It is 1 mm thick with fixed temperatures of 60°C and 10°C, respectively, at the leftmost ends of its hot arm and cold arm, as shown below. Suppose the bimorph has a uniform internal heat generation at 100 W/m³, and ignore the effects of convective heat transfer. (1) Determine the steady-state temperature distribution in the thermal bimorph. (2) Determine the thermally induced deformation and stresses in the bimorph if the two arms are fixed on their left ends.

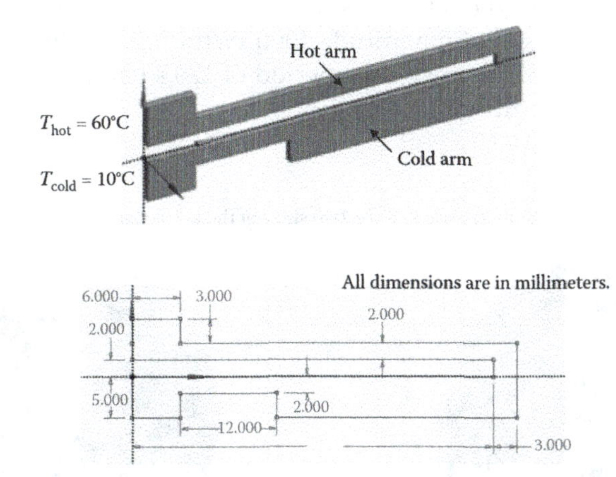

9.6 The heat sink shown below has a square base of size 22 mm × 22 mm × 2 mm and consists of an array of square pin-fins, each of size 2 mm × 2 mm × 8 mm. The pin-fins are separated by a gap distance of 4 mm in each direction. The heat sink is initially at a temperature of 28°C. Suppose a heat flux input is prescribed on the bottom face of the heat sink. The heat flux is of a symmetrical square wave pattern with a period of 60 s (equal time for each half-cycle) and peak magnitudes transitioning from 800 W/m² to 0 W/m². A fan forces air over all surfaces of the heat sink except for the bottom face, where the heat flux is prescribed. The surrounding air is 28°C with a film coefficient of 30 W/(m²°C). Assume that the heat sink is made of aluminum with thermal conductivity $k = 170$ W/(m K), density $\rho = 2800$ kg/m³, specific heat $c = 870$ J/(kg K), Young's modulus $E = 70$ GPa, Poisson's ratio $\nu = 0.3$, and thermal expansion coefficient $\alpha = 22 \times 10^{-6}$/°C. Study the transient thermal response of the heat sink during the first three cycles.

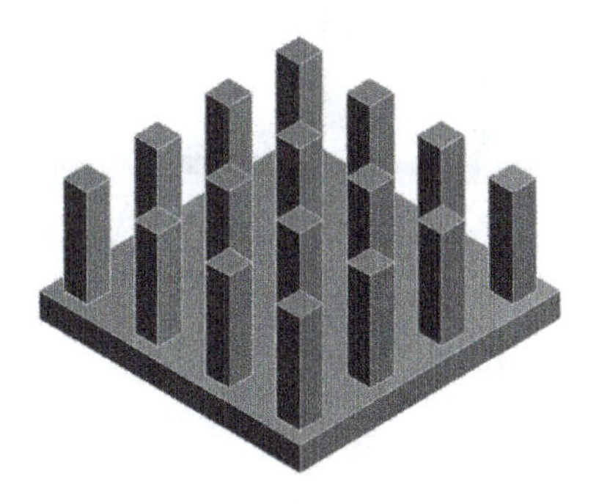

9.7 The electric kettle base shown below has a thickness of 0.5 cm and is made of structural steel with thermal conductivity $k = 60.5$ W/(m K), density $\rho = 7850$ kg/m^3, and specific heat $c = 434$ J/(kg K). Suppose a heat flux condition is defined through a serpentine heating element embedded at the bottom of the kettle base. Except for the imprinted heating element face, all other faces of the kettle base are subjected to convective heat transfer with a bulk air temperature of 40°C and a film coefficient of 80 W/(m^2°C). (1) Study the steady-state thermal response of the kettle base if a constant heat flux of 800 W/(m^2°C) is applied through the heating element. (2) Suppose the heating element is cycled on and off with a period of 60 s (30 s on-time followed by 30 s off-time). During on-time, the heat flux is of magnitude 1000 W/(m^2°C). Study the transient thermal response of the kettle base for a period of 120 s using the steady-state solution as the initial condition.

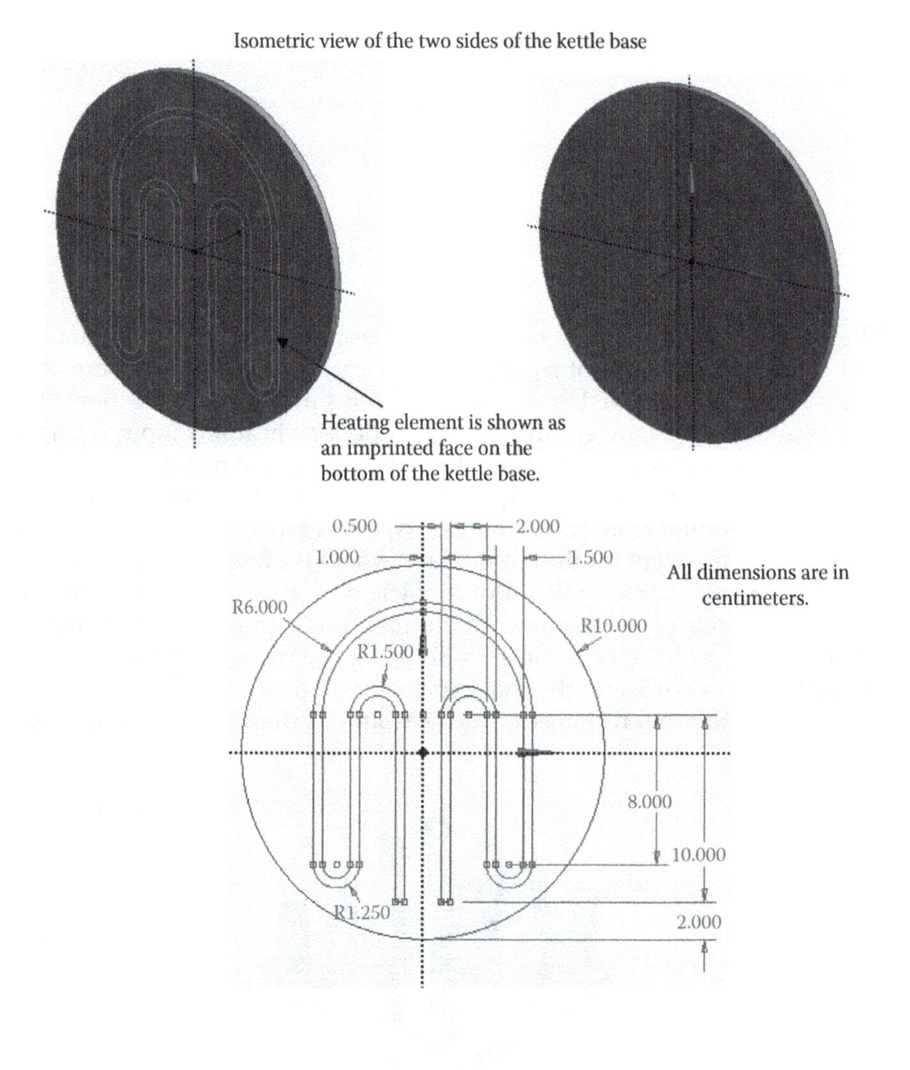

9.8 The air to water heat exchanger shown below is made of copper with thermal conductivity $k = 400$ W/(m K), Young's modulus $E = 100$ GPa, Poisson's ratio $\nu = 0.3$, and thermal expansion coefficient $\alpha = 18 \times 10^{-6}$/°C. The exterior surfaces are in contact with cold water with a film coefficient of 30 W/(m^2°C) and a bulk temperature of 20°C. The interior surfaces are in contact with hot air with a film coefficient of 100 W/(m^2°C) and a bulk temperature of 80°C. (1) Determine the steady-state thermal response of the heat exchanger. (2) Suppose the two annulus faces at the ends of the heat exchanger are fixed. Determine the thermal deformation and stresses induced in the exchanger.

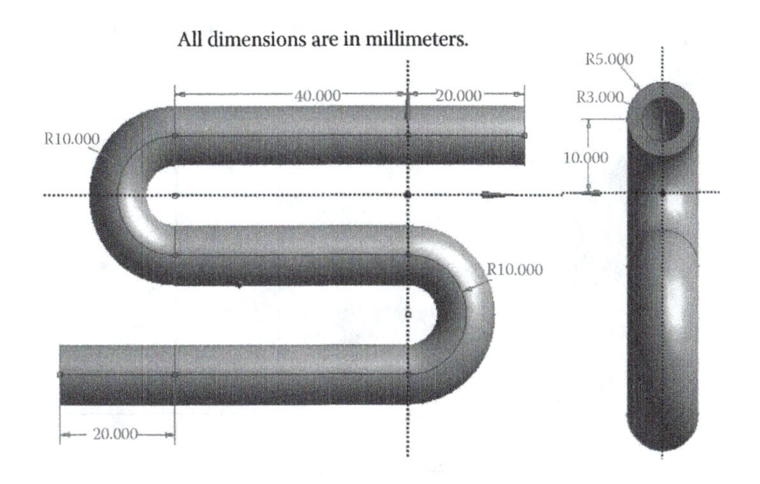

9.9 Thermal expansion mismatch is of great concern in high-density semiconductor devices. The model shown below consists of a copper dye on a silicon substrate. Suppose the dye and substrate are perfectly bounded and the substrate is fixed on the bottom face. Determine the thermally induced deformation and stresses if the dye is assigned a uniform temperature of 60°C and the substrate is assigned a uniform temperature of 40°C. For copper, use Young's modulus $E = 100$ GPa, Poisson's ratio $\nu = 0.3$, and thermal expansion coefficient $\alpha = 18 \times 10^{-6}$/°C, and for silicon, use $E = 200$ GPa, $\nu = 0.27$, and $\alpha = 2.6 \times 10^{-6}$/°C.

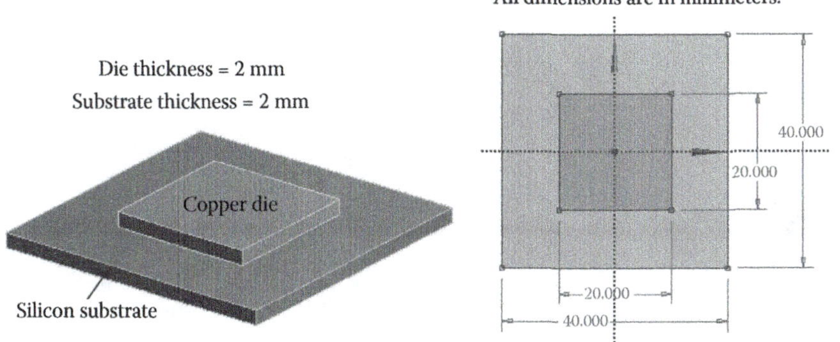

9.10 The following square platform with two arms has a thickness of 1 mm and is made of silicon with a thermal conductivity $k = 130$ W/(m°C), Young's modulus $E = 200$ GPa, Poisson's ratio $\nu = 0.27$, and thermal expansion coefficient $\alpha = 2.6 \times 10^{-6}$/°C. Ignore the effects of convective heat transfer. (1) Determine the temperature distribution in the platform when the temperatures of the upper and lower ends are set uniform at 40°C and 20°C, respectively, as shown below. (2) Suppose the ends of the two arms are fixed. Determine the thermally induced deformation and stresses in the platform.

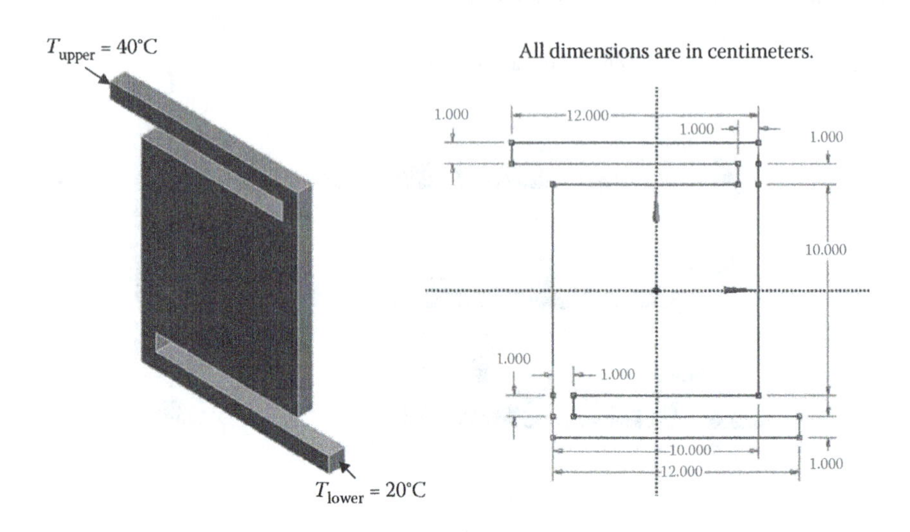

10

Introduction to Fluid Analysis

10.1 Introduction

Flow simulations are widely used in engineering applications ranging from flow around airplane wings and hydraulic turbines to flow in blood vessels and other circulatory systems (see Figure 10.1). We may gain a better understanding of the motion of fluid around objects as well as the fluid behavior in complex circulatory systems by conducting fluid analysis. Computational fluid dynamics (CFD) simulation complements experimental testing, helps reduce cost and turnaround time for design iterations, and has become an indispensable tool whenever practical design involving fluids is required. In this chapter, we will discuss fluid analysis using *ANSYS® Workbench*.

10.2 Review of Basic Equations

We begin by briefly reviewing the fundamentals of fluid mechanics.

10.2.1 Describing Fluid Motion

In fluid dynamics, the motion of a fluid is mathematically described using physical quantities such as the flow velocity $\mathbf{u}$, flow pressure p, fluid density ρ, and fluid viscosity ν. The flow velocity or flow pressure is different at a different point in a fluid volume. The objective of fluid simulation is to track the fluid velocity and pressure variations at different points in the fluid domain.

10.2.2 Types of Fluid Flow

There are many different types of fluid flow. A flow can be compressible ($\rho \neq$ constant) or incompressible ($\rho =$ constant), viscous ($\nu \neq 0$) or inviscid ($\nu = 0$), steady ($\partial \mathbf{u}/\partial t = 0$) or unsteady ($\partial \mathbf{u}/\partial t \neq 0$) and laminar (streamline) or turbulent (chaotic). Furthermore, a fluid can be Newtonian (if the viscosity depends only on temperature and pressure, not on forces acting upon it; in other words, shear stress is a linear function of the fluid strain rate) or non-Newtonian (if the viscosity depends on forces acting upon it, i.e., shear stress is a nonlinear function of the fluid strain rate).

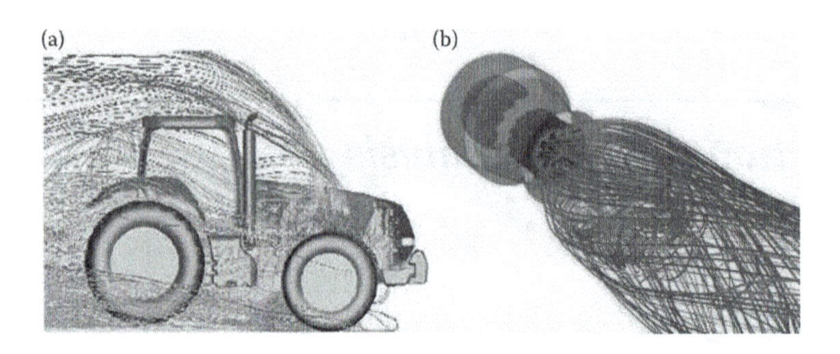

FIGURE 10.1
Examples of CFD: (a) airflow around a tractor (Courtesy ANSYS, Inc., http://www.ansys.com/Industries/Automotive/Application+Highlights/Body), and (b) streamlines inside a combustion chamber (Courtesy ANSYS, Inc., http://www.edr.no/en/courses/ansys_cfd_advanced_modeling_ reacting_flows_and_combustion_in_ans ys_fluent).

10.2.3 Navier–Stokes Equations

For the purpose of this chapter, we limit ourselves to the study of incompressible Newtonian flows. All fluids are compressible to some extent, but we may consider most common liquids as incompressible, whose motion is governed by the following Navier–Stokes (N–S) equations:

$$\frac{\partial \mathbf{u}}{\partial t} = -\mathbf{u}\nabla\mathbf{u} + \nu\nabla^2\mathbf{u} - \frac{\nabla p}{\rho} + \mathbf{f} \tag{10.1}$$

which shows that the acceleration $\partial\mathbf{u}/\partial t$ of a fluid particle can be determined by the combined effects of advection $\mathbf{u}\nabla\mathbf{u}$, diffusion $\nu\nabla^2\mathbf{u}$, pressure gradient $\nabla p/\rho$, and body force f.

The N–S equations can be derived directly from the conservation of mass, momentum, and energy principles. Note that for each particle of a fluid field we have a set of N–S equations. A particle's change in velocity is influenced by how the surrounding particles are pushing it around, how the surrounding resists its motion, how the pressure gradient changes, and how the external forces such as gravity act on it [15].

In 3-D Cartesian coordinates, the N–S equations become:

$$\rho\left(\frac{\partial u}{\partial t} + u\frac{\partial u}{\partial x} + v\frac{\partial u}{\partial y} + w\frac{\partial u}{\partial z}\right) = \nu\left(\frac{\partial^2 u}{\partial x^2} + \frac{\partial^2 u}{\partial y^2} + \frac{\partial^2 u}{\partial z^2}\right) - \frac{\partial p}{\partial x} + f_x \tag{10.2}$$

$$\rho\left(\frac{\partial v}{\partial t} + u\frac{\partial v}{\partial x} + v\frac{\partial v}{\partial y} + w\frac{\partial v}{\partial z}\right) = \nu\left(\frac{\partial^2 v}{\partial x^2} + \frac{\partial^2 v}{\partial y^2} + \frac{\partial^2 v}{\partial z^2}\right) - \frac{\partial p}{\partial y} + f_y \tag{10.3}$$

$$\rho\left(\frac{\partial w}{\partial t} + u\frac{\partial w}{\partial x} + v\frac{\partial w}{\partial y} + w\frac{\partial w}{\partial z}\right) = \nu\left(\frac{\partial^2 w}{\partial x^2} + \frac{\partial^2 w}{\partial y^2} + \frac{\partial^2 w}{\partial z^2}\right) - \frac{\partial p}{\partial z} + f_z \tag{10.4}$$

where u, v, and w are components of the particle's velocity vector $\mathbf{u}$.

In CFD modeling, the N–S equations for particle motion are numerically solved, along with specified boundary conditions, on a 3-D grid that represents the fluid domain to be analyzed. For more details on numerical solution techniques adopted in CFD, please refer to the theory guide in *ANSYS* online documentation.

10.3 Modeling of Fluid Flow

Practical aspects of CFD modeling are discussed next. The topics include fluid domain specification, meshing, boundary condition assignments, and solution visualization.

10.3.1 Fluid Domain

A fluid domain is a continuous region with respect to the fluid's velocity, pressure, density, viscosity, and so on. Figure 10.2 illustrates examples of an internal flow and an external flow. For an internal flow (see Figure 10.2a), the fluid domain is confined by the wetted surfaces of the structure in contact with the fluid. For an external flow (see Figure 10.2b), the fluid domain is the external fluid region around the immersed structure.

10.3.2 Meshing

In CFD analysis, mesh quality has a significant impact on the solution time and accuracy as well as the rate of convergence. A good mesh needs to be fine enough to capture all relevant flow features, such as the boundary layer and shear layer and so on (see Figure 10.3), without overwhelming the computing resources. A good mesh should also have smooth and gradual transitions between areas of different mesh density, to avoid adverse effect on convergence and accuracy.

10.3.3 Boundary Conditions

Appropriate boundary conditions are required to fully define the flow simulation, as the flow equations are solved subject to boundary conditions. The common fluid boundary conditions include the inlet, outlet, opening, wall, and symmetry plane [16]. An inlet condition is used for boundaries where the flow enters the domain. An outlet condition is for where the flow leaves the domain. An opening condition is used for boundaries where the fluid can enter or leave the domain freely. A wall condition represents a solid boundary of

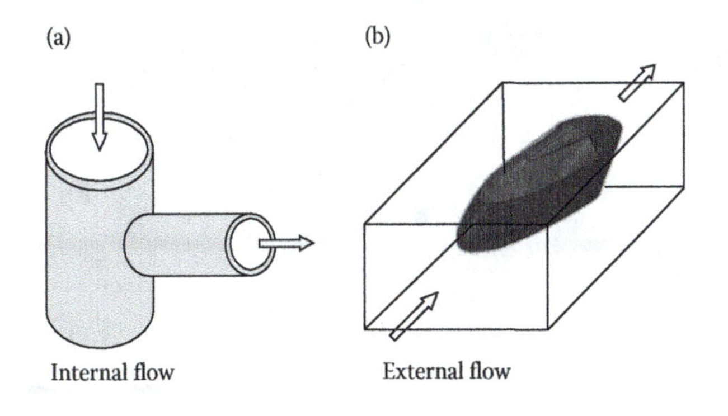

FIGURE 10.2
Flow region definition: (a) internal flow and (b) external flow.

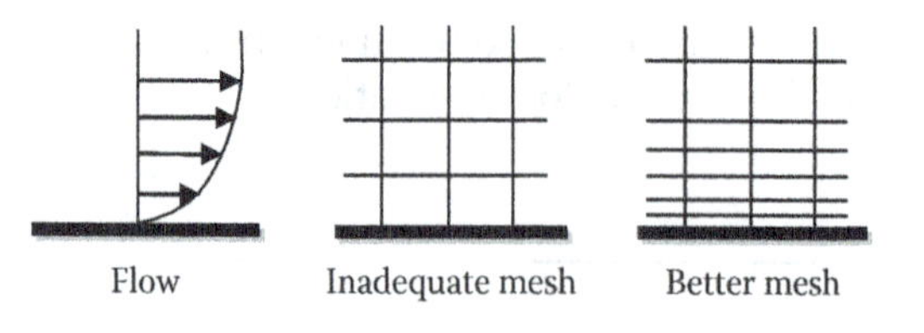

FIGURE 10.3
The resolution of a mesh needs to adequately represent the flow feature.

the flow model. For fixed walls, this typically means a no-slip boundary condition for the flow velocity. A symmetry plane condition is used for planes exhibiting both geometric and flow symmetry.

A pressure boundary is a boundary where the flow pressure is defined. A velocity boundary is a boundary where the velocity profile is given. Figure 10.4 gives two simulated airflow streamline plots for a city block. The boundary conditions for (a) include a velocity inlet on the left edge, a pressure outlet on the right edge, and wall conditions on all buildings (depicted in gray lines). The boundary conditions for (b) include pressure inlets on the four sides of the city block, a pressure outlet at the city's circular center, and wall conditions on all buildings. As we can see from Figure 10.4, different boundary conditions result in different flow patterns. Assigning realistic boundary conditions that truly reflect the flow conditions is crucial in CFD analysis.

10.3.4 Solution Visualization

Once a problem is setup with the fluid domain meshed and the boundary conditions specified, it can be submitted to the solver for computation of a solution. Flow motion can be described by plotting the flow variables on a section plane or in 3-D space. The commonly used plots in CFD are contour plots of pressure, shear strain rate and turbulence kinetic

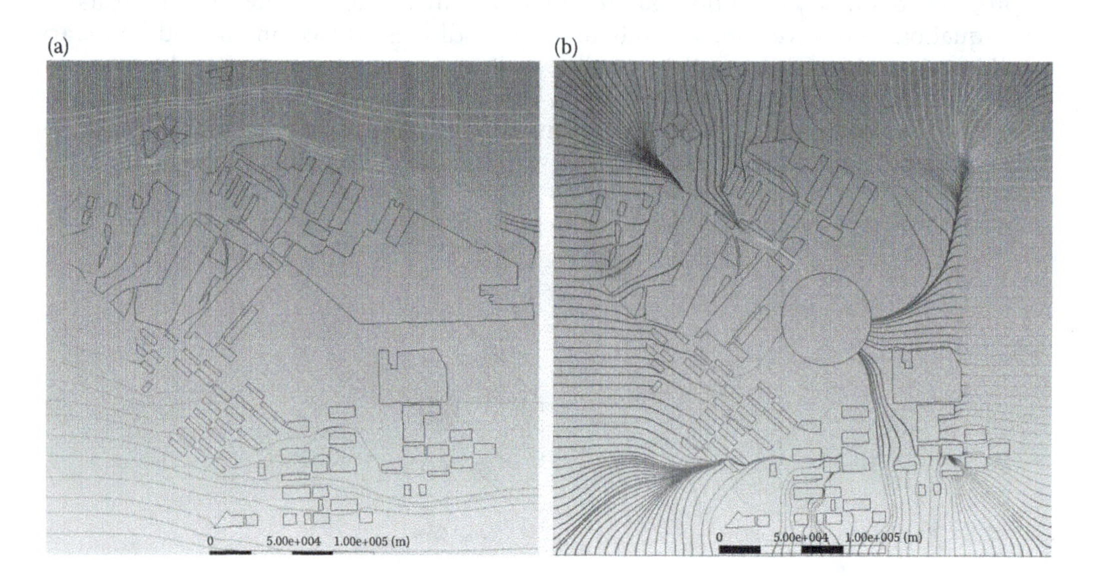

FIGURE 10.4
Airflow over a city block: (a) crosswind blowing from left to right; (b) a circular low-pressure area at the city center.

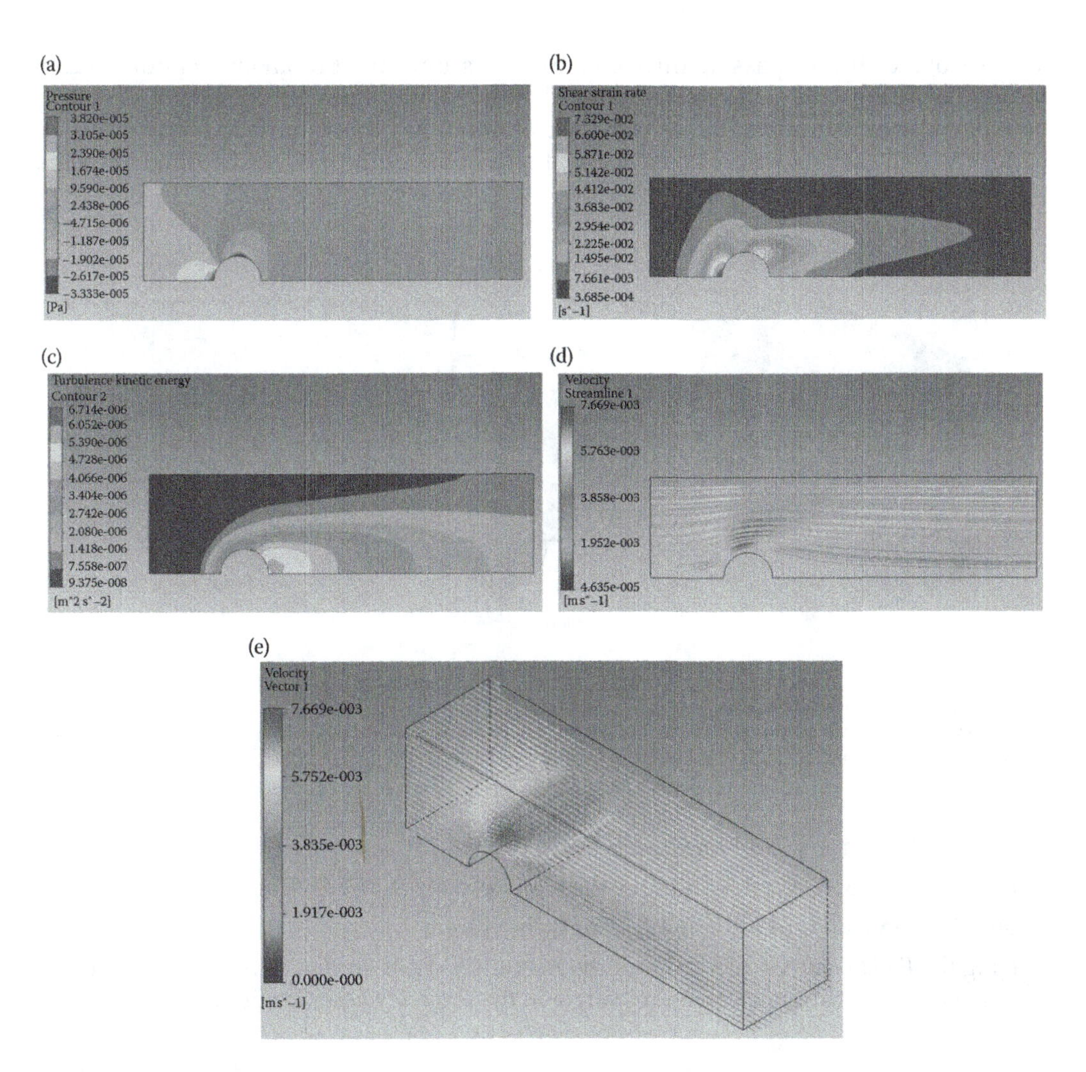

FIGURE 10.5
Solution plots of fluid flow past a single cylinder (a half-model using symmetry). (a) Pressure distribution; (b) shear strain rate distribution; (c) turbulence kinetic energy distribution; (d) streamline; and (e) velocity vector plot.

energy distributions, streamline plot, and vector plot of velocity field. Figure 10.5 shows the related solution plots of a flow passing over a single cylinder. A symmetric half-model is used in the example, with the flow pressure, velocity, shear stress, turbulence intensity, and particle trajectories plotted as shown above.

10.4 Case Studies with *ANSYS Workbench*

Problem Description: The aerodynamic performance of vehicles can be improved by utilizing computational fluid dynamics simulation. In this case study, we conduct fluid analysis of the air flow passing through a truck. Assume air at room temperature of 25°C for the flow field with an air velocity of 40 km/h blowing from left to right. Use nonslip boundary conditions along the walls of the truck and the ground surface. Find the airflow pattern as well as the pressure and velocity distributions of the flow field around the truck.

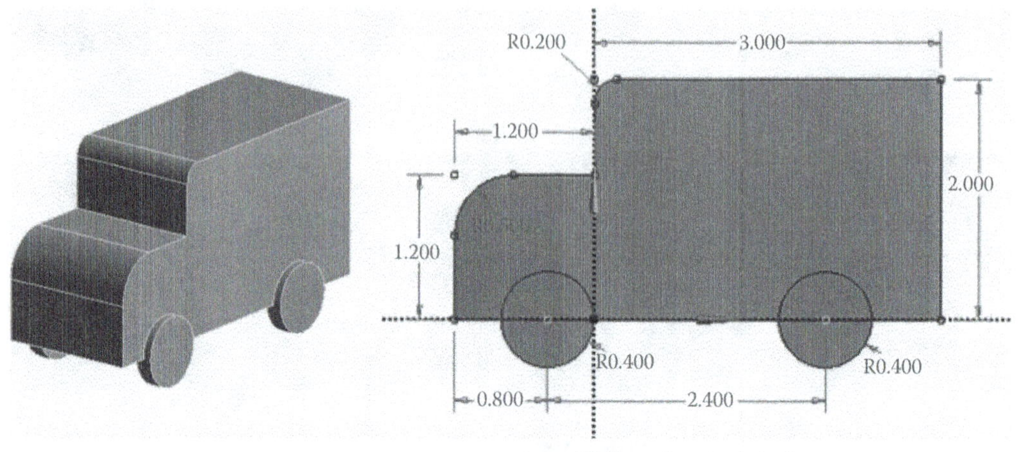

Flow field: Incompressible air at 25°C.
Boundary conditions: Air is blowing from left to right at a velocity of 40 km/h.
Nonslip boundaries along the walls of the truck and the ground surface.

Step 1: Start an *ANSYS Workbench* Project

Launch *ANSYS Workbench* and save the blank project as *"CFD_Truck.wbpj."*

Step 2: Create a *Fluid Flow (CFX)* Analysis System

Drag the *Fluid Flow (CFX)* icon from the *Analysis Systems Toolbox* window and drop it inside the highlighted green rectangle in the *Project Schematic* window.

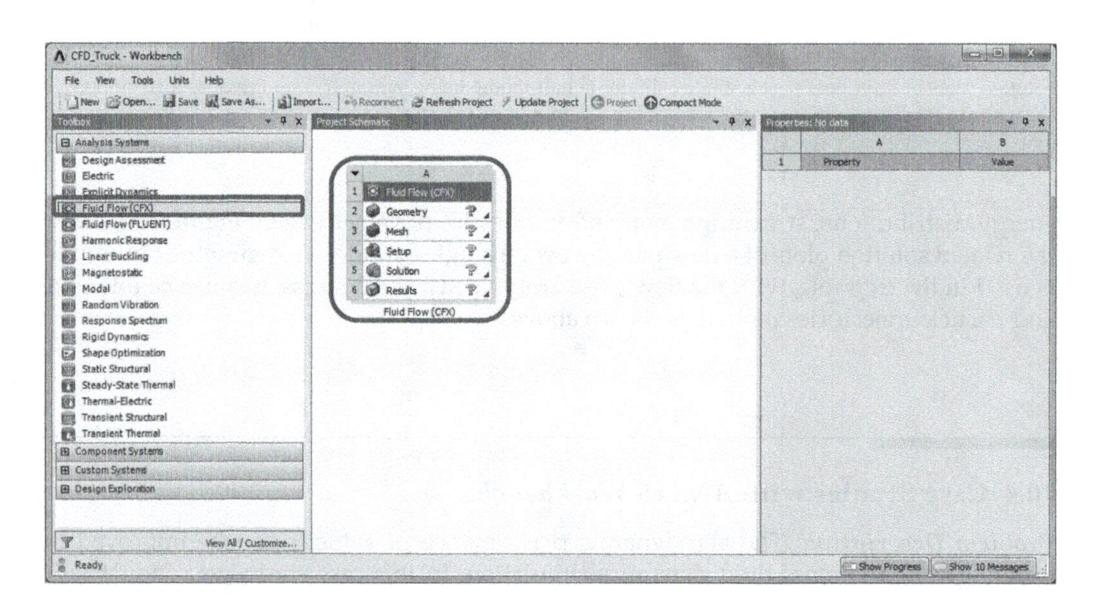

Step 3: Launch the *DesignModeler* Program

Double-click the *Geometry* cell to launch *DesignModeler,* and select *"Meter"* in the *Units* pop-up window.

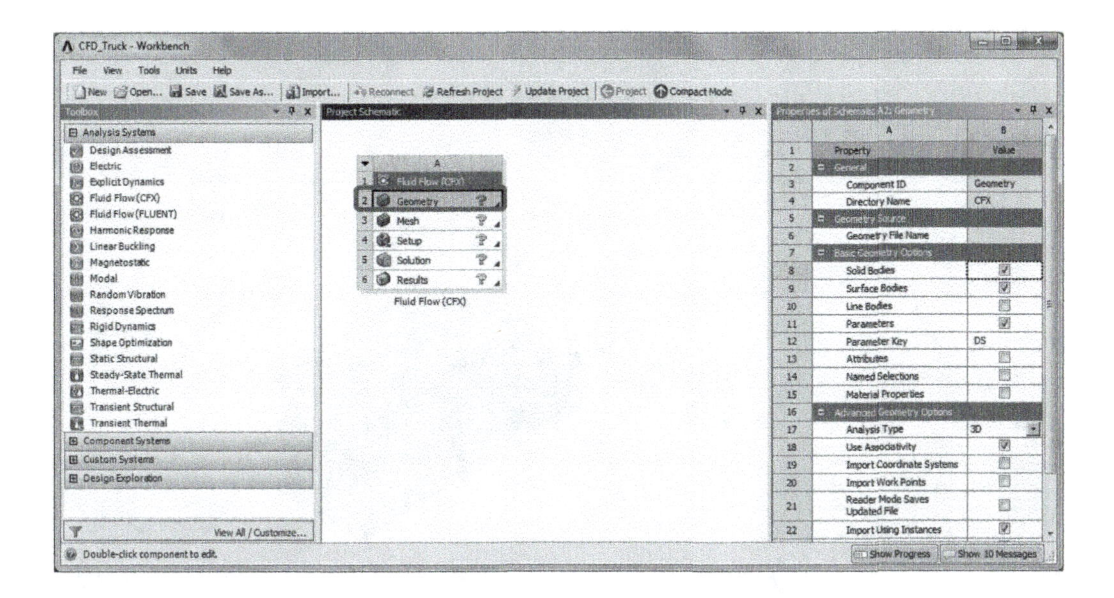

Step 4: Create a Truck Model

Create a Sketch—Click on *Sketching*. Select *Draw* and draw a 2D profile as shown below.

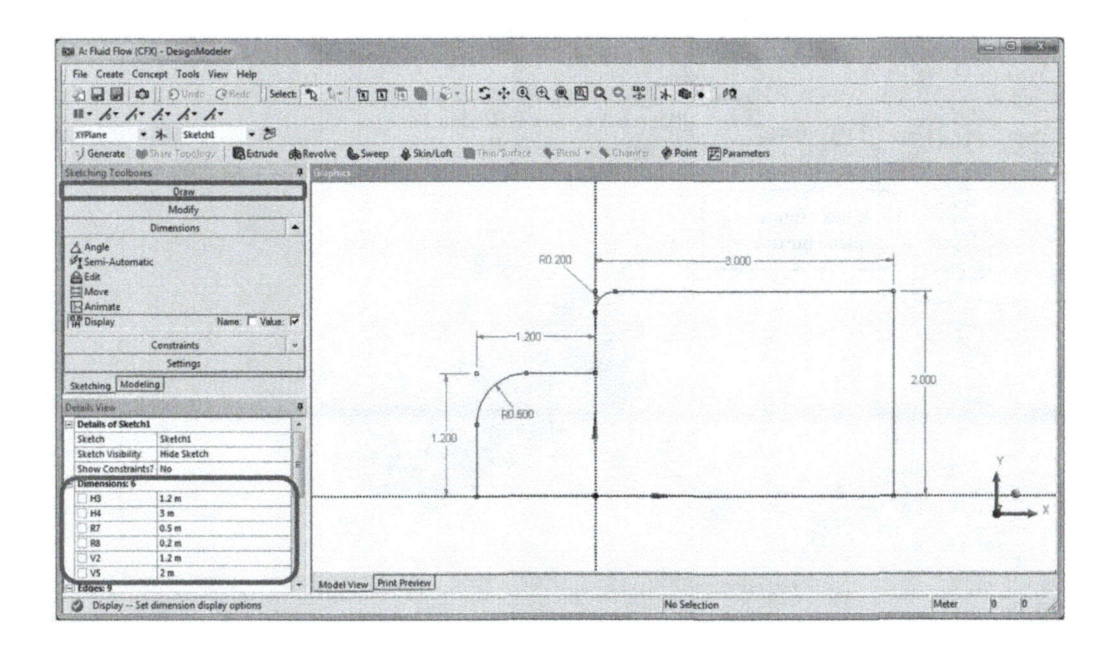

Create an Extruded Body—Switch to the *Modeling* tab and click on *Extrude*. In the *Details of Extrude1*, use *Sketch1* as the *Base Object* and change the *Direction* to *Both-Symmetric*. Set the extrusion depth to *0.8 m* in the field of *FD1, Depth,* and click *Generate* to create a solid body.

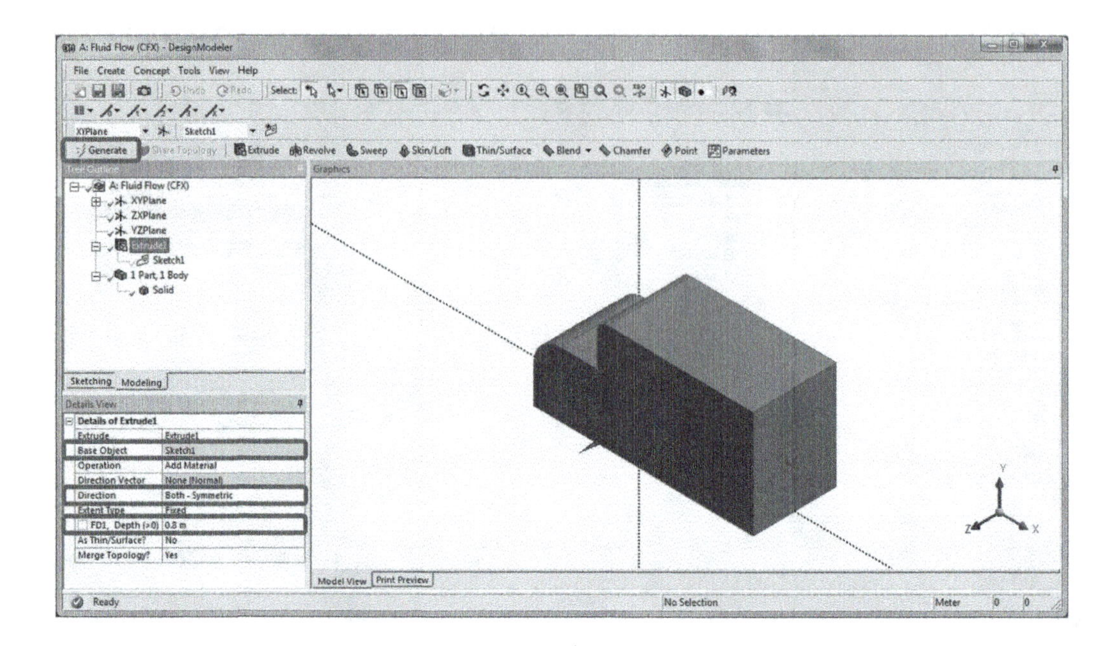

Create Sketch of Two Wheels—Click on the New Plane button to create a new plane (Plane4) on the front face of the truck. Then click on the New Sketch button to create a new sketch (Sketch2) on the plane as shown below.

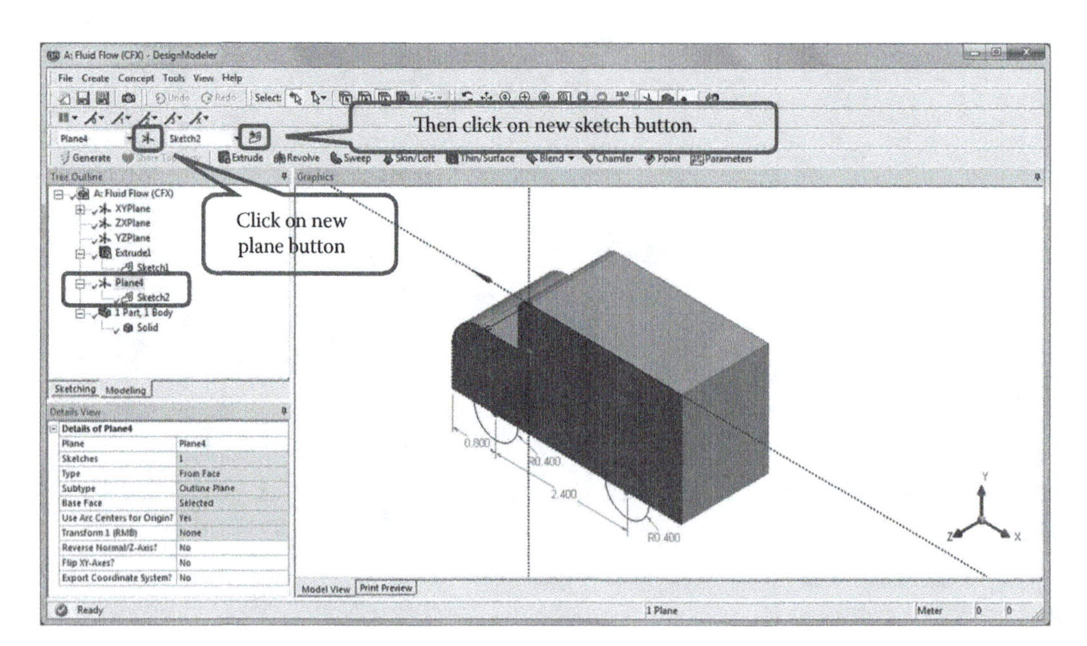

Create Two Solid Wheels—Switch to the *Modeling* tab and click on *Extrude*. In the *Details of Extrude2*, use *Sketch2* as the *Base Object* and change the *Direction* to *Both-Symmetric*. Set *FD1, Depth* to *0.1 m*. Click *Generate* to create two wheels.

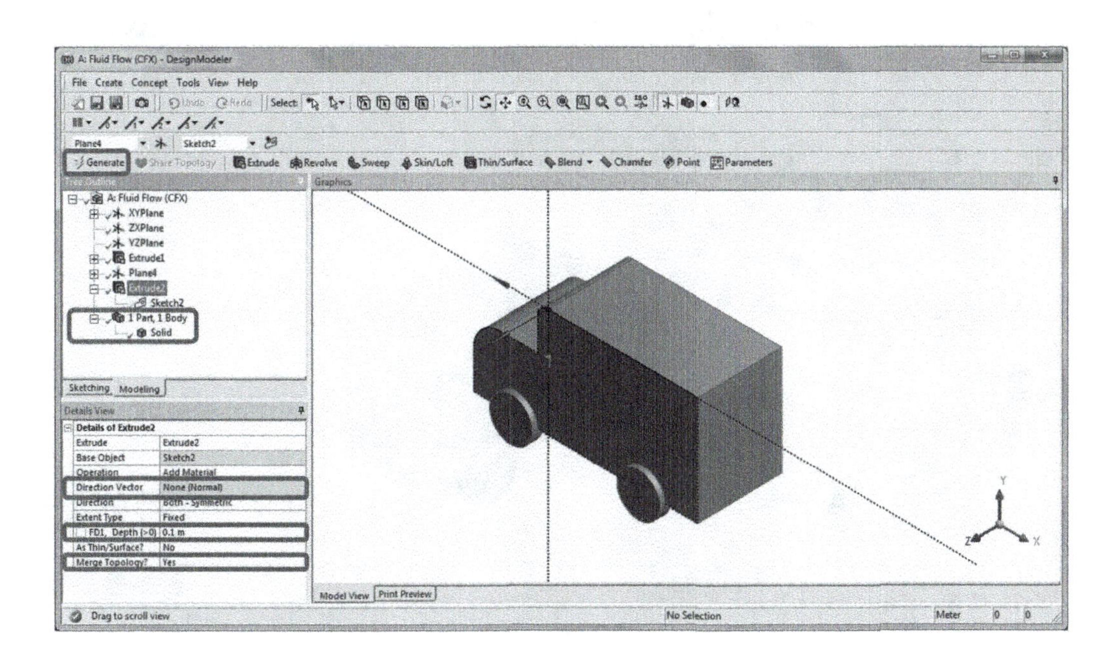

Create the Other Two Wheels—Repeat the previous two steps to create two wheels on the other side of the truck model as shown below.

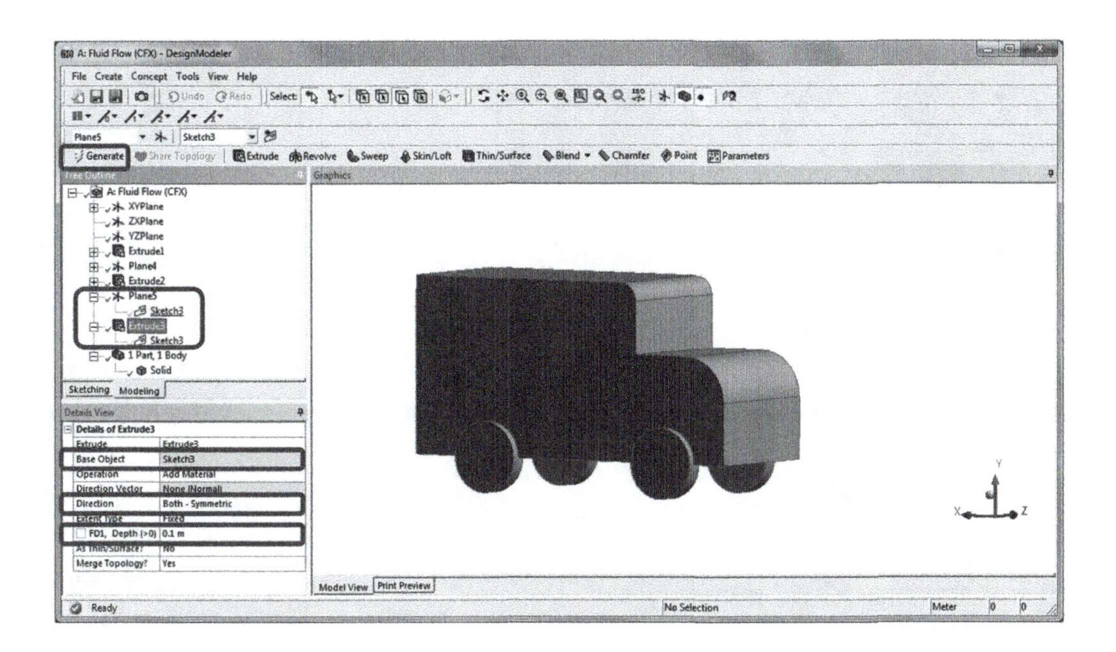

Step 5: Create a Box Enclosure

Select *Enclosure* from the drop-down menu of *Tools* to create a box enclosure.

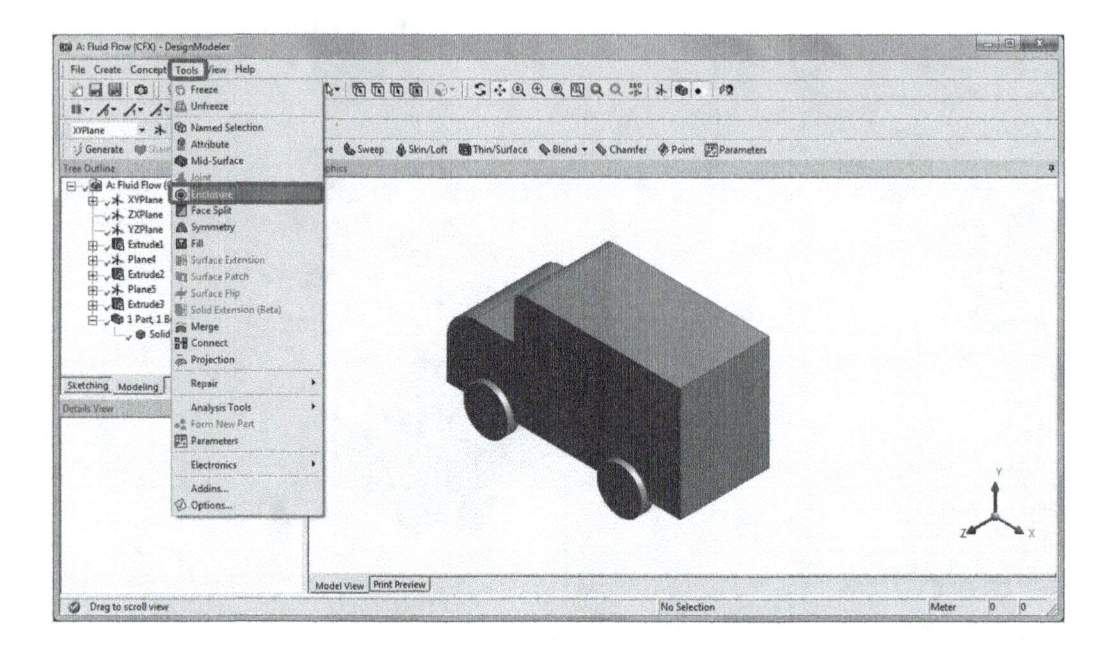

In the *Details of Enclosure1*, enter the cushion dimensions as shown below, and click *Generate*.

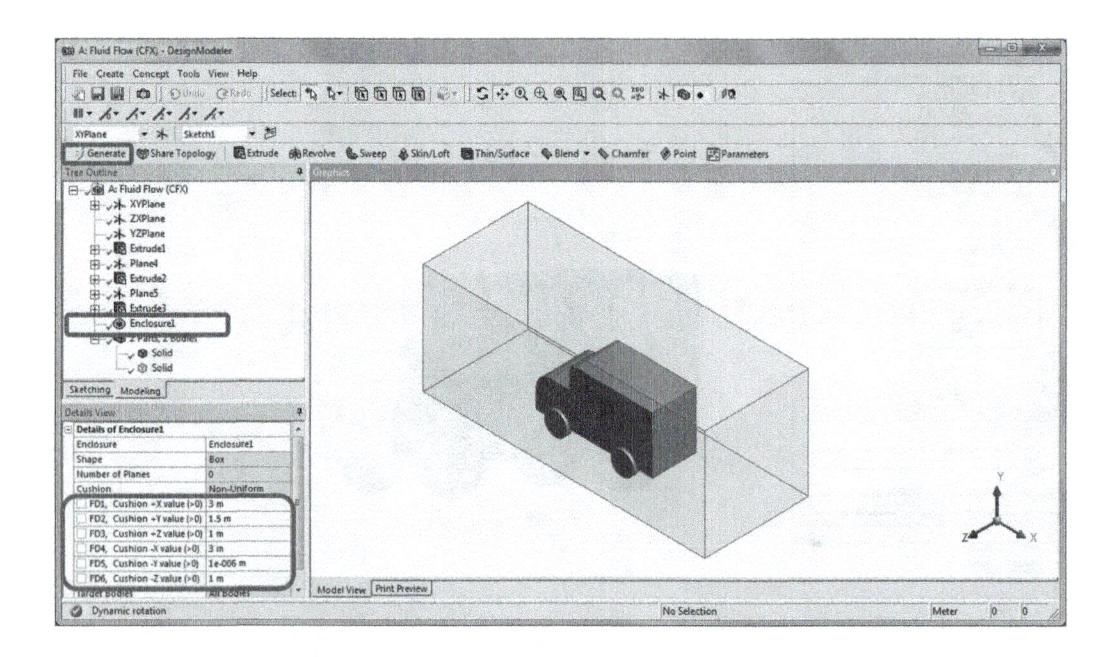

In the *Tree Outline*, right click on the first *Solid* body, and select *Suppress Body*. The first solid body represents the truck. When suppressed, the truck body will not be meshed later on.

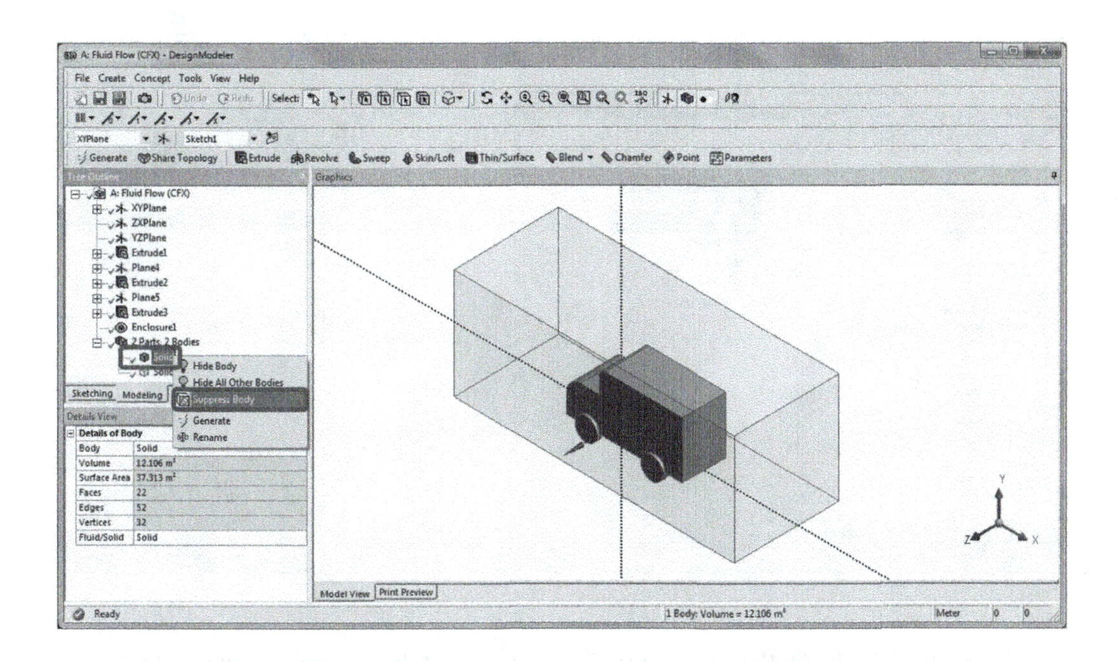

From the *Tree Outline* click on the second *Solid* body, which represents the enclosing fluid domain. Ensure that *Fluid* is selected for the *Fluid/Solid* field in the *Details of Body*.

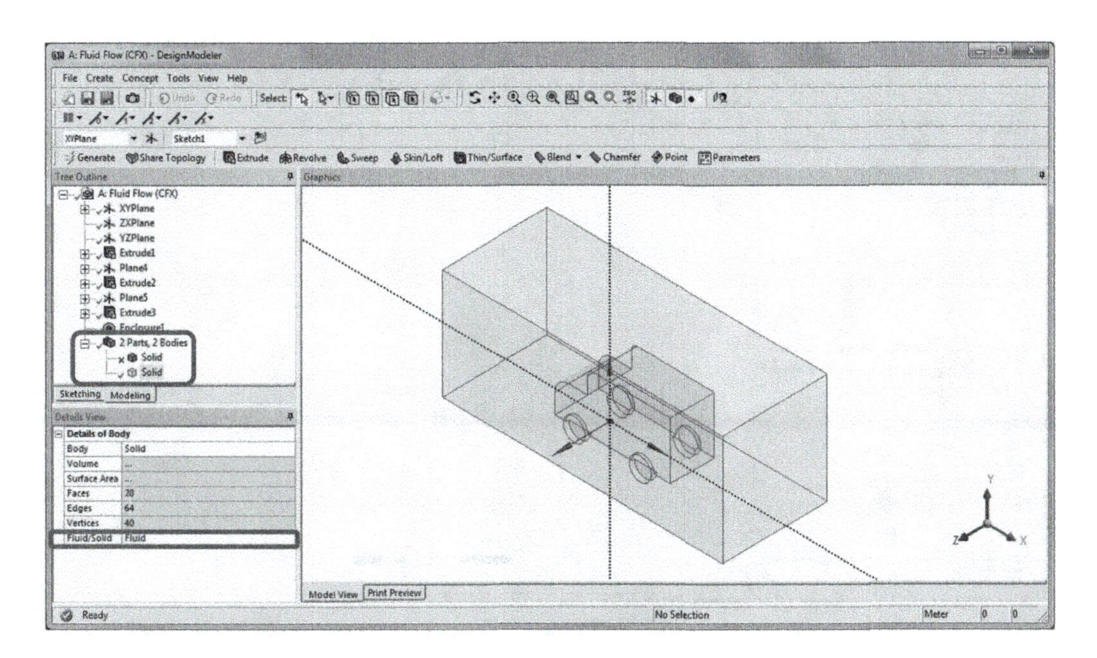

Step 6: Generate Mesh

Double-click on the *Mesh* cell to launch the *Meshing* program.

Click on *Mesh* in the *Outline*. In the *Details of "Mesh,"* set the *Relevance Center* to *Medium* under *Sizing*, and the *Element Midside Nodes* to *Dropped*. This helps reduce the total number of nodes to an acceptable level not exceeding the requested resources of educational licenses. Click on *Update* to generate the mesh.

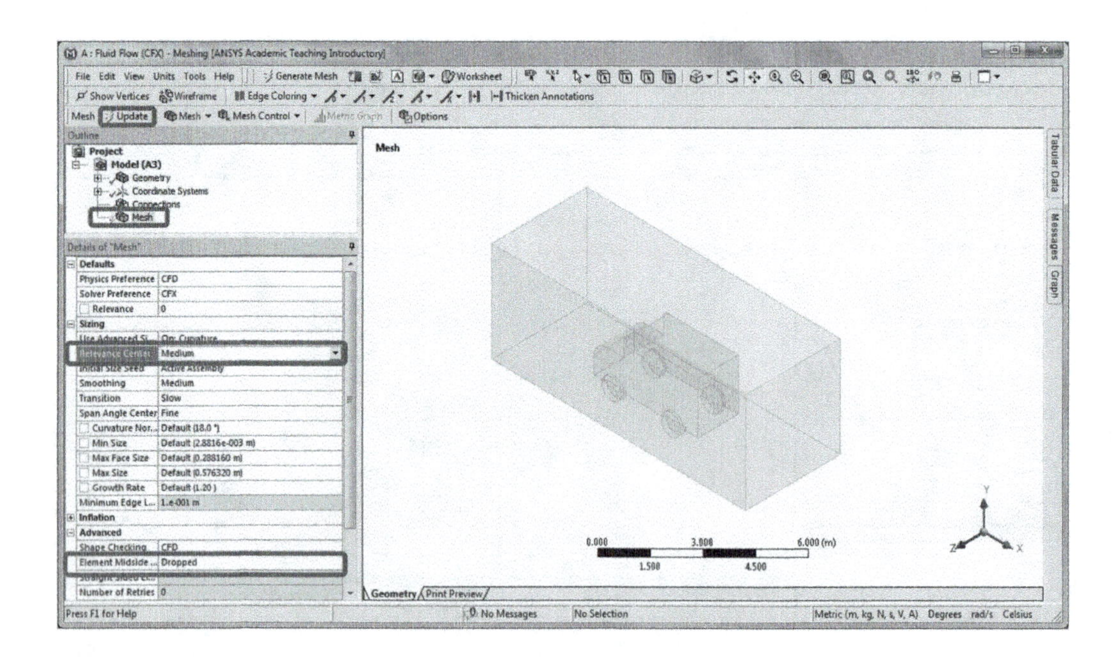

The following figure shows the generated mesh for the fluid domain.

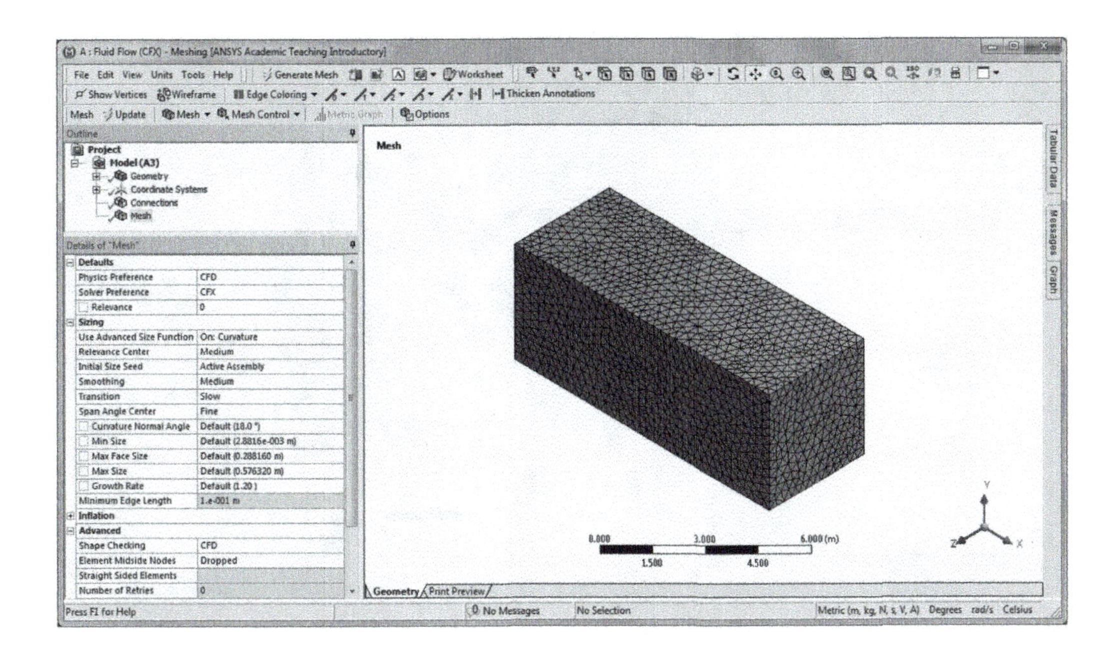

Step 7: Set Up Flow Domain

Double-click on the *Setup* cell to launch the *Fluid Flow (CFX)—CFX-Pre* program.

Double-click on *Analysis Type* in the *Outline* tree.

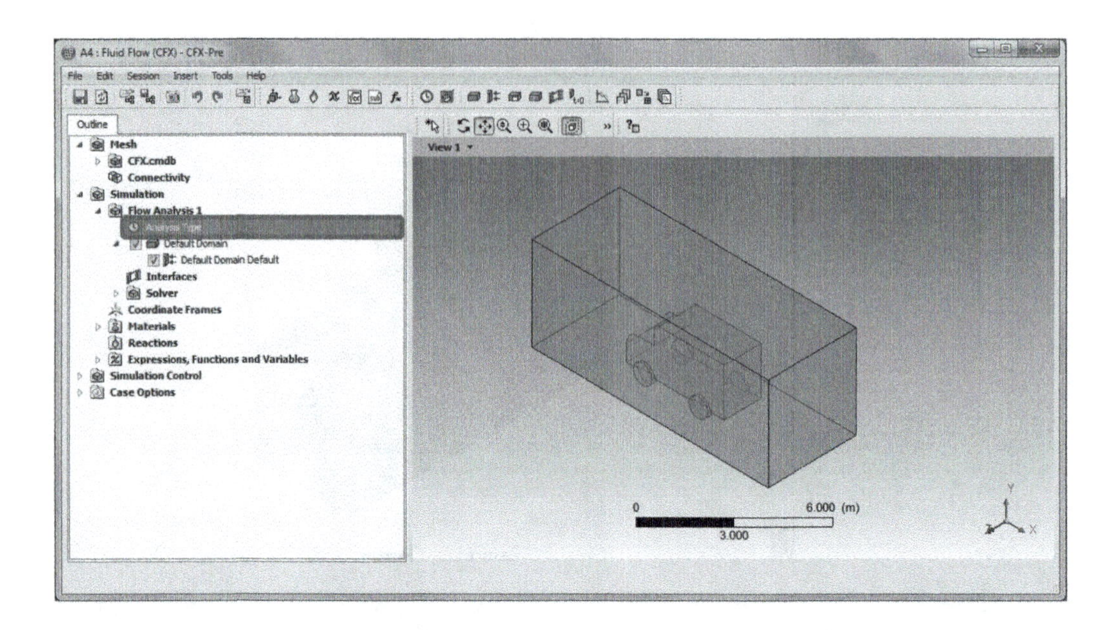

In the *Details of Analysis Type in Flow Analysis 1*, set the *Analysis Type* to *Steady State*. Click *Ok* to exit the menu.

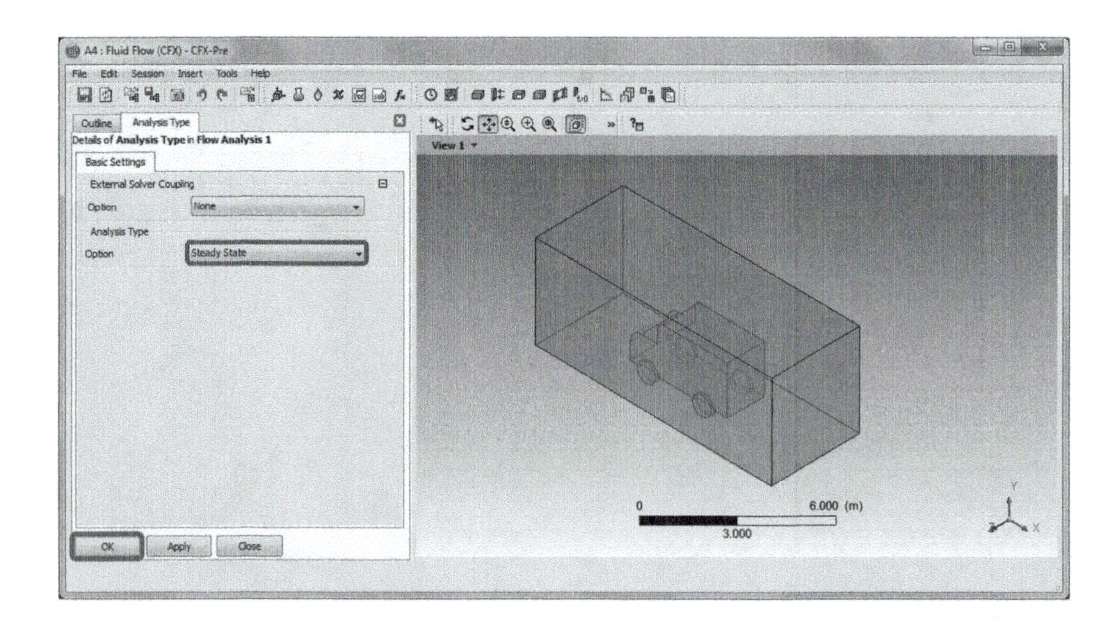

Next, double-click on *Default Domain* in the *Outline* tree.

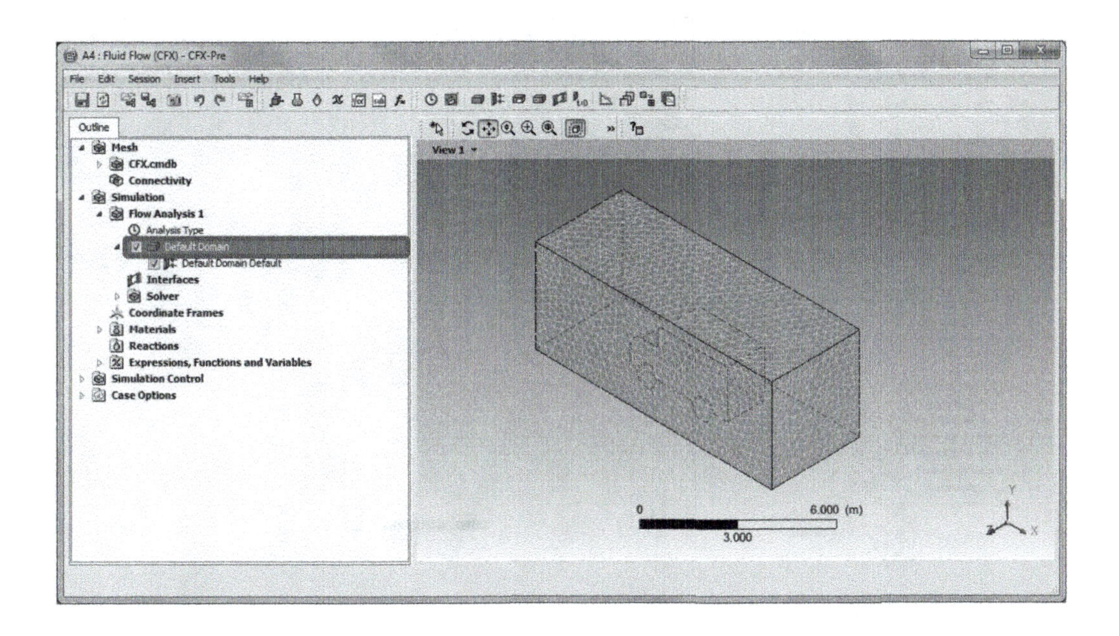

In the *Details of Default Domain in Flow Analysis 1,* set the *Domain Type* to *Fluid Domain, Material* to *Air at 25°C,* and *Reference Pressure* to *1[atm].* Click *Ok* to exit the menu.

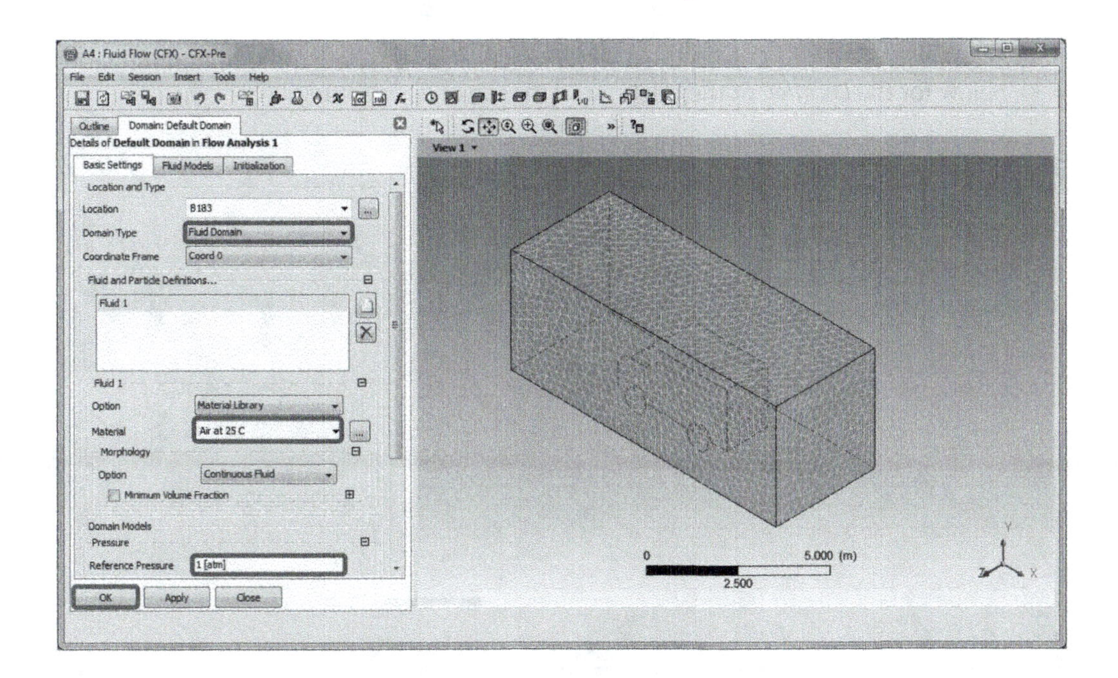

Step 8: Set Up Boundary Conditions

Add Inlet—Right-click on *Default Domain* in the *Outline*. Select *Insert* and then *Boundary*.

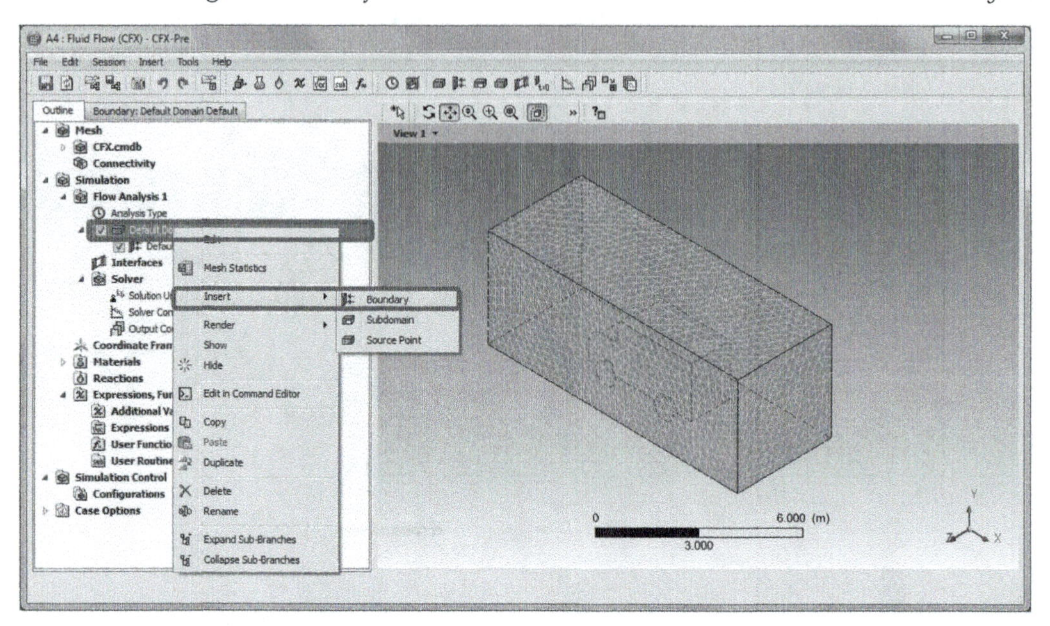

Change the *Name* to *Inlet* in the pop-up menu shown below. Click *Ok*.

Insert Boundary

Name Inlet

OK Cancel

In the *Details of Inlet in Default Domain in Flow Analysis 1*, select the face highlighted below for the *Location* of the inlet under the tab of *Basic Settings*.

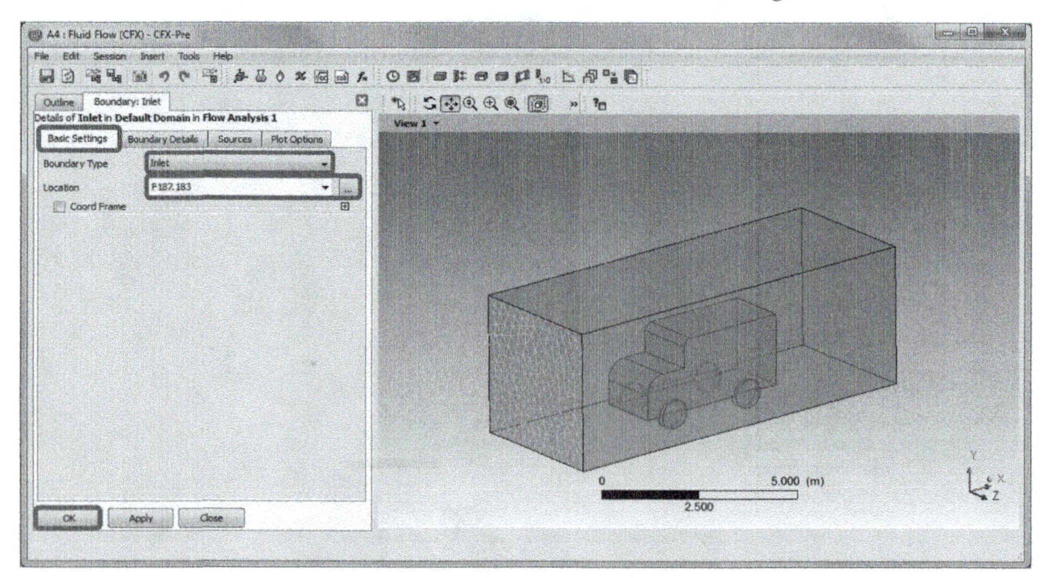

Click on the *Boundary Details* tab, and set the *Normal Speed* to *40 [Km Hr ^ –1]*. Click **OK**.

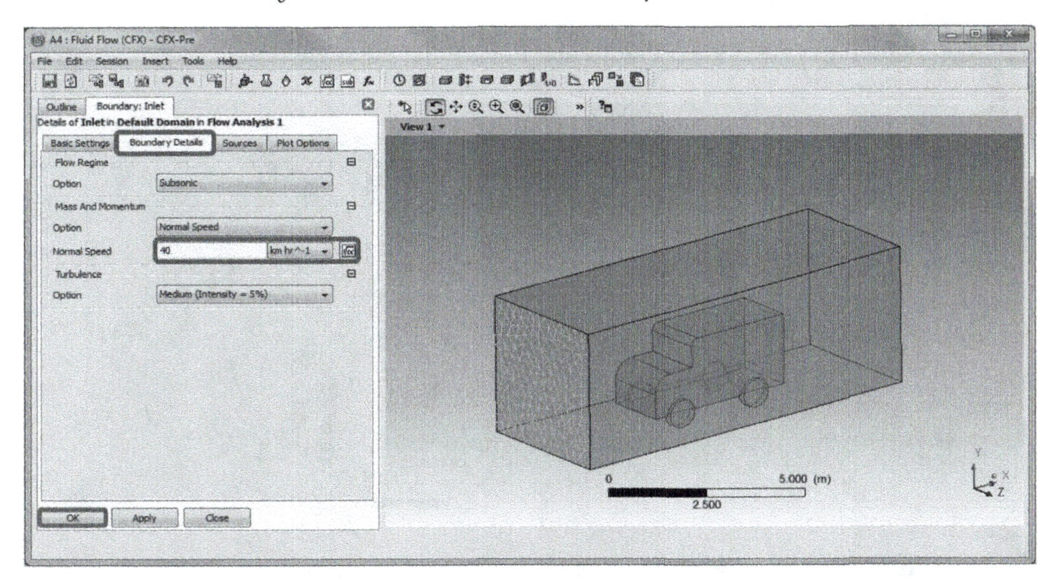

Add Outlet—Right-click on *Default Domain* in the *Outline*. Select *Insert* and then *Boundary*. In the pop-up menu shown below, set the *Name* to *Outlet*. Click *Ok*.

In the *Details of Outlet in Default Domain in Flow Analysis 1*, select the face highlighted below for the *Location* of the outlet under the tab of *Basic Settings*.

Click on the *Boundary Details* tab, and set the *Relative Pressure* to *0 [Pa]*. Click *Ok*.

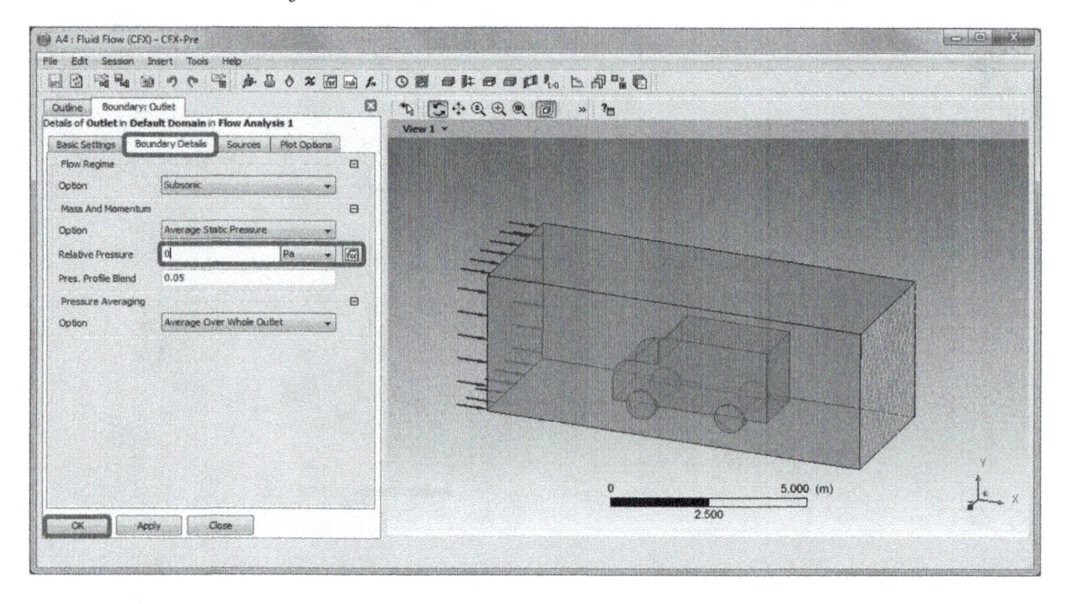

Add Wall—Right-click on *Default Domain* in the *Outline*. Select *Insert* and then *Boundary*. In the pop-up menu shown below, set the *Name* to *Wall*. Click *Ok*.

In the *Details of Wall in Default Domain in Flow Analysis 1*, click on the button to the right of the Location field under the tab of *Basic Settings*.

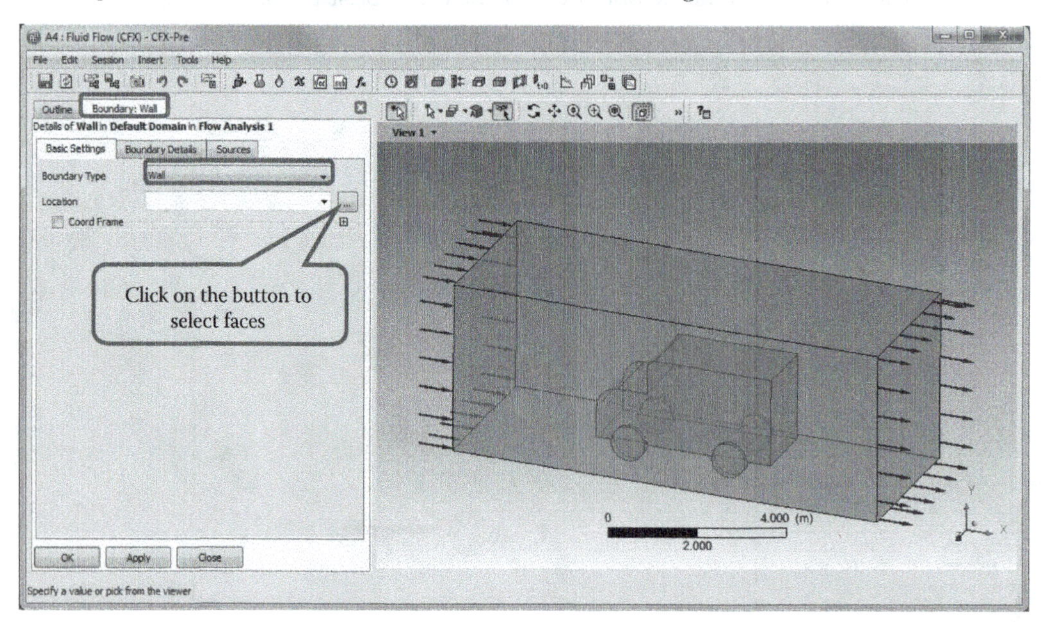

In the pop-up menu of *Selection Dialog*, select multiple faces (all surfaces of the truck and the ground surface) with Shift-Click. Click *Ok* to accept the selection.

Selection Dialog
F184.183
F185.183
F186.183
F187.183
F188.183
F189.183
F190.183
F191.183
F192.183
F193.183
F194.183
F195.183
F196.183
F197.183
F198.183
F199.183
F200.183
F201.183
F202.183
F203.183
F204.183
F205.183
F206.183
F207.183
F208.183
F209.183
F210.183
F211.183
OK Cancel

Switch to the tab of *Boundary Details*, select *No Slip Wall* as the *Option* for *Mass and Momentum*, and *Smooth Wall* for *Wall Roughness*. Click *Ok* to exit the menu.

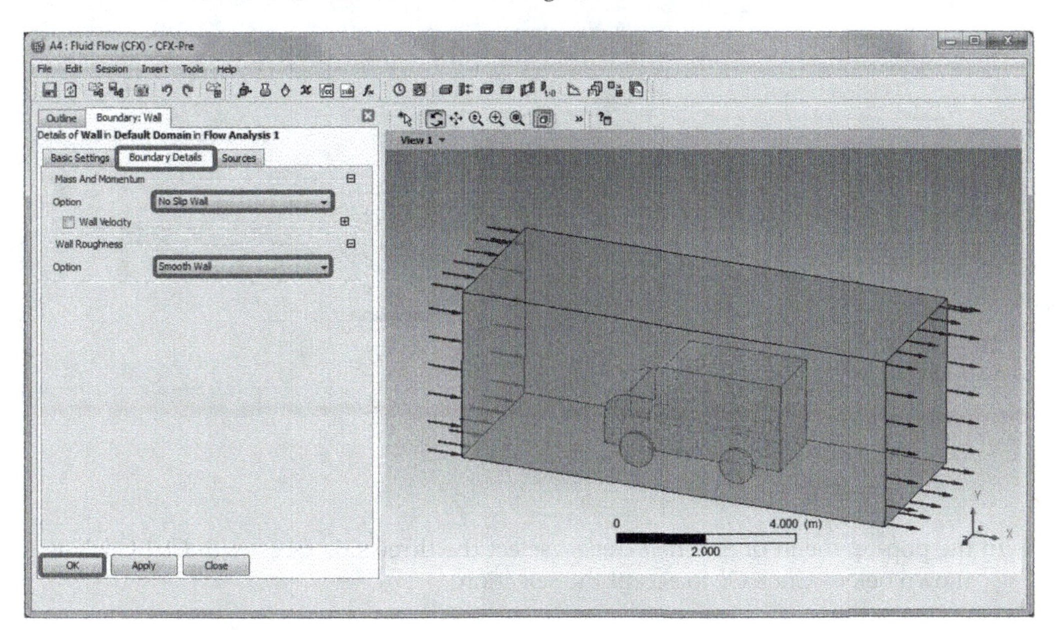

Add Opening—Right-click on *Default Domain* in the *Outline*. Select *Insert* and then *Boundary*. In the pop-up menu shown below, set the *Name* to *Opening*. Click *Ok*.

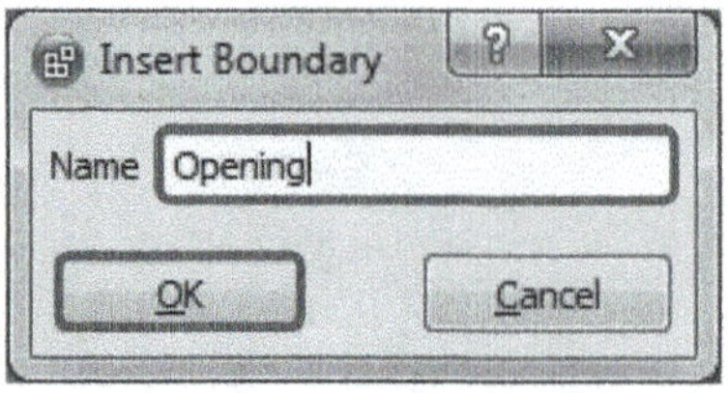

In the *Details of Opening in Default Domain in Flow Analysis 1*, click on the button to the right of the Location field under the tab of *Basic Settings*.

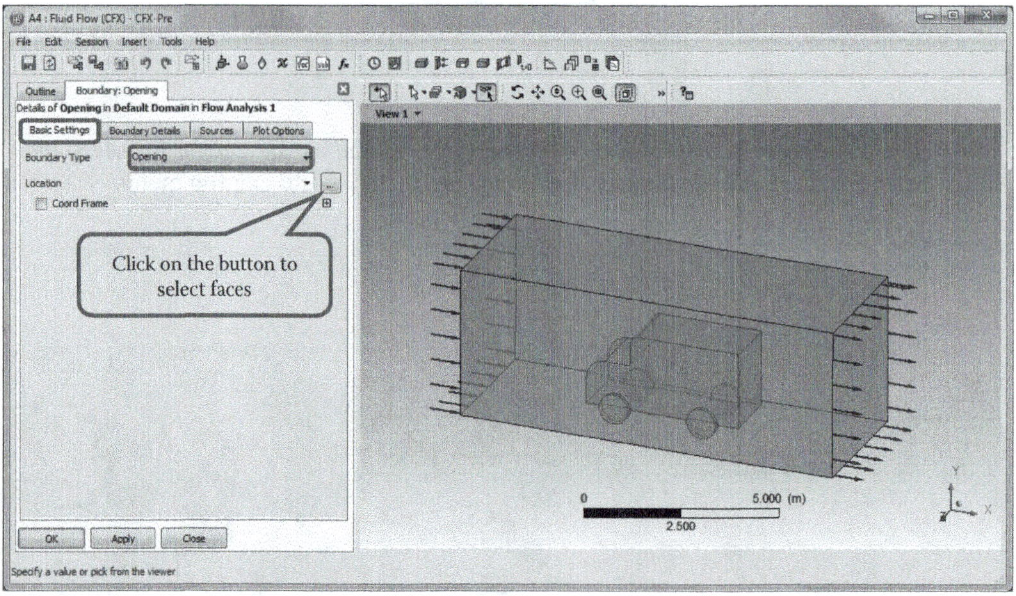

In the pop-up menu of *Selection Dialog*, select the three side faces with Ctrl-Click as shown below. Click *Ok* to accept the selection.

Selection Dialog

F184.183
F185.183
F186.183
F187.183
F188.183
F189.183
F190.183
F191.183
F192.183
F193.183
F194.183
F195.183
F196.183
F197.183
F198.183
F199.183
F200.183
F201.183
F202.183
F203.183
F204.183
F205.183
F206.183
F207.183
F208.183
F209.183
F210.183
F211.183

OK Cancel

Switch to the tab of *Boundary Details*, set the *Relative Pressure* to *0 [Pa]*. Click *Ok* to exit.

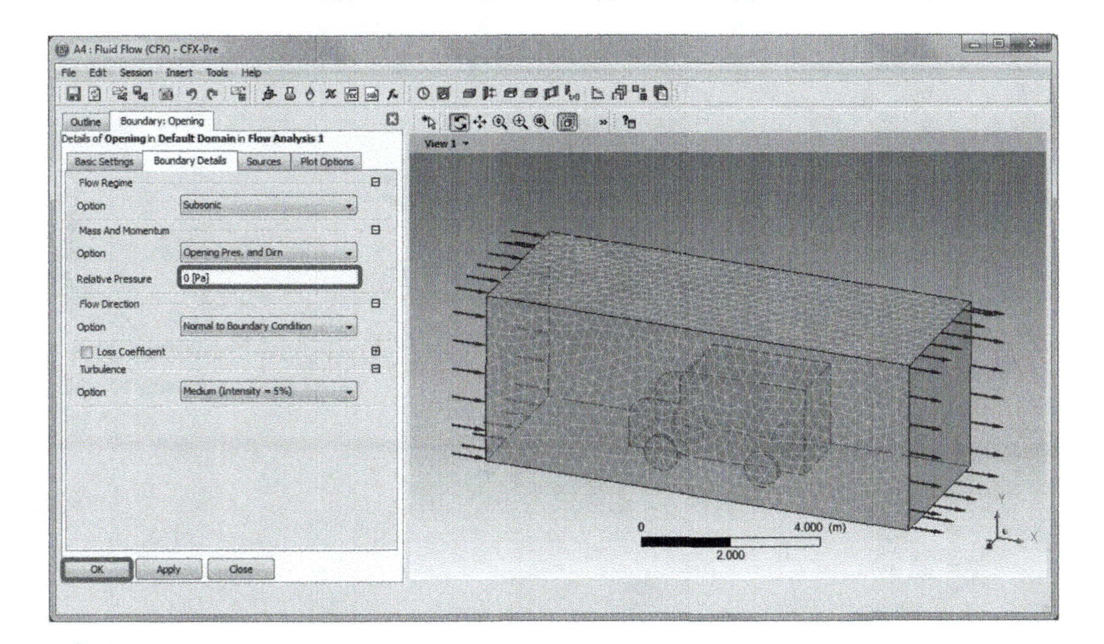

After completion, the finished domain boundaries should look like the following. Exit *CFX-Pre.*

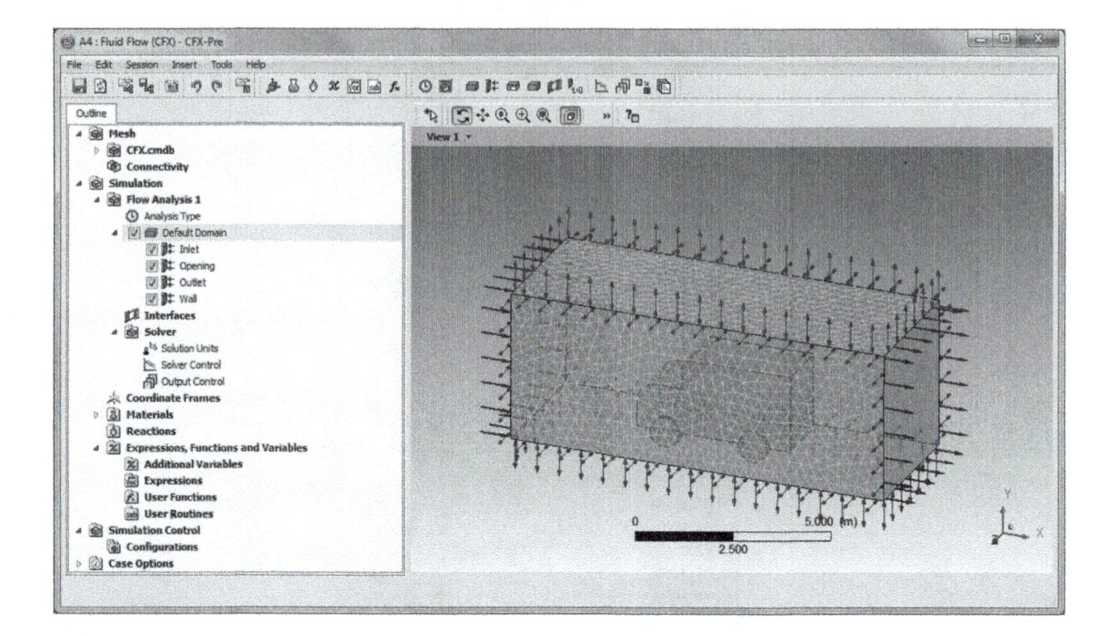

Step 9: Start a Solution

Double-click on the *Solution* cell to obtain a solution of the fluid flow problem.

In the pop-up menu of *Define Run*, click the *Start Run* button to launch the *CFX-Solver*.

When a menu pops up to indicate that the *Solver Run* has finished normally, click *Ok*. Then exit the *CFX-Solver Manager* program.

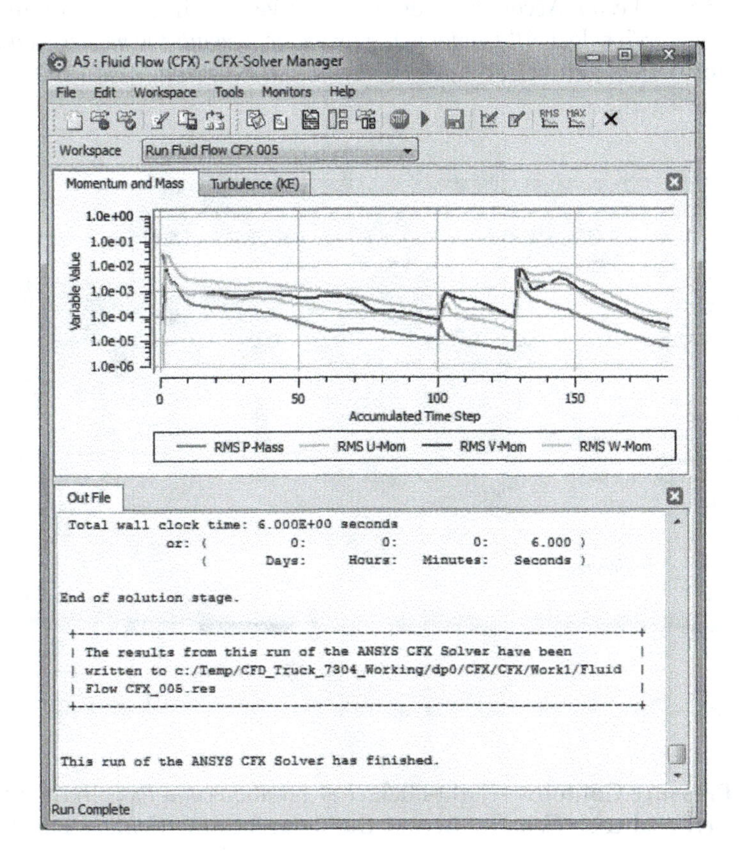

Step 10: Retrieve Results

Double-click on the *Results* cell to visualize the results.

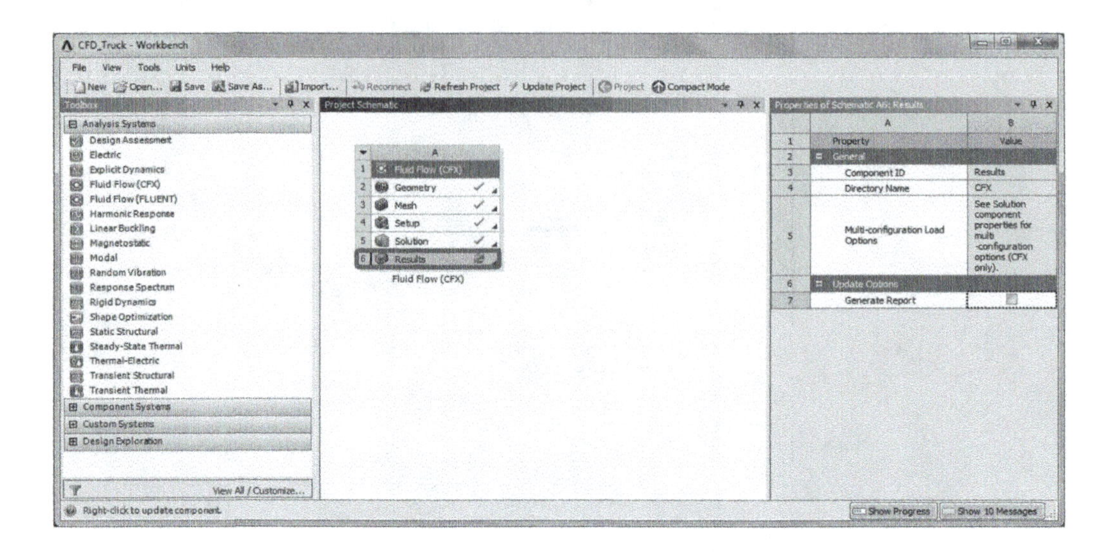

Plot Velocity Vector—Right-click *User Locations and Plots* in the *Outline,* and select *Insert* and then *Vector.* Accept the default name *Vector 1* in the pop-up menu. In the *Details of Vector 1,* select *All Domains* for *Domains, Default Domain* for *Locations* and *Velocity* for *Variable,* and click *Apply.* Check *Default Legend View 1* in the *Outline* to display the legend.

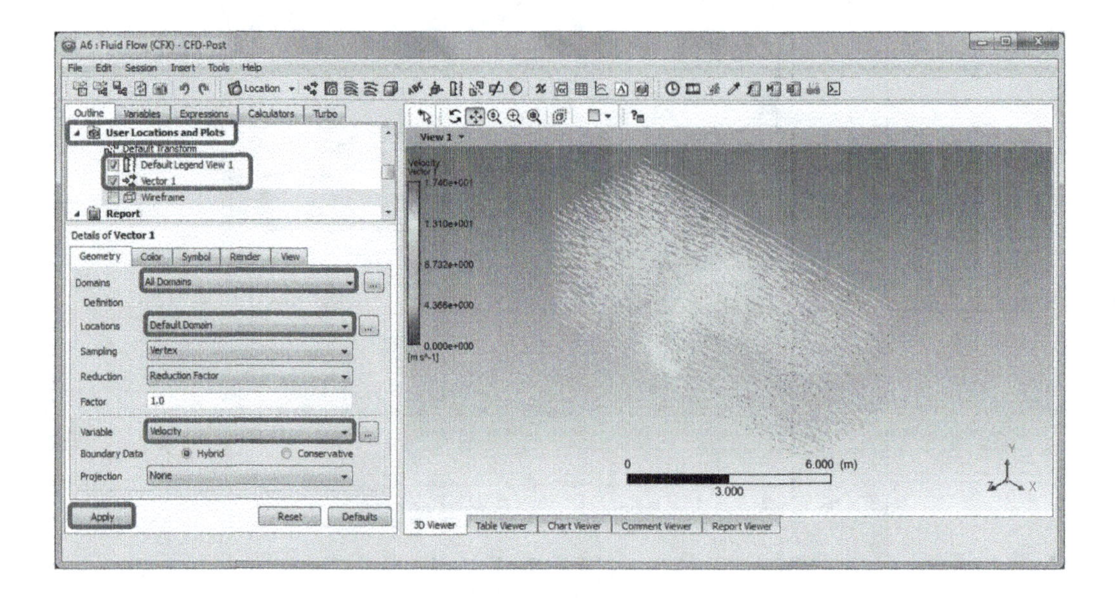

Plot Wall Pressure Contour—Right-click *User Locations and Plots* in the *Outline,* and select *Insert* and then *Contour.* Accept the default name *Contour 1* in the pop-up menu. In the *Details of Contour 1*, select *All Domains* for *Domains, Wall* for *Locations* and *Pressure* for *Variable*, and click *Apply*. Check *Default Legend View 1* and *Contour1* in the *Outline* to view the pressure distribution on the wall.

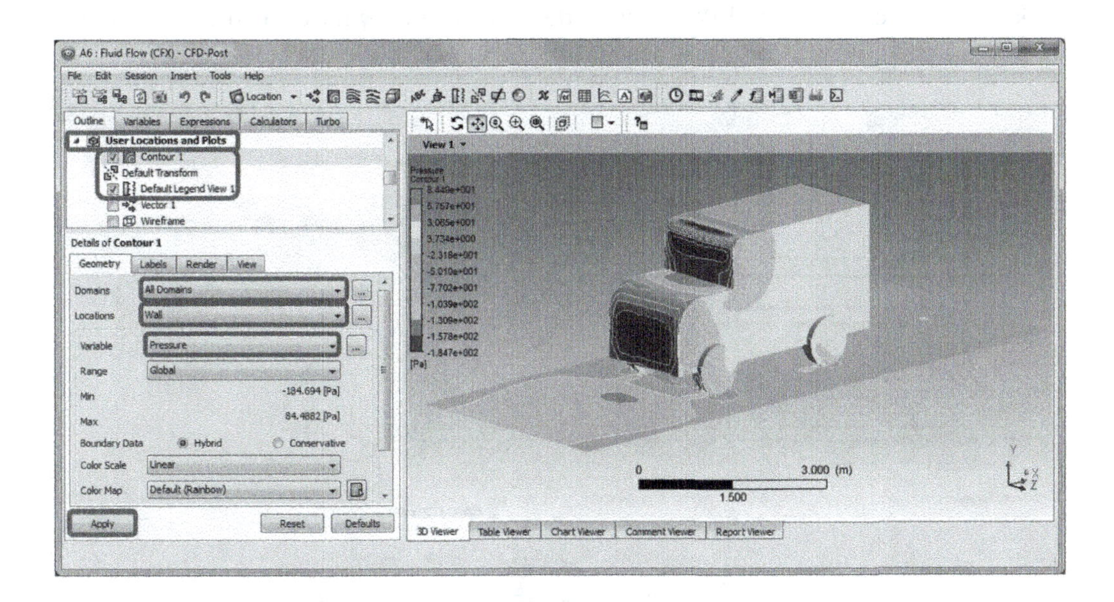

Plot Streamline—Right-click *User Locations and Plots* in the *Outline*, and select *Insert* and then *Streamline*. Accept the default name *Streamline 1* in the pop-up menu. In the *Details of Streamline 1*, select *3D Streamline* for *Definition*, *All Domains* for *Domains*, *Inlet* for *Start From*, and *Velocity* for *Variable*, and click *Apply*. Check *Default Legend View 1*, *Streamline 1*, and *Wireframe* in the *Outline* to view streamlines.

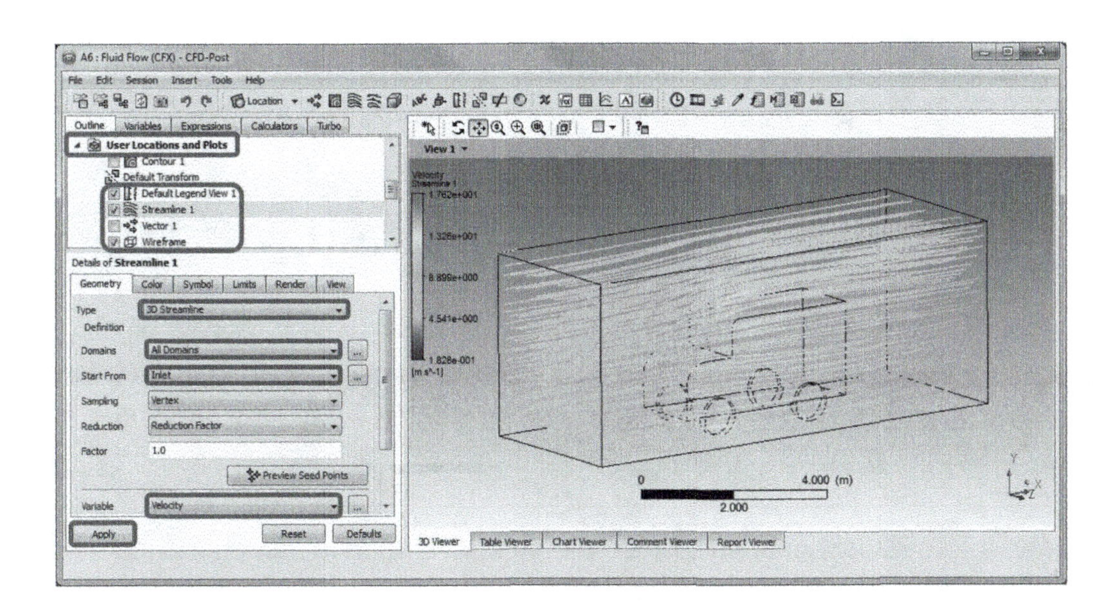

Modeling tips: Follow the steps below to visualize flow results on a section plane. First, right-click *User Locations and Plots* in the *Outline*, then select *Insert* and *Location* and then *Plane* in the context menu. Accept the default name *Plane 1* in the pop-up menu. In the *Details of Plane 1*, select *All Domains* for *Domains*, *XY Plane* for *Method*, and *0.0 [m]* for Z, and click *Apply*. Check *Plane 1* and *Wireframe* in the *Outline* to view the created section plane.

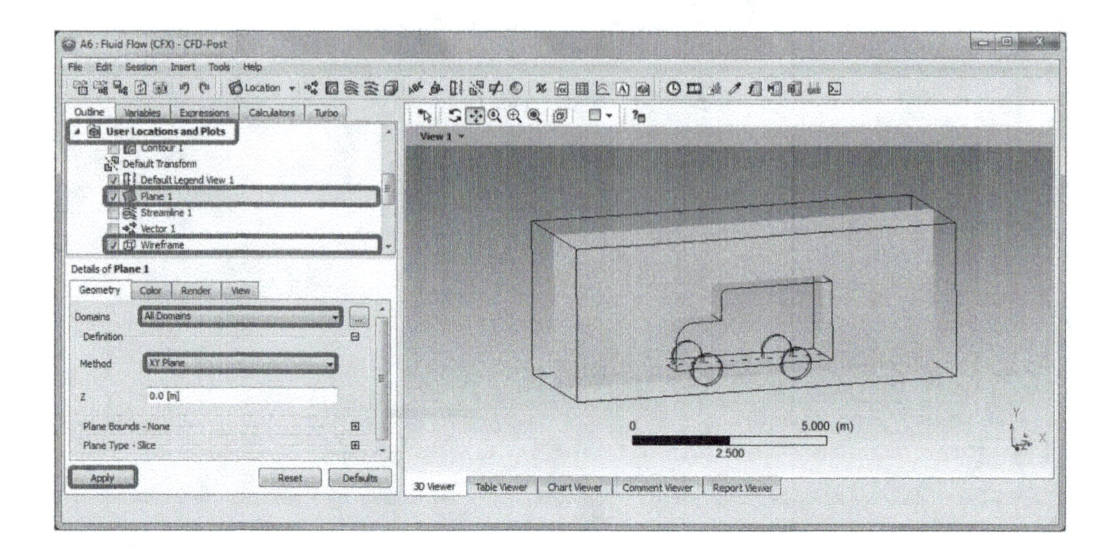

To generate a contour plot on the plane, right-click *User Locations and Plots*, and select *Insert* and then *Contour*. Accept the default name *Contour 2* in the pop-up menu. In the *Details of Contour 2*, select *All Domains* for *Domains*, *Plane 1* for *Locations*, and *Velocity* for *Variable*, and click *Apply*. Check *Default Legend View 1*, *Contour2*, and *Wireframe* in the *Outline* to retrieve the velocity distribution on the section plane.

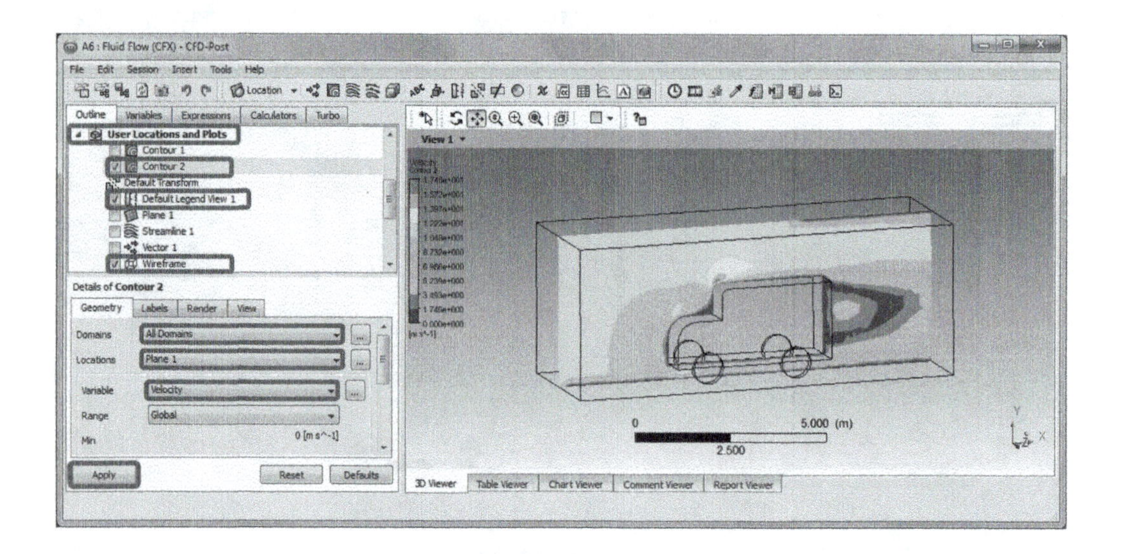

To generate a streamline plot on the plane, right-click *User Locations and Plots*, and select *Insert* and then *Streamline*. Accept the default name *Streamline 2* in the pop-up menu. In the *Details of Streamline 2*, select *All Domains* for *Domains*, *Plane 1* for *Start From*, *Rectangular Grid* for *Sampling*, *0.03* for *Spacing*, and click *Apply*. Check *Default Legend View 1*, *Streamline2*, and *Wireframe* in the *Outline* to view the streamlines on the section plane.

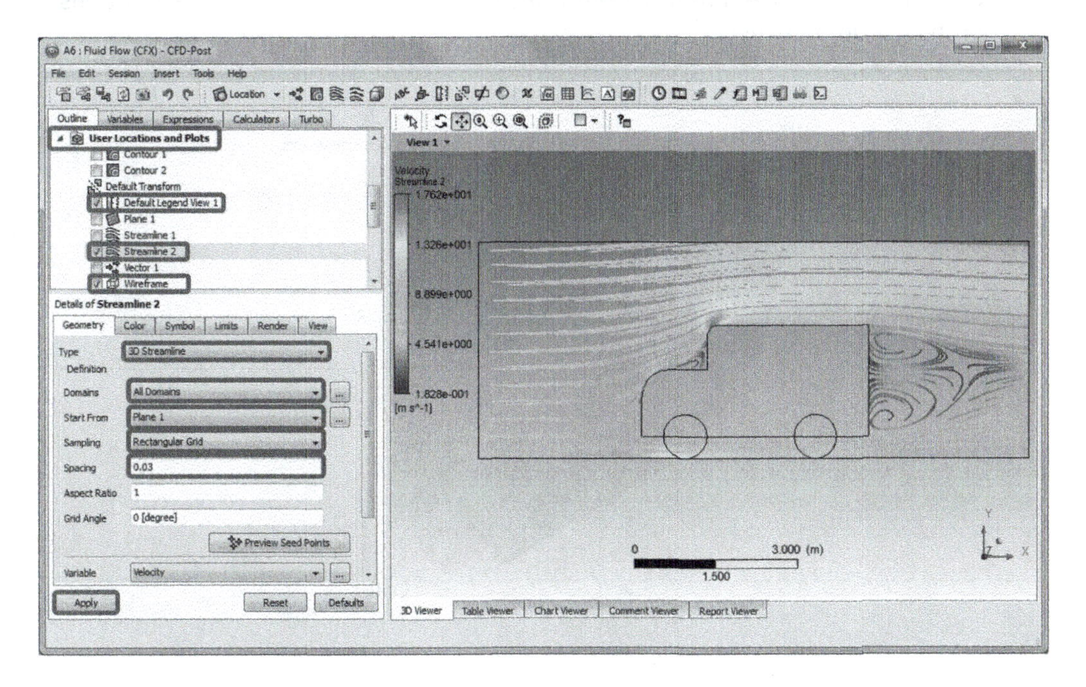

10.5 Summary

In this chapter, we briefly discussed the governing equations in fluid dynamics, main variables of concern, boundary conditions, and procedures in conducting CFD modeling and analysis. A case study using *ANSYS Workbench* is demonstrated using a truck model.

10.6 Review of Learning Objectives

Now that you have finished this chapter you should be able to

1. Understand the unique nature of fluid dynamics problems, including its governing equations, variables, and boundary conditions
2. Know how to create quality mesh for CFD analysis
3. Know how to apply the boundary conditions for CFD analysis
4. Perform CFD modeling and analysis of flows in complex domains using *ANSYS Workbench*

PROBLEMS

10.1 Conduct fluid analysis of the water flow passing through a U-tube. The tube has an inner and outer diameter of 4 and 4.2 cm, respectively. Assume nonslip boundary conditions on the surface walls. The inlet flow velocity is 5 mm/s and the outlet flow velocity is 4 mm/s. Determine the pressure and velocity distributions of the interior flow field.

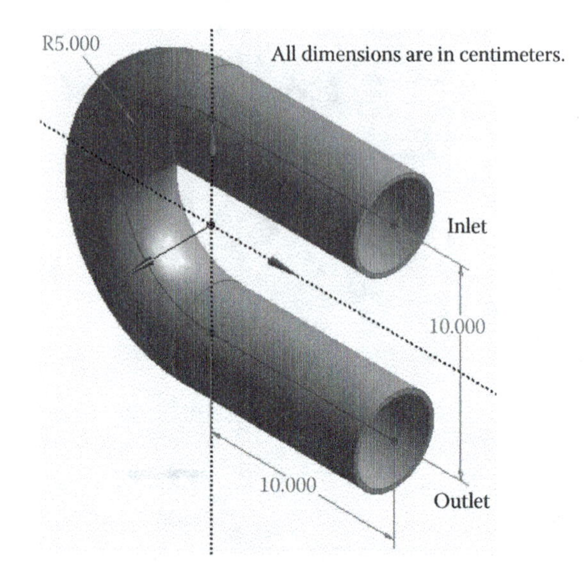

10.2 Determine the pressure and velocity distributions of the flow of air over a solid half-cylinder. The 20 cm long half-cylinder has a radius of 5 cm, and is placed on the ground. The upstream air has a velocity of 5 mm/s, and the air in the downstream passageway is at atmospheric pressure. Assume nonslip boundary conditions at the air–solid interface and the air is at a constant room temperature of 25°C.

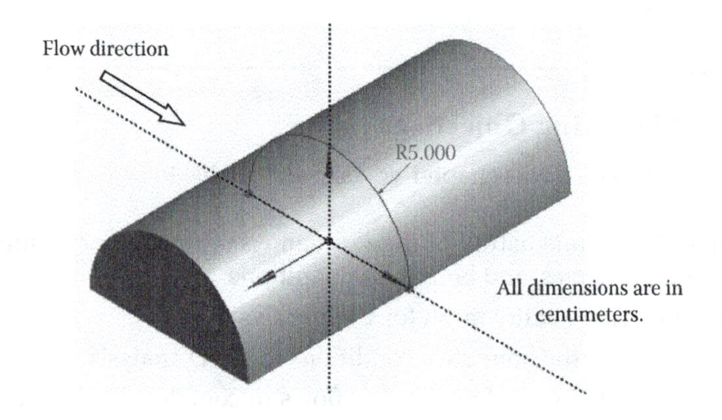

10.3 Determine the pressure and velocity distributions of the flow of air over a solid sphere. The sphere has a radius of 5 cm. The upstream air has a velocity of 5 mm/s, and the air in the downstream passageway is at atmospheric pressure. Assume nonslip boundary conditions at the spherical surface and the air is at a constant room temperature of 25°C.

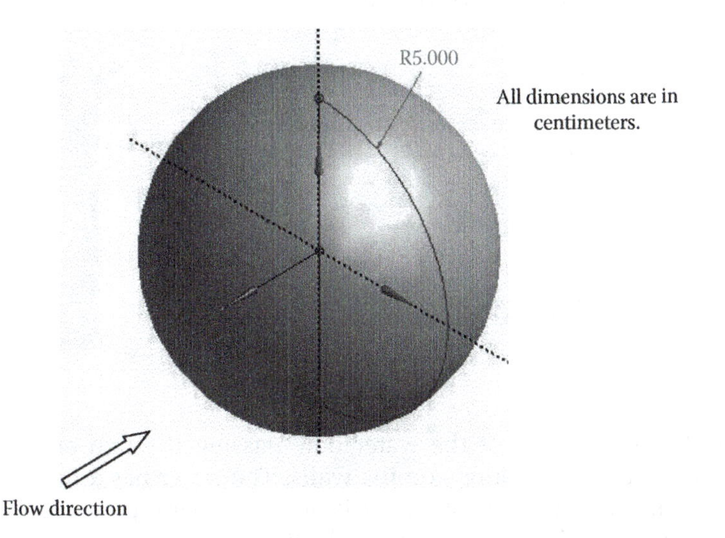

10.4 The heat sink shown below has a square base of size 22 mm × 22 mm × 2 mm and consists of an array of square pin-fins, each of size 2 mm × 2 mm × 8 mm. The pin-fins are separated by a gap distance of 4 mm in each direction. Determine the pressure and velocity distributions of the flow of air over the heat sink at room temperature of 25°C. Suppose the square base of the heat sink is in contact with the ground. The upstream air has a velocity of 10 mm/s, and the air in the downstream passageway is at atmospheric pressure.

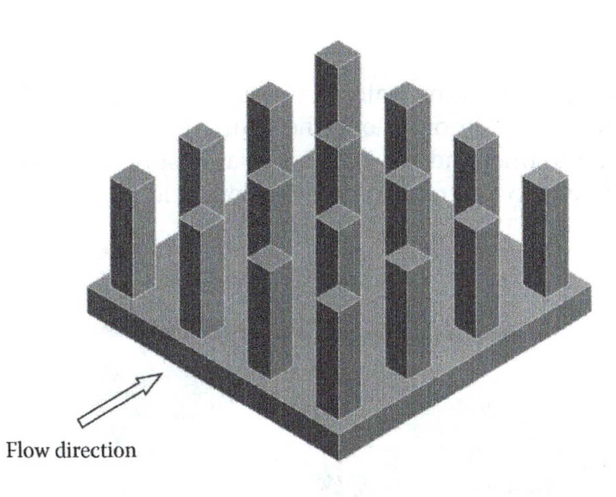

10.5 Conduct fluid analysis of the air flow passing through an S-pipe. Assume nonslip boundary conditions along the pipe walls and the air is at a room temperature of 25°C. The velocity of the air is 5 mm/s at the inlet and the outlet is at atmospheric pressure. Determine the pressure and velocity distributions of the interior flow field.

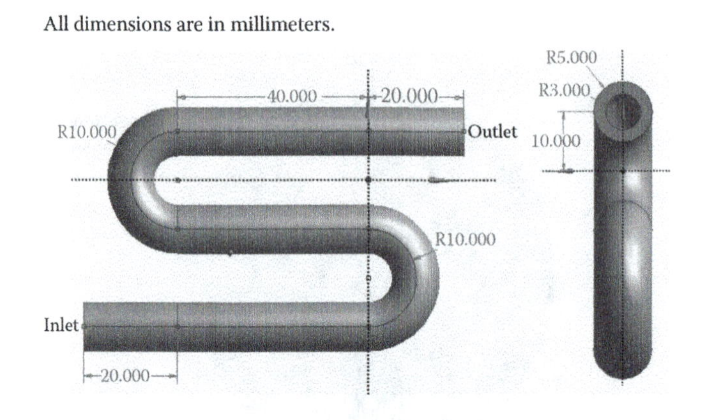

10.6 Conduct fluid analysis of the water flow passing through an orifice. Assume nonslip boundary conditions on the walls. The water has a uniform velocity of 15 mm/s at the inlet and the outlet is at atmospheric pressure. Determine the pressure and velocity distributions of the flow field.

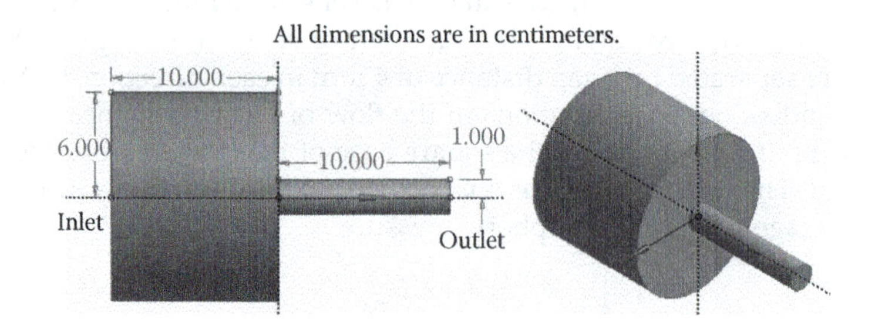

10.7 Conduct fluid analysis of the water flow passing through a diffuser pipe. Assume nonslip boundary conditions along the walls. The water has a uniform velocity of 10 mm/s at the inlet and the outlet is at atmospheric pressure. Determine the pressure and velocity distributions of the flow field.

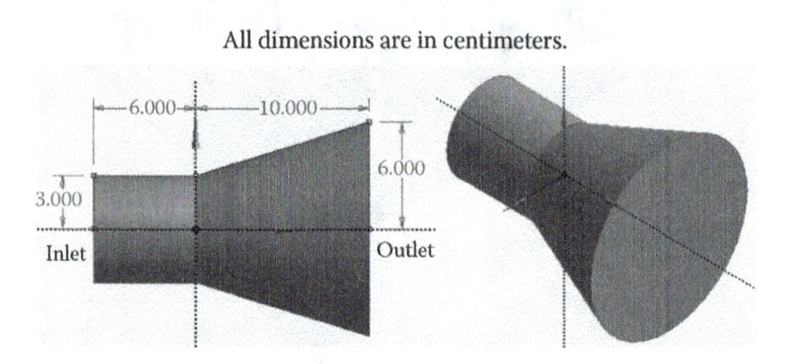

10.8 Conduct fluid analysis of the water flow passing through a transition pipe. Assume nonslip boundary conditions along the walls. The water has a uniform velocity of 25 mm/s at the inlet and the outlet is at atmospheric pressure. Determine the pressure and velocity distributions of the flow field.

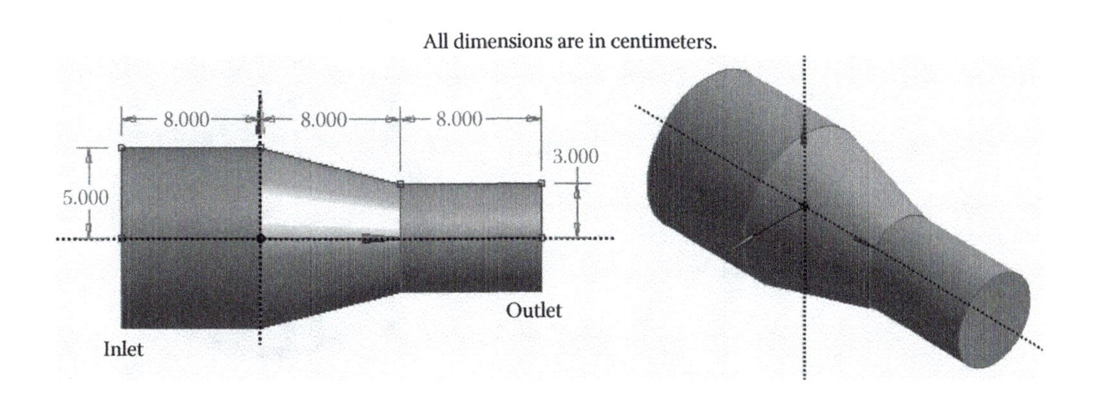

10.9 Determine the pressure and velocity distributions of the flow of water inside the following circular chamber. The chamber has a depth of 2 m. Assume nonslip boundary conditions at the walls. The velocity of water at the inlets is 10 cm/s, and the outlets are at atmospheric pressure.

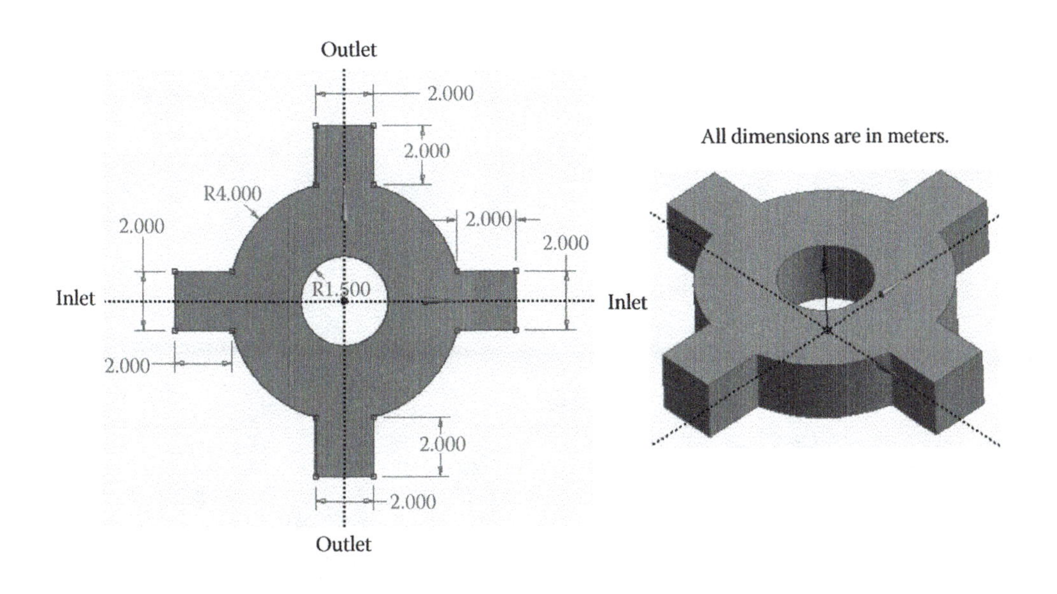

10.10 Determine the pressure and velocity distributions of the flow of water inside the following circular chamber. The chamber has a depth of 5 mm. Assume nonslip boundary conditions at the walls. The velocity of water at the inlets is 5 mm/s, and the outlets are at atmospheric pressure.

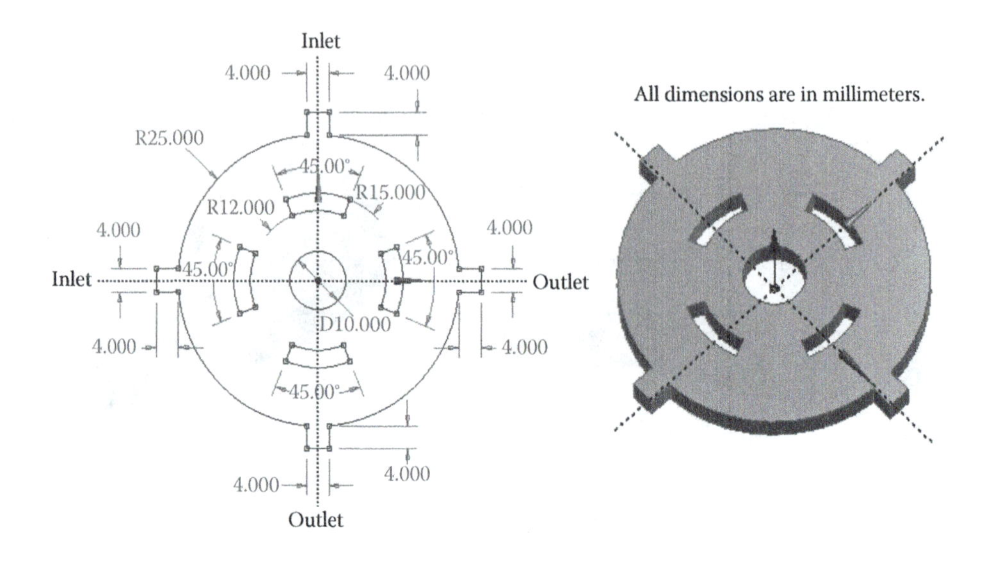

11

Design Optimization

11.1 Introduction

Optimization is an integral part of engineering design. Traditionally, optimization tasks have been carried out mostly by trial and error, when unexpected failure of a design to meet certain criteria is identified. Often, a best design is not obtained after many design iterations. Only a feasible design meeting all the requirements is created. The conventional way of changing a design when it is found to be deficient in certain criteria can be incredibly costly and time-consuming. For today's engineers, a more productive and cost-effective practice is to use numerical optimization techniques to guide in the evaluation of design trade-offs. Often, a best design is put forward after running simulation-based optimization a few times. In this chapter, we will cover materials relating to the topics of design optimization via simulation. The concepts of topology optimization, parametric optimization, and design space exploration will be introduced, along with optimization examples using *ANSYS® Workbench*.

11.2 Topology Optimization

A design's performance should be optimized from different perspectives as the design process evolves. In the early stage of a design, topology optimization can be used to help designers arrive at a good initial design concept. The goal of topology optimization is to find the ideal distribution of material within a predefined design space for a given set of loading and boundary conditions. The regions that contribute the least to the load bearing are identified and taken out from the design to minimize the weight. As a result, an optimal material layout is determined, from which a good design concept can be derived. Figure 11.1 illustrates topology optimization of a bridge structure. The 3-D design space of the bridge is shown as a solid rectangle box in Figure 11.1a. The bridge is fixed on the bottom two edges and applied a surface load on the top face. In Figure 11.1b, an arch is clearly suggested as the ideal layout by the topology optimization study aiming at 80% weight reduction. Materials are removed from the least stressed regions in the simulation model, that is, regions contributing the least to the overall stiffnesss of the structure. This optimization study perhaps helps elucidate why the arch continues to play an important part in bridge design after thousands of years of architectural use.

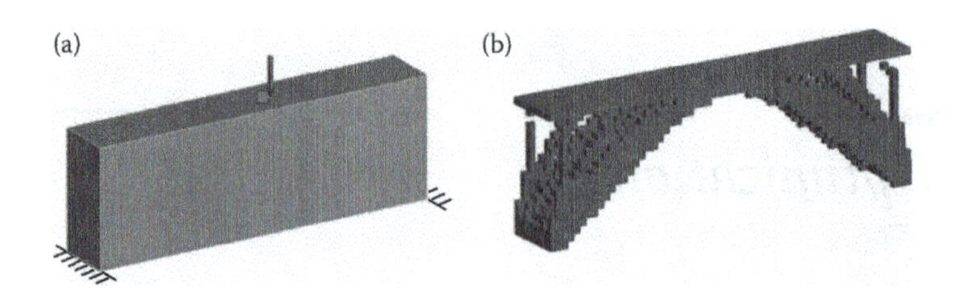

FIGURE 11.1
Topology optimization of a bridge structure: (a) The original design space and (b) the optimized layout with an 80% weight reduction target.

11.3 Parametric Optimization

In the final design stage, a design's performance is greatly influenced by its shape and size. Parametric optimization can be used to help designers determine the optimal shape and dimensions of a structure. In parametric optimization, the independent variables whose values can be changed to improve a design are called *design variables*. Design variables are usually geometric parameters such as length, thickness, or control point coordinates that control a design's shape. Responses of the design to applied loads are known as *state variables*, which are functions of the design variables. Examples of state variables are stresses, deformations, temperatures, frequencies, and so on. The restrictions placed on the design are *design constraints*. In general, parametric optimization involves minimizing an objective function of the design variables subject to a given set of design constraints [17].

For example, consider the parametric optimization of a stiffened aluminum panel with clamped edges. Stiffened panels are suited for weight-sensitive designs and are widely used in ship decks, air vehicles, and offshore structures. The main drawback is that they are light-weight structures with low natural frequencies, leading to a greater risk of resonance. For the design shown in Figure 11.2, the variables that are allowed to change, that is, the design variables, are the stiffener height h, the plate thickness t, the longitudinal stiffener thickness t_{long}, and the lateral stiffener thickness t_{lat}. To reduce the panel's vulnerability to vibration-induced movement, its fundamental frequency f_{base}, that is, the state variable in the study, is to be set above, say, 20 Hz. Suppose your objective is to minimize the panel's overall weight. You may set up a parametric optimization study solving for the optimum values of the four design variables so as to minimize the panel's weight while satisfying the design constraint of $f_{base} > 20$ Hz.

11.4 Design Space Exploration for Parametric Optimization

A "black box" model shown in Figure 11.3 can be used to illustrate the parametric optimization schematic, as a direct relationship between the design variables and responses is generally unknown. The schematic includes the use of a parametric finite element model for finding the effects of design variables (inputs) to the responses (outputs). After multiple response datasets are collected from finite element simulations, a mapping relationship can

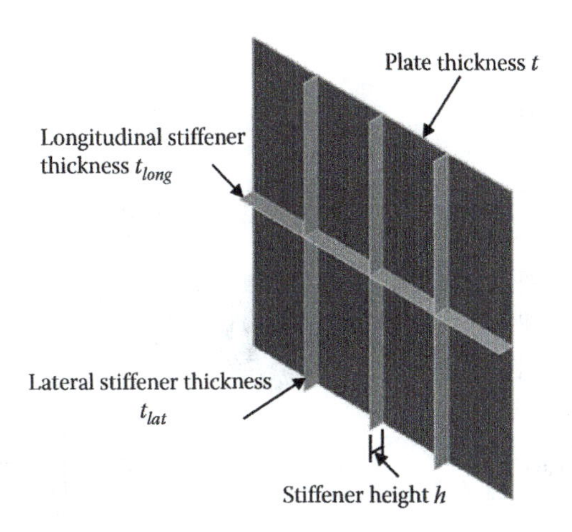

FIGURE 11.2
A stiffened panel with clamped edges.

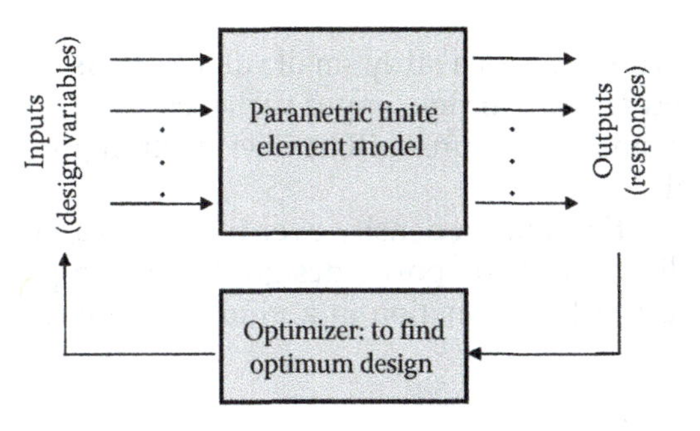

FIGURE 11.3
"Black Box" optimization schematic.

be established between the design variables and responses. The optimizer then explores the mapping of the parametric design space to find the optimal set of design variables that fulfill the given response criteria.

In the design space exploration, design of experiments are usually used in combination with response surface modeling to efficiently map out a parametric design space through simulating a minimum number of design scenarios.

11.4.1 Design of Experiments

Design of experiments (DOE) is a technique originally developed for model fitting with experimental data. In design optimization, DOE can be used to fit the simulated response data to mathematical equations. These equations, also referred to as *response surface* equations, serve as models (see Figure 11.4) to predict the responses for any combination of design variable values.

In DOE, each design variable is viewed as one dimension in a design space. In general, we have an n-dimensional space if there are n design variables. Each design variable has many possible discrete values or *levels*. An array can be constructed by considering all the combinations of design variable levels. In most cases, we cannot afford the response data

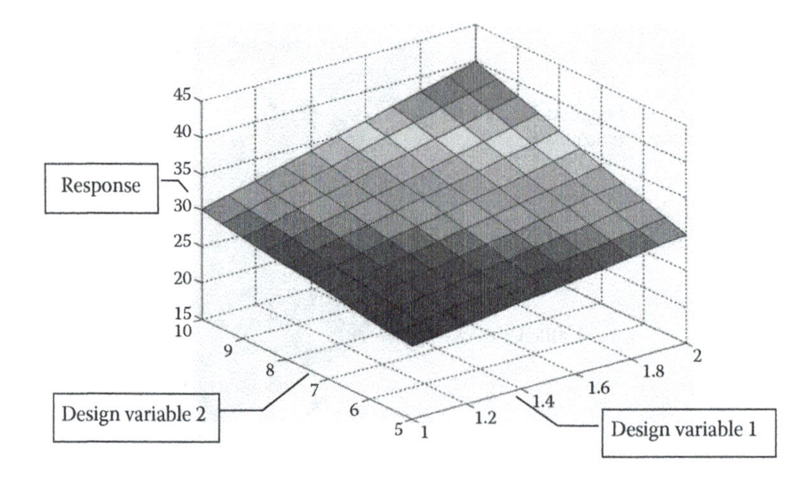

FIGURE 11.4
A 3-D response surface representation.

collection for such an array even for a fairly small value of n. As Table 11.1 shows, 16 experimental runs are needed already for four input variables (x_1, x_2, x_3, and x_4) at two levels (0 and 1). The number of simulation runs will be overwhelmingly large for multiple inputs at many different levels.

A grid of data points of minimum density over the multivariable design space is often used instead. Especially, central composite design (CCD) is a popular design, due to its efficiency in providing much information in a minimum number of required numerical

TABLE 11.1
Design of Experiments: 16 Experimental Runs for 4 Inputs at 2 Levels

	Inputs			
Experimental runs	x_1	x_2	x_3	x_4
1	0	0	0	0
2	0	0	0	1
3	0	0	1	0
4	0	0	1	1
5	0	1	0	0
6	0	1	0	1
7	0	1	1	0
8	0	1	1	1
9	1	0	0	0
10	1	0	0	1
11	1	0	1	0
12	1	0	1	1
13	1	1	0	0
14	1	1	0	1
15	1	1	1	0
16	1	1	1	1

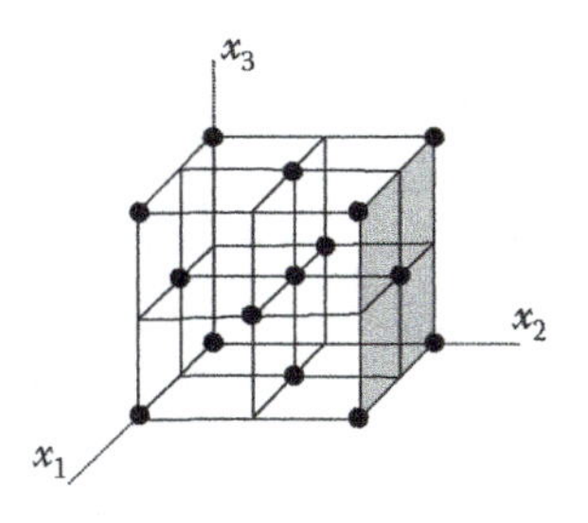

FIGURE 11.5
A face-centered central composite design for three design variables.

experiments [17]. CCD enables an efficient construction of a second-order fitting model. As illustrated in Figure 11.5, a typical face-centered CCD for three design variables (x_1, x_2, and x_3) at 3 levels suggests the use of 15 design points, compared to 3^3 possible combinations of the full design. The 15 design points are marked as back dots in Figure 11.5. Each design point corresponds to a design scenario. By using the CCD, the maximum amount of information can be extracted while requiring a significantly reduced number of numerical experiments.

11.4.2 Response Surface Optimization

After the design space is sampled through an experimental design such as CCD, a response dataset (e.g., the maximum deformation and stress results) can be readily obtained for a design scenario through the finite element simulation. The response datasets, with each dataset corresponding to a simulation scenario, are then used to fit the response surface models. These models are interpolation models that can provide continuous variation of the responses with respect to the design variables (see Figure 11.4).

In the optimizer, a designer sets up a design objective and constraints. For optimization with multiple objectives, the relative importance of different objectives and constraints can be specified. Using the fitted response surface models, the *feasible region*, that is, the region satisfying all design constraints can be identified in the design space. The best design candidate is then determined by searching for the best available value of the objective function over the entire design space's feasible region.

In the next section, we will use *ANSYS Workbench* to optimize an L-shaped structure using the above-mentioned techniques.

11.5 Case Studies with *ANSYS Workbench*

Problem Description: Determine if weight reduction pockets can be generated in the L-shaped structure shown below. The structure is 2 mm thick and is made of structural steel. The boundary and loading conditions are specified as follows: A downward force of 300 N is applied at the bottom edge of the free end, and the top face of the L-shape is fixed. The allowed maximum deformation in the structure is 0.3 mm. (a) Perform topology optimization to achieve 60% weight reduction. (b) Redesign the structure based on the results from topology optimization, and conduct parametric optimization to minimize weight subject to the deformation constraint.

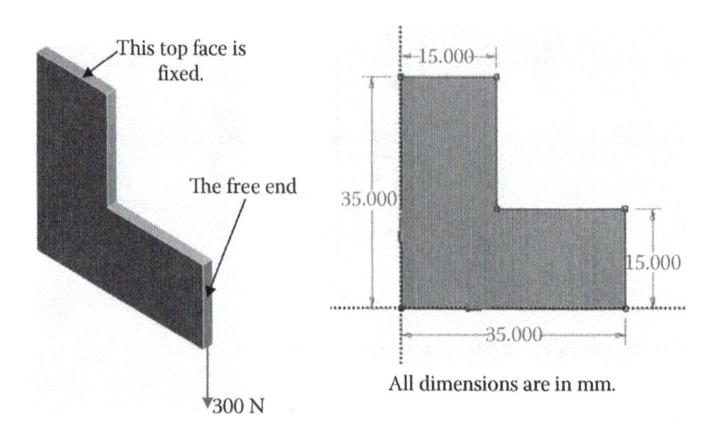

Solution steps for portion (A):

Step 1: Start an *ANSYS Workbench* Project

Launch *ANSYS Workbench* and save the blank project as *"Lshape.wbpj."*

Step 2: Create a *Shape Optimization* Analysis System

Drag the *Shape Optimization* icon from the *Analysis Systems Toolbox* window and drop it inside the highlighted green rectangle in the *Project Schematic* window.

Note: If *Shape Optimization* does not appear in the *Toolbox* of *Analysis Systems,* you may have to turn on the beta option. Go to *Tools* > *Options…* > *Appearance,* and check the box next to *Beta Options.* If beta options are on and the analysis system still is not coming up, go to the bottom of the *Toolbox* in *Workbench* and click *"View All/Customize"* and make sure that there is a check mark next to *"Shape Optimization (Beta)."*

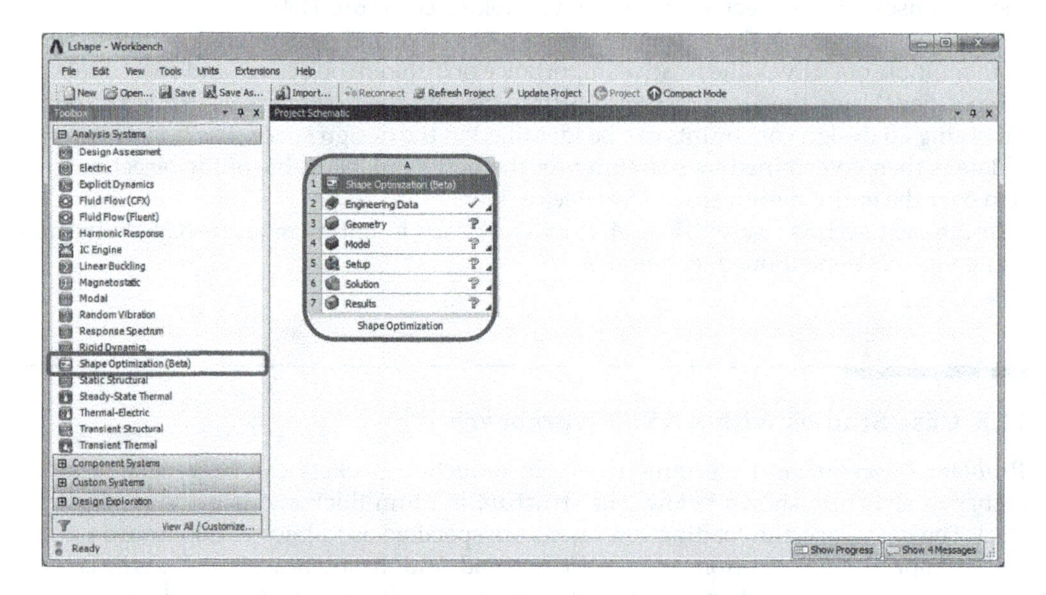

Step 3: Launch the *DesignModeler* Program

Double-click the *Geometry* cell to launch *DesignModeler,* and select *"Millimeter"* in the *Units* pop-up window.

Step 4: Create the Geometry

Click on the *Sketching* tab. Draw a sketch of the L-shape on the *XY Plane*, as shown below. An entity named *Sketch1* will be shown underneath *XY Plane* of the model's *Tree Outline*.

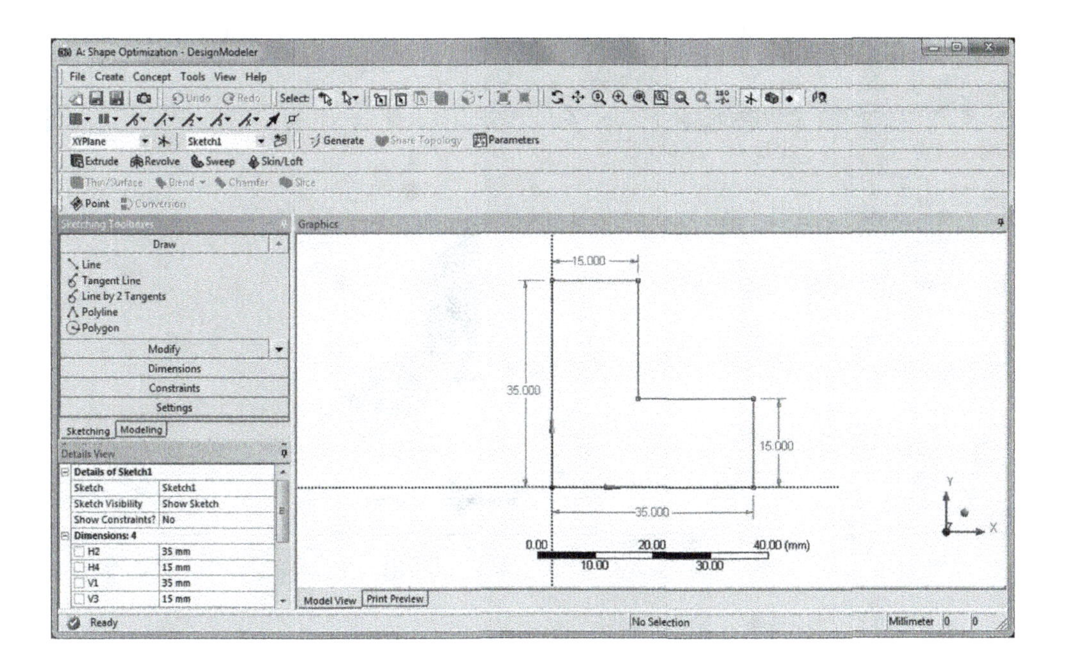

Extrude *Sketch1* to create a *2 mm* thick solid body, as shown below.

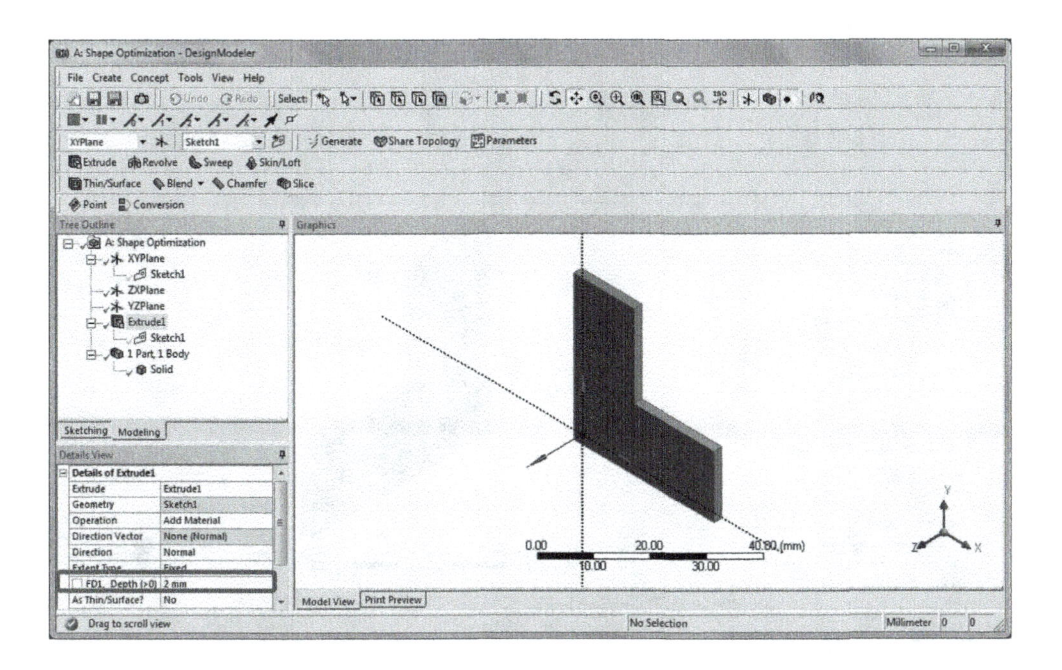

Step 5: Launch the *Shape Optimization* Program

Double-click on the *Model* cell to launch the *Shape Optimization* program. Change the *Units* to *Metric (mm, kg, N, s, mV, mA)*.

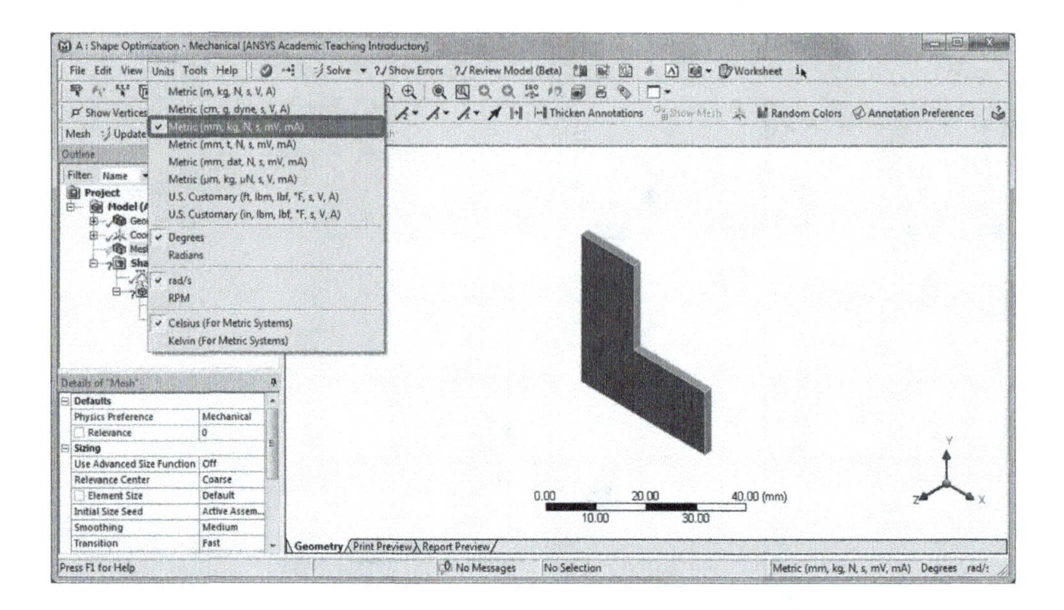

Step 6: Generate Mesh

Click on *Mesh* in the *Outline* tree. In the *Details of "Mesh,"* enter *"1.0 mm"* for the *Element Size*. Right-click on *Mesh* and select *Generate Mesh*.

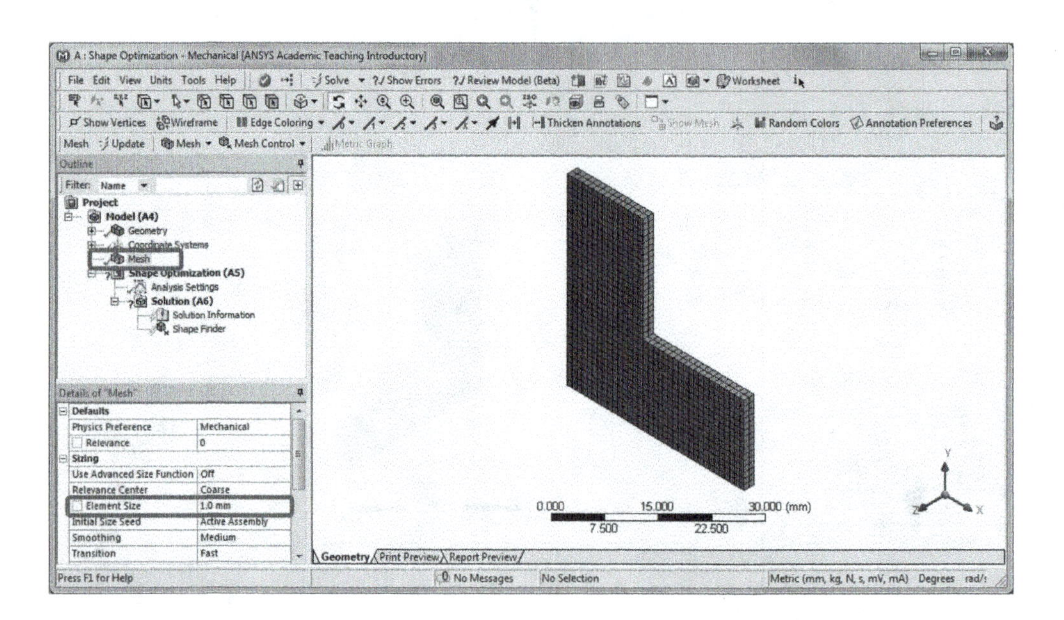

Step 7: Apply Boundary Conditions

Right-click on *Shape Optimization (A5)*. Choose *Insert* and then *Fixed Support* from the context menu. Click on the top face, and apply it to the *Geometry* selection in the *Details of "Fixed Support."* The top face of the L-shape is now fixed as shown below.

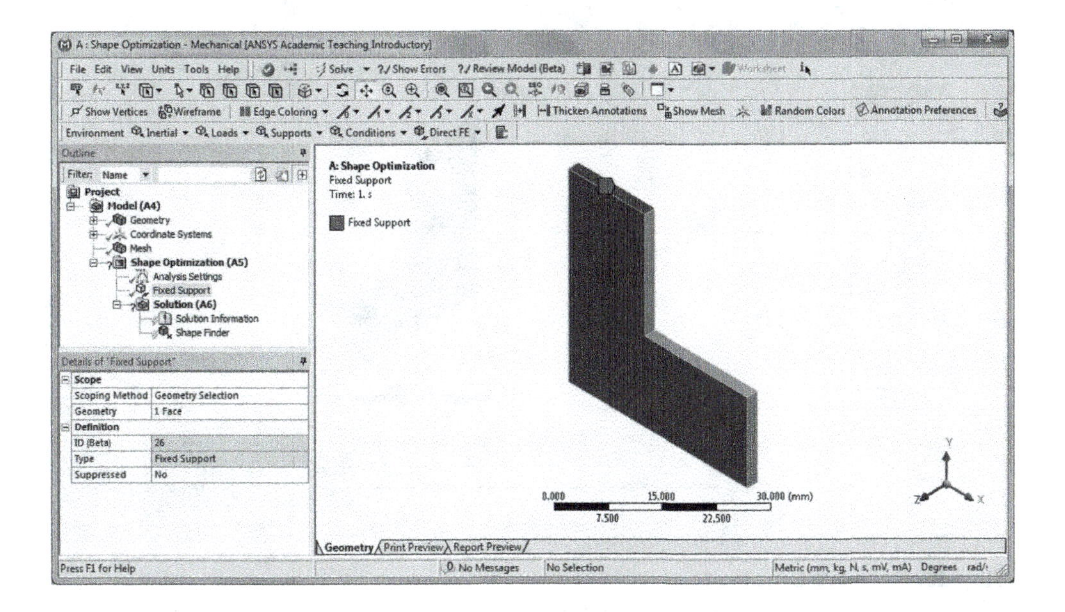

Step 8: Apply Loads

Right-click on *Shape Optimization (A5)*. Choose *Insert* and then *Force*. In the *Details of "Force,"* apply a *300* N force to the top edge of the free end in the *Y*-direction, as shown below.

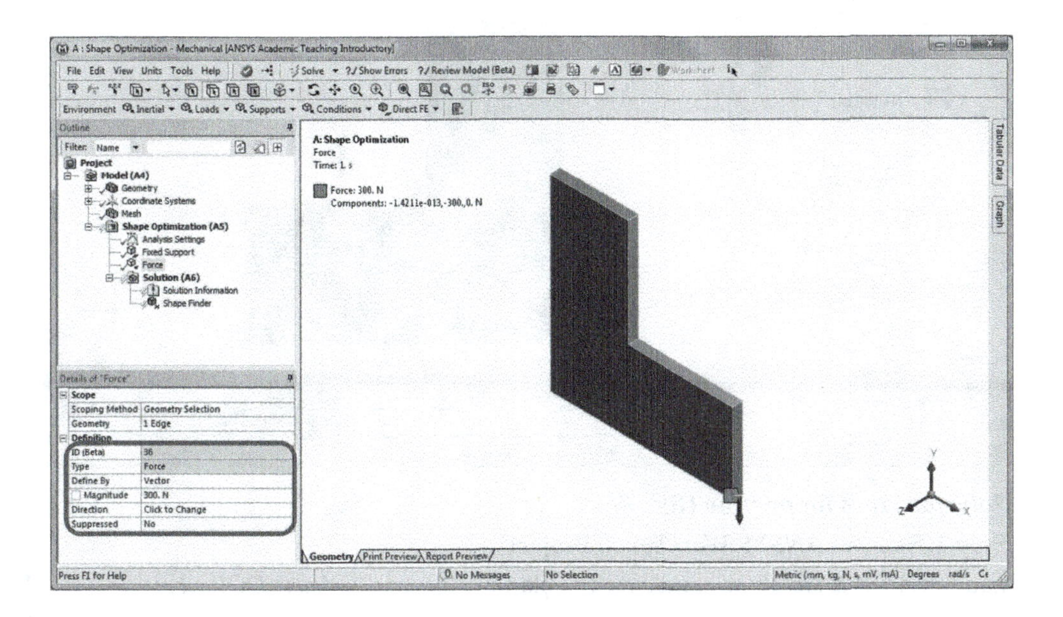

Step 9: Define Target Weight Reduction

Click on *Shape Finder* underneath *Solution (A6)* in the *Outline* tree. In the *Details of "Shape Finder,"* enter "60%" for the *Target Reduction*. Right-click on *Solution (A6)* and select *Solve*.

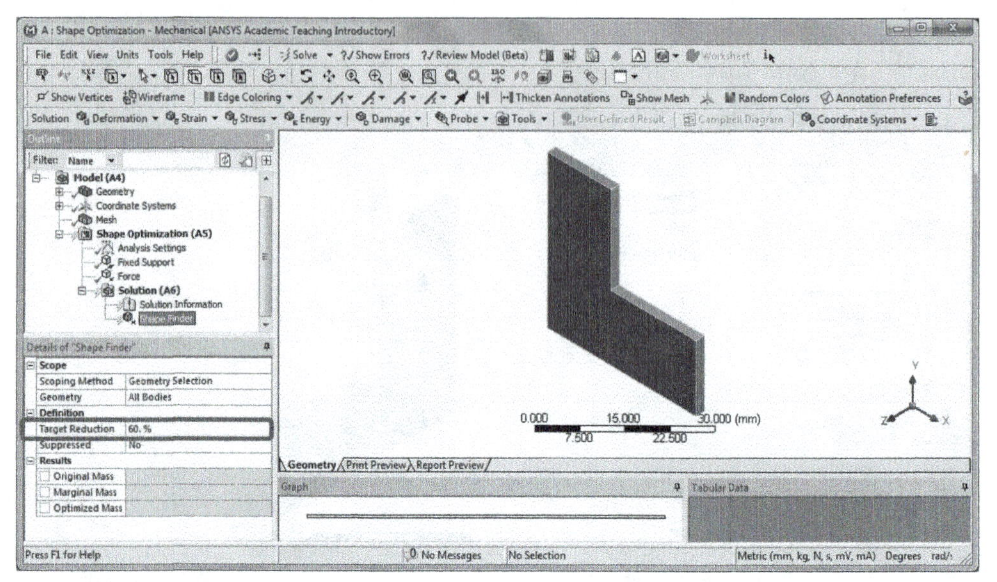

Step 10: Retrieve Solution

Click on *Shape Finder* underneath *Solution (A6)* in the *Outline* tree to review results. The region marked in gray is suggested as the optimal shape for the design under the given loads. Save project and exit the *ANSYS Workbench*.

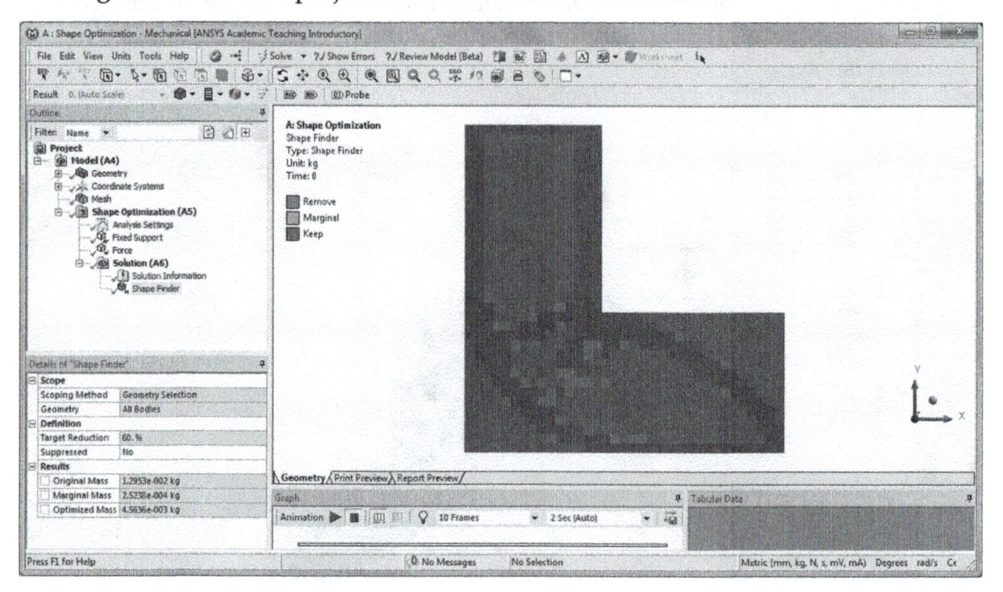

Solution steps for portion (B):

Step 1: Start an *ANSYS Workbench* Project

Launch *ANSYS Workbench* and save the blank project as *"Lshape_DOE.wbpj."*

Step 2: Create a *Static Structural* Analysis System

Drag the *Static Structural* icon from the *Analysis Systems Toolbox* window and drop it inside the highlighted green rectangle in the *Project Schematic* window.

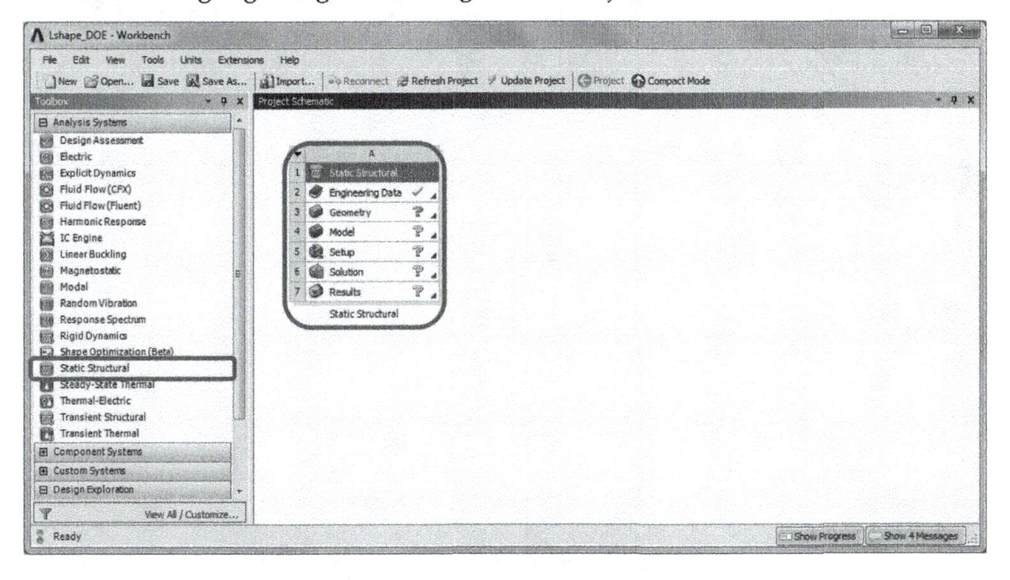

Step 3: Launch the *DesignModeler* Program

Double-click the *Geometry* cell to launch *DesignModeler*, and select *"Millimeter"* in the *Units* pop-up window.

Step 4: Create the Geometry

Click on the *Sketching* tab. Draw a sketch of the L-shape on the *XY Plane*, as shown below. An entity named *Sketch1* will be shown underneath *XY Plane* of the model's *Tree Outline*.

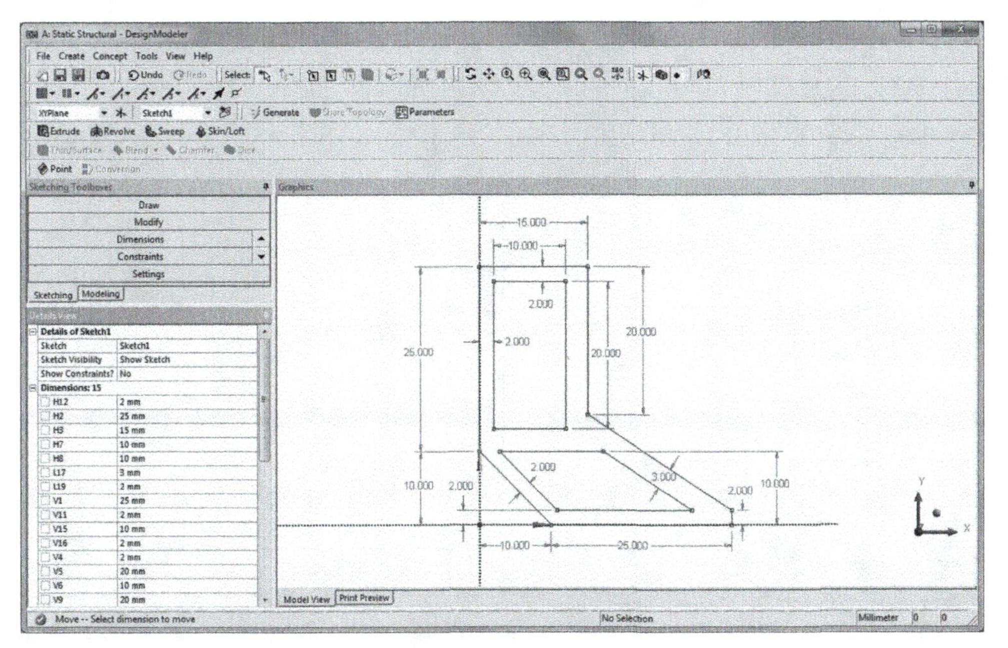

Apply fillets of radius *1 mm* to the six interior corners as shown below. Apply the *Equal Radius* constraint to all six fillets. Turn on the *Name* option for the *Display* of *Dimensions.*

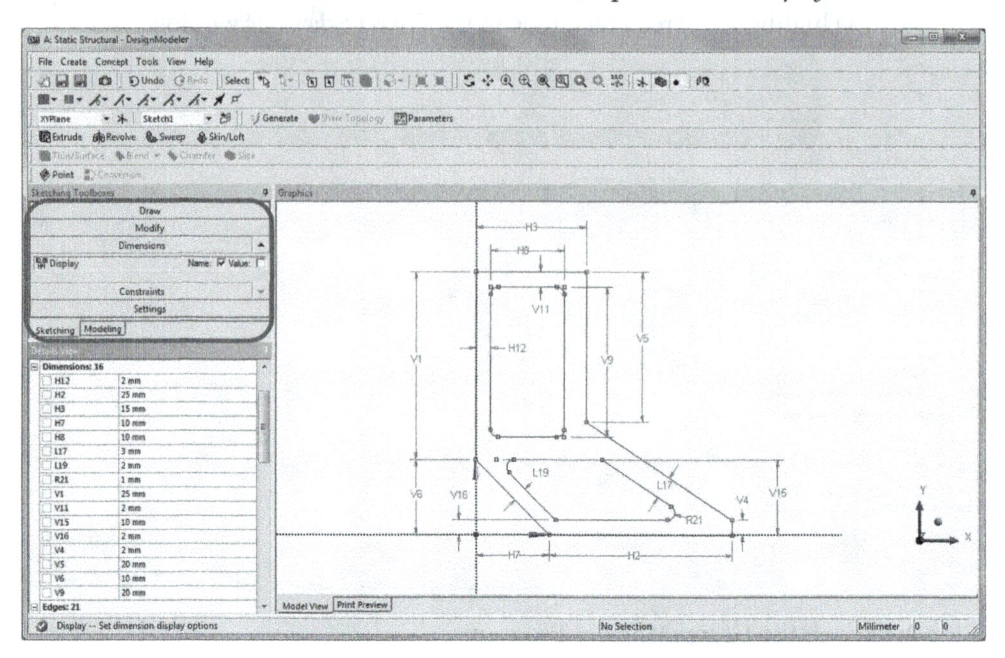

In the *Details View,* click on the box to the left of the dimension name *V5.* In the pop-up window, enter *"height1"* as the parameter name and then click *OK.* Repeat the steps for the dimension name *V15* and enter *"height2"* as the corresponding parameter name (note your dimension names may not be the same as they appear here). A letter *D* will be marked on the box to the left of the two selected dimensions in the *Details View* as shown below.

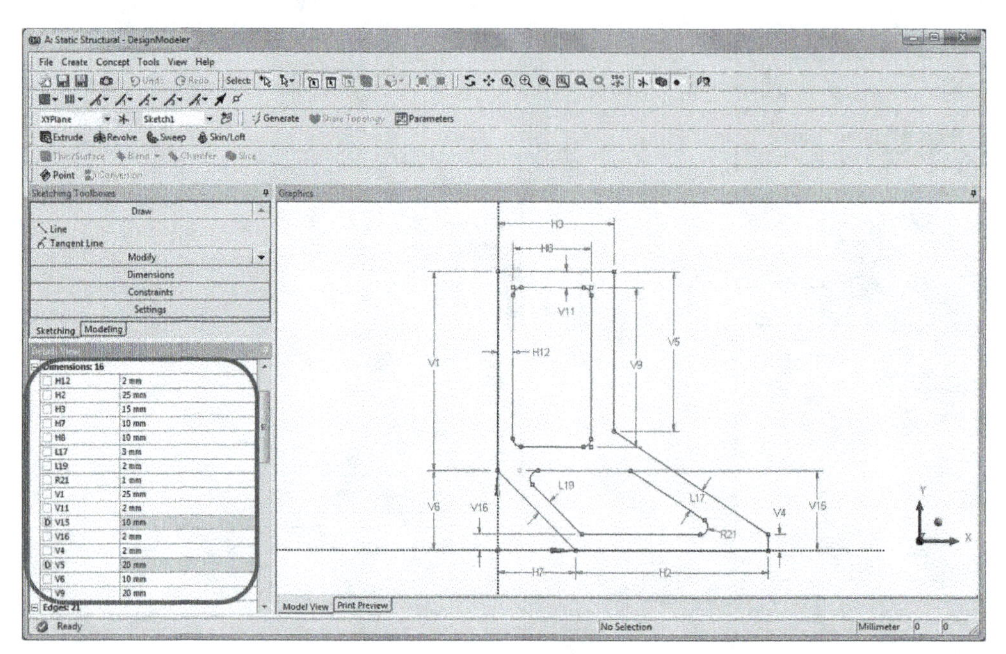

Extrude *Sketch1* to create a *2 mm* thick solid body, and close *DesignModeler.*

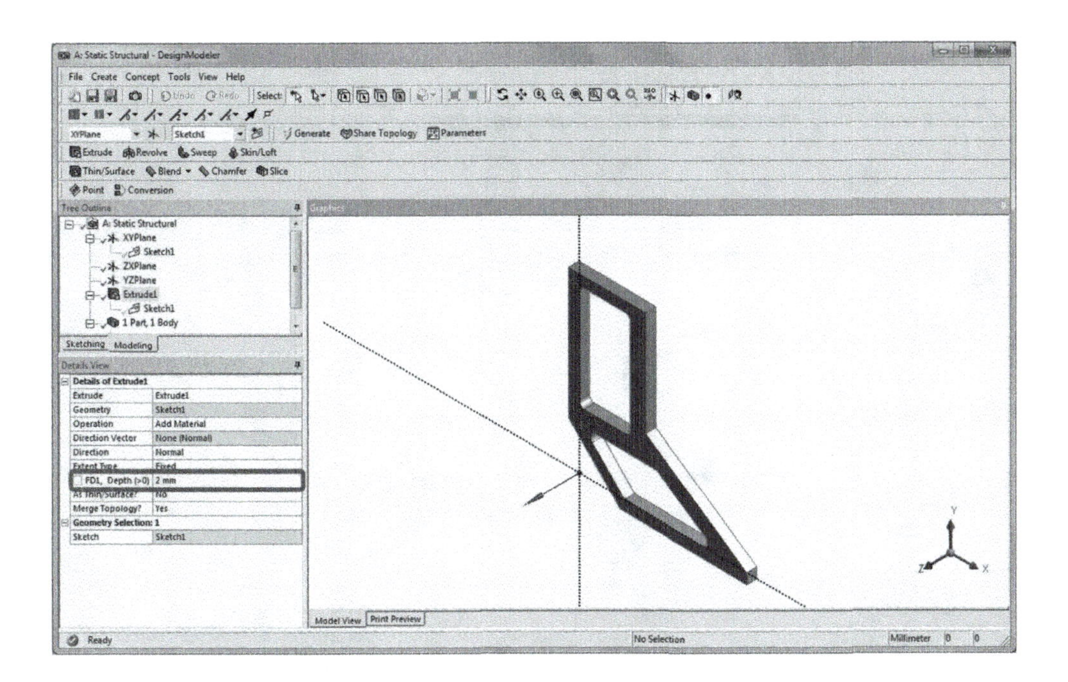

Note that a *Parameter Set* bus is added to the project schematic.

Double-click on the *Parameter Set*. Note that the selected dimension parameters, namely, *height1* and *height2*, are now included as two design variables in the *Outline* window.

	A	B	C	D
1	ID	Parameter Name	Value	Unit
2	Input Parameters			
3	Static Structural (A1)			
4	P5	height1	20	
5	P6	height2	10	
*	New input parameter	New name	New expression	
7	Output Parameters			
*	New output parameter		New expression	
9	Charts			

Step 5: Launch the *Static Structural* Program

Double-click on the *Model* cell to launch the *Static Structural* program. Change the *Units* to *Metric (mm, kg, N, s, mV, mA)*.

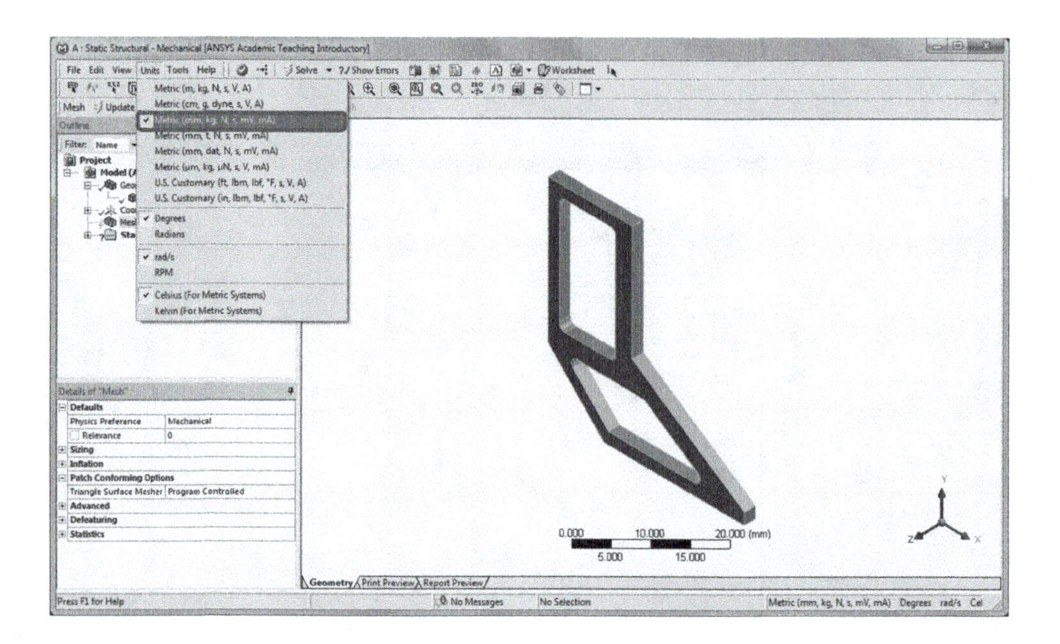

Step 6: Generate Mesh

Click on *Mesh* in the *Outline* tree. In the *Details of "Mesh,"* enter *"1.0 mm"* for the *Element Size*. Right-click on *Mesh* and select *Generate Mesh*.

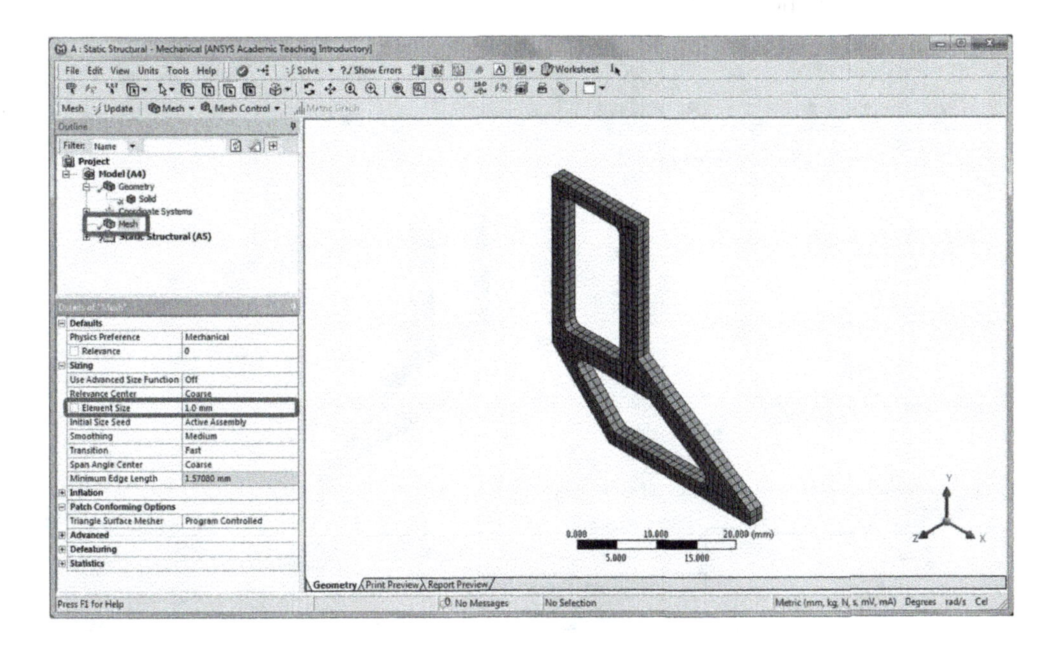

Step 7: Apply Boundary Conditions

Right-click on *Static Structural (A5)*. Choose *Insert* and then *Fixed Support* from the context menu. Click on the top face, and apply it to the *Geometry* selection in the *Details of "Fixed Support."* The top face of the L-shape is now fixed as shown below.

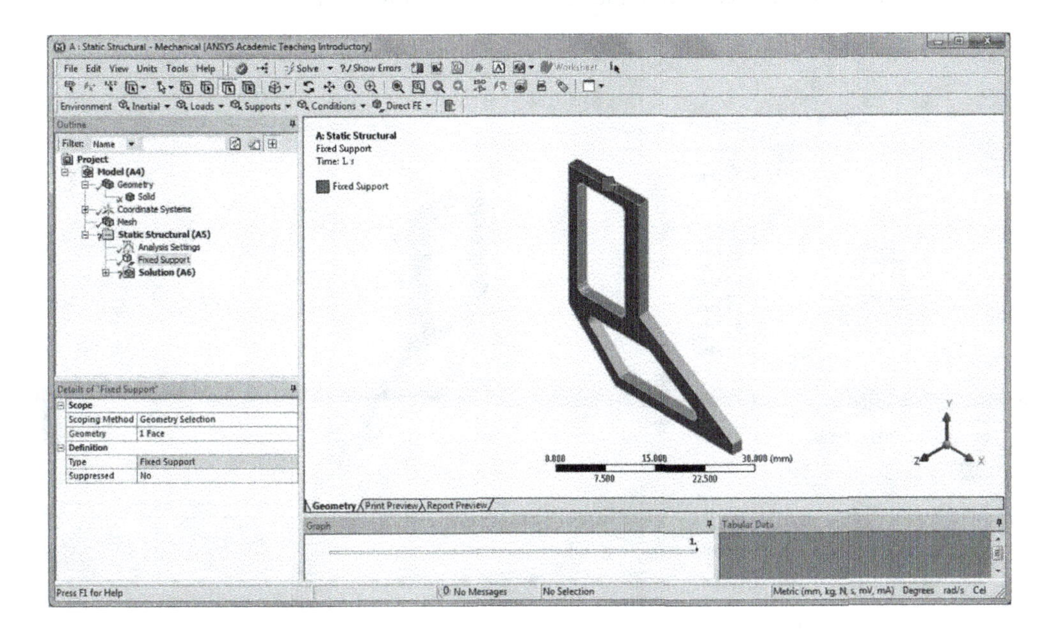

Step 8: Apply Loads

Right-click on *Static Structural (A5).* Choose *Insert* and then *Force.* In the *Details of "Force,"* apply a *300* **N** force to the bottom edge on the right along negative *Y*-direction, as shown below.

Step 9: Retrieve Solution and Define Response Parameters

Insert *Total Deformation* by right-clicking *Solution (A6)* in the *Outline* tree. Right-click on *Solution (A6)* and select *Solve.* Click on *Total Deformation* in the *Outline* to review deformation results. Next, click on the box to the left of *Maximum* in the *Details of "Total Deformation."* Note that a letter "**P**" is added to the box.

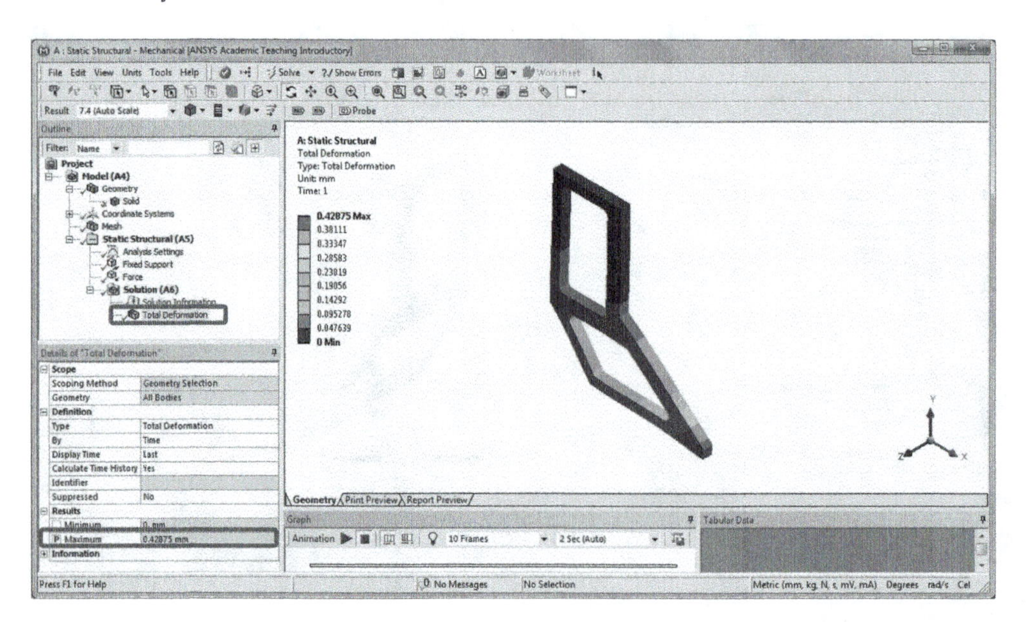

In the *Details of "Solid,"* click to the left of *Mass* under *Properties*. Note that a letter "**P**" is added to the box. Exit the *Static Structural* window.

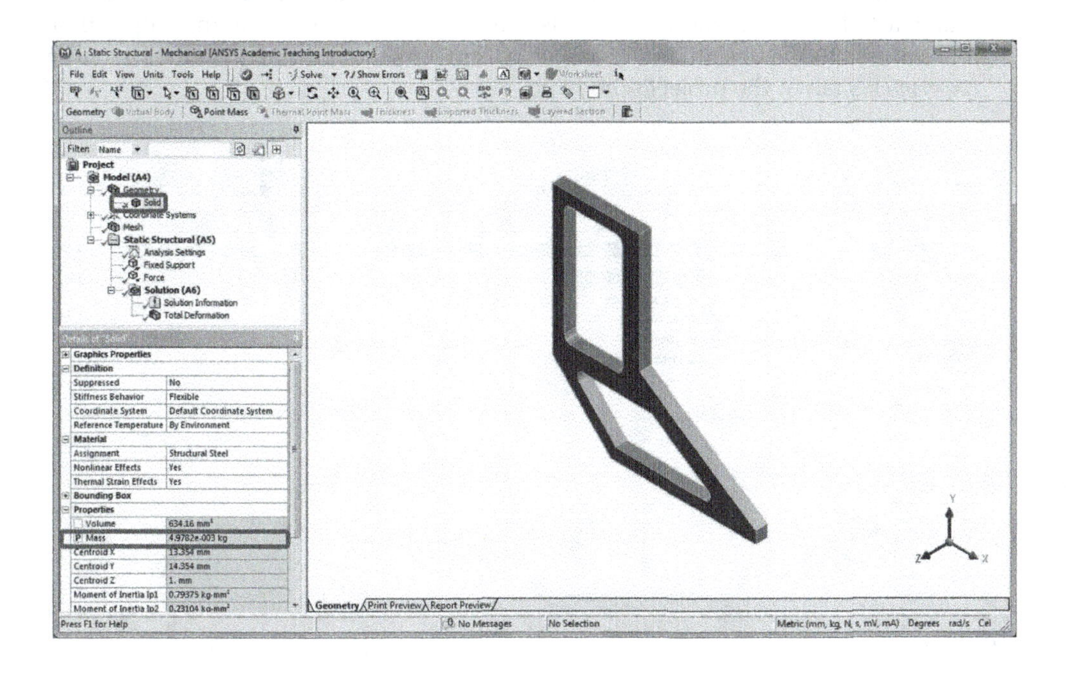

Step 10: Explore the Parameter Set

Double-click the *Parameter Set* in the project schematic. The input parameters (design variables) and output parameters (responses) are listed with their corresponding base values in the *Outline* window. In the *Table of Design Points*, the current design is shown as the only design point, click *Return to Project*.

Step 11: Create Response Surface Optimization

Double-click the *Response Surface Optimization* underneath *Design Exploration* from the *Toolbox*. A *Response Surface Optimization* system will be added to the project schematic. The optimization system is linked to the *Parameter Set* indicating that it has access to the design and response parameters. Double-click on the *Design of Experiments* tab in the project schematic.

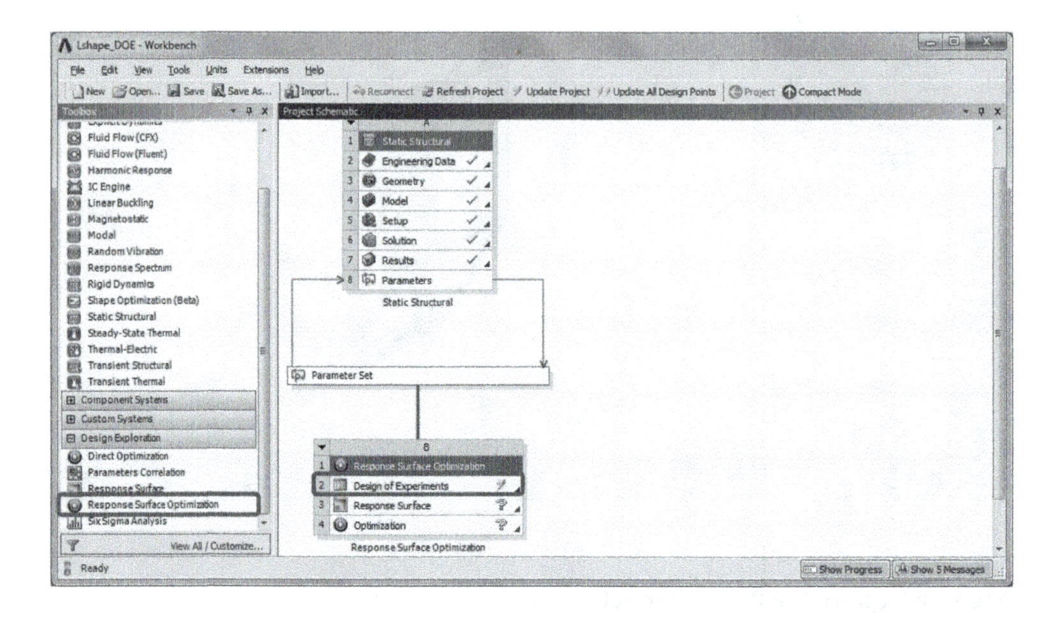

Step 12: Generate DOE Table

In the *Outline of Schematic B2: Design of Experiments,* click on the *P1-height1* tab. Enter the *Lower Bound* and *Upper Bound* values as shown below.

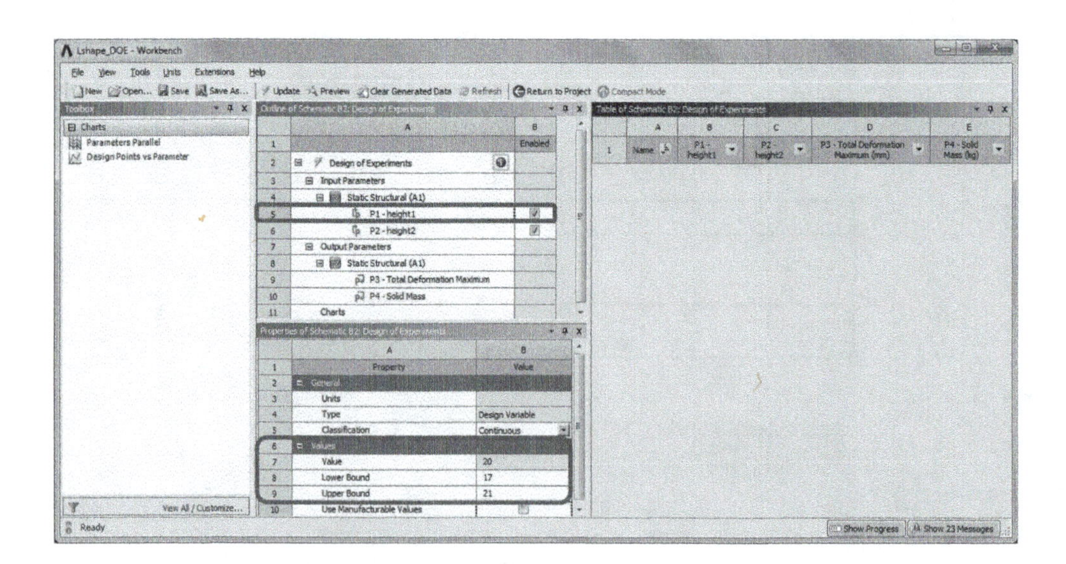

In the *Outline of Schematic B2: Design of Experiments,* click on the *P2-height2* tab. Enter the *Lower Bound* and *Upper Bound* values as shown below in the *Properties of Schematic B2: Design of Experiments.*

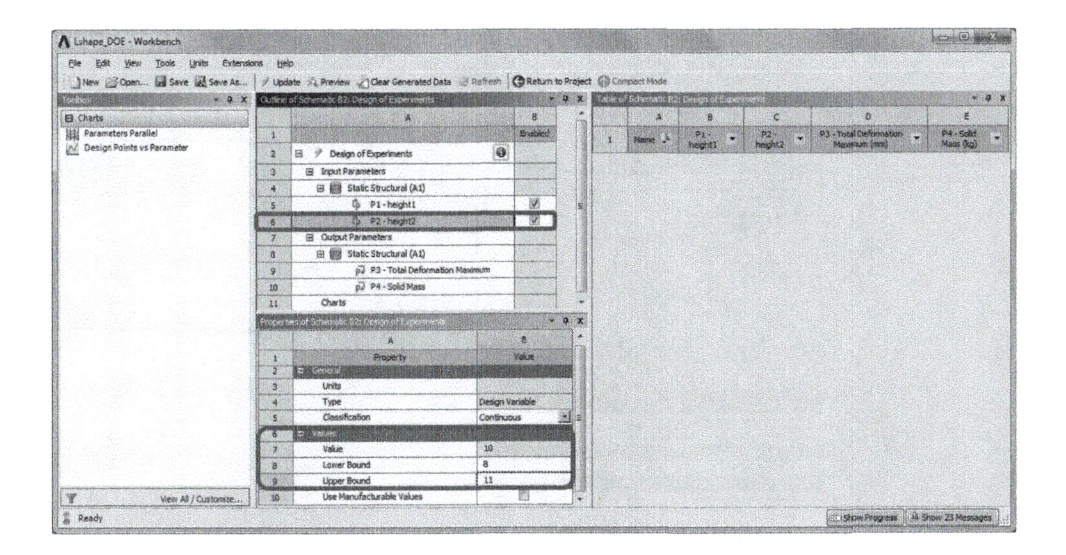

In the *Outline of Schematic B2: Design of Experiments,* click on the *Design of Experiments* tab. Turn on *Preserve Design Points After DX Run.* Change the *Design Type* to *Face-Centered.*

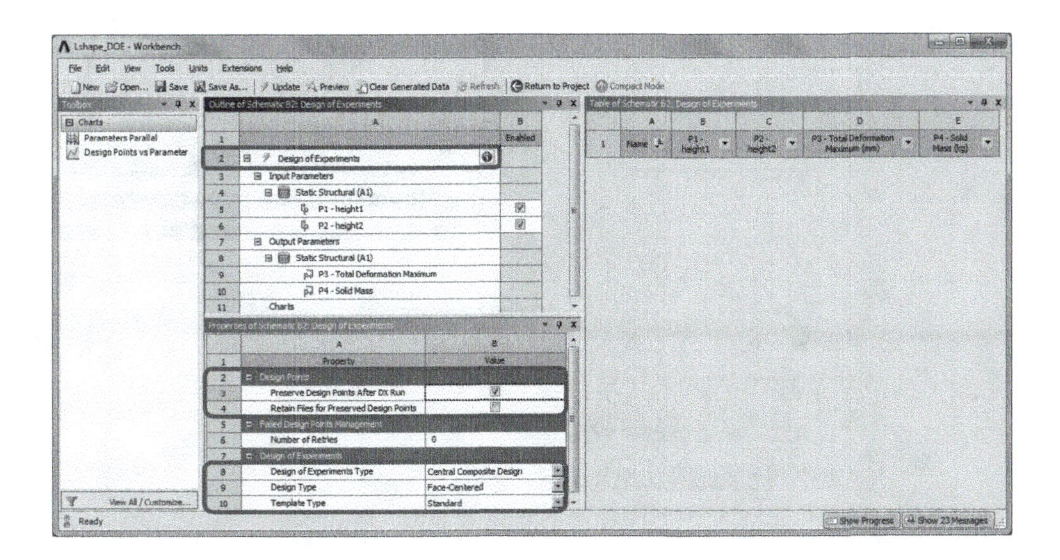

In the toolbar, click *Preview*. A DOE table listing nine design points is generated. Then click *Update* in the toolbar to run finite element simulations for the nine different design scenarios. It will take a while to finish running all the simulations. After completion, review the nine design points (simulation results) listed in the *Table of Schematic B2*, and click *Return to Project*.

	A	B	C	D	E
1	Name	P1 - height1	P2 - height2	P3 - Total Deformation Maximum (mm)	P4 - Solid Mass (kg)
2	1	19	9.5	0.36531	0.0051654
3	2	17	9.5	0.34983	0.0053454
4	3	21	9.5	0.39794	0.0050334
5	4	19	8	0.273	0.0055433
6	5	19	11	0.60622	0.0048026
7	6	17	8	0.26056	0.0057618
8	7	21	8	0.29837	0.0053594
9	8	17	11	0.59091	0.0049379
10	9	21	11	0.63551	0.004731

Step 13: Build Response Surface

Double-click the *Response Surface* tab in the project schematic. Turn on the two options underneath *Design Points* in the *Properties of Schematic B3: Response Surface*. Click *Update* to create the response surface.

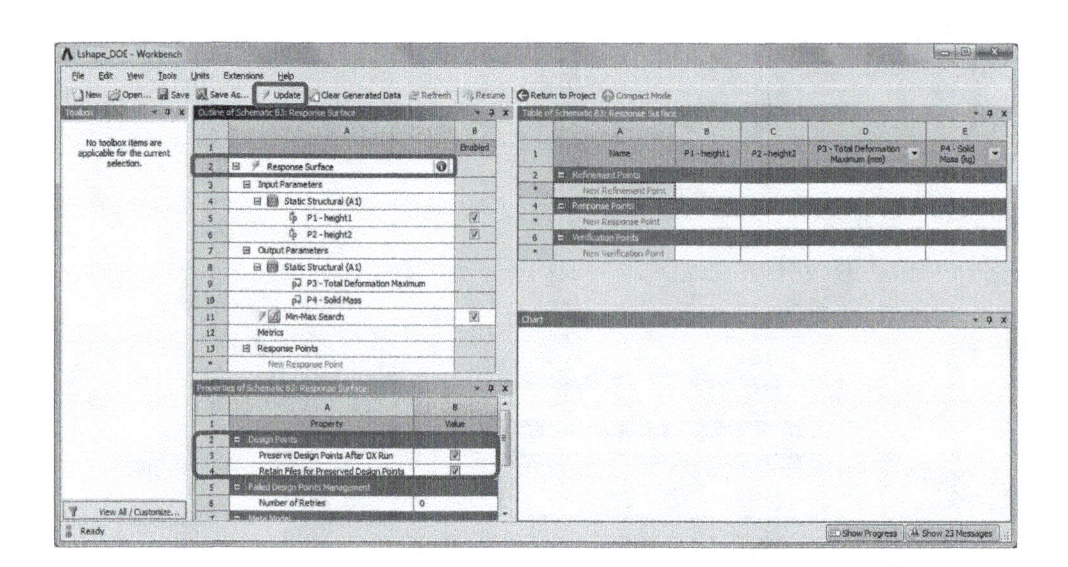

To view the response surface, click on the *Response* tab in the *Outline of Schematic B3: Response Surface*. Select *3D* for *Mode* in the *Properties of Outline: Response*. The response surface of the maximum total deformation as a function of height1 and height2 is shown in the *Response Chart*. Review the response surface chart, and click *Return to Project*.

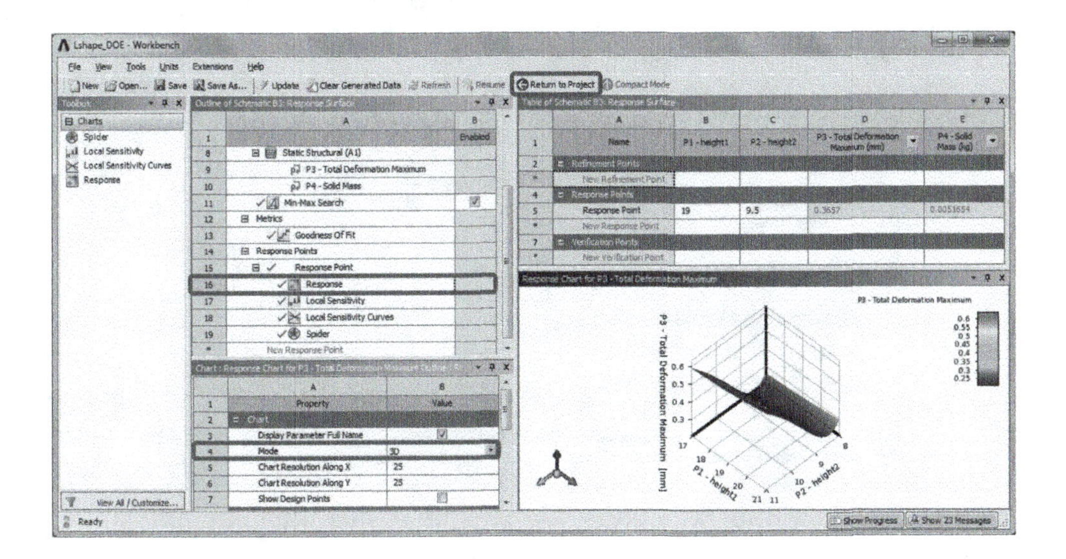

Step 14: Set Up Objectives and Constraints

Double-click the *Optimization* tab in the project schematic. Click on the *Objectives and Constraints* tab in the *Outline of Schematic B4: Optimization*. Set up the objective and constraints as shown below in the *Table of Schematic B4: Optimization*.

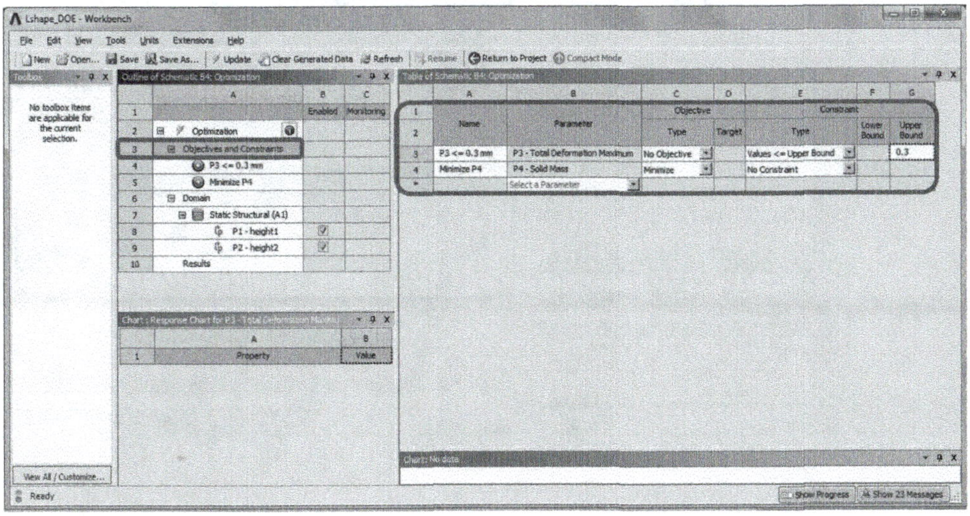

Click on the *Optimization* tab in the *Outline of Schematic B4: Optimization*. Select *Screening* for *Optimization Method* in the *Chart: Response Chart for P3*. Click *Update*.

Three candidates for best design are now listed in the *Table of Schematic B4: Optimization*. Review the best design candidates in the *Candidate Points* table. Click *Return to Project*.

Step 15: Update Current Design

Double-click the *Parameter Set* in the project schematic. In the *Table of Design Points* shown below, *DP1, DP2,* and *DP3* are feasible solutions that meet the design requirements. Among them, *DP3* achieves the best result in minimizing the total weight while meeting the maximum total deformation requirement. It has a value of *21 mm* for *height1* and *8 mm* for *height2*.

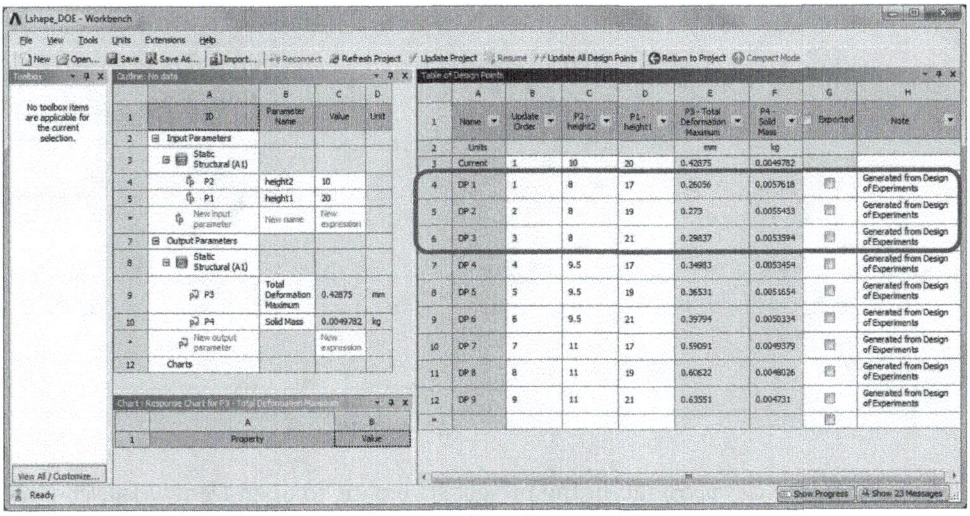

We will now update the current design with this design point (*DP3*), which can then be served as the base case for another round of optimization iteration if necessary. Right-click on the *DP3* tab in the *Table of Design Points*, and select *Copy inputs to Current* in the drop-down menu. This will update the *Current* design point with the selected design point *DP3*.

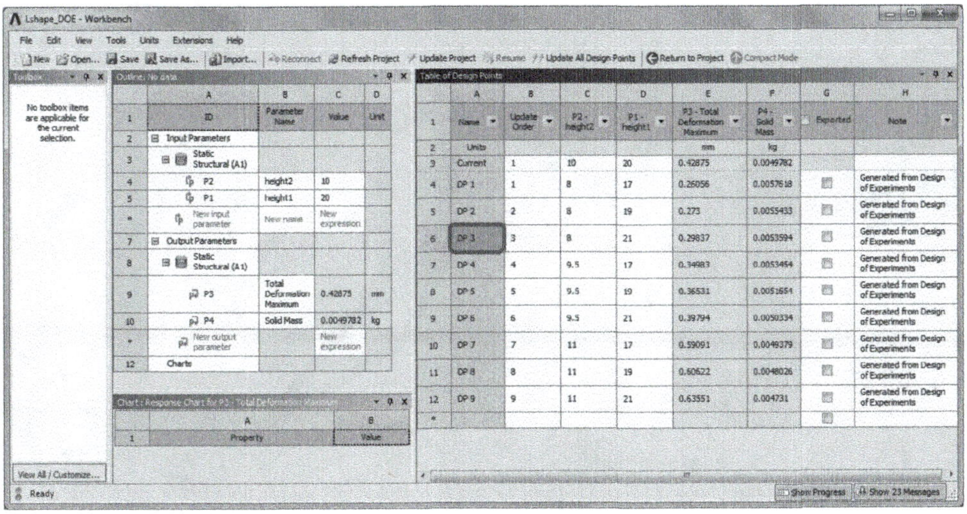

Right-click on the *Current* tab in the *Table of Design Points*, and select *Update Selected Design Points*. This will update the results of the *Current* design point. Click *Return to Project*.

Lshape_DOE - Workbench

File Edit View Tools Units Extensions Help

New Open... Save Save As... Import... Reconnect Refresh Project Update Project Resume Update All Design Points Return to Project Compact Mode

Toolbox: No toolbox items are applicable for the current selection.

Outline: No data

	A	B	C	D
1	ID	Parameter Name	Value	Unit
2	Input Parameters			
3	Static Structural (A1)			
4	P2	height2	8	
5	P1	height1	21	
*	New input parameter	New name	New expression	
7	Output Parameters			
8	Static Structural (A1)			
9	P3	Total Deformation Maximum	0.29837	mm
10	P4	Solid Mass	0.0053594	kg
*	New output parameter		New expression	
12	Charts			

	A	B
1	Property	Value

Table of Design Points

	A	B	C	D	E	F	G	H
1	Name	Update Order	P2 - height2	P1 - height1	P3 - Total Deformation Maximum	P4 - Solid Mass	Exported	Note
2	Units				mm	kg		
3	Current	1	8	21	0.29837	0.0053594		
4	DP 1	1	8	17	0.26056	0.0057618		Generated from Design of Experiments
5	DP 2	2	8	19	0.273	0.0055433		Generated from Design of Experiments
6	DP 3	3	8	21	0.29837	0.0053594		Generated from Design of Experiments
7	DP 4	4	9.5	17	0.34983	0.0053454		Generated from Design of Experiments
8	DP 5	5	9.5	19	0.36531	0.0051654		Generated from Design of Experiments
9	DP 6	6	9.5	21	0.39794	0.0050334		Generated from Design of Experiments
10	DP 7	7	11	17	0.59091	0.0049379		Generated from Design of Experiments
11	DP 8	8	11	19	0.60622	0.0048026		Generated from Design of Experiments
12	DP 9	9	11	21	0.63551	0.004731		Generated from Design of Experiments
*								

View All / Customize... Ready Show Progress Show 23 Messages

Double-click the *Geometry* tab in the project schematic to open *DesignModeler*. Note that the geometry displayed in the *Graphics* window is now updated based on the current design point (*21 mm* for *height1* and *8 mm* for *height2*). Close the *DesignModeler*.

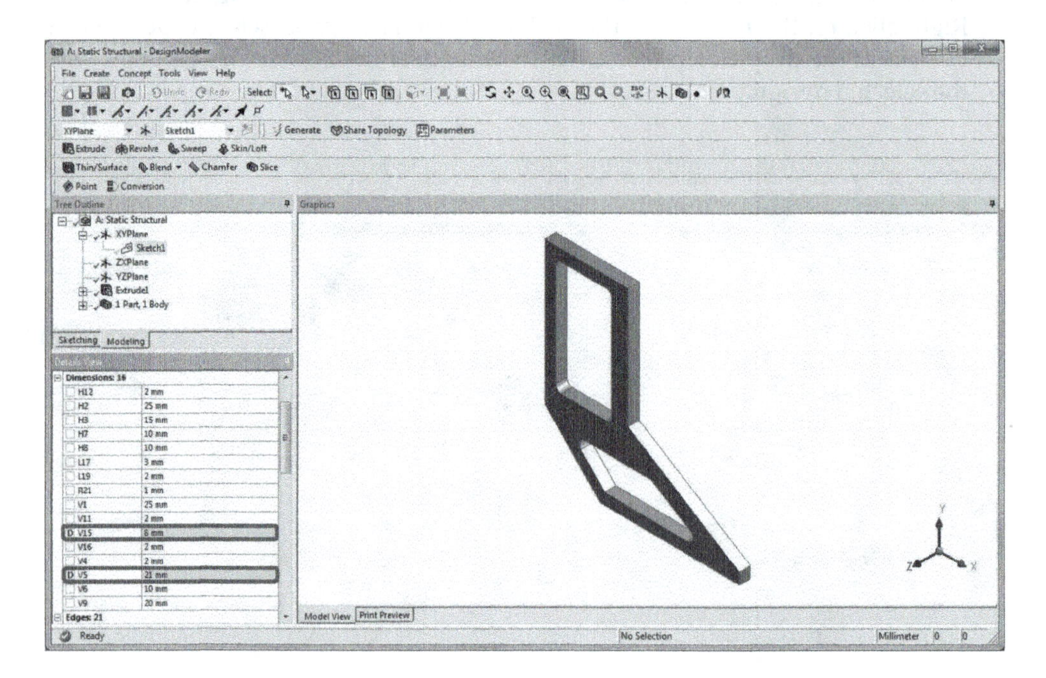

Double-click the *Results* tab in the project schematic to view the updated deformation results from the *Static Structural–Mechanical* program.

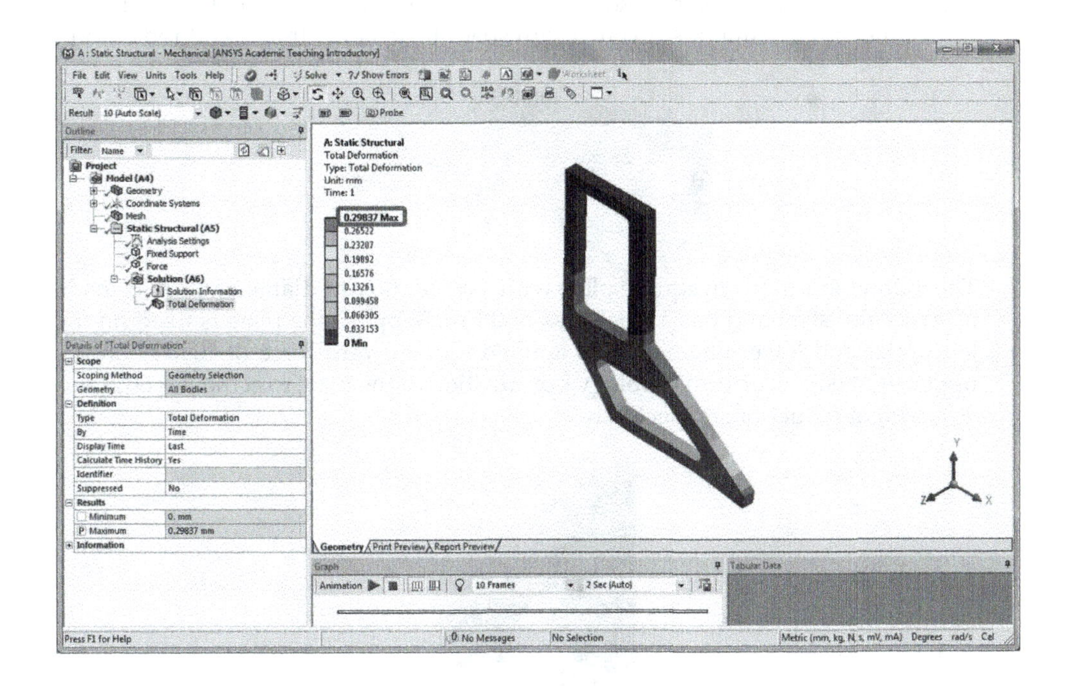

11.6 Summary

In this chapter, we briefly discussed design optimization problems and the type of optimizations. We also learnt how to apply FEA in design optimization studies. A case study is demonstrated using *ANSYS Workbench* to reduce the weight of an L-shaped structure.

11.7 Review of Learning Objectives

Now that you have finished this chapter you should be able to

1. Understand the basic concepts in design optimization and the steps in conducting an optimization of a design with the FEA
2. Perform optimization of structure designs using *ANSYS Workbench*

PROBLEMS

11.1 The 3-D design space of a steel bridge structure is 6 m long, 2 m tall, and 1 m wide, as shown below. Suppose the structure is fixed on the bottom two edges and is applied a uniform pressure load on the top surface. Perform topology optmization of the bridge using the given design space to achieve 80% weight reduction.

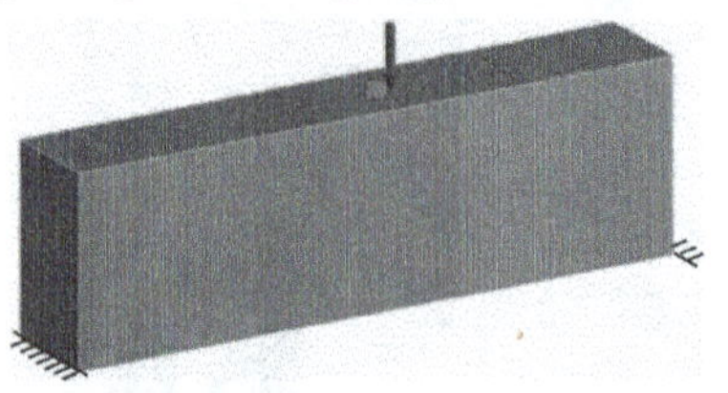

11.2 The following 1 m × 1 m square plate with a center hole of diameter 0.4 m is made of structural steel and has a thickness of 0.1 m. Suppose the plate is fixed on the left upper and lower edges, and is applied a downward force of 1000 N on the upper right edge. Perform topology optmization of the plate to achieve a 60% and 90% weight reduction, respectively.

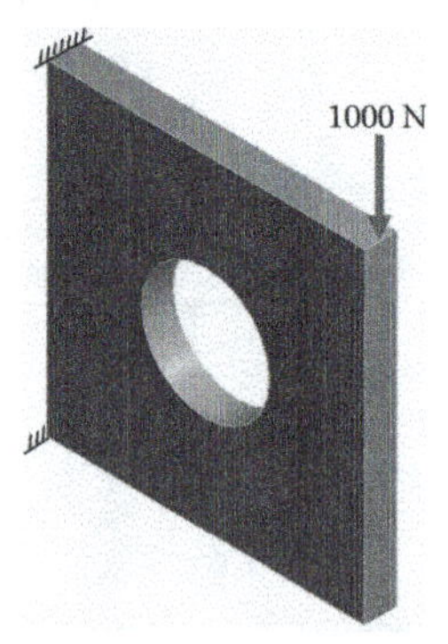

11.3 Optimize the stiffened aluminum panel design with clamped edges as shown below. In the initial design, the 100 in. × 100 in. panel has a thickness of 0.5 in. The stiffeners are each 100 in. long, 5 in. tall, and 0.25 in. thick, and divide the panel evenly into eight blocks of the same area. The design variables, that is, the panel thickness t, the longitudinal stiffener thickness t_{long}, and the lateral stiffener thickness t_{lat}, have the following range of variations: 0.4 in. $\leq t \leq$ 0.6 in., 0.2 in. $\leq t_{long} \leq$ 0.3 in., and 0.2 in. $\leq t_{lat} \leq$ 0.3 in. The optimization aims to minimize the panel's weight subject to the constraint of the panel's fundamental frequency $f_{base} \geq 130$ Hz.

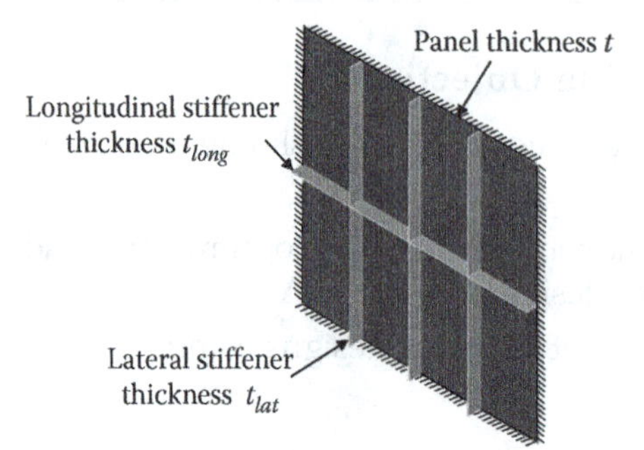

11.4 Optimize the design of the following L-shaped bracket made of structural steel with Young's modulus $E = 200$ GPa and Poisson's ratio $\nu = 0.3$. In the initial design, the L-shape has a length $L = 8$ cm, a height $h = 3$ cm, and an extrusion depth $d = 4$ cm, and is rigidly connected to a rectangular block of size 3 cm × 8 cm × 1 cm, as shown below. The bracket is fixed on the right end. A uniform pressure of 100 N/cm^2 is applied on the protruding top face on the left. The design variables, that is, L, h, and d, have the following range of variations: 6 cm $\leq L \leq$ 8 cm, 2.5 cm $\leq h \leq$ 3 cm, and 3 cm $\leq d \leq$ 4 cm. The optimization aims to minimize the bracket weight subject to the maximum allowed displacement limit of 0.1 mm and stress limit of 100 MPa.

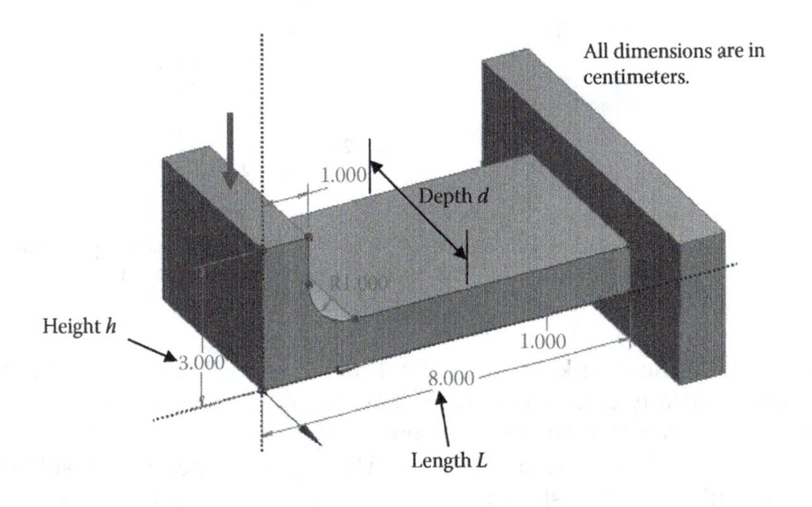

11.5 Perform topology and size optimization of the following bookshelf bracket design. The bracket is 2 mm thick and is made of structural steel. The vertical side of the bracket is fixed, and the horizontal side is applied a uniform pressure of 25 MPa. (a) The objective of the topology optimization is to achieve a weight reduction of 60%. (b) Redesign the bracket based on the results from topology optimization. For the parametric optimization of the redesigned bracket, minimize the bracket weight subject to the maximum allowed deformation constraint of 0.5 mm and stress constraint of 100 MPa.

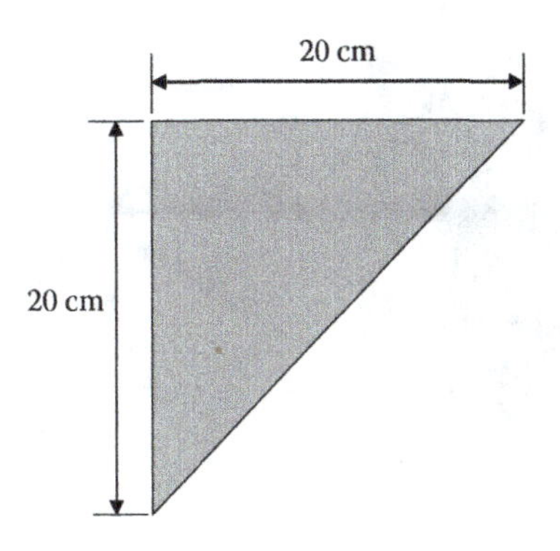

11.6 The original design space along with the constraints and loading conditions is given below for structural optimization of a bike frame. The coordinate dimensions are all in millimeters. Assume a 5 mm thickness for the design space and perform topology optimization for the frame structure to achieve a weight reduction target of 60% and 80%, respectively. Use structural steel for the frame and discuss on the results of optimization.

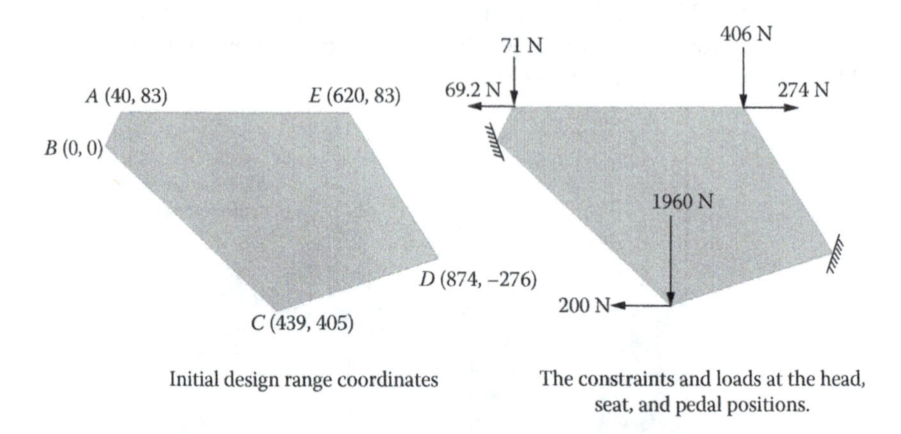

Initial design range coordinates

The constraints and loads at the head, seat, and pedal positions.

11.7 The following heat sink consists of a 1 m tall central tube and 12 radial fins that are uniformly distributed every 30°C around the central tube. The central tube has an outer diameter of 0.4 m and an inner diameter of 0.36 m. Each fin is 0.048 m wide, 0.4 m long, and 1 m tall. The vapor temperature inside the tube is 315°C with a heat transfer coefficient of 2.94×10^5 W/(m^2°C). The surrounding air temperature is 0°C with a heat transfer coefficient of 40.88 W/(m^2°C). The heat sink is made of aluminum alloy with a thermal conductivity of 175 W/(m K), a density of 2770 kg/m^3, and a specific heat of 870 J/(kg K). Optimize the fin width W and the fin length L to minimize the use of material while achieving a minimum heat dissipation of 75 kW per unit length. The design variables W and L have the following range of variations: 0.03 m < W < 0.052 m, and 0.2 m < L < 0.45 m.

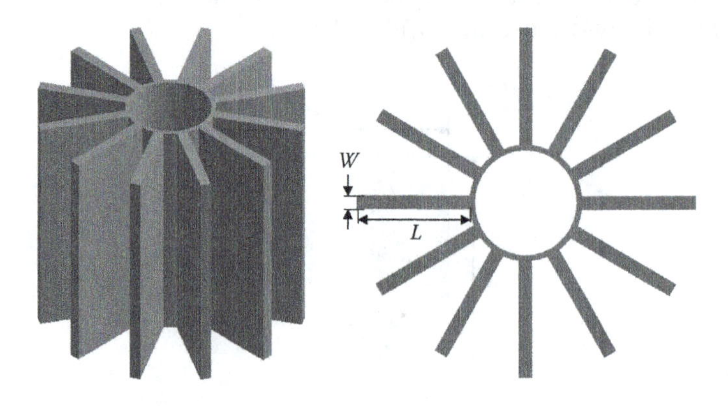

11.8 The original design space along with the constraints and loading conditions is given below for structural optimization of a chair. Assume that the chair is made of Douglas fir with $E = 10^4$ MPa and $\nu = 0.3$. The chair is subjected to two surface loads with force magnitudes of 980 and 98 N, respectively, as shown in the figure. The four corners at the bottom of the chair are fixed. Optimize the chair design to achieve a 90% weight reduction.

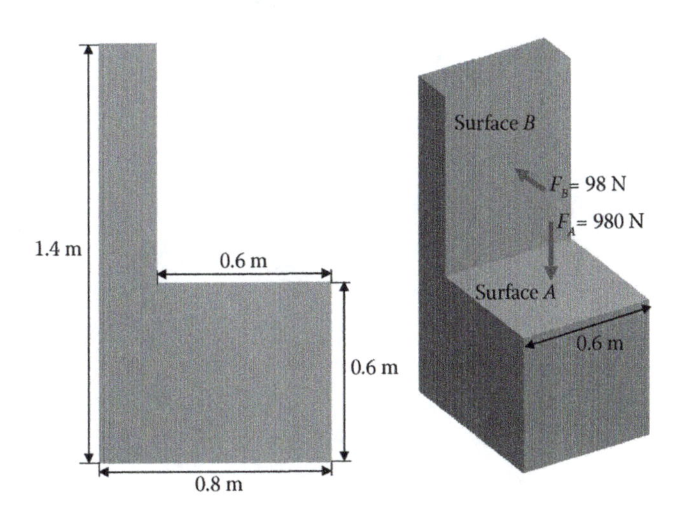

12

Failure Analysis

12.1 Introduction

Structures such as bridges, aircraft, and machine components can fail in many different ways. An overloaded structure may experience permanent deformation, which can lead to compromised function or failure of the entire structure. When subjected to millions of small repeated loads, a structure may have a slow growth of surface cracks that can cause material strength degradation and a sudden failure. When a slender structure is loaded in compression, it may undergo an unexpected large deformation and lose its ability to carry loads. Failure analysis plays an important role in improving the safety and reliability of an engineering structure. In this chapter, we will discuss some of the topics related to structural failure analysis. The concepts of static, fatigue, and buckling failures will be introduced, along with examples using *ANSYS® Workbench*.

12.2 Static Failure

Under static loading conditions, material failure can occur when a structure is stressed beyond the elastic limit. There are two types of material failures from static loading, namely, ductile failure and brittle failure. The main difference between the two failure types is the amount of plastic deformation a material experiences before fracture. As illustrated in Figure 12.1, ductile materials tend to have extensive plastic deformation before fracture, while brittle materials are likely to experience no apparent plastic deformation before fracture. For more information on ductile and brittle failure, please see Reference [18].

12.2.1 Ductile Failure

In this section, two common theories on ductile failure, that is, the maximum shear stress theory (or Tresca Criterion) and the distortion energy theory (or von Mises Criterion), are reviewed.

12.2.1.1 Maximum Shear Stress Theory (Tresca Criterion)

According to the maximum shear stress theory, ductile failure occurs when the maximum shear stress τ^{max} exceeds one half of the material yield strength S_y. Suppose that a factor of safety n is considered in the design. The design equation is given as

$$\tau^{max} < \frac{S_y}{2n} \tag{12.1}$$

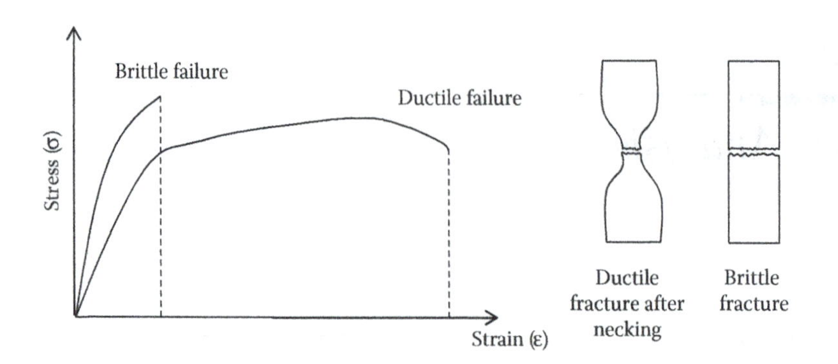

FIGURE 12.1
Stress–strain curves for brittle and ductile materials.

12.2.1.2 Distortion Energy Theory (von Mises Criterion)

Based on the distortion energy theory, ductile failure occurs when the maximum von Mises stress σ_e^{max} exceeds the material yield strength S_y. Applying a factor of safety n to the design, the design equation becomes:

$$\sigma_e^{max} < \frac{S_y}{n} \tag{12.2}$$

12.2.2 Brittle Failure

For brittle materials, the two common failure theories are the maximum normal stress theory and the Mohr–Coulomb theory.

12.2.2.1 Maximum Normal Stress Theory

The maximum normal stress theory states that brittle failure occurs when the maximum principal stress exceeds the ultimate tensile (or compressive) strength of the material. Suppose that a factor of safety n is considered in the design. The safe design conditions require that

$$-\frac{S_{uc}}{n} < \{\sigma_1, \sigma_2, \sigma_3\} < \frac{S_{ut}}{n} \tag{12.3}$$

where σ_1, σ_2, σ_3 are the principal stresses in the three directions, S_{ut} and S_{uc} are the ultimate tensile strength and the ultimate compressive strength, respectively. To avoid brittle failure, the principal stresses at any point in a structure should lie within the square failure envelope illustrated in Figure 12.2 based on the maximum normal stress theory.

12.2.2.2 Mohr–Coulomb Theory

The Mohr–Coulomb theory predicts brittle failure by comparing the maximum principal stress with the ultimate tensile strength S_{ut} and the minimum principal stress with the

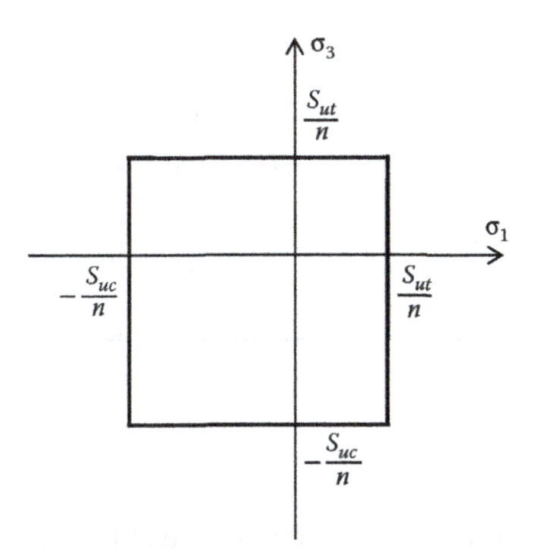

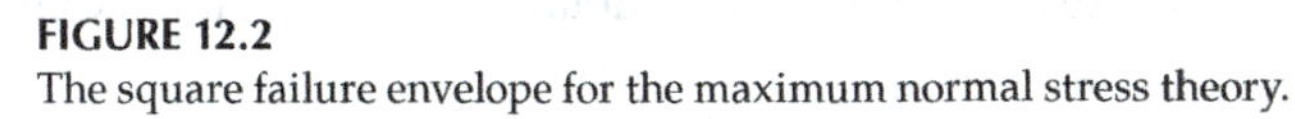
FIGURE 12.2
The square failure envelope for the maximum normal stress theory.

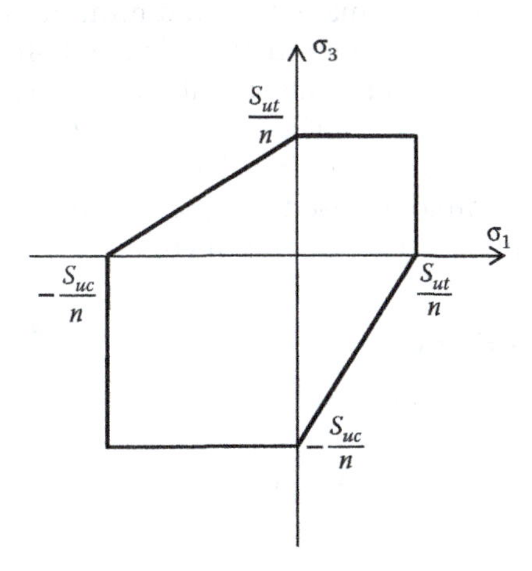

FIGURE 12.3
The hexagonal failure envelope for the Mohr–Coulomb theory.

ultimate compressive strength S_{uc}. Suppose that a factor of safety n is considered in the design. The safe design conditions require that the principal stresses lie within the hexagonal failure envelope illustrated above in Figure 12.3.

12.3 Fatigue Failure

Under dynamic or cyclic loading conditions, fatigue can occur at stress levels that are considerably lower than the yield or ultimate strengths of the material. Fatigue failures are

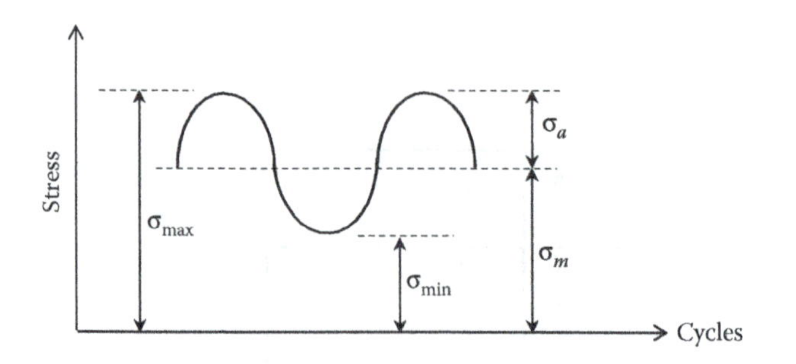

FIGURE 12.4
A typical fatigue stress cycle.

often sudden with no advanced warning and can lead to catastrophic results. In this section, theories related to fatigue failure are reviewed. For more details, we refer the reader to Reference [18].

Design of components subjected to cyclic load involves the concept of mean and alternating stresses (see Figure 12.4 for an illustration of a typical fatigue stress cycle). The mean stress, σ_m, is the average of the maximum and minimum stresses in one cycle, that is, $\sigma_m = (\sigma_{max} + \sigma_{min})/2$. The alternating stress, σ_a, is one-half of the stress range in one cycle, that is, $\sigma_a = (\sigma_{max} - \sigma_{min})/2$. For most materials, there exist a fatigue limit, and parts having stress levels below this limit are considered to have infinite fatigue life. The fatigue limit is also referred to as the endurance limit, S_e.

In the following, three commonly used fatigue failure theories, namely, Soderberg, Goodman, and Gerber failure criteria, are presented.

12.3.1 Soderberg Failure Criterion

This theory states that the structure is safe if

$$\frac{\sigma_a}{S_e} + \frac{\sigma_m}{S_y} < \frac{1}{n} \tag{12.4}$$

where σ_a is the alternating stress, σ_m the mean stress, S_e the endurance limit, S_y the yield strength, and n the factor of safety. When the alternating stress σ_a is plotted versus the mean stress σ_m, the Soderberg line can be drawn between the points of $\sigma_a = S_e/n$ and $\sigma_m = S_y/n$. If the stress is below the line, then the design is safe. This is a conservative criteria based on the material yield strength S_y (see Figure 12.5).

12.3.2 Goodman Failure Criterion

This theory states that the structure is safe if

$$\frac{\sigma_a}{S_e} + \frac{\sigma_m}{S_{ut}} < \frac{1}{n} \tag{12.5}$$

where S_{ut} is the ultimate tensile strength and n is the factor of safety. When the alternating stress σ_a is plotted versus the mean stress σ_m, the Goodman line can be drawn between

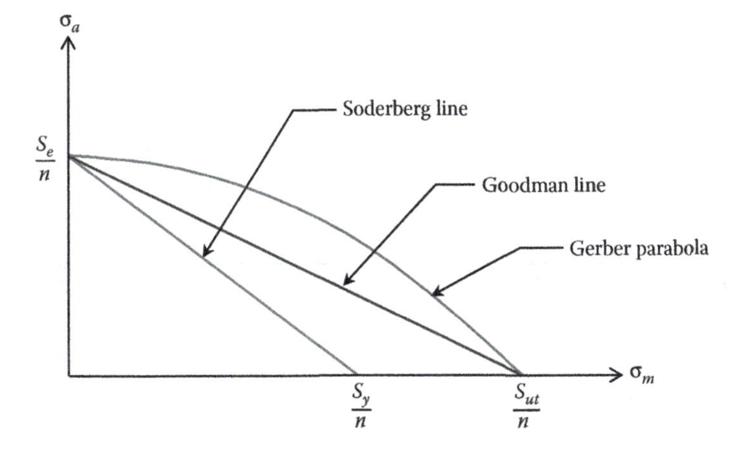

FIGURE 12.5
Fatigue diagrams showing various criteria of failure.

the points of $\sigma_a = S_e/n$ and $\sigma_m = S_{ut}/n$. If the stress is below the line, then the design is safe. This is a less conservative criteria based on the material ultimate strength S_{ut} (see Figure 12.5).

12.3.3 Gerber Failure Criterion

This theory states that the structure is safe if

$$\frac{\sigma_a}{S_e} + \left(\frac{\sigma_m}{S_{ut}}\right)^2 < \frac{1}{n} \tag{12.6}$$

where S_{ut} is the ultimate tensile strength and n is the factor of safety. When the alternating stress σ_a is plotted versus the mean stress σ_m, the Gerber parabola can be drawn between the points of $\sigma_a = S_e/n$ and $\sigma_m = S_{ut}/n$. If the stress is below the parabola, then the design is safe. This is a less conservative criteria based on the material ultimate strength S_{ut} (see Figure 12.5).

In general, most structures will fail eventually after a certain number of cycles of repeated loading. In fatigue life analysis, *S–N* curves are widely used for the life estimation, especially in high-cycle fatigue situations where the cyclic loading is elastic and the number of cycles to failure is large. An *S–N* curve (see Figure 12.6) relates the alternating stress, σ_a, to the number of cycles to failure, N_f, and is generated from fatigue tests on samples of the given material.

Most *S–N* curves are developed based on zero mean stress. However, many loading applications involve a nonzero mean stress. The Soderberg, Goodman, and Gerber theories mentioned above can be used to estimate the mean stress effects on fatigue life. Note that there is little difference in the three theories when the mean stress σ_m is relatively small compared to the alternating stress σ_a.

Take the Gerber theory as an example. To estimate a structure's fatigue life for a given combination of mean stress σ_m and alternating stress σ_a, one may first solve for the effective alternating stress σ_e, based on the following modified Gerber equation:

$$\frac{\sigma_a}{\sigma_e} + \left(\frac{\sigma_m}{S_{ut}}\right)^2 = \frac{1}{n} \tag{12.7}$$

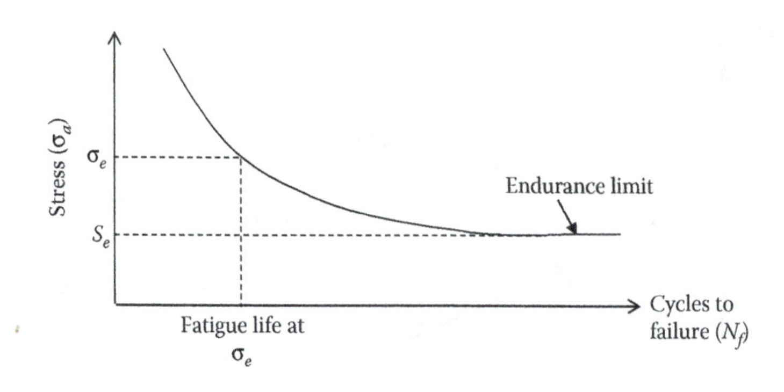

FIGURE 12.6
Use of the *S–N* curve for fatigue life estimation.

Once the effective alternating stress, σ_e, is obtained from Equation 12.7, the fatigue life for the corresponding effective alternating stress σ_e can be determined from the *S–N* curve of the given material (see Figure 12.6). If the effective alternating stress, σ_e, is kept below threshold of the endurance limit, S_e, the structure is considered to have an infinite life.

12.4 Buckling Failure

Under compressive loads, slender members such as columns can become structurally unstable, leading to a sudden failure. The sudden loss of stability under compressive loads is referred to as buckling [19]. Buckling failure usually occurs within the elastic range of the material. The actual compressive stress in the structure at failure is often smaller than the ultimate compressive strength of the material.

In buckling analysis, one may be interested in finding the critical load, that is, the maximum compressive load a structure can support before buckling occurs, or the buckling mode shapes, that is, shapes the structure takes while buckling. The critical load is a structure's elastic stability limit, which is adequate to force the structure to buckle into the first mode shape. A structure can potentially have infinitely many buckling modes, and higher critical load values are required to trigger higher-order buckling modes (see Figure 12.7 for an illustration of buckling failure of a slender column, and Figure 12.8 for different buckling mode shapes). Because buckling failure can lead to catastrophic results, a high factor of safety (at least >3) is often used for critical load calculations.

When developing a model for buckling, one should pay attention to the boundary conditions as they have a considerable effect on a structure's buckling behavior. For example, a column with both ends pinned can buckle more freely than a column with both ends fixed. Also note that bending moments acting on a structural member, for example, due to eccentricity of an axial load, may play a role in affecting buckling and should not be neglected in the analysis in general.

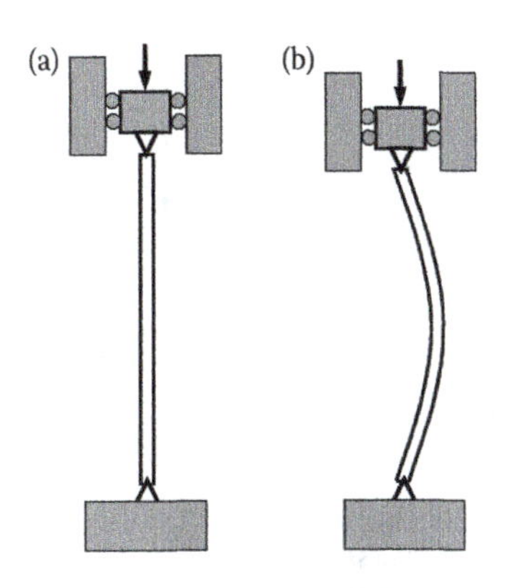

FIGURE 12.7
Buckling of a slender column: (a) load < critical load, (b) load > critical load. (http://en.wikipedia.org/wiki/Buckling.)

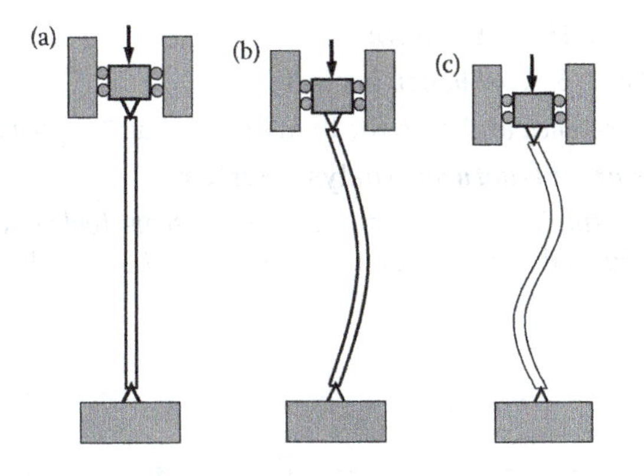

FIGURE 12.8
Buckling of a slender column: (a) unbuckled, (b) first buckling mode, and (c) second buckling mode.

12.5 Case Studies with *ANSYS Workbench*

Problem Description: A dog-bone shaped specimen is examined for static, fatigue, and buckling failures. The specimen is made of structural steel with geometric dimensions shown below. The bottom face of the specimen is fixed, and the top face of the specimen is applied a static pressure load of 50 MPa. (a) Determine whether or not the specimen undergoes plastic deformation under the given static pressure load. (b) If the static pressure load is changed into a fully reversed cyclic load with a magnitude of 50 MPa, find the life of the specimen, and also determine whether or not fatigue failure occurs in the specimen assuming a design life of 10^6 cycles. (c) Determine whether or not the specimen buckles under the given static pressure load, and obtain the first three buckling mode shapes.

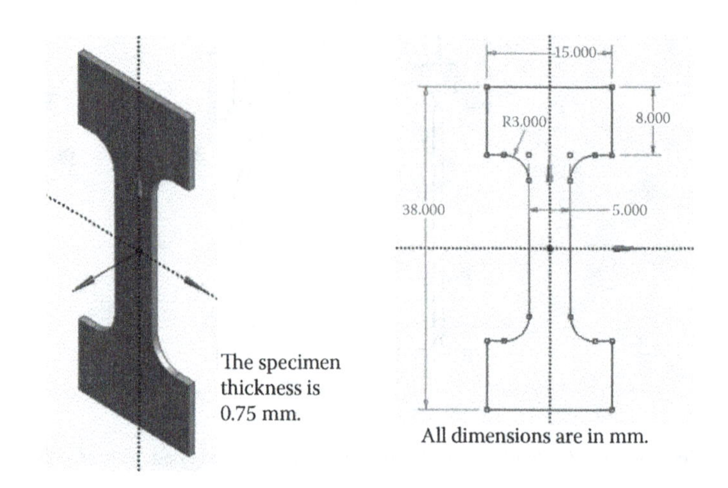

Solution steps for portion (A and B):

Step 1: Start an *ANSYS Workbench* Project

Launch *ANSYS Workbench* and save the blank project as *"Dogbone.wbpj."*

Step 2: Create a *Static Structural* Analysis System

Drag the *Static Structural* icon from the *Analysis Systems Toolbox* window and drop it inside the highlighted green rectangle in the *Project Schematic* window.

Step 3: Launch the *DesignModeler* Program

Double-click the *Geometry* cell to launch *DesignModeler,* and select *"Millimeter"* in the *Units* pop-up window.

Step 4: Create the Geometry

Click on the *Sketching* tab. Draw a sketch of the dog bone shape on the *XY Plane*, as shown below. An entity named *Sketch1* will be shown underneath *XY Plane* of the model's *Tree Outline*.

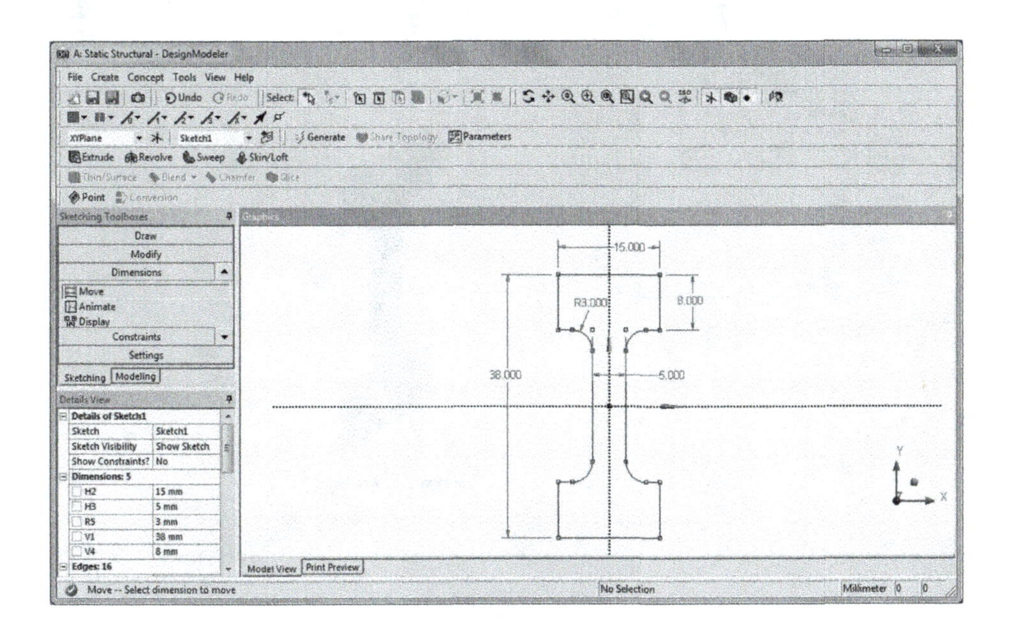

Extrude *Sketch1* to create a *0.75 mm* thick solid body, as shown below.

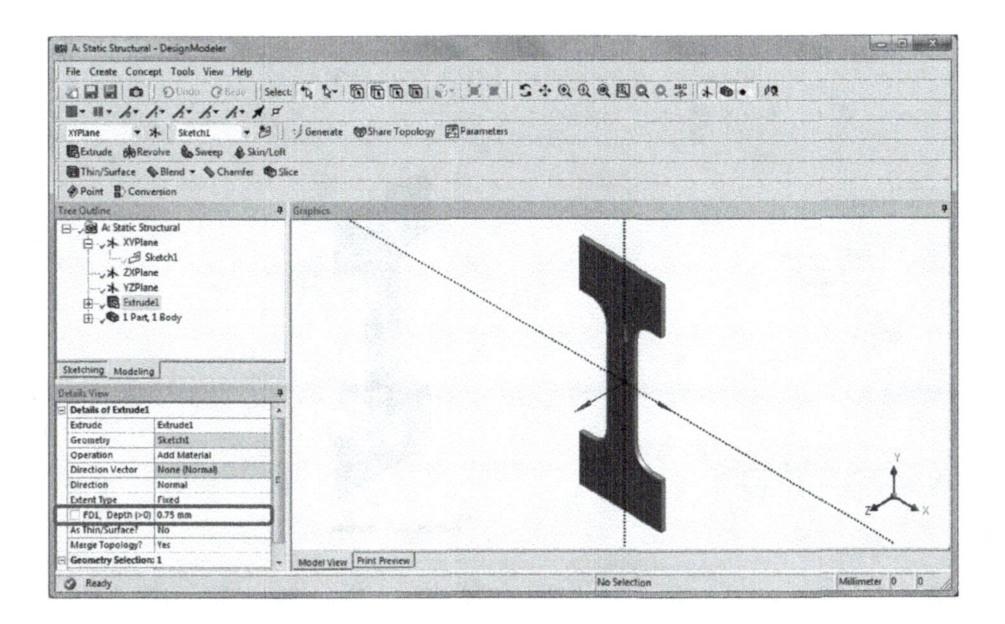

Step 5: Launch the *Static Structural* Program

Double-click on the *Model* cell to launch the *Static Structural* program. Change the *Units* to *Metric (mm, kg, N, s, mV, mA).*

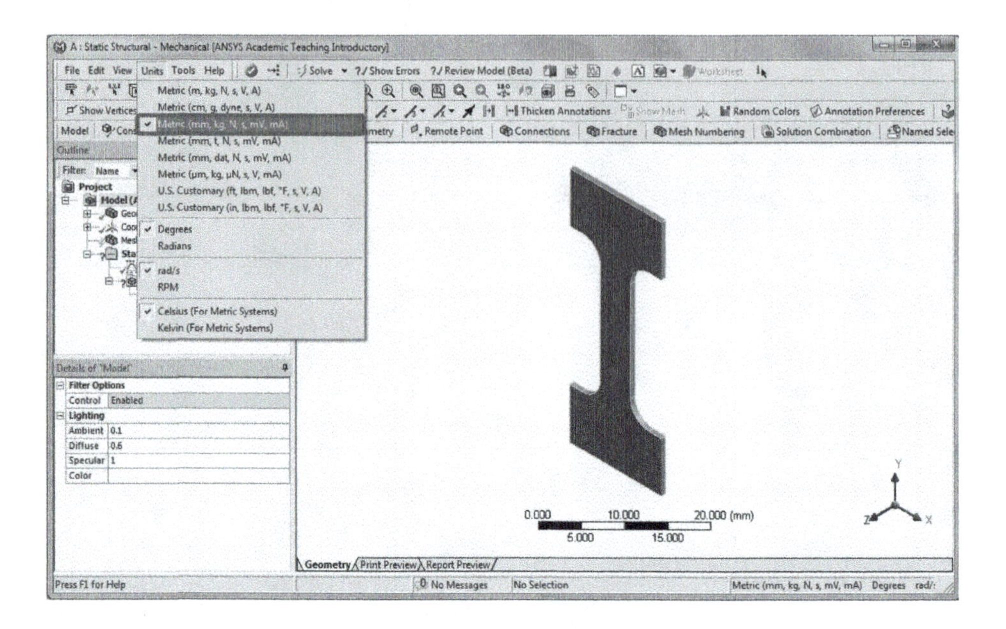

Step 6: Generate Mesh

Click on *Mesh* in the *Outline* tree. In the *Details of "Mesh,"* enter *"0.5 mm"* for the *Element Size*. Right-click on *Mesh* and select *Generate Mesh*.

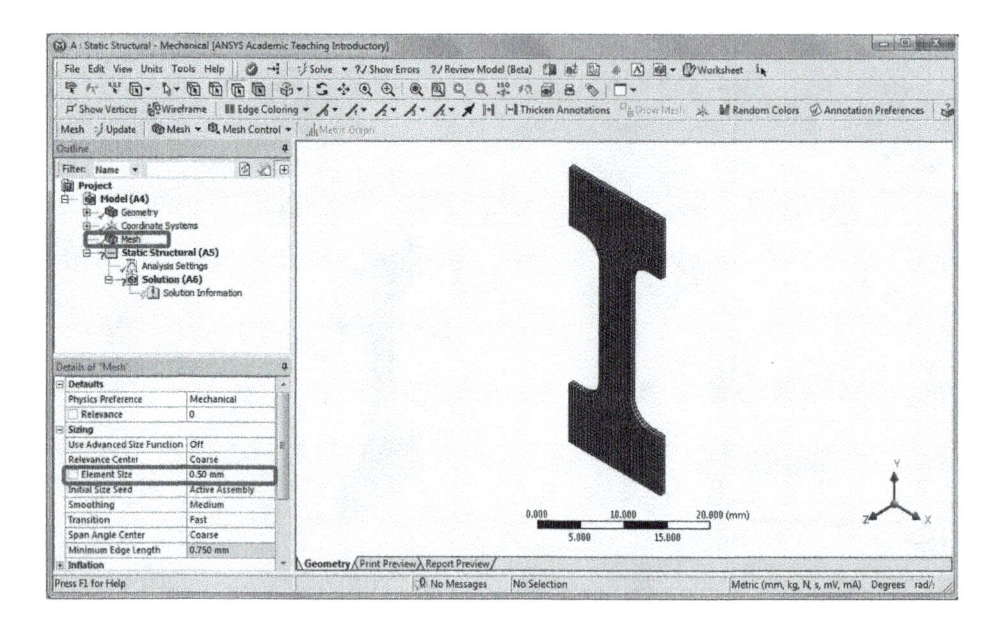

Step 7: Apply Boundary Conditions

Right-click on *Static Structural (A5)*. Choose *Insert* and then *Fixed Support* from the context menu. Click on the bottom face, and apply it to the *Geometry* selection in the *Details of "Fixed Support."* The bottom face of the dog bone shape is now fixed as shown below.

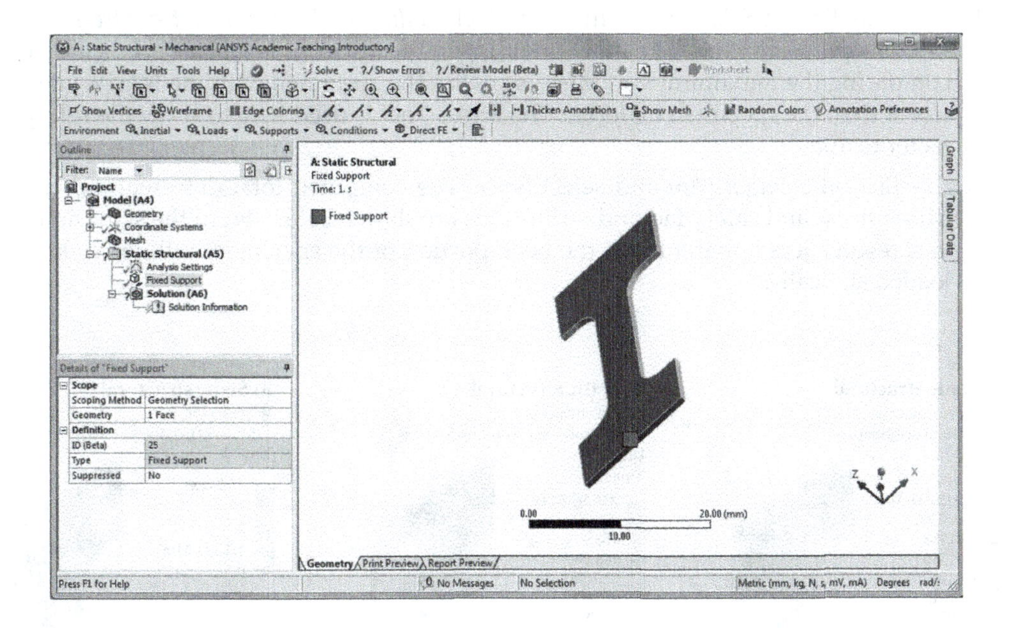

Step 8: Apply Loads

Right-click on *Static Structural (A5)*. Choose *Insert* and then *Pressure*. In the *Details of "Pressure,"* apply a *50 MPa* pressure to the top face, as shown below.

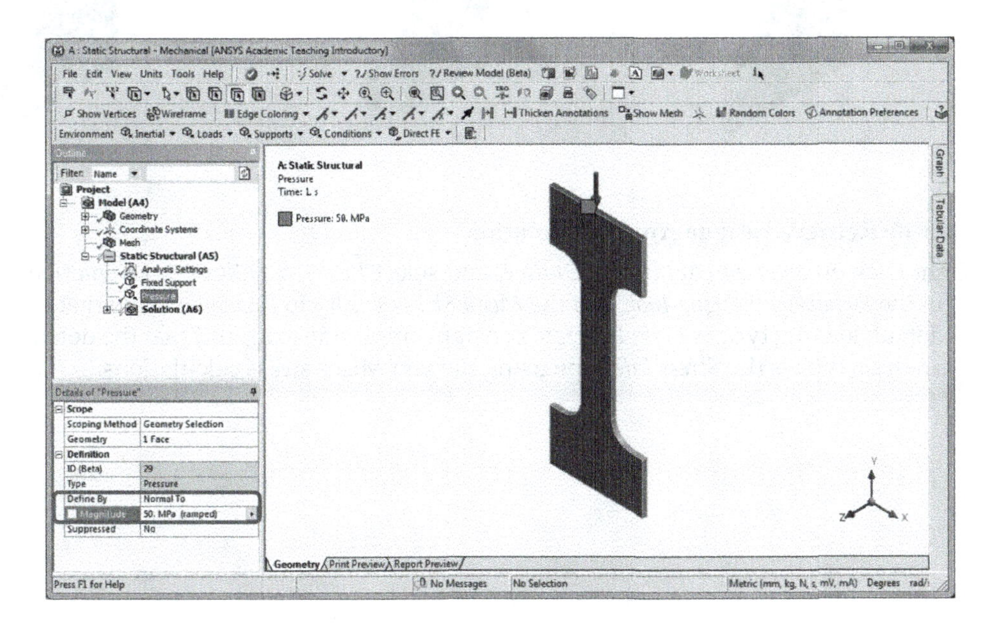

Step 9: Retrieve Static Analysis Results

First, insert a *Total Deformation* item by right-clicking on *Solution (A6)* in the project *Outline.*

Then, insert an *Equivalent Stress* item by right-clicking on *Solution (A6)* in the project *Outline.*

Next, right-click on *Solution (A6)* in the project *Outline,* and select *Insert -> Stress Tool -> Max Equivalent Stress.* The initial yielding in the test sample may be predicted by comparing the maximum von Mises stress in the specimen with the tensile yield strength of the specimen material. The Stress Tool is used here to show the safety factor results.

Right-click on *Solution (A6)* and select *Solve.* The computed total deformation, von Mises stress and safety factor distributions are shown below. From the static analysis results, it is apparent that the neck portion of the specimen will not yield if loaded statically.

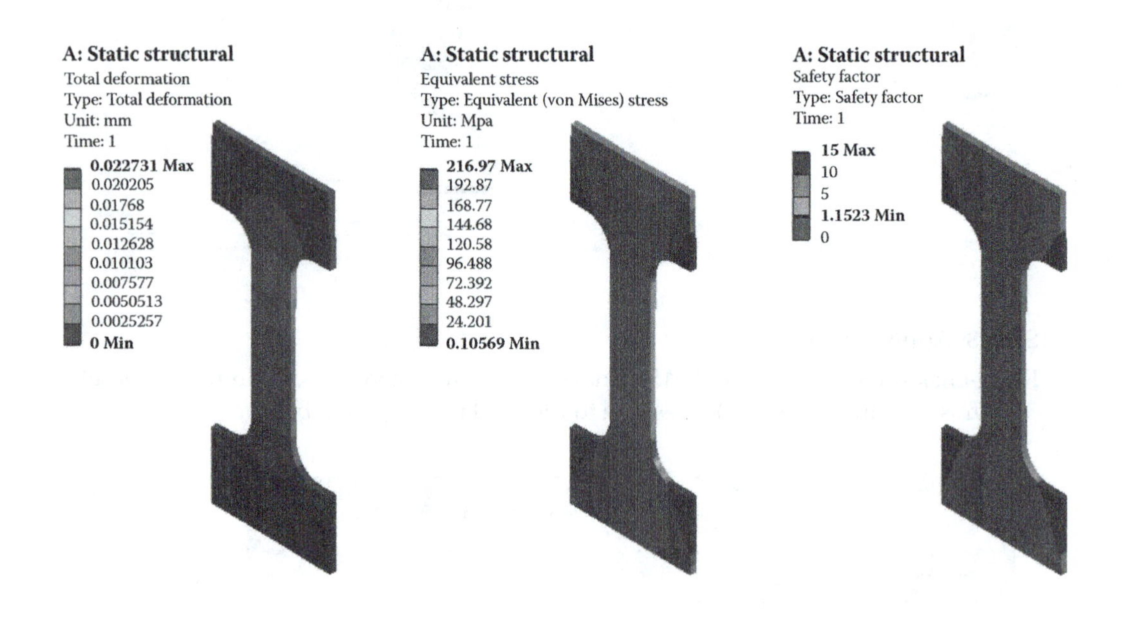

Step 10: Retrieve Fatigue Analysis Solution

Right-click on *Solution (A6)* in the *Outline,* and select *Insert -> Fatigue -> Fatigue Tool.* In the *Details of "Fatigue Tool,"* set the *Mean Stress Theory* to *Goodman.* Note that the default loading type is *Fully Reversed* constant amplitude load, and that the default analysis type is the *Stress Life* type using the von Mises stress calculations.

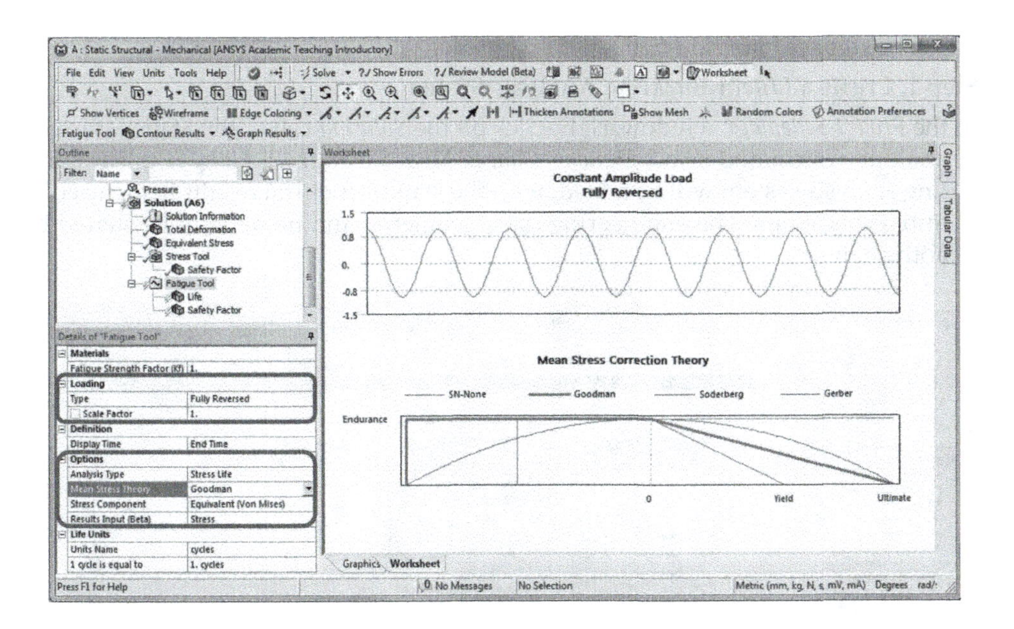

Right-click on the *Fatigue Tool* in the *Outline*, and select *Insert -> Life*. Next, right-click on the *Fatigue Tool* and select *Insert -> Safety Factor*. In the *Details of "Safety Factor,"* change the *Design life* from the default value of *10^9 cycles* to *10^6 cycles*. Finally, right-click on the *Fatigue Tool* and select *Evaluate All Results*. From the fatigue analysis results, the shortest life is at the undercut fillets (19,079 cycles) followed by the neck portion of the specimen. The neck portion of the specimen has a fairly small safety factor with a minimum value of 0.3973. The results show that the specimen will not survive the fatigue testing assuming a design life of 10^6 cycles.

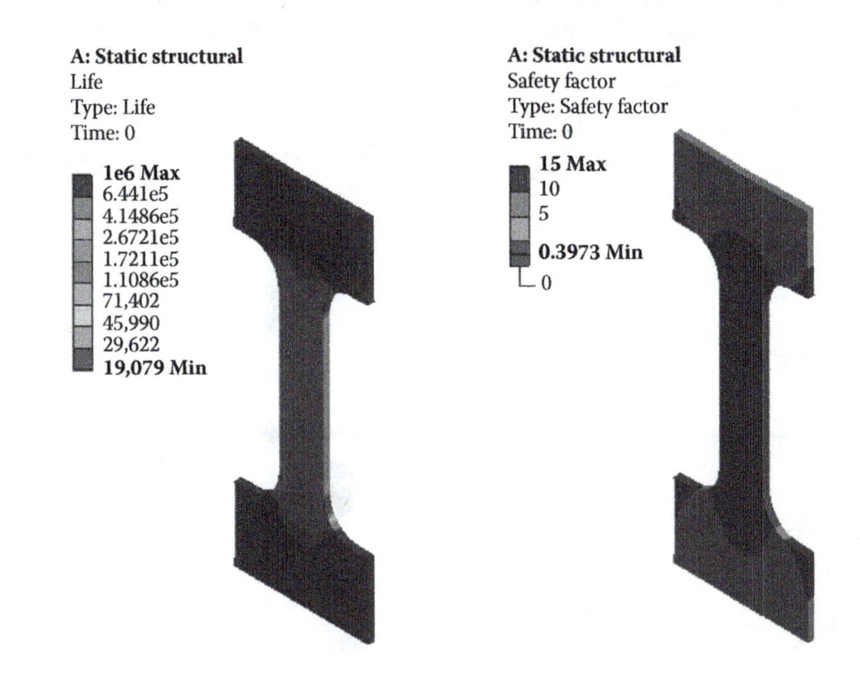

Solution steps for portion (C):

Step 1: Create a *Linear Buckling* Analysis System

In the *Project Schematic* window, right-click on the *Solution* cell of the *Static Structural* analysis system and select *Transfer Data to New -> Linear Buckling.* A linear buckling analysis system will be added, with the static structural results being used as initial conditions. The engineering data, geometry, and model will be shared by both analyses.

Step 2: Launch the *Multiple Systems–Mechanical* Program

Double-click the *Setup* cell of the *Linear Buckling* system to launch the *Multiple Systems–Mechanical* program. Click on the *Analysis Settings* under *Linear Buckling (B5)* in the *Outline*. In the *Details of "Analysis Settings,"* set the *Max Modes to Find* to 3.

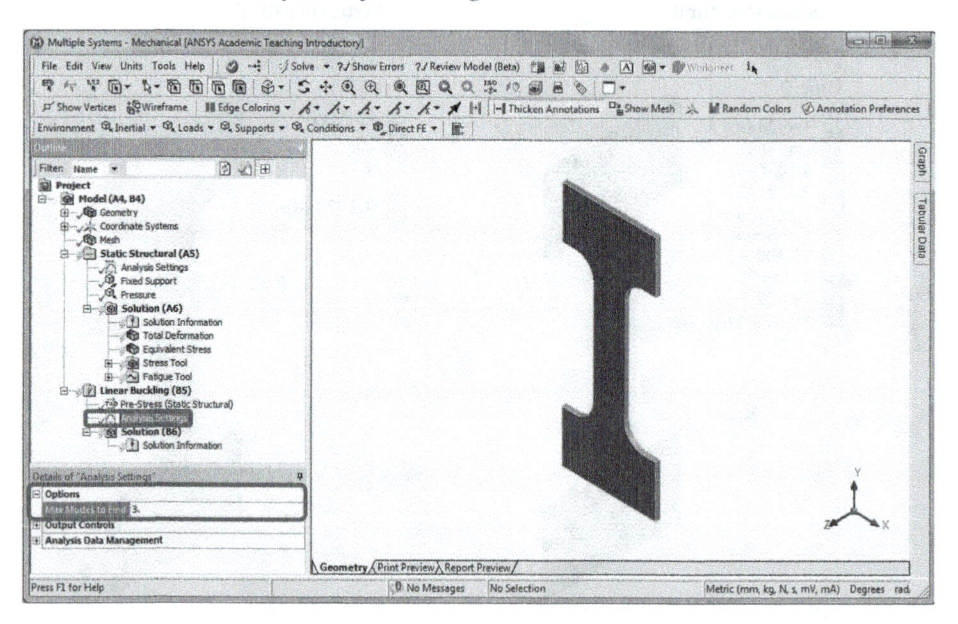

Step 3: Retrieve Linear Buckling Analysis Results

Insert three *Total Deformation* items by right-clicking on *Solution (B6)* in the *Outline*. In the *Details of "Total Deformation,"* set *Mode* to 1. In the *Details of "Total Deformation 2,"* set *Mode* to 2. In the *Details of "Total Deformation 3,"* set *Mode* to 3.

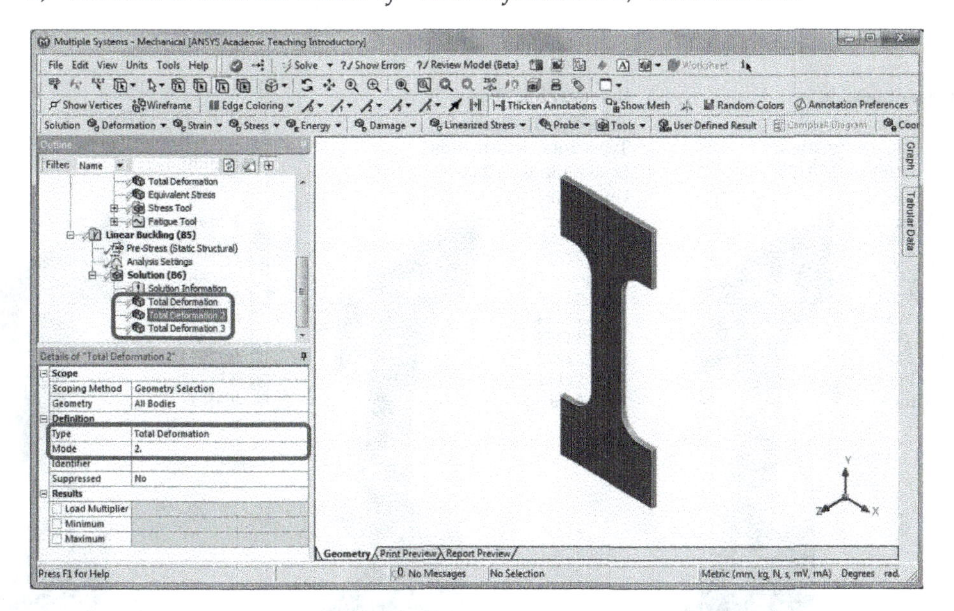

Right-click on *Solution (B6)* and select *Solve* to view the buckling modes. To use the default window layout as shown below, select *View* -> *Windows* -> *Reset Layout* from the top menu bar. Note that the load modifier for the first buckling mode is found to be 0.78173. To find the load required to buckle the structure, multiply the applied load by the load multiplier. For example, the first buckling load will be 39.0865 MPa (0.78173 × 50 MPa), thus the applied pressure of 50 MPa will cause the specimen to buckle. In the *Graph* window, you can play the buckling animation.

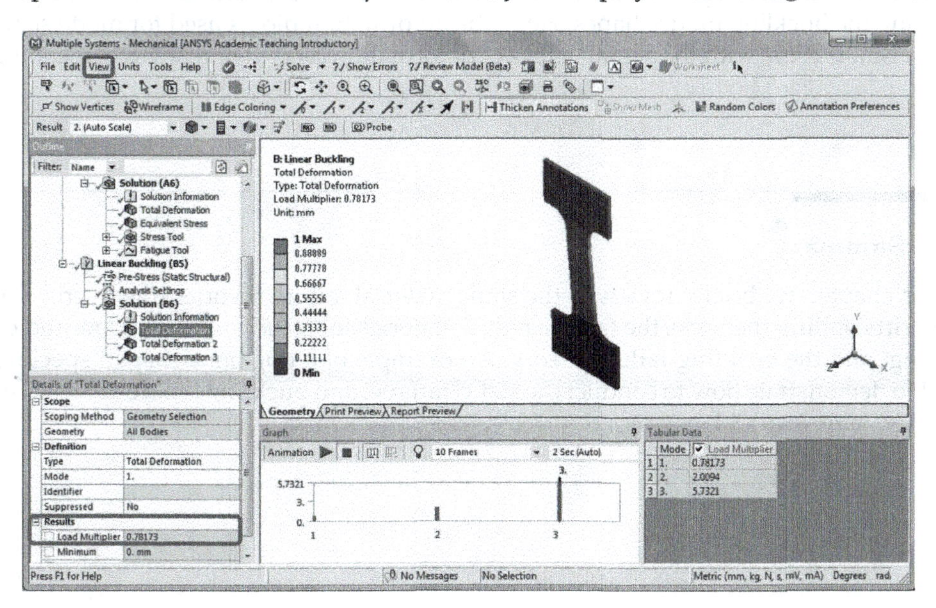

The following figures show the first three buckling mode shapes. The corresponding load multipliers for the first, second, and third mode shapes are 0.78173, 2.0094, and 5.7321, respectively.

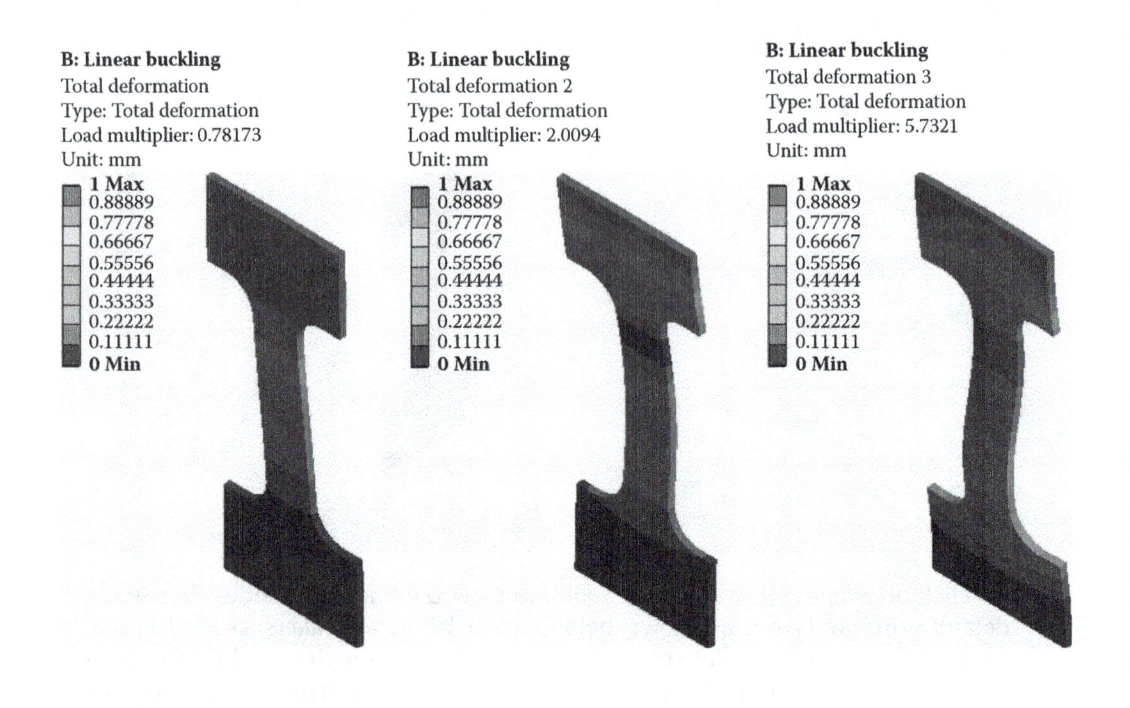

Note that the max value in the total deformation plots is scaled to 1 when displaying the buckling mode shapes. Here, the deformation plot is used for mode shape visualization, with the actual values of deformation carrying no physical meaning.

12.6 Summary

In this chapter, we briefly reviewed the static material failure theories (the ductile failure and brittle failure theories), the fatigue failure theories for structures under low and cyclic loading, and the buckling failure theory. An example of a dog-bone shaped specimen is used to demonstrate how to conduct the static, fatigue, and buckling failure analyses using the FEM with *ANSYS Workbench*.

12.7 Review of Learning Objectives

Now that you have finished this chapter you should be able to

1. Understand the failure theories for structures under static load
2. Understand the fatigue failure theories for structures under cyclic load
3. Understand the buckling failure of a slender structure under static compressive load
4. Perform static failure analysis of structures using *ANSYS Workbench*
5. Perform fatigue failure analysis of structures using *ANSYS Workbench*
6. Perform buckling failure analysis of structures using *ANSYS Workbench*

PROBLEMS

12.1 The following figure shows a simple bracket made of structural steel with Young's modulus $E = 200$ GPa and Poisson's ratio $\nu = 0.3$. The bracket has a front face profile shown below and an extrusion depth of 12 cm. A through-hole of diameter 1.5 cm is centered on the bracket's flat ends, one on each side. Suppose that the bracket is fixed on the bottom, and is subjected to a fully reversed cyclic pressure load of magnitude 10 MPa on the top face. Find the life of the bracket and determine whether or not fatigue failure occurs in the part assuming a design life of 10^6 cycles.

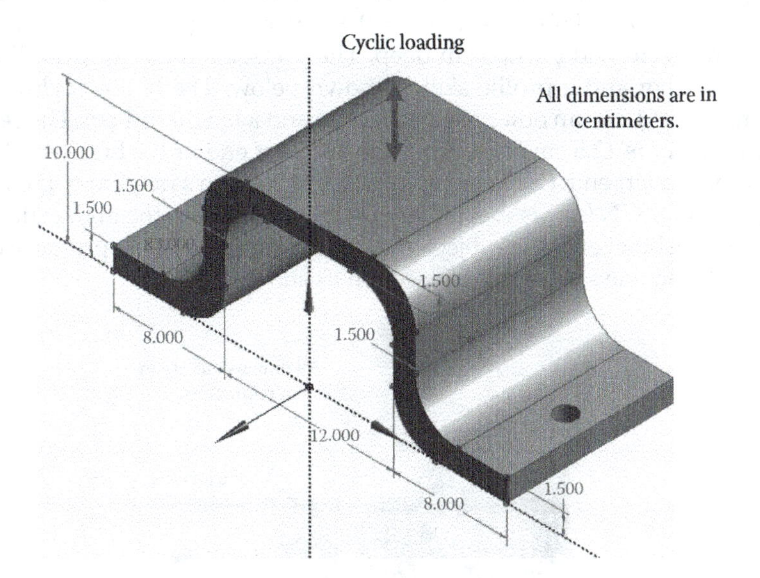

12.2 The following figure shows a lifting bracket made of structural steel with Young's modulus $E = 200$ GPa and Poisson's ratio $\nu = 0.3$. The base portion of the bracket is 20 cm long, 12 cm wide, and 3 cm deep. The fillet radius at the base corners is 1.75 cm. Four through-holes each of radius 1 cm are located symmetrically both length wise and width wise in the base, with the hole centers being 12 cm apart by 6 cm apart. The upper portion of the base has a profile sketch shown below and is 3 cm thick. The bracket is fixed on the bottom. (1) Suppose the bracket is lifted by a bearing load of 100 kN. Determine if the bracket can safely carry the given load. (2) Find the life of the bracket if it is subjected to a fully reversed cyclic bearing load of magnitude 50 kN.

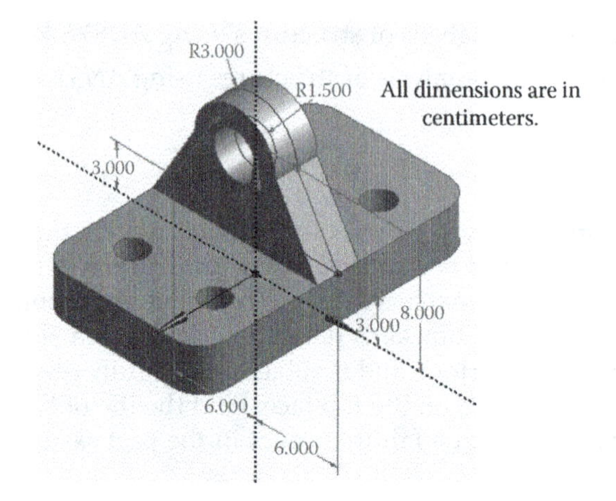

12.3 The figure below shows a bracket made of structural steel with Young's modulus $E = 200$ GPa and Poisson's ratio $\nu = 0.3$. The rectangular base of the bracket is 12 cm long, 6 cm wide, and 3 cm deep. The midsection has an overall extrusion depth of 1.6 cm and a profile sketch shown below. The hollow cylinder has an inner radius of 1 cm, an outer radius of 2 cm, and a height of 4 cm. The central axis of the cylinder is 12.5 cm away from the leftmost end of the bracket. The bracket is fixed on the left end. (1) Determine if the bracket can safely carry the load when a pressure of 800 N/cm^2 is exerted on the inclined face of the midsection. (2) Find the life of the bracket if it is subjected to a fully reversed cyclic pressure of magnitude 5 MPa on one side of the midsection as shown below.

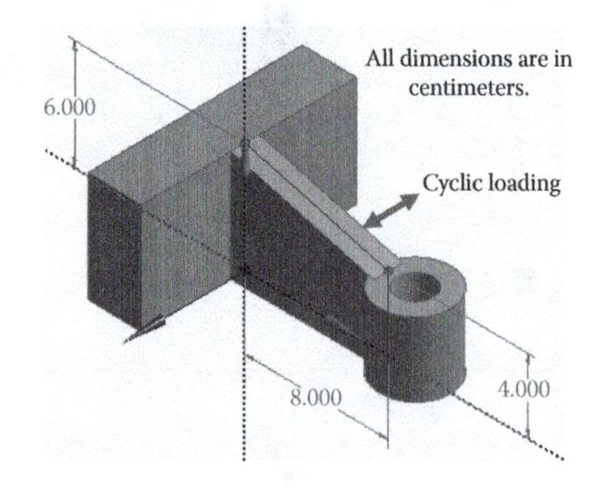

12.4 The following figure shows an L-shaped bracket made of structural steel with Young's modulus $E = 200$ GPa and Poisson's ratio $\nu = 0.3$. The 4 cm thick L-shape with the cross-sectional profile shown in the figure is rigidly connected to a rectangular base of size 3 cm × 8 cm × 1 cm. The bracket is fixed on the right end at the base. A surface pressure of 10 MPa is exerted on the left end as shown below. (1) Determine if static failure or buckling failure occurs in the bracket. (2) Find the life of the bracket if it is subjected to a fully reversed cyclic pressure of magnitude 10 MPa on the left end.

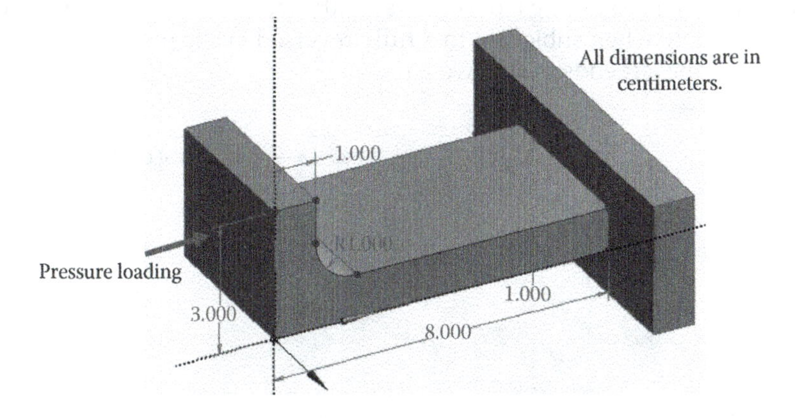

12.5 The following figure shows the cross-sectional profile of a 39 ft long straight railroad track. Use structural steel for the track with Young's modulus $E = 200$ GPa and Poisson's ratio $\nu = 0.3$. Assume the track is fixed on one end and subjected to a pressure load along the axis of the track on the other end, that is, the cross section at a distance of 39 ft from the fixed end. Determine the critical buckling load for the railroad track.

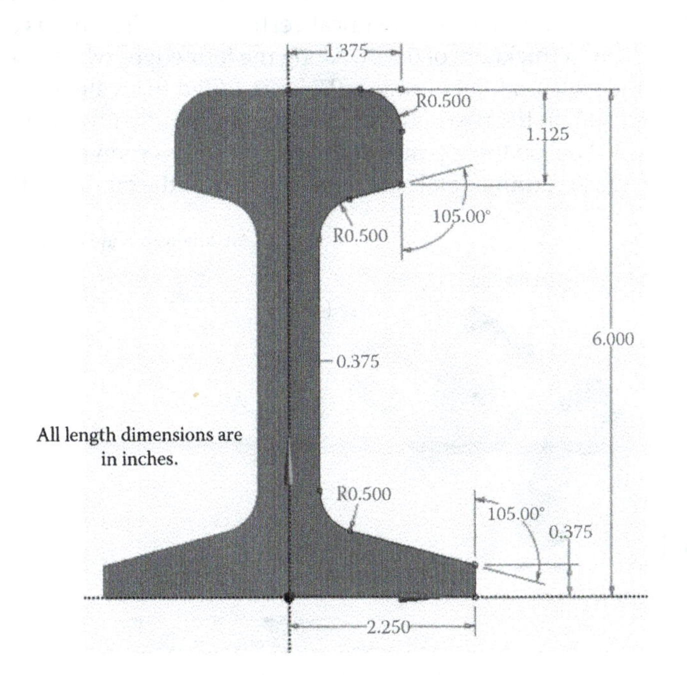

12.6 The following two-bar assembly consists of four components made of structural steel with Young's modulus $E = 200$ GPa, Poisson's ratio $\nu = 0.3$, and density $\rho = 7850$ kg/m^3. As shown in the figure below, the two bars are identical in size. Each bar has a diameter of 2 cm and a length of 16 cm. The two identical end blocks each has a thickness of 2 cm and a fillet radius of 1 cm at the round corners. The distance of the gap in between the two blocks is 8 cm. The two bars are fixed on the right ends, and subjected to a pressure load of 100 MPa on the left ends as shown below. (1) Determine if static failure or buckling failure occurs in the assembly. (2) Assuming a design life of 10^6, determine if fatigue failure occurs in the assembly when subjected to a fully reversed cyclic pressure of magnitude 100 MPa on the left ends of the two bars.

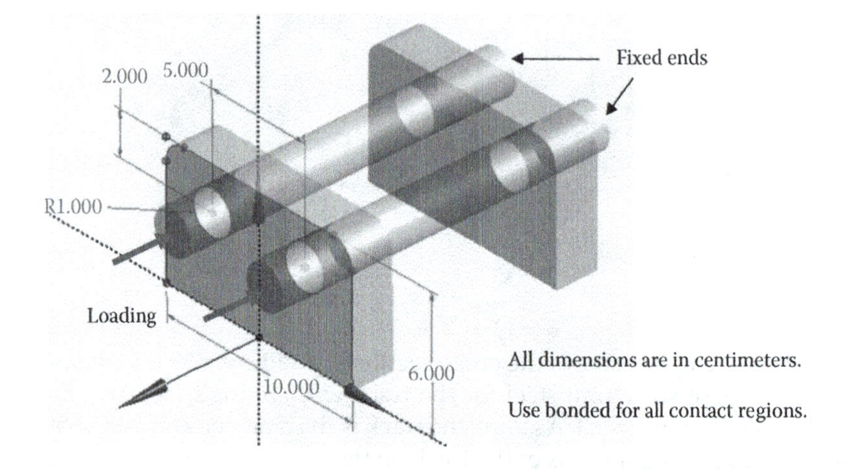

12.7 The following figure shows a crank shaft made of structural steel with Young's modulus $E = 200$ GPa and Poisson's ratio $\nu = 0.3$. The crank shaft consists of three identical horizontal rods and two identical vertical pieces. The rod is each 4 in. long. The plate each has a thickness of 0.5 in. Along the four edges where the rods are connected with the plates, a fillet of radius 0.2 in. is added to create round corners. The crank shaft is fixed on the two ends and subjected to a fully reversed cyclic pressure of magnitude 300 psi on the top face of the central rod as shown below. Assuming a design life of 10^6, determine if fatigue failure occurs in the crank shaft assembly.

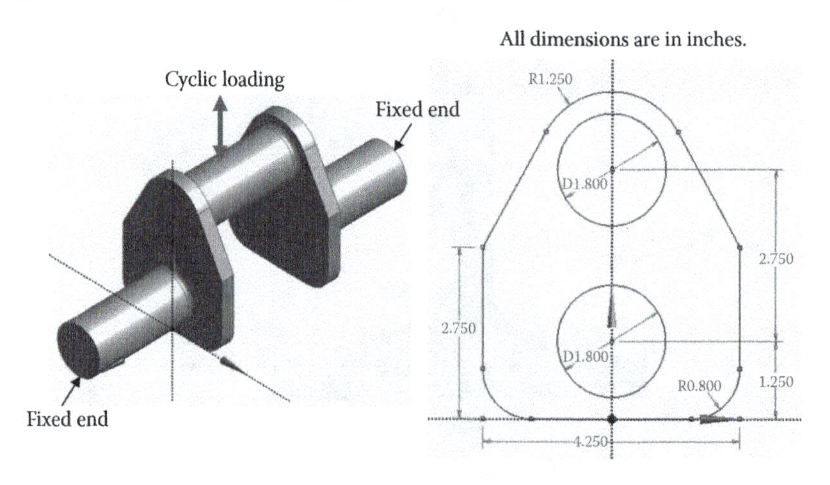

Appendix 1: Review of Matrix Algebra

A.1 Linear System of Algebraic Equations

$$
\begin{aligned}
a_{11}x_1 + a_{12}x_2 + \cdots + a_{1n}x_n &= b_1 \\
a_{21}x_1 + a_{22}x_2 + \cdots + a_{2n}x_n &= b_2 \\
\cdots\cdots & \\
a_{n1}x_1 + a_{n2}x_2 + \cdots + a_{nn}x_n &= b_n
\end{aligned}
\tag{A.1}
$$

where $x_1, x_2, \ldots, x_n$ are the unknowns.

In *matrix form*:

$$
\mathbf{Ax} = \mathbf{b} \tag{A.2}
$$

where

$$
\mathbf{A} = [a_{ij}] = \begin{bmatrix} a_{11} & a_{12} & \cdots & a_{1n} \\ a_{21} & a_{22} & \cdots & a_{2n} \\ \cdots & \cdots & \cdots & \cdots \\ a_{n1} & a_{n2} & \cdots & a_{nn} \end{bmatrix}, \quad \mathbf{x} = \{x_i\} = \begin{Bmatrix} x_1 \\ x_2 \\ \vdots \\ x_n \end{Bmatrix}, \quad \mathbf{b} = \{b_i\} = \begin{Bmatrix} b_1 \\ b_2 \\ \vdots \\ b_n \end{Bmatrix} \tag{A.3}
$$

A is called an $n \times n$ (square) matrix, and **x** and **b** are (column) vectors with dimension n.

A.2 Row and Column Vectors

$$
\mathbf{v} = [v_1 \quad v_2 \quad v_3] \qquad \mathbf{w} = \begin{Bmatrix} w_1 \\ w_2 \\ w_3 \end{Bmatrix} \tag{A.4}
$$

A.3 Matrix Addition and Subtraction

For two matrices **A** and **B**, both of the *same size* ($m \times n$), the addition and subtraction are defined by

$$\begin{aligned} \mathbf{C} = \mathbf{A} + \mathbf{B} \quad &\text{with} \quad c_{ij} = a_{ij} + b_{ij} \\ \mathbf{D} = \mathbf{A} - \mathbf{B} \quad &\text{with} \quad d_{ij} = a_{ij} - b_{ij} \end{aligned} \tag{A.5}$$

A.4 Scalar Multiplication

$$\lambda \mathbf{A} = [\lambda a_{ij}] \tag{A.6}$$

A.5 Matrix Multiplication

For two matrices **A** (of size $l \times m$) and **B** (of size $m \times n$), the product of **AB** is defined by

$$\mathbf{C} = \mathbf{AB} \quad \text{with } c_{ij} = \sum_{k=1}^{m} a_{ik} b_{kj} \tag{A.7}$$

where $i = 1, 2, \ldots, l$; $j = 1, 2, \ldots, n$.

Note that, in general, $\mathbf{AB} \neq \mathbf{BA}$, but $(\mathbf{AB})\mathbf{C} = \mathbf{A}(\mathbf{BC})$ (associative).

A.6 Transpose of a Matrix

If $\mathbf{A} = [a_{ij}]$, then the transpose of **A** is

$$\mathbf{A}^T = [a_{ji}] \tag{A.8}$$

Notice that

$$(\mathbf{AB})^T = \mathbf{B}^T \mathbf{A}^T \tag{A.9}$$

A.7 Symmetric Matrix

A *square* $(n \times n)$ matrix $\mathbf{A}$ is called symmetric, if

$$\mathbf{A} = \mathbf{A}^T \quad \text{or} \quad a_{ij} = a_{ji} \tag{A.10}$$

A.8 Unit (Identity) Matrix

$$\mathbf{I} = \begin{bmatrix} 1 & 0 & \dots & 0 \\ 0 & 1 & \dots & 0 \\ \dots & \dots & \dots & \dots \\ 0 & 0 & \dots & 1 \end{bmatrix} \tag{A.11}$$

Note that $\mathbf{AI} = \mathbf{A}$, $\mathbf{Ix} = \mathbf{x}$.

A.9 Determinant of a Matrix

The determinant of *square* matrix $\mathbf{A}$ is a scalar number denoted by det $\mathbf{A}$ or $|\mathbf{A}|$. For 2×2 and 3×3 matrices, their determinants are given by

$$\det\begin{bmatrix} a & b \\ c & d \end{bmatrix} = ad - bc \tag{A.12}$$

and

$$\det\begin{bmatrix} a_{11} & a_{12} & a_{13} \\ a_{21} & a_{22} & a_{23} \\ a_{31} & a_{32} & a_{33} \end{bmatrix} = a_{11}a_{22}a_{33} + a_{12}a_{23}a_{31} + a_{21}a_{32}a_{13} - a_{13}a_{22}a_{31} - a_{12}a_{21}a_{33} - a_{23}a_{32}a_{11} \tag{A.13}$$

A.10 Singular Matrix

A *square* matrix $\mathbf{A}$ is *singular* if det $\mathbf{A} = 0$, which indicates problems in the system (non-unique solutions, degeneracy, etc.).

A.11 Matrix Inversion

For a *square* and *nonsingular* matrix $\mathbf{A}$ (det $\mathbf{A} \neq 0$), its *inverse* $\mathbf{A}^{-1}$ is constructed in such a way that

$$\mathbf{AA}^{-1} = \mathbf{A}^{-1}\mathbf{A} = \mathbf{I} \tag{A.14}$$

The *cofactor matrix* $\mathbf{C}$ of matrix $\mathbf{A}$ is defined by

$$C_{ij} = (-1)^{i+j} M_{ij} \tag{A.15}$$

where M_{ij} is the determinant of the smaller matrix obtained by eliminating the ith row and jth column of $\mathbf{A}$.

Thus, the inverse of $\mathbf{A}$ can be determined by

$$\mathbf{A}^{-1} = \frac{1}{\det \mathbf{A}} \mathbf{C}^T \tag{A.16}$$

We can show that $(\mathbf{AB})^{-1} = \mathbf{B}^{-1}\mathbf{A}^{-1}$.

EXAMPLE A.1

(A) $$\begin{bmatrix} a & b \\ c & d \end{bmatrix}^{-1} = \frac{1}{(ad-bc)} \begin{bmatrix} d & -b \\ -c & a \end{bmatrix}$$

Checking:

$$\begin{bmatrix} a & b \\ c & d \end{bmatrix}^{-1} \begin{bmatrix} a & b \\ c & d \end{bmatrix} = \frac{1}{(ad-bc)} \begin{bmatrix} d & -b \\ -c & a \end{bmatrix} \begin{bmatrix} a & b \\ c & d \end{bmatrix} = \begin{bmatrix} 1 & 0 \\ 0 & 1 \end{bmatrix}$$

(B) $$\begin{bmatrix} 1 & -1 & 0 \\ -1 & 2 & -1 \\ 0 & -1 & 2 \end{bmatrix}^{-1} = \frac{1}{(4-2-1)} \begin{bmatrix} 3 & 2 & 1 \\ 2 & 2 & 1 \\ 1 & 1 & 1 \end{bmatrix}^T = \begin{bmatrix} 3 & 2 & 1 \\ 2 & 2 & 1 \\ 1 & 1 & 1 \end{bmatrix}$$

Checking:

$$\begin{bmatrix} 1 & -1 & 0 \\ -1 & 2 & -1 \\ 0 & -1 & 2 \end{bmatrix} \begin{bmatrix} 3 & 2 & 1 \\ 2 & 2 & 1 \\ 1 & 1 & 1 \end{bmatrix} = \begin{bmatrix} 1 & 0 & 0 \\ 0 & 1 & 0 \\ 0 & 0 & 1 \end{bmatrix}$$

If det $\mathbf{A} = 0$ (i.e., $\mathbf{A}$ is *singular*), $\mathbf{A}^{-1}$ does not exist.

The solution of the linear system of equations (Equation 1.1) can be expressed as (assuming the coefficient matrix $\mathbf{A}$ is nonsingular)

$$\mathbf{x} = \mathbf{A}^{-1}\mathbf{b}$$

Thus, the main task in solving a linear system is to find the inverse of the coefficient matrix.

A.12 Solution Techniques for Linear Systems of Equations

- Gauss elimination methods
- Iterative methods

The two methods are briefly reviewed in Chapter 5.

A.13 Positive-Definite Matrix

A *square* ($n \times n$) matrix **A** is said to be *positive definite*, if for all nonzero vector **x** of dimension n,

$$\mathbf{x}^T \mathbf{A} \mathbf{x} > 0$$

Note that positive-definite matrices are nonsingular. We can show that all stiffness matrices are positive definite and the above condition is a statement that the strain energy in a structure should be positive if the structure is constrained and the stiffness matrix is nonsingular.

A.14 Differentiation and Integration of a Matrix

Let $\mathbf{A}(t) = [a_{ij}(t)]$, then the differentiation is defined by

$$\frac{d}{dt}\mathbf{A}(t) = \left[\frac{da_{ij}(t)}{dt} \right] \tag{A.17}$$

and the integration by

$$\int \mathbf{A}(t)dt = \left[\int a_{ij}(t)dt \right] \tag{A.18}$$

Appendix 2: Photo Credits

Chapter 1

Figure 1.1(a) A train engine built with LEGO®.

(http://lego.wikia.com/wiki/10020_Santa_Fe_Super_Chief)

Figure 1.3(a) Wind load simulation of an offshore platform.

(Courtesy ANSYS, Inc. http://www.ansys.com/Industries/Energy/Oil+&+Gas);

Figure 1.3(b) Modal response of a steel frame building with concrete slab floors.

(http://www.isvr.co.uk/modelling/)

Figure 1.3(c) Underhood flow and thermal management.

(Courtesy ANSYS, Inc. http://www.ansys.com/Industries/Automotive/Application+Highlights/Underhood)

Figure 1.3(d) Electric field pattern of antenna mounted on helicopter.

(Courtesy ANSYS, Inc. http://www.ansys.com/Industries/Electronics+&+Semiconductor/Defense+&+Aerospace+Electronics)

Chapter 2

Figure 2.1(a) Montreal Biosphere Museum.

(http://en.wikipedia.org/wiki/Montreal_Biosph%C3%A8re)

Figure 2.1(b) Betsy Ross Bridge.

(http://en.wikipedia.org/wiki/Betsy_Ross_Bridge)

Chapter 3

Figure 3.1(a) A car frame.

(http://www.carbasics-1950.com/)

Figure 3.1(b) An exercise machine.

(http://www.nibbledaily.com/body-by-jake-cardio-cruiser/)

Chapter 5

Figure 5.1(a) Reflective symmetry.
(http://www.thinkingfountain.org/s/symmetry/butterflypattern.gif)
Figure 5.1(c) Rotational symmetry.
(http://csdt.rpi.edu/na/pnwb/symmetry2a.html)
Figure 5.1(d) Translational symmetry.
(http://library.thinkquest.org/16661/background/symmetry.1.html)

Chapter 6

Figure 6.1 An airplane made of numerous plate and shell structures.
(http://en.wikipedia.org/wiki/Boeing_787_Dreamliner)

Chapter 8

Figure 8.7 Two equations for determining the proportional damping coefficients.
(From R. D. Cook, *Finite Element Modeling for Stress Analysis* (John Wiley & Sons, Inc., Hoboken, NJ, 1995).)
Figure 8.13 Car crash analysis using the FEA (from LS-Dyna website).
(http://www.dynaexamples.com/)

Chapter 10

Figure 10.1(a) Airflow around a tractor.
(Courtesy ANSYS, Inc. http://www.ansys.com/Industries/Automotive/Application+Highlights/Body)
Figure 10.1(b) Streamlines inside a combustion chamber.
(Courtesy ANSYS, Inc. http://www.edr.no/en/courses/ansys_cfd_advanced_modeling_reacting_flows_and_combustion_in_ansys_fluent)

Chapter 12

Figure 12.7 Buckling of a slender column.
(http://en.wikipedia.org/wiki/Buckling)

References

1. O. C. Zienkiewicz, The birth of the finite element method and of computational mechanics, *International Journal for Numerical Methods in Engineering*, **60**, 3–10, 2004.
2. *ANSYS Workbench User's Guide*, Release 14.5, ANSYS, Inc., 2012.
3. C. Felippa, *Introduction to Finite Element Methods*, Lecture Notes, University of Colorado at Boulder (http://www.colorado.edu/engineering/cas/courses.d/IFEM.d/).
4. J. N. Reddy, *An Introduction to the Finite Element Method*, 3rd ed. (McGraw-Hill, New York, 2006).
5. O. C. Zienkiewicz and R. L. Taylor, *The Finite Element Method*, 5th ed. (McGraw-Hill, New York, 2000).
6. S. P. Timoshenko and S. Woinowsky-Krieger, *Theory of Plates and Shells*, 2nd ed. (McGraw-Hill, New York, 1987).
7. R. D. Cook, *Finite Element Modeling for Stress Analysis* (John Wiley & Sons, Inc., Hoboken, NJ, 1995).
8. K. J. Bathe, *Finite Element Procedures* (Prentice-Hall, Englewood Cliffs, NJ, 1996).
9. T. R. Chandrupatla and A. D. Belegundu, *Introduction to Finite Elements in Engineering*, 3rd ed. (Prentice-Hall, Upper Saddle River, NJ, 2002).
10. Y. C. Fung, *A First Course in Continuum Mechanics*, 3rd ed. (Prentice-Hall, Englewood Cliffs, NJ, 1994).
11. P. Timoshenko and J. N. Goodier, *Theory of Elasticity*, 3rd ed. (McGraw-Hill, New York, 1987).
12. ANSYS Help System, *Mechanical Applications—Mechanical User Guide*, Release 14.5, ANSYS, Inc., 2012.
13. R. D. Cook, D. S. Malkus, M. E. Plesha, and R. J. Witt, *Concepts and Applications of Finite Element Analysis*, 4th ed. (John Wiley & Sons, Inc., Hoboken, NJ, 2002).
14. S. Moaveni, *Finite Element Analysis—Theory and Application with ANSYS*, 3rd ed. (Prentice-Hall, Upper Saddle River, NJ, 2007).
15. B. R. Munson, D. F. Young, and T. H. Okiishi, *Fundamentals of Fluid Mechanics*, 5th ed. (John Wiley & Sons, Inc., Hoboken, NJ, 2006).
16. *ANSYS Help System, CFX—Modeling Guide*, Release 14.5, ANSYS, Inc., 2012.
17. *ANSYS Help System, Design Exploration User Guide*, Release 14.5, ANSYS, Inc., 2012.
18. J. E. Shigley, C. R. Mischke, and R. G. Budynas, *Mechanical Engineering Design*, 7th ed. (McGraw-Hill, New York, 2003).
19. F. P. Beer, E. R. Johnston, J. T. DeWolf,x and D. F. Mazurek, *Mechanics of Materials*, 6th ed. (McGraw-Hill, New York, 2011).

Index

U

V